Engineering Materials

This series provides topical information on innovative, structural and functional materials and composites with applications in optical, electrical, mechanical, civil, aeronautical, medical, bio- and nano-engineering. The individual volumes are complete, comprehensive monographs covering the structure, properties, manufacturing process and applications of these materials. This multidisciplinary series is devoted to professionals, students and all those interested in the latest developments in the Materials Science field.

More information about this series at http://www.springer.com/series/4288

Victor A. Karkhin

Thermal Processes in Welding

Springer

Victor A. Karkhin
Department of Welding and Laser Technologies
St. Petersburg Polytechnic University
St. Petersburg, Russia

ISSN 1612-1317 ISSN 1868-1212 (electronic)
Engineering Materials
ISBN 978-981-13-5964-4 ISBN 978-981-13-5965-1 (eBook)
https://doi.org/10.1007/978-981-13-5965-1

Library of Congress Control Number: 2018966399

This Springer imprint is published by the registered company Springer Nature Singapore Pte Ltd.
The registered company address is: 152 Beach Road, #21-01/04 Gateway East, Singapore 189721, Singapore

Dedicated to
my parents Joachim and Anne

Preface to the English Edition

Welding as a technological tool has been widely used for 100 years. This is caused by notable advantages of welding over other methods of material joining.

The ultimate goal of welding is to obtain the weld which would be equivalent to the base metal by all the properties. In reality, this goal is not achieved. A vast majority of failures of welded structures are initiated in welds. The main factors of these failures are degradation of the weld metal and the heat-affected zone, and welding residual stresses and distortion, occurred under a thermal effect.

Thermal processes in welding include the processes that lead to the change of temperature in a workpiece and filler metals due to heat of a welding source, heat propagation across the workpiece and heat removal outward.

Almost all welding methods and related technologies are accompanied by metal heating in a limited region of a joint. The range of heat temperature varies from the initial ambient temperature to the temperature of metal evaporation, with heating and cooling rates up to thousands of degrees per second. Besides, this can lead to base metal melting, chemical reactions in a weld pool, solidification of liquid metal, volume changes, microstructure transformations in a solidified metal, elastic and plastic deformations, etc. These processes determine local properties of metal and, consequently, fitness-for-purpose of the joint and the whole welded structure. In order to optimise a welding technology, it is required to know how welding conditions (base and filler metals, welding method and its parameters) affect these processes.

Thermal processes in welding came into focus in the 20–30s of the previous century after welding had been widely adopted in the industry. A great contribution to the theory as applied to welding was made by D. Rosenthal, who published his research work in 1935. In his paper D. Rosenthal provided solutions of heat conduction problems for moving concentrated heat sources in quasi-stationary conditions, when temperature fields do not change with time in a moving coordinate system related to a heat source. In 1938, there were published research papers by N. N. Rykalin, who suggested solutions of a more generalised problem including transient thermal processes. Later, it was revealed, which is common for the history of science, that the fundamental solutions to the problem of heat and mass transfer in the same mathematical form had been gained by H. A. Wilson before, in 1904.

Foundations of the general theory of thermal processes in welding were laid and published in three monographs by N. N. Rykalin in 1941, 1947 and 1951. Due to its logic and simple physical interpretation, the theory, based on the analytical method of sources, became accessible to welding researchers and engineers. Analytical approaches were developed in papers of K. M. Gatovsky, V. I. Makhnenko, N. T. Nguyen and C. L. Tsai.

The technological breakthrough resulted in working out numerical methods of solving heat conduction problems, i.e. finite difference method and finite element method. Such methods provide the foundation for computational codes for modelling thermal processes in welding (Abaqus, Ansys, Sysweld, etc.). A great contribution to the development of the numerical methods of solving welding problems was made by T. DebRoy, V. F. Demchenko, J. Goldak, A. S. Kurkin, V. I. Makhnenko, L.-E. Lindgren, V. A. Sudnik and Y. Ueda.

Within the recent two decades, the interest in modelling welding processes has shown a dramatic rise, which is proved by regular international seminars "Mathematical Modelling of Weld Phenomena". However, in spite of numerous publications, there are no monographs that systematise and generalise methods of computing thermal processes and predicting their effects to optimise microstructure, local mechanical properties and residual stresses in various zones of welds.

The present book considers main welding heat sources, provides thermophysical properties of metals, formulates welding heat conduction problems and presents analytical and numerical methods of solving heat conduction problems for fusion and pressure welding. Particularly, the book analyses temperature fields under programmed heat input. The author formulates and solves inverse problems of heat conduction in welding and exemplifies optimisation of welding conditions for different objective functions. It is demonstrated that knowledge of thermal cycles allows predicting local microstructure state and assessing local mechanical properties of metal.

The book is aimed at engineers, lecturers and researchers in the field of welding and related technologies, strength of welded structures, surface heat treatment by a concentrated heat source. Besides, the book can be of particular interest for graduate and postgraduate students majoring in welding.

In order to study thermal processes in welding, it is required to know general physics, higher mathematics, materials science and other general subjects. Besides, thermal processes in welding are the foundation for studying fundamentals of welding metallurgy, fusion and pressure welding technologies, theory of welding stresses and distortion, theory of strength of welded structures, modelling and optimisation of welding processes. They are worth studying for related technologies, i.e. cutting, soldering, brazing, hot strengthening of structures, thermal surfacing, material treatment with high-concentrated heat sources.

The book is based on my regular lectures at the Peter the Great St. Petersburg Polytechnic University for more than 30 years.

The manuscript was translated by M. N. Sosnina. The diagrams were meticulously drawn by P. N. Khomich. All this help is gratefully acknowledged. I am particularly indebted to A. M. Levchenko for his permanent support.

I am wholeheartedly grateful to my beloved wife, Vera Brosset, whose encouragement and care inspired me to work on this book.

St. Petersburg, Russia
September 2018

Victor A. Karkhin

Contents

Symbols

A	Area, m^2; amplitude of oscillations, m; austenite
A_{C1}	Ferrite-to-austenite transformation start temperature, K
A_{C3}	Ferrite-to-austenite transformation end temperature, K
A_m	Cross-sectional area of molten pool (weld), m^2
a	Thermal diffusivity, $m^2\ s^{-1}$
B	Bainite
Bi	Biot number
b	Coefficient of heat loss for thin plate and rod, s^{-1}
C	Concentration factor of normal distribution heat source, m^{-2}; concentration
CE	Carbon equivalent
c	Specific heat capacity, $J\ g^{-1}\ K^{-1}$
$c\rho$	Volume-specific heat capacity, $J\ m^{-3}\ K^{-1}$
D	Diffusion coefficient, $m^2\ s^{-1}$
d	Diameter, m
d_e	Diameter of electrode, m
d_n	Weld nugget diameter, m
Ei	Exponential integral (5.1.32)
e	Euler number (e = 2.718282…)
F	Ferrite; force, N; function; function accounting for influence of source pulsation; objective function
Fo	Fourier number
f	Frequency, s^{-1}; function
f_L	Liquid fraction
f_S	Solid fraction
G	Green's function
grad$T$	Temperature gradient, $K\ m^{-1}$
H	Hardness; specific enthalpy, $J\ g^{-1}$
ΔH	Latent heat of phase transformation, $J\ g^{-1}$

H_m	Depth of fused metal or weld pool, m; specific enthalpy at melting temperature, J g^{-1}
HV	Vickers hardness
h	Thickness, m
I	Electric current, A
$\mathrm{I_n}$	Modified Bessel function of the first kind of order n (5.2.115)
$\mathrm{i}\Phi^*$	Integral error function (5.1.37)
$\mathrm{J_n}$	Bessel function of the first kind of order n (4.2.44)
j	Current density, A m^{-2}
$\mathrm{K_n}$	Modified Bessel function of the second kind of order n (5.2.116)
L	Length, m; latent heat of melting and solidification, J g^{-1}
L_m	Length of molten pool, m
l	Length, m
M	Martensite
M_S	Temperature at which martensite transformation begins, K
n	Outward normal to the surface, m
P	Pearlite; point under study
Pe	Peclet number
p	Perimeter, m; parameter
Q	Heat amount, J
Q_0	Heat amount of point source, J
Q_1	Heat density per unit length, J m^{-1}
Q_2	Area-specific heat density, J m^{-2}
Q_3	Volume-specific heat density, J m^{-3}
q	Heat source power, W
q_0	Mean value of pulsed power, power of point heat source, W
q_1	Power per unit length, W m^{-1}
$q_{1\max}$	Maximum value of normal distribution heat power density (per unit length), W m^{-1}
q_2	Area-specific heat source power density, heat flow density, W m^{-2}
$q_{2\max}$	Maximum value of normal distribution heat source power density (area-specific), W m^{-2}
q_{2S}	Heat flow density at body surface, W m^{-2}
q_3	Volume-specific heat source power density, W m^{-3}
q_b	Low pulse (background) power, W
q_{gross}	Gross heat source power, W
q_p	High pulse (peak) power, W
R	Distance in space from centre (three-dimensional radius vector), radius of circular cylinder, m; solidification rate, m s^{-1}; electrical resistance, Ω
r	Distance in plane from centre (two-dimensional radius vector), m
r_0	Radius of heat source with constant distribution, m
$r_{0.05}$	Radius of heat source with normal distribution, m
r_e	Normal radius of heat source, m
r_e	Radius of electrode, m

r_m	Weld radius in cross section, m
r, θ, z	Cylindrical system of coordinates
S	Surface area, m^2
T	Temperature, K
$\{T\}$	Column vector of nodal point temperatures, K
ΔT	Temperature difference, temperature increment, K
T_0	Initial temperature, K
T_1	One-dimensional temperature field (in rod), K
T_2	Two-dimensional temperature field (in plate), K
T_3	Three-dimensional temperature field (in infinite body), K
T_∞	Ambient temperature, K
T_a	Austenitising temperature, K
T_{av}	Average temperature, K
T_L	Liquidus temperature, K
T_m	Melting temperature, K
$T_{\max}$	Peak temperature, K
$T_{\min}$	Minimum temperature, K
T_S	Solidus temperature, temperature at surface (boundary), K
t	Time, s
t_0	Time interval, s
t_b	Low pulse time (background duration), s
t_c	Cycle duration (total pulse time), s
t_{eq}	Equivalent time, s
$t_{\max}$	Time at which maximum temperature occurs, s
t_p	High pulse time (peak duration), s
t_s	Slope-up and slope-down pulse time, s
t_w	Time after start of welding, s
$\Delta t_{8/5}$	Cooling time from 1073 K (800 °C) to 773 K (500 °C), s
$\Delta t_{8.5/5}$	Cooling time from 1123 K (850 °C) to 773 K (500 °C), s
U	Voltage, V
V	Volume, m^3
v	Travel (welding) speed of heat source, feed rate of electrode (filler material), m s^{-1}
W, w	Width or half-width, m
W_m	Width of molten pool (weld), m
w	Cooling rate, $w = -\partial T/\partial t$, K s^{-1}; weight function
x	Longitudinal axis (in welding direction), m
x, y, z	Cartesian frame, m
x_e, y_e, z_e	Parameters of source with normal distribution, m
y	Space coordinate in weld transverse direction, m
y_m	Weld half-width, m
z	Space coordinate in thickness direction of body, m
α	Coefficient of surface heat transfer, W m^{-2} K^{-1}
α_c	Coefficient of convective heat transfer, W m^{-2} K^{-1}
α_r	Coefficient of radiative heat transfer, W m^{-2} K^{-1}

δ	Dimensionless thickness; Dirac delta function (1.2.6); elongation
ε	Dimensionless peak power; error
ζ	Coordinate of source on z-axis, m; dimensionless coordinate in z-direction
η	Coordinate of source on y-axis, m; dimensionless coordinate in y-direction
η_h	Heat source efficiency
η_T	Thermal efficiency of base metal melting
θ	Dimensionless temperature
θ_1	Dimensionless temperature in rod
θ_2	Dimensionless temperature in plate
θ_3	Dimensionless temperature in infinite body
λ	Thermal conductivity, W m^{-1} K^{-1}
λ_x, λ_y, λ_z	Thermal conductivity in the x-, y- and z-directions, W m^{-1} K^{-1}
μ	Coefficient of heat transfer influence; root of equation
ξ	Dimensionless coordinate in x-direction; source coordinate on x-axis, m
ξ, η, ζ	Cartesian system of dimensionless coordinates; coordinate of heat source, m
π	Archimedes constant (π = 3.141593…)
ρ	Density, g m^{-3}
ρ_1	Dimensionless distance in line from centre
ρ_2	Dimensionless distance in plane from centre
ρ_3	Dimensionless distance in space from centre
ρ_e	Specific electrical resistance, Ω m
σ	Normal stress, Pa
$\sigma_{0.2}$	0.2% offset yield strength, Pa
σ_u	Ultimate tensile strength, Pa
τ	Dimensionless time
$\Delta\tau_{8/5}$	Dimensionless cooling time from 1073 K (800 °C) to 773 K (500 °C)
Φ	Error function (3.7.4)
Φ^*	Complementary error function (4.2.16)
ϕ	Polar angle
X, χ	Functional
Ψ	Heat saturation function; reduction in area
BEM	Boundary element method
BM	Base (parent) metal
FDM	Finite difference method
FEM	Finite element method
GMAW	Gas metal arc welding
GTAW	Gas tungsten arc welding
HAZ	Heat-affected zone
SAW	Submerged arc welding
WM	Weld metal

Abstract

Welding is the most important method of joining metal components. The theory of the thermal dynamics of welding is one of the means by which welding processes are studied and controlled.

This book discusses welding heat sources, the thermophysical properties of different metals, formulation of heat conduction problems and presents analytical and numerical methods for solving the heat conduction problems in fusion and pressure welding. Temperature fields under programmed heat input are analysed. Inverse heat conduction problems are stated and solved. Optimisation of welding conditions using different objective functions is demonstrated.

This text shows that a knowledge of thermal cycles makes it possible to predict the local microstructure and to assess the local mechanical properties of weld and heat-affected zone metals.

The book is aimed at engineers, researchers, lecturers and students, specialising in welding and related technologies.

The book is based on the author's regular lectures at the Peter the Great St. Petersburg Polytechnic University.

Revised edition of the original Russian edition.

Chapter 1
Energy Characteristics of Welding Heat Sources

1.1 Requirements for Welding Heat Sources

In most cases welding is carried out with local heating of bodies up to the temperature which is determined by the type of welding and properties of the materials to be welded. During fusion welding, this temperature T must be greater than the melting temperature T_m ($T > T_m$), while during pressure welding it can be either higher or lower than T_m. In order to use the energy efficiently, its maximally localised input must be ensured so that the welded material can be heated in a minimally required volume.

For welding heating, different kinds of energy are converted into heat, in particular, electric, beam, chemical and mechanical energy and also their combinations.

During fusion welding, the power density of the source is within the range of $1 - 10^5$ W mm^{-2} (Fig. 1.1). In case of a lower density the material cannot be welded. If the density is higher, evaporation prevails.

1.2 Welding Arc

In terms of the volume of application, arc welding exceeds all other welding methods.

A welding arc represents a stable electric discharge through a gas gap. Current goes between two electrodes. A positive electrode is called an anode, a negative one is called a cathode (Fig. 1.2). During alternating current welding the polarity of electrodes changes together with the frequency of the power supply. Typical welding arcs have currency from 1 to 1500 A with the voltage from 8 to 50 V.

The electric field intensity in the arc is very uneven. In the narrow cathode and anode regions it is by orders of magnitude higher than in the arc column, where it does not exceed several volts per one millimeter (Fig. 1.2). So, the volumetric density of the heat generating in the arc is very uneven, which results in a non-uniform temperature

V. A. Karkhin, *Thermal Processes in Welding*, Engineering Materials,
https://doi.org/10.1007/978-981-13-5965-1_1

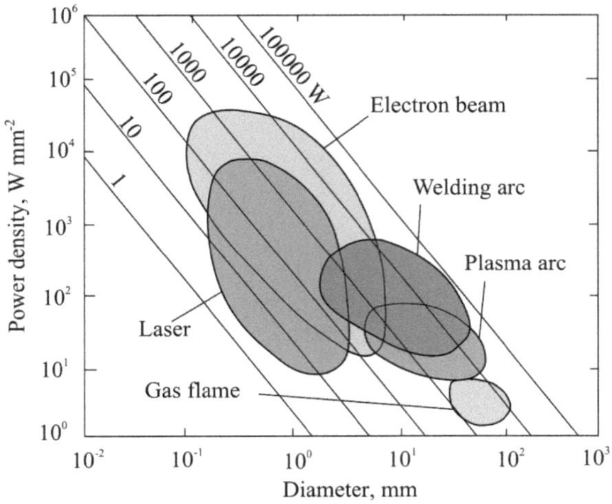

Fig. 1.1 Power densities and characteristic dimensions of heat sources for fusion welding (Lindgren 2007)

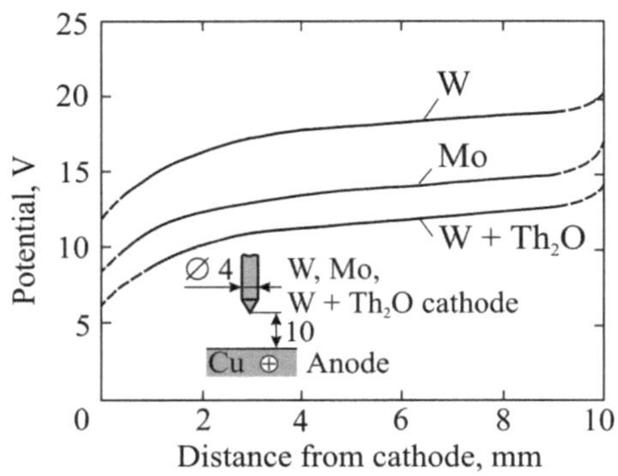

Fig. 1.2 Experimental axial potential distribution in low pressure argon arcs with different cathode materials (current is 60 A, argon pressure is 4 kPa, electrode negative) (Matsunawa and Nishiguchi 1979; Lancaster 1986)

field with big gradients. It follows from the figure that the voltage falls mostly in narrow electrode regions, being higher in the cathode region than in the anode one.

Tungsten and carbon are used as a non-consumable electrode. Molybdenum and tantalum are also refractory, but they are used in exceptional cases. The work function (the lowest energy needed to remove an electron from metal) of thoriated tungsten is lower than that of pure tungsten. So, the cathode fall of voltage and, consequently, the temperature of the cathode region are relatively low. The tip of the electrode does not melt and keeps its initial sharpened form.

The state of the arc (temperature field, plasma density, electric potential, speed, pressure, current density, etc.) depends on the material of electrodes, gas environment (composition and pressure), distance between electrodes, current and many other factors. It is determined by experimental methods (Farmer et al. 1986; Gick et al. 1973; Haddad and Farmer 1984; Haddad et al. 1985; Hiraoka et al. 1997; Hsu et al. 1983; Jackson 1960; Key et al. 1983; Kobayashi and Suga 1979; Nestor 1962; Olsen 1963; Smaars and Acinger 1968; Yushchenko et al. 2006) and by numerical methods

(Choo et al. 1990; Gick et al. 1973; Glickstein 1981; Haddad et al. 1985; Kovitya and Lowke 1982, 1985; Lee and Na 1996; Schnick et al. 2010; Ushio and Matsuda 1982).

Despite the fact that welding arc has been researched for quite a long time, it is still a physical object which has not been studied enough (Finkelnburg and Maecker 1961; Lancaster 1987; Leskov 1970; Dresvin 1972; Rykalin 1974; Sosnin et al. 2008; Lancaster 1986; Tikhodeyev 1961; Yampolsky 1972). It is especially true with regard to its interaction with a consumable electrode and the surface of the weld pool.

Let us consider the steady axisymmetric arc, which burns in argon between a thoriated tungsten electrode and a water-cooled copper anode (Fig. 1.3a). The temperature field of the anode depends on the state of the arc, which, in turn, depends on the conditions on the contact arc-anode and arc-cathode surfaces. Thus, with a strict approach, the object of research must include all three bodies simultaneously (electrode, arc, workpiece) (Murphy and Thomas 2017). Five regions are distinguished when solving this complex problem: cathode, cathode region of the arc, arc column, anode region of the arc and anode. Equations of mass and current continuity, of momentum and energy conservation are written for the column. It is necessary to consider the complicated physical phenomena in the electrode region (electronic current and ionic current, thermal conductivity, etc.) and surface effects on the electrodes. Thermal conductivity and electric heating are considered in the electrodes. Some relevant boundary conditions must be set. The formulated problem is strongly non-linear and can be solved only in a numerical way. The conditions on the cathode-arc and arc-anode boundaries are finalised by an iteration method (Wendelstorf et al. 1997). Thus, even in case of a simple methodological example given in Fig. 1.3a, it is extremely difficult to solve the problem with quite a strict statement and a lot of preparatory and computer time is required.

The steady-state temperature fields in the electrode, arc and workpiece are shown in Fig. 1.3a. The temperature on the arc axis reaches its peak in the proximity to the tungsten cathode, where the volumetric power density is maximal due to a great cathode voltage drop. The temperature reaches its minimum close to the water-cooled copper anode, which is proven by experiments (Haddad and Farmer 1984).

An important result of the problem solution is the heat flow distribution from the arc to the anode and its components (Fig. 1.3b). In terms of its form, the radial distribution curve of the total heat flow is close to a normal curve (Gaussian curve). It is seen that in the proximity to the arc axis, the input of electron energy in the heat flow in the anode is the main one. The anode is heated considerably due to heat conduction, especially where current is weak. Plasma radiation heating is low. If current increases, heat transfer from plasma to the anode grows. Let us emphasise that in order to calculate the temperature field of the workpiece (anode), the distribution of the heat flow from the arc to the workpiece must be known.

The calculation of the heat flux from the arc to the workpiece becomes much more difficult if the reaction of the liquid metal of the weld pool is considered during fusion welding (surface deflection of the pool due to the arc pressure, formation of metal vapour in the arc, etc.). For example, with aluminium vapour in helium the radial distribution of the arc temperature and, consequently, the heat flow in the workpiece

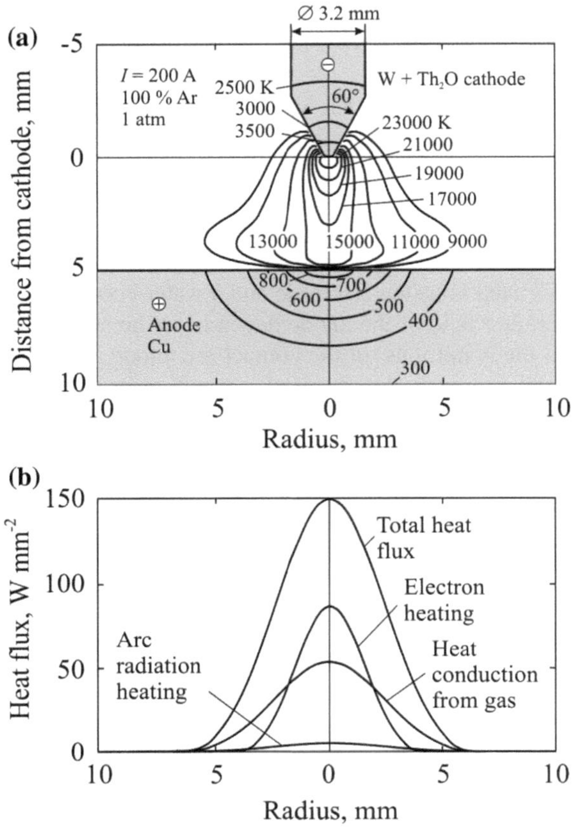

Fig. 1.3 Calculated temperature fields in electrode, arc and anode (**a**) and heat fluxes from the arc to the anode (**b**) (Wendelstorf et al. 1997; Zhu et al. 1995)

change considerably (Glickstein 1981). The presence of iron vapour in argon results in abnormal radial distribution of temperature and current density with minimum on the arc axis (Schnick et al. 2010). If the impact of the workpiece surface is considered during multi-pass butt welding and welding of T-joints, the axial symmetry of the arc is disturbed and the arc model becomes three-dimensional. More difficulties are there because the temperature dependencies of many properties of materials are not studied enough, such as viscosity, electrical conductivity, radiation coefficient, and thermal electron emission.

Some parameters of the axisymmetric arc on the surface of the heated body (radial distributions of heat flow density, current density, pressure) are determined experimentally by the split-anode technique (Fig. 1.4a). Let us look into the technique by the example of determining the radial distribution of the power density $q_2(r)$. Let us assume that the function $q_2(r)$ is known. Then the power $q(x)$ falls on the right part of the anode:

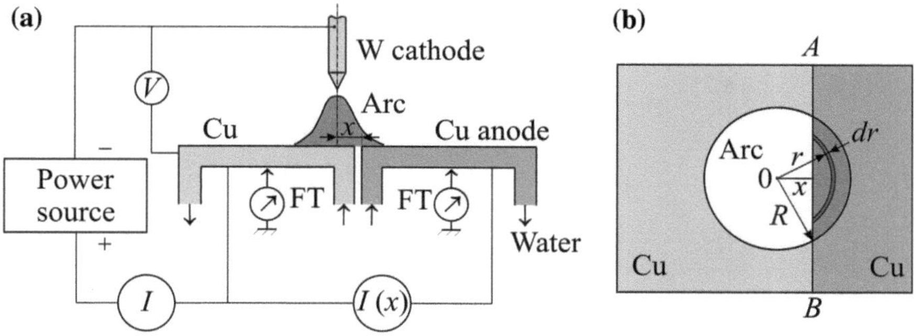

Fig. 1.4 Schematic illustration of the split-anode technique (**a**) and the split-anode coordinate system (**b**) (Lu and Kou 1988): FT is the force transducer

$$q(x) = 2 \int_{x}^{R} q_2(r) r \arccos(x/r) dr, \qquad (1.2.1)$$

where x is the distance from the centre of the arc O to the heat-impermeable plane of separation AB, R is the radius of the arc (Fig. 1.4b). In the figure the arc on the right part of the anode is darkened. Integration is taken along the radius. In reality the power $q(x)$ is known (it is measured by a direct method of calorimetry). The unknown function $q_2(r)$ is the solution to the integral Eq. (1.2.1) (Lu and Kou 1988):

$$q_2(r) = \frac{1}{\pi} \int_{x}^{R} \frac{q''(x)}{\sqrt{x^2 - r^2}} dx, \qquad (1.2.2)$$

where $q''(x)$ is the second-order derivative on the right part of the anode in reference to the position of arc x.

The function $q''(x)$ is very sensitive to measurement error, so the experimental curve $q(x)$ should be smoothed over in advance. If the law $q_2(\mathbf{p}, r)$ is assigned beforehand, the parameter vector $\mathbf{p}$ of this law can be found from the condition of minimal deviation of the calculated curve $q(\mathbf{p}, x_n)$ at the points x_n $(n = 1, \ldots N)$ from the experimental values of the function at these points. There is an algorithm for solving this optimisation problem (Lu and Kou 1988).

The problems referring to the radial distribution of current density and arc pressure are stated in a similar way. The distribution of the current $I(x)$ and force $F(x)$ is measured directly by an ammeter and a force transducer FT (Fig. 1.4a), while the radial distribution of the arc pressure is measured by a micromanometer (Kopayev et al. 2006).

In the further calculations of heat processes in the workpiece, it is convenient to present the obtained radial distribution of the power density $q_2(r)$ as a known function. Numerous experiments give evidence that the radial distribution of the arc

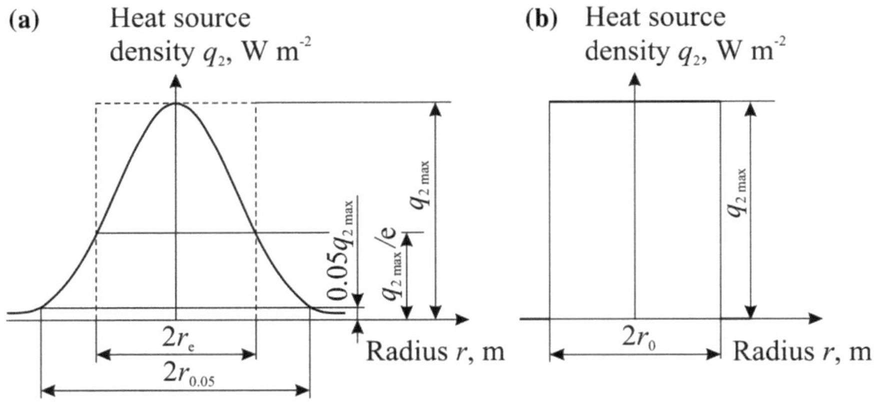

Fig. 1.5 Circular source with normal distribution of heat source density q_2 (**a**) and a circular source with constant heat source density q_2 (**b**)

heat flow density over the plane surface of the body located perpendicularly to the arc axis is described satisfactorily by a Gaussian curve (normal distribution law):

$$q_2(r) = q_{2\max} \exp(-Cr^2), \tag{1.2.3}$$

where $q_{2\max}$ is the heat flow density in the heat centre (with $r = 0$), W m^{-2} (Fig. 1.5a); C is the concentration factor of heat flow of the arc, m^{-2}. Such heat sources are called *normally distributed circular sources*. The distribution parameters $q_{2\max}$ and C in (1.2.3) are chosen to meet the condition of minimal deviation from the experimental distribution $q_2(r)$.

The connection of the effective power of the source q (W) with the distribution parameters $q_2(r)$ is established through integration of the heat flow density q_2 (1.2.3) over the entire plane:

$$q = \int_0^\infty q_2(r) 2\pi r dr = \frac{\pi}{C} q_{2\max} \text{ or } q_{2\max} = \frac{C}{\pi} q. \tag{1.2.4}$$

In theory, the source (1.2.3) operates on the entire surface, $0 < r < \infty$. In practice, it is useful to introduce the notion of conventional heat spot, which radius is taken up from the following conditions.

1. At the distance r_e from the centre, the flow density falls by e ≈ 2.718282 times. Hence, Eq. (1.2.3) suggests $r_e = 1/\sqrt{C}$. This radius is called *normal* (Rykalin et al. 1988).
2. At the distance $r_{0.05}$ from the centre, the flow density falls by e$^3 \approx 20$ times and is equal to 0.05 $q_{2\max}$. Then $r_{0.05} = \sqrt{3/C} = \sqrt{3} r_e$. For instance, for the arc in argon medium $r_{0.05} = 4.5$–6.5 mm, for the submerged arc of low and medium

power $r_{0.05} = 5$–10 mm, high power $-r_{0.05} = 10$–18 mm, for the open arc (by a carbon or metallic electrode) of small and medium power $r_{0.05} = 10$–15 mm, high power $-r_{0.05} = 15$–25 mm (Makhnenko and Kravtsov 1976).

The bigger factor C is, the quicker the power density decreases with distance away from the arc centre and the lower the radii r_e and $r_{0.05}$ are. With $C = \infty$ a point source on the surface of the body is obtained from a plane source:

$$q_2(r) = q\delta(0), \tag{1.2.5}$$

where $\delta\ (u)$ is the δ-Dirac function (delta-function):

$$\delta(u) = \begin{cases} \infty \text{ if } u = 0; \\ 0 \text{ if } u \neq 0, \end{cases} \tag{1.2.6}$$

where $\int\limits_{-\infty}^{\infty} \delta(u)du = 1$ (infinite spike of unit intensity). The δ-function has the fundamental property:

$$\int\limits_{-\infty}^{\infty} f(u)\delta(u - a)du = f(a), \tag{1.2.7}$$

i.e. allows setting down the space density of a physical quantity (for example, power density), concentrated or applied at point a of the space. It is widely used in mathematical physics.

A normally distributed source can be substituted for a uniformly distributed source with the same power q and radius r_0, $r_0 = r_e = r_{0.05}/\sqrt{3}$ (Fig. 1.5b). In the calculations of heat processes during arc welding it is more common to accept the hypothesis of normal distribution (Fig. 1.5a), rather than uniform distribution (Fig. 1.5b). The two independent parameters of the arc out of several ones (q, $q_{2\max}$, C, r_e, $r_{0.05}$) are usually determined experimentally.

If the arc axis is inclined to the x-axis on the surface of the body at angle β, then the source can be approximately considered as a *normally distributed elliptic source*:

$$q_2(x, y) = q_{2\max} \exp(-C_x x^2 - C_y y^2), \tag{1.2.8}$$

where

$$q_{2\max} = \frac{\sqrt{C_x C_y}}{\pi} q; \tag{1.2.9}$$

$$C_x = \frac{1}{r_{ex}^2}; C_y = \frac{1}{r_{ey}^2}; r_{ex} = \frac{r_{ey}}{\sin \beta}.$$

Here C_x and C_y are the concentration factors of arc power by the axes x and y, r_{ex} and r_{ey} are the radii (major and minor semi-axes) of the ellipse.

The line of the constant density of the power $q_2 = \text{const}$ is described by an equation of ellipse

$$\frac{x^2}{r_{ex}^2} + \frac{y^2}{r_{ey}^2} = \ln\left(\frac{\text{const}}{q_{2\max}}\right), \tag{1.2.10}$$

which gave the name to this source. In a special case, when $C_x = C_y = C$, formulas (1.2.8) and (1.2.9) turn into (1.2.3) and (1.2.4), i.e. the normally distributed circular source is a special case of a normally distributed elliptical one.

Equation (1.2.8) is true for a cylinder-shaped arc (power density is maximal in the centre of the ellipse and the plane $x = 0$ is the plane of symmetry, Fig. 1.6b). For a cone-shaped arc (the concentration factor decreases with distance away from the electrode) the plane $x = 0$ is not the plane of symmetry.

Further on more complicated laws of power distribution will be looked into for spatially limited and unlimited heat sources (see Sect. 5.2.1).

Figure 1.7 shows experimental distributions of power density for two arcs with constant current but with a different length of the arc and shape of the tungsten electrode. The experimental data was approximated with a normal distribution curve (1.2.3). The parameters $q_{2\max}$ and C for each arc were defined by the two conditions: the effective power q is constant (the volume under the surface $q_2(r)$ is constant) and the area under the curves $q_2(r)$ is constant (Lu and Kou 1988). Figure 1.7 shows that with an increase in the length (voltage) of the arc and diameter of the electrode, the arc becomes less concentrated ($q_{2\max}$ and C go down).

The shielding gas and iron vapour it contains exert a great impact on the temperature field of the arc and distribution of power and current density on the surface of the anode made of austenitic steel SUS 304 (≤0.08% C, 8–10.5% Ni, 18–20% Cr) (Fig. 1.8). The researched two-dimensional axisymmetric region included a tungsten electrode, 3.2 mm in diameter and a tip angle equal to 60°, an arc, 5 mm in length at

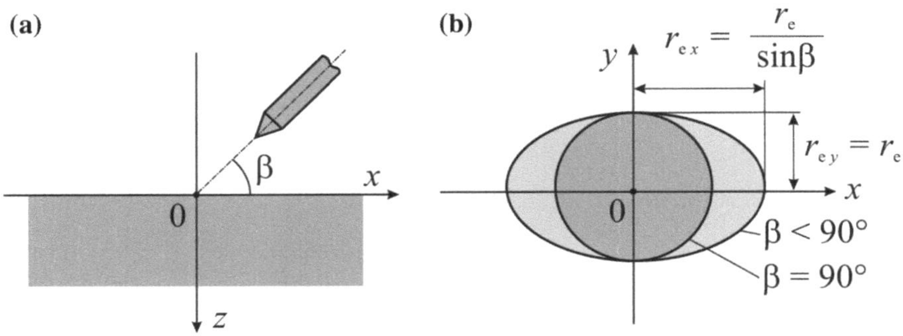

Fig. 1.6 Schematic of electrode angle (**a**) and heat source sizes (**b**) (Sudnik et al. 2005)

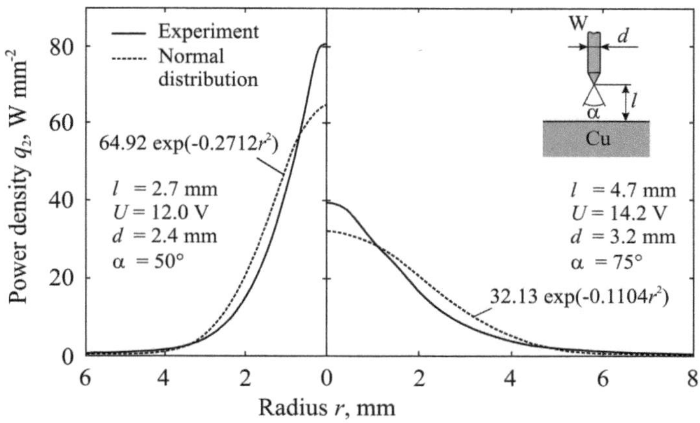

Fig. 1.7 Effective power-density distribution in a 100 A arc with different arc length l, voltage U, W electrode diameter d and tip angle α (Lu and Kou 1988)

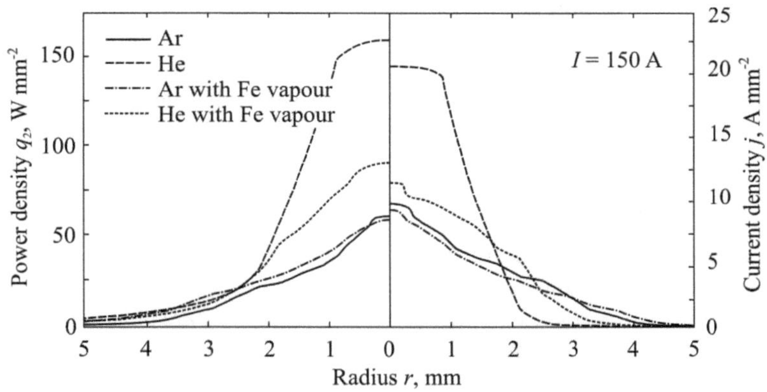

Fig. 1.8 Calculated radial dependence of power density q_2 and current density j at the anode for arcs in Ar and He with SUS 304 austenitic steel anode (Murphy et al. 2009)

the pressure of 1 atm and current of 150 A and a water-cooled anode with a liquid pool. The arc is stationary relative to the heated body. It has been revealed that the volumetric heat capacity of gas makes a profound effect on the compression of the arc. The iron vapour in the arc increases electrical conductivity at a low temperature, which causes a considerable decrease in the heat flow and electric current on the surface of the weld pool and, consequently, a reduced depth of the pool (Murphy et al. 2009). With the current equal to 200 A, iron vapour in argon reduces the temperature in the centre of the arc from 18000 K down to 8000 K (Hertel et al. 2013).

Different techniques are used to define the parameters of the welding arc (Rykalin and Kulagin 1953; Tsarkov and Orlik 2001). By comparing the experimental thermal cycles and the thermal cycles calculated with a non-linear model of 3D heat transfer, the following linear equation was obtained for the arc radius $r_{0.05}$ during welding of austenitic steel 12Kh18N10T ($\leq$0.12% C, 17.0–19.0% Cr, 9.0–11.0% Ni, 0.6–0.8% Ti) with a non-consumable electrode under the conditions: current $I = 160$–320 A, arc length $l = 2$–6 mm, welding speed $v = 20$–40 mm s^{-1}, 100% Ar (Sudnik and Rybakov 1992; Sudnik et al. 2005):

$$r_{0.05} = 8.43 - 0.007I + 0.043l. \quad (1.2.11)$$

The dimension $r_{0.05}$ and l–mm, current I–A. It follows from the formula that the arc radius grows with reduction in current and increase in the length of the arc. For more powerful arcs the tendency remains the same (Yushchenko et al. 2006).

The welding arc exerts a force impact on the body that is being welded. The pressure force is described by a parabolic dependence on current. During welding with a coated electrode at the current of 150 A it reaches 0.017 N (Doan and Lorentz 1941), while during welding with a consumable electrode, 5 mm in diameter at the current of 700 A it is 0.12 N (Petrunichev 1960). The radial pressure distribution of the arc is close to normal (Fig. 1.9), and its parameters are determined by gas medium (Yamauchi and Taka 1979; Zaehr et al. 2010), current level, electrode tip angle and arc length (Yamauchi and Taka 1979), electrode diameter, constriction in the additional magnetic field (Nomura et al. 2010) and other factors. In turn, the arc pressure makes the surface of the weld pool deflect and the characteristics of the arc change (Fig. 1.10). The higher the current is, the greater the deflection of the pool surface is and the higher the temperature is (Fig. 1.10a). In the proximity to the relatively cold anode, the temperature gradients are high, and isotherms are parallel to the surface of the anode. On the radial distributions of power and current density their sharp falls are seen on the arc axis (Fig. 1.10b). Differently from the plane anode (Fig. 1.3b) the distributions are far from being normal, they are bimodal (Choo et al. 1990). These curves can be described by an equation (Yerofeyev and Maslennikov 2005)

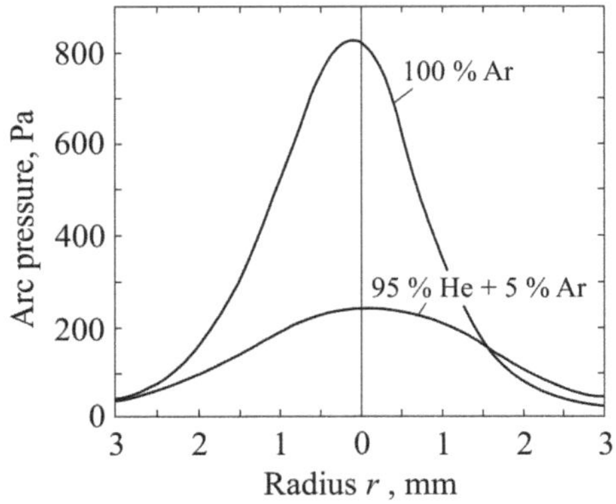

Fig. 1.9 The effect of shielding gas on arc force (experiment: current 200 A, W electrode negative, electrode diameter 3.2 mm, tip angle 30°, arc length 3 mm) (Radaj 1992; Lancaster 1986; Yamauchi 1979)

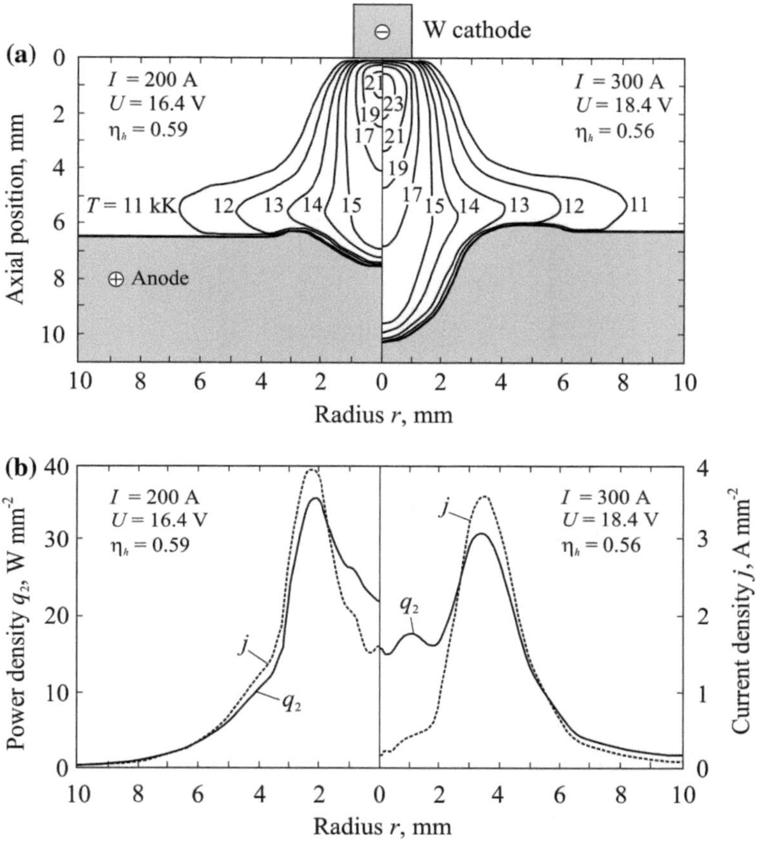

Fig. 1.10 Calculated radial distributions of temperature (**a**), heat flux q_2 and current density j on the anode (**b**) (Choo et al. 1990)

$$q_2(r) = \frac{2q}{\pi r_d^2}\left[a_d + (1 - a_d)\frac{2r^2}{r_d^2}\right]\exp\left(\frac{-2r^2}{r_d^2}\right), \qquad (1.2.12)$$

where r_d is the radius of the arc (the radius of the circle, where 86% of the arc heat flow goes); a_d is a coefficient which considers the deviation of flow distribution from the normal law and is determined by the ratio between the depth h_c and the radius r_c of the crater under the arc:

$$a_d = \frac{r_c}{r_c + h_c}. \qquad (1.2.13)$$

Thus, the heat flow depends, to a large extent, on the shape of the free surface of the weld pool.

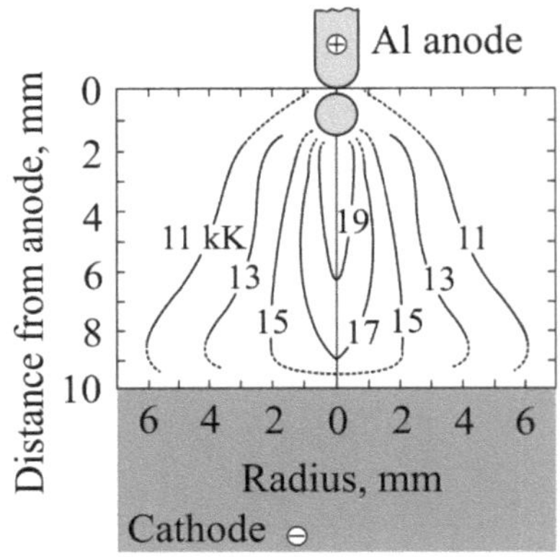

Fig. 1.11 Experimental temperature distribution in gas metal arc welding of aluminium at 250 A (Smaars and Acinger 1968; Lancaster 1986)

The temperature field of the arc during welding with a consumable electrode in argon is similar to the one during welding with a non-consumable electrode (Figs. 1.11 and 1.2a). In the first case the isotherms on the axis are more elongated (Smaars and Acinger 1968). An additional heat flow on the surface is created due to liquid drops of the consumable electrode.

With the voltage of the arc growing from 30 up to 38 V $q_{2\max}$ decreases from 300 down to 250 W mm^{-2}, and $r_{0.05}$ increases from 8 to 9.5 mm (Fig. 1.12b). Such type of the change in the power density distribution can be explained by a change in the arc length with a change in the arc voltage. With a growth in the length, the heat exchange of the arc column and the surface of the workpiece increases, but, at the same time, the heat exchange with the arc spot on the electrode reduces. Thus, with the reducing arc voltage, the heat flow becomes more concentrated close to the arc axis.

The flux grade and current density exert no considerable effect on the distribution of the heat flow on the surface of the workpiece (Fig. 1.12c).

If the arc burns in the gap between the edges of plates during butt welding or between the surfaces of the workpiece when T-joints are being welded, it becomes non-axisymmetric and its temperature field, distribution of current and power density on the surface of the workpiece are three-dimensional.

There are numerous publications where the distributions of arc power density are given for welding with a non-consumable electrode (Belousov 2002; Kopayev et al. 2006; Lee and Na 1996; Lowke and Tanaka 2006; Lu and Kou 1988; Mochizuki et al. 2010; Murphy et al. 2009; Nestor 1962; Rykalin 1951, 1957; Sudnik et al. 2005; Szekely 1989; Wendelstorf et al. 1997) and with a consumable electrode (Rykalin and Kulagin 1953; Sudnik and Yerofeyev 1988).

The gross power of the arc q_{gross} is determined by current I and arc voltage U, but only a part of it goes into the welded body. This part is called the *effective power* q (W):

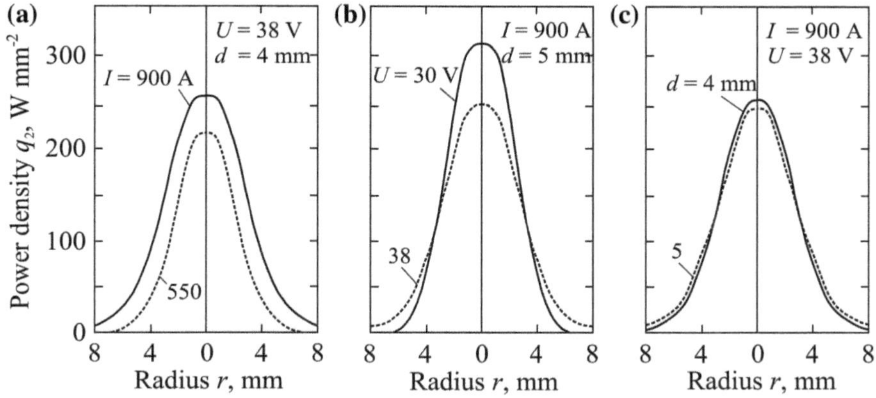

Fig. 1.12 Heat source power density of a submerged arc over the radial distance from centre r, dependency on amperage I (**a**), voltage U (**b**) and electrode diameter (**c**) (alternating current, low-carbon steel electrode, fused flux FTs-4) (Rykalin and Kulagin 1953)

$$q_{gross} = UI; \tag{1.2.14}$$

$$q = \eta_h q_{gross} = \eta_h UI, \tag{1.2.15}$$

where η_h is the arc efficiency (also called the efficiency of heating the workpiece by the arc (Rykalin 1951, 1957) and the heat transfer efficiency of the process (Messler 1999)).

The losses of power q_{gross}–q are conditioned by the radiation of the high temperature plasma of the arc, convection of the ambient gas, liquid drop spattering and other factors. The arc efficiency η_h depends on many factors: material to be welded, welding technique (consumable or non-consumable electrode), shielding gas and its consumption, welding conditions (current, voltage, welding speed), etc. The experimental values η_h also depend on the measuring method (calorimetry, method of a split-anode calorimeter, etc.).

For welding with a non-consumable electrode, the experimental values η_h have a range from 0.21 (if aluminium is alternating current welded) (Christensen et al. 1965) to 0.91 (if the copper anode is water-cooled) (Tsai and Eagar 1985). The dependence of the arc efficiency on the arc current is shown in Fig. 1.13. The wide range of values η_h can be explained by the difference in the welding conditions and measuring method. The arc length makes a considerable impact. The longer the arc is, the lower the arc efficiency is, which is shown by band 3: the upper values correspond to the arc length of 2 mm, the lower ones to that of 6 mm.

With a non-linear model of three-dimensional heat transfer and the design of the experiment, a formula was obtained for arc efficiency under the same conditions as Eq. (1.2.11) (Sudnik and Rybakov 1992; Sudnik et al. 2005):

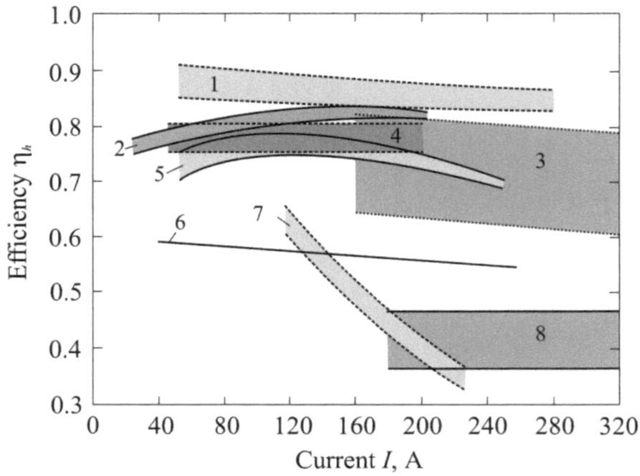

Fig. 1.13 Dependence of gas tungsten arc efficiency η_h on current I: 1—water-cooled anode (Tsai and Eagar 1985); 2—304 stainless steel (Giedt et al. 1989); 3—12Kh18N10T stainless steel (Sudnik and Rybakov 1992); 4—Zircaloy-4 (Glickstein and Friedmann 1983); 5—stainless steel (Smartt et al. 1986); 6—steel (Eagar and Tsai 1983); 7—steel (Niles and Jackson 1975); 8—mild steel (Christensen et al. 1965)

$$\eta_h = 0.947 - 0.00017\,I - 0.057\,l + 0.002\,v. \tag{1.2.16}$$

The dimension I–A, l–mm, v–mm s^{-1}. It follows from the equation that with the growing current I and the arc length l, the arc efficiency falls down while with the growing speed v it rises. Welding processes with non-consumable electrodes have an efficiency up to 25% higher than described in the current standard EN 1011-1 (Haelsig et al. 2017).

If the pressure of the shielding gas (argon and helium) increases, the arc efficiency falls dramatically due to growing radiation losses. For example, if the current is 100 A $\eta_h = 0.7$ with the pressure $p = 0.1$ MPa and $\eta_h = 0.4$ with $p = 30$ MPa (Katsaounis 1993).

During welding with a consumable electrode, the components of effective power have a more complicated nature and the arc efficiency is, as a rule, higher (Figs. 1.14 and 1.15) than during welding with a non-consumable electrode (Fig. 1.13). The higher values η_h are explained by the fact that some heat, which melts the electrode, goes with the drops to the pool and contributes to heating the base metal (Rykalin 1951, 1957).

The value η_h is, to a larger degree, dependent on the position of the arc against the welded body. For example, during multipass narrow gap welding almost all heat goes into the welded joint. During submerged arc welding, most heat losses are accounted for by flux melting (Fig. 1.14b). During submerged arc welding of low-carbon steel with alternating current the values η_h reach 0.90–0.99 (Christensen et al. 1965) and depend on the arc voltage, welding speed and flux composition (Rykalin and Kulagin 1953).

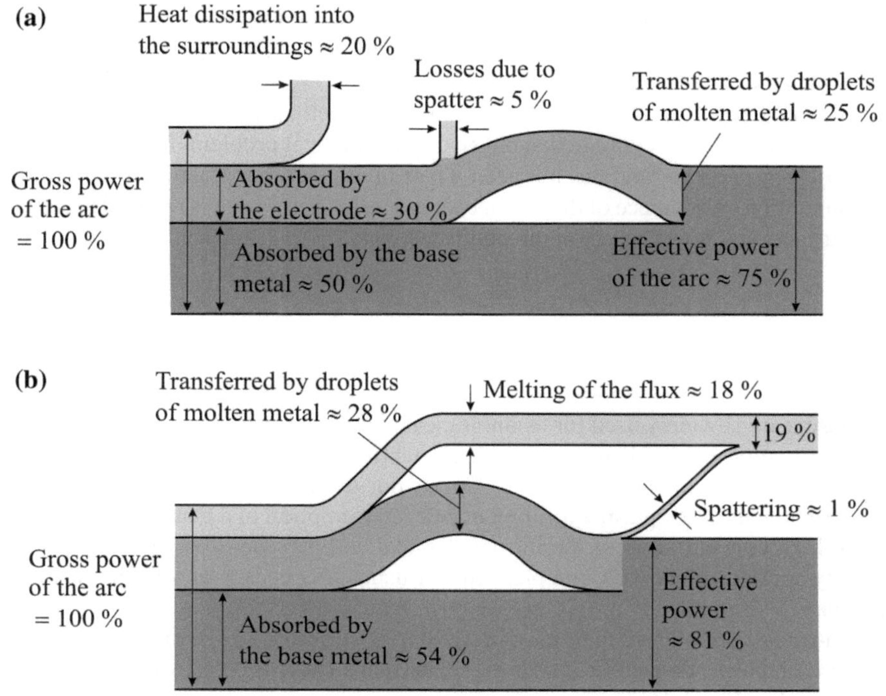

Fig. 1.14 Heat balance of welding arcs: an open metallic arc ($I = 150$–250 A, $U = 20$–25 V) and a submerged metallic arc ($I = 1000$ A, $U = 36$ V, $v = 6.67$ mm s^{-1}) (Radaj 1992; Rykalin 1951, 1957)

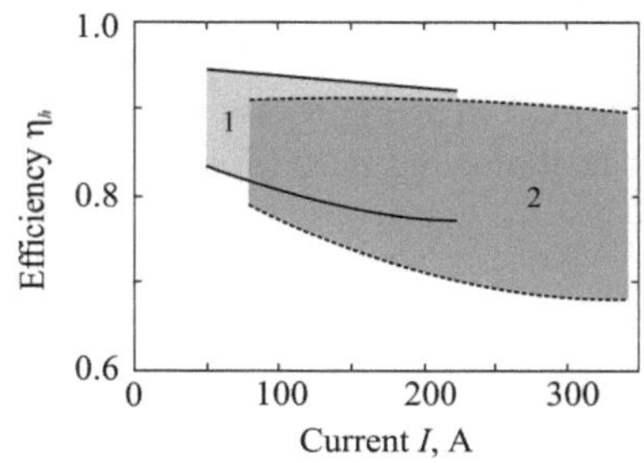

Fig. 1.15 Dependence of arc efficiency η_h on current I with arc length 1–2 mm: (1) wire diameter 0.89 mm, Ar + 8% CO_2; (2) wire diameter 1.2 mm, Ar + 18% CO_2 (Boswort 1991; Sudnik and Ivanov 1998; Watkins et al. 1990)

Welding processes with consumable electrodes should be categorised depending on the type of arc (short arc $\eta_h = 0.85$, pulsed arc $\eta_h = 0.77$, spray arc $\eta_h = 0.70$) (Haelsig et al. 2017).

It should be noted that physical phenomena modelling in the arc, electrode and welded joint is a very complicated physico-mathematical problem. It is mostly done for research purposes and has not been a part of engineering practice so far. So the consideration of influence of the arc on heat processes in the joint is reduced to setting the surface heat flows based on the published data.

1.3 Plasma Arc

A plasma arc is widely used for technological metal treatment due to the advantages it has in comparison with an open arc: high stability allowing performing the heating process by a long arc; versatile controllability due to some additional process parameters (diameter of the plasma forming nozzle, consumption of a plasma forming gas, etc.); high concentration of the introduced heat and pressure allow obtaining high quality welded joints with complete fusion using less energy and having a lower residual distortion.

The most common arc to be used in welding is a transferred plasma arc (the workpiece is an electrode and the nozzle is isolated) at a straight polarity (the workpiece is an anode) (Fig. 1.16).

The plasma arc thanks to being constricted by a gas streaming at an orifice in the nozzle is marked with a temperature higher than that of the conventional (open) arc (Fig. 1.17), a higher rate of plasma flow and, consequently, a higher pressure on the surface of the workpiece. These qualities of the plasma arc make it possible to weld, for example, 10 mm thick aluminium alloy AMg6 (91.1–93.7% Al, 5.8–6.8% Mg) plates at the speed of 8 mm s^{-1} in a single pass and obtain relatively narrow welds with a small width of the heat affected zone (Sosnin et al. 2008).

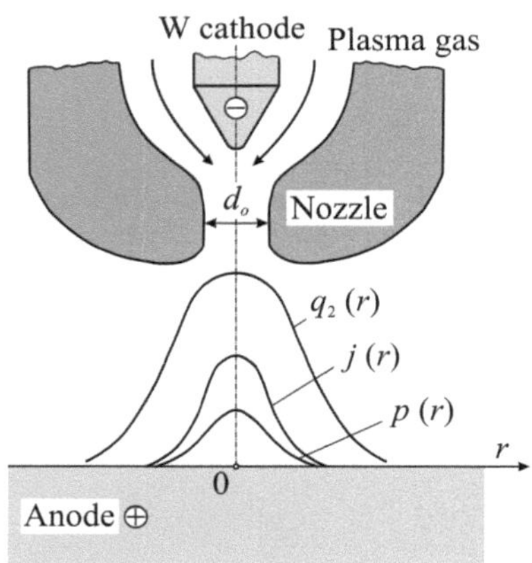

Fig. 1.16 Schematic of plasma torch and radial distributions of power density q_2, current density j and pressure p at the workpiece (Sosnin et al. 2008)

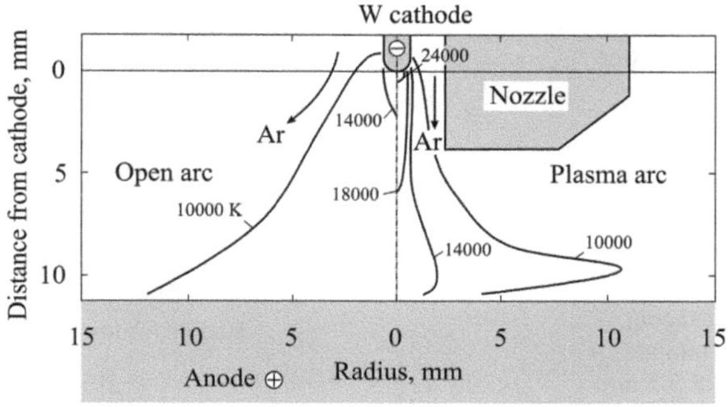

Fig. 1.17 Temperature fields of the open arc (current 200 A, voltage 14.5 V) and plasma arc (current 200 A, voltage 29 V, orifice diameter 4.8 mm) (Gage 1959)

Plasma torches are used to obtain a plasma jet. Their main part is a nozzle having an orifice with diameter d_o (Fig. 1.16). The plasma jet is formed in the orifice and stabilised by its walls and a cold plasma-forming gas, separating the arc from the walls. Argon, helium and other gases that are inert to the electrode (tungsten, copper, etc.) are used as a plasma gas.

A plasma arc is used a lot for surfacing including hardfacing and cladding. The principle of these plasma techniques is coating a metal surface with a melted powder or wire filler material in order to restore the dimensions of worn pieces or to make new products with wear and corrosion resistant properties of the surface. In gas metal arc welding, the plasma arc is considered to be a normally distributed heat source (Alaluss et al. 2010).

Low-amperage arcs can be used to join high-temperature materials that are hard to weld. Their feature is low heat input introduced to a workpiece, which improves its weldability (Prokhorov 1976).

When open and plasma arcs are compared at the same current, it can be seen that the open welding arc in argon is characterised by a higher concentration of the heat flow (Fig. 1.18a). The heat flow concentration factor C noticeably reduces with the growing current of the open arc against the plasma arc, whose C virtually does not depend on the current (Fig. 1.18b). The heat efficiency η_h of the open arc is 2–10% higher than that of the plasma arc. When the current grows, the efficiency reduces, which is caused by a higher heat loss due to a larger cross-section of the arc (Yushchenko et al. 2006). This tendency is common for more powerful arcs as well (Eq. 1.2.16).

During plasma arc welding of the aluminium alloy 6061 (0.8–1.2% Mg, 0.4–0.8% Si), thickness 6.3 mm under the typical conditions (current 100–130 A, voltage 22–32 V, welding speed 3.0 mm s^{-1}, orifice diameter 3.2 mm, plasma-forming argon consumption 0.94 l min^{-1}, consumption of shielding helium 19 l min^{-1}) the dependence of arc efficiency η_h on the current reduces, but on the voltage (arc length) increases. For example, when the current is 110 A and the voltage increases from 22 to 32 V (the distance from the orifice to the workpiece grows from 5.6 to 10.5 mm) the values η_h rise from 0.49 to 0.60, which is explained

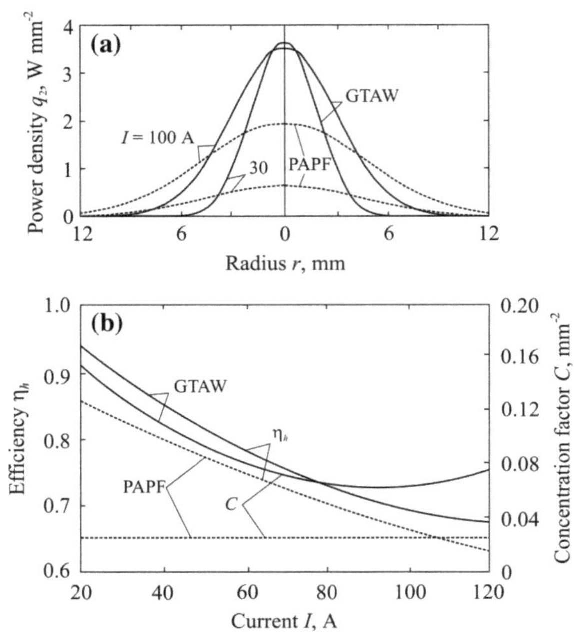

Fig. 1.18 Experimental radial distributions of power density $q_2\ (r)$ (**a**) and dependence of concentration factor C and arc efficiency η_h on current I (**b**) in gas tungsten arc welding with filler wire (GTAW) and a 5 mm long arc and plasma-arc powder facing (PAPF) with a 7.5 mm arc length and nozzle of 3.5 mm in diameter (Yushchenko et al. 2006)

by a growth in the heat transfer caused by the shielding gas (Evans et al. 1998). For open arcs the latter relation is inverse (see Eq. (1.2.16)).

In order to set the heat and force impact on the weld pool, it is necessary to know the radial distribution of the temperature and the velocity of the plasma-forming gas, which can be determined by experimental and computational methods (Carlson et al. 2013; Turichin et al. 2013). The arc efficiency values of plasma arcs are presented in the papers (DuPont and Marder 1995; Evans et al. 1998; Fuerschbach 1995; Metcalfe and Quingley 1975; Yushchenko et al. 2006).

It should be noted that in order to apply refractory coatings and hot straightening of welded structures, it is common to use an arc which burns in a plasma torch between the electrode and the plasma gas nozzle (a non-transferred arc). The density of the plasma jet heat flow is satisfactorily described by the Gaussian distribution law. The flow concentration factor C and the efficiency of jet η_h decrease drastically with distancing from the plasma torch (Kudinov and Ivanov 1981).

1.4 Electron Beam

An electron beam and a laser beam are highly concentrated energy sources. The power density of both sources is 1–2 orders of magnitude greater than that of conventional welding arcs (Fig. 1.1).

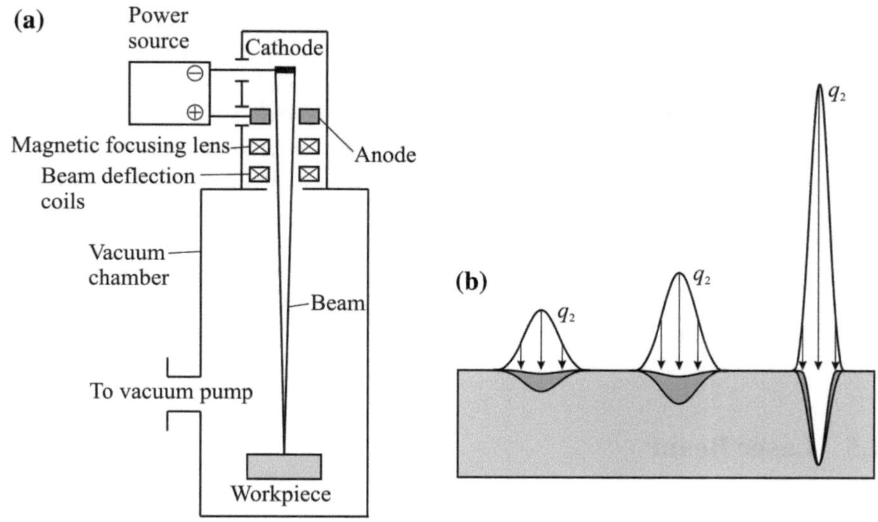

Fig. 1.19 Schematic representation of electron beam welding (**a**) and effect of power density on melt-in (or conduction) and keyhole modes (**b**) (Frolov 1988; Nerovny 2016)

An electron beam as a process tool allows heating, melting and evaporating virtually all materials. The source of electrons is a cathode, which has a high emission ability (Fig. 1.19a). A high accelerating voltage (20–150 kV) is applied to the cathode and an annular anode. In the main gap, electrons accelerate up to a high velocity. At a voltage equal to 150 kV, the electron velocity in high vacuum reaches 2×10^8 m s^{-1} (2/3 of the velocity of light) and the electrons have a relativistic mass increase of 35% (Messler 1999). By using a magnetic lens, the electron beam is focused to the required size. The deflecting magnetic lens allows moving the beam on the processed surface along virtually any trajectory. The electron flow meets the workpiece and its energy transforms into thermal one.

The dependence of the relative current density on the relative radius corresponds to the Gaussian distribution (Fig. 1.20).

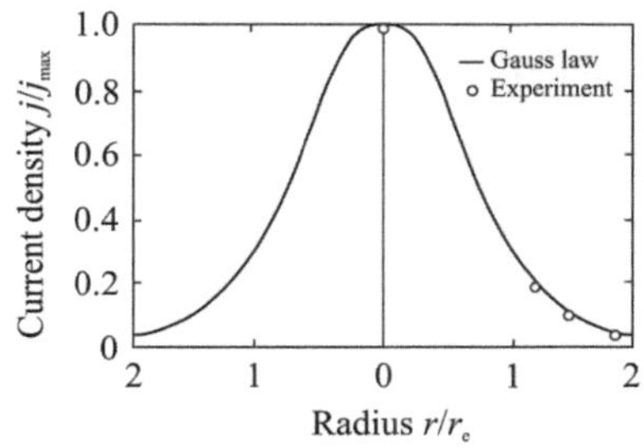

Fig. 1.20 Radial distribution of the electron beam current (Rykalin et al. 1988)

The electron beam control system allows obtaining any power density required for welding. The power distribution in-depth is determined by the way the beam interacts with the surface of the welded body. If the beam is defocused and has a low power density, it acts as a surface source (Fig. 1.19b, on the left). If the beam is focused and has a high power density, a vapour cavity emerges and the beam acts as a 3D source, distributed both in the plane of the plate and along its depth (Fig. 1.19b, on the right). Heat models for the beam are discussed in numerous publications.

The heat efficiency of the beam η_h is determined by losses (X-radiation, secondary electron emission, electron reflection and thermal electron emission from the processed surface). It changes from 0.70 for the surface beam to 0.95 and above for the submerged beam (Messler 1999).

1.5 Laser Beam

Laser technologies are widely used in industry. Differently from electron beam welding, laser welding does not need vacuum environment and at the same time it allows obtaining very high densities of the heat flow (Fig. 1.1).

Depending on the type of the applied active medium, lasers are divided into solid-state, gas, liquid and semiconductor ones. In welding Nd:YAG solid state, CO_2 gas and semiconductor lasers are used.

Nd: YAG (neodymium: yttrium-aluminium-garnet) solid-state lasers generate a laser with a wavelength of 1.06 μm, have power output in general equal to 0.1–3 kW (Messler 1999). They consist of a rod made of an active medium (Fig. 1.21). Mirror 1 is fully reflecting, and mirror 2 is partially reflecting. A gas discharge lamp, fed from a high voltage capacity battery, serves for energisation. The rod and the lamp are located in a reflecting case. Focusing optics allows obtaining high power density on the surface of the workpiece.

CO_2 gas lasers generate laser light with a wavelength of 10.6 μm. They have high power of 0.5–25 kW (Messler 1999) with a high energy conversion efficiency (0.08–0.3) (Frolov 1988). Similarly to electron beam welding, the power distribution of the laser beam in the plane of the workpiece and along its depth is determined

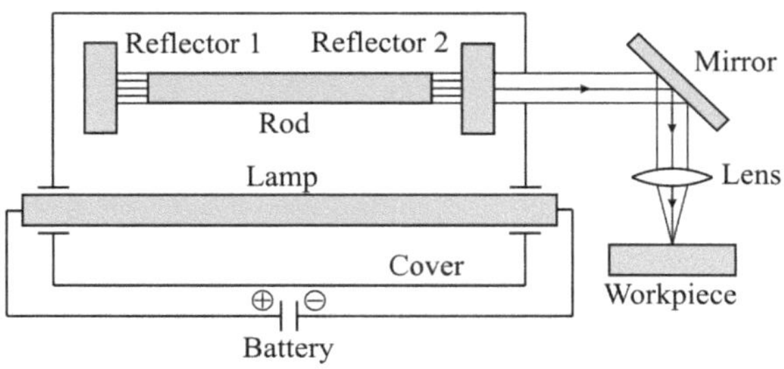

Fig. 1.21 Schematic of a solid-state laser (Messler 1999; Frolov 1988)

Table 1.2 Heat source efficiency η_h for various fusion welding processes (Messler 1999)

Process	Efficiency η_h
1	2
Oxy-fuel gas	
Low combustion intensity fuel	0.25–0.50
High combustion intensity fuel	0.50–0.80
Gas-tungsten arc	
Low current, straight polarity	0.40–0.60
High current, straight polarity	0.60–0.80
Reverse polarity	0.20–0.40
Alternating current	0.20–0.50
Plasma arc	
Melt-in mode	0.70–0.85
Keyhole mode	0.85–0.95
Gas-metal arc	
Globular transfer mode	0.60–0.75
Spray transfer mode	0.65–0.85
Shielded-metal or flux-cored arc	0.65–0.85
Submerged arc	0.85–0.99
Electroslag	0.55–0.85
Electron beam	
Melt-in mode	0.70–0.85
Keyhole mode	0.85–0.95
Laser beam	
Reflective surfaces or vapours	0.005–0.50
Keyhole mode	0.50–0.75

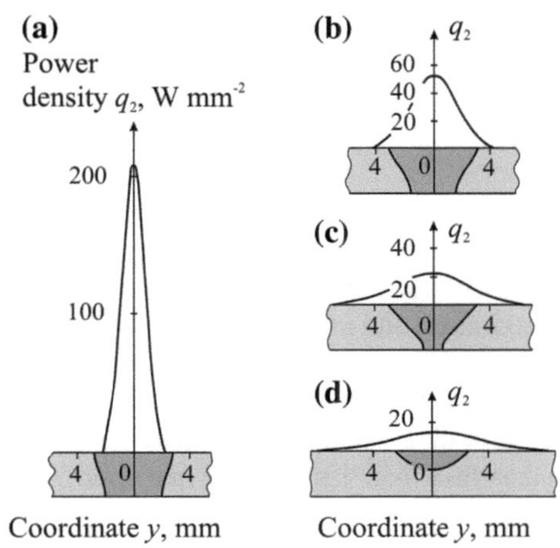

Fig. 1.25 The effect of power density distribution on the weld penetration and cross-sectional shape in welding of AA 6061 aluminium alloy 3.2 mm thick (speed 4.23 mm s^{-1}, power 880 W): electron beam welding (**a**); plasma arc welding (**b**); gas tungsten arc welding, electrode negative (**c**) and tungsten arc welding, electrode positive (**d**) (Kou and Le 1984; Messler 1999)

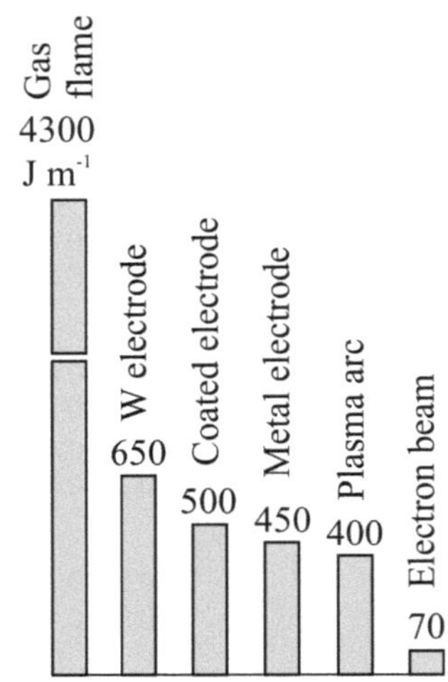

Fig. 1.26 Energy consumption for butt welding of a 3 mm thick steel plate using different welding methods (Sandvik 1977)

1.8 Resistance Spot Welding

In terms of the scope of application, resistance spot welding exceeds all other methods of welding with pressure combined. Resistance spot welding remains the main method of joining in passenger car and light truck production, with 5,000 spot welds per car. The principle of resistance spot welding is shown in Fig. 1.27a. Sheets are held by water-cooled copper electrodes, with force F. Current I for a time t_w flows through sheets with generation of heat:

$$Q = \int_{0}^{t_w} I^2(t)R(t)dt, \quad (1.8.1)$$

where R is the electrical resistance of circuit components: basic materials (electrodes and welded sheets) and interfaces (electrode-sheet and sheet-sheet).

Heat is generated mostly in the contact points of the plates (where resistance is maximum) and melts the metal, forming the future weld spot, or nugget (Fig. 1.27b). After current I is switched off, sheets are pressed with the same or stronger force and the liquid metal solidifies (Fig. 1.27c). It is common to use direct or alternating current of 1–100 kA. Power consumption is several dozens or hundreds of kVA. It takes less than 1 s to weld one spot. An ideal nugget diameter between sheets of equal thickness is 0.6–0.7 of the joint thickness (Messler 1999).

Resistance seam welding is carried out in the same way as resistance spot welding, ensuring a continuous tight joint. It is achieved by a subsequent positioning of a series

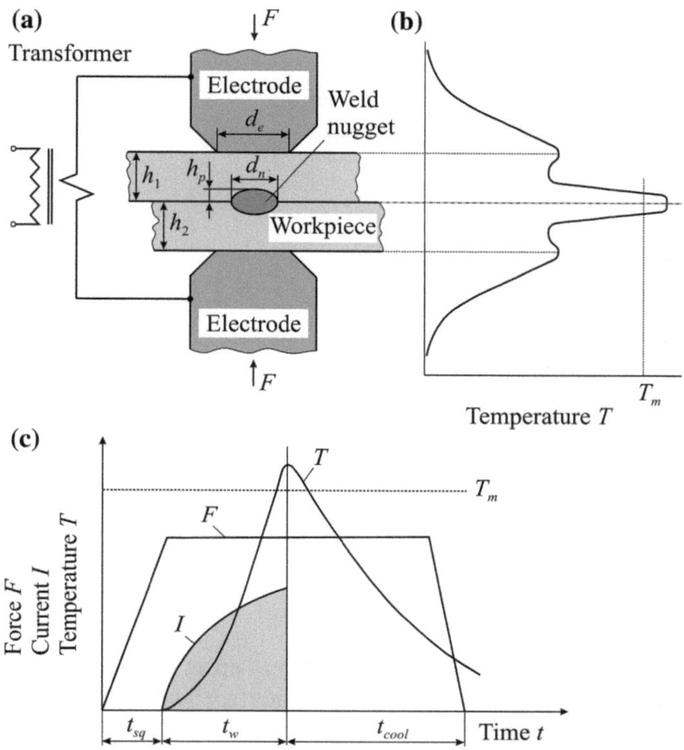

Fig. 1.27 Schematic of resistance spot welding (**a**), axial temperature distribution at the time when maximum temperature is reached (**b**) and the welding cycle: t_{sq} is the squeeze time, t_w is the weld time, t_{cool} is the cooling time and T_m is the melting temperature (Sandvik 1977)

of spots with partial overlapping of the previous spot by the next one. Force is applied continuously and current continuously or intermittently to produce a linear weld.

The heat source in resistance spot welding is three-dimensional, its power density is determined by the product of $\rho_e \, j^2$ (W m^{-3}), where ρ_e is the specific electric resistance and j is the current density. In engineering heat calculation, the source form is simplified to a plane circle and positioned between sheets on the interface with the contact resistance R_c. Hence, an essential problem emerges: how big is the resistance R_c? It depends on the surface state, which, in turn, depends on the pressure and unknown temperature, i.e. the problem is non-linear in the temperature. There are many models for interfaces (Galler et al. 2010; Gelman 1970; Kochergin 1987).

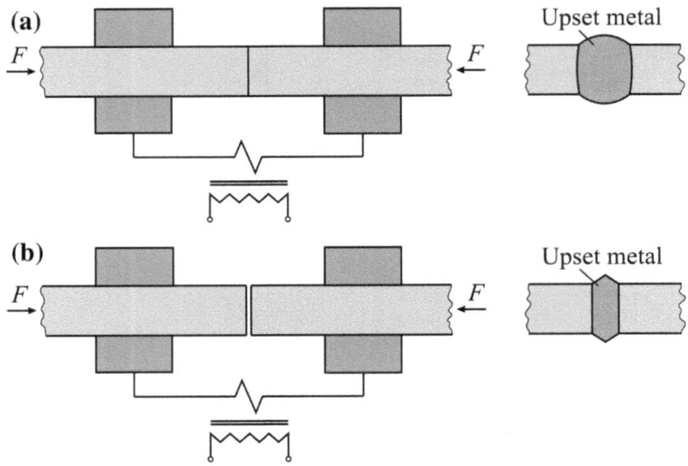

Fig. 1.28 Schematic diagrams of resistance butt welding (**a**) and flash welding (**b**)

1.9 Resistance Butt Welding

During resistance butt welding (designated also as upset welding), the heating process is ensured by the heat generated at the interface and in welded components (Fig. 1.28a). The components are butted under pressure before heating is started. Since the contact resistance R_c is considerably greater than that of the components, heat is concentrated at the interface:

$$Q_c = \int_0^{t_w} I^2(t)R_c(t)dt, \tag{1.9.1}$$

where t_w is the time of welding.

The volumetric power density of the heat source in the pieces q_3 is determined by the current density j and specific electric resistance ρ_e, $q_3 = \rho_e j^2$.

In resistance butt welding, a considerable non-uniformity in heating is common for components over their cross-section due to the non-uniform (rough contacting surfaces) resistance distribution R_c. Contact resistance depends on the material and initial state of the surface as well as on the upsetting force F and current I. These dependencies are the most difficult to determine when setting heat sources for calculating the temperature fields in the welding process. This problem is discussed in the special literature (Gelman 1970; Kochergin 1987).

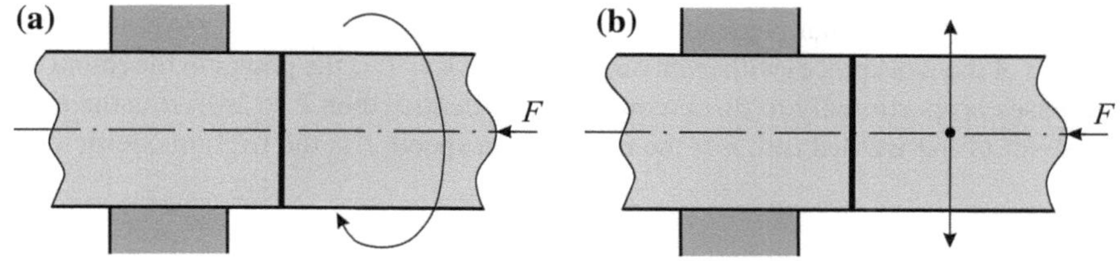

Fig. 1.29 Schematic diagrams of rotational (**a**) and linear (**b**) friction welding (Lebedev et al. 1987)

1.10 Flash Welding

A flash welding scheme is shown in Fig. 1.28b. As the initial gap between two parts is reduced under voltage, high points of the faying surface of each part begin to cause arcing. The metal in the contact quickly heats and explodes, which results in continuous sparking, common for flash welding. With flashing-off of the entire section of the workpieces for an equal value, the average density of heat generated at the faying surface becomes uniform (Gelman 1970). With sufficient flashing-off of the workpieces, the force F is applied and, consequently, molten metal is expelled, the hot metal is plastically upset, forming a weld.

In the steady state, the frequency of contact forming at the faying surface (spark generation frequency) is large. So the heat source can be considered as plane and uniformly distributed. The boundary conditions on the interface can be set in reference to the temperature or heat flow (see Sect. 6.2).

1.11 Friction Welding

In friction welding the surfaces of the workpieces are heated by direct conversion of mechanical energy into heat due to friction.

In rotational friction welding two workpieces are placed in a clamping device, one of them is static and the other one is brought to rotation (Fig. 1.29a). The interface is heated to the required temperature (when steel is welded the temperature in the butt joint reaches 1300–1600 K) (Olshansky 1978). During friction the heated softened metal is pressed and upset under an axial force, scrubbing away any foreign oxides and inclusions and forming a weld. Heating is stopped by an abrupt cessation of rotation. The theory of friction welding is discussed in the special literature (Lebedev et al. 1987; Vill 1962).

The main parameters of the rotational friction welding conditions are the rotation speed n, the axial force F and the welding time t_w, which set the power of the heat source (Olshansky 1978)

$$q = k\pi\, dnf F, \tag{1.11.1}$$

where k is the coefficient depending on the type of power distribution over the cross-section of the workpiece (with even distribution of $k = 1$; if the power in the round rod increases proportionally to distancing from the centre, then $k = 2/3$), d is the outer diameter of the welded rod, n is the rod rotation speed, f is the friction coefficient.

1.12 Friction Stir Welding

The process of friction stir welding, developed at The Welding Institute, UK (Thomas et al. 1991), allows making butt, corner, angle and lap joints. It has spread mainly for aluminium alloy joints (Ishchenko et al. 2007). The principle of the method is relatively simple: a cylindrical rotating tool of a special shape with a shoulder and a pin (tool probe) is placed in the butt plane of workpieces that are closely pressed together (Fig. 1.30a). Herewith the metal is heated due to friction and plastic deformation. When the tool traverses along the butt plane of the joint, the heated metal moves from the area in front of the pin to the area behind it, is formed by the shoulder and forms a welded joint (Fig. 1.30b). The tool is pressed by the force F_z and inclined at a tilt angle $\alpha < 3°$ (Fig. 1.30d). After welding there is a hole, which is welded up or brought to run-off tabs (Fig. 1.30c).

The weld seam turns out asymmetrical in reference to its longitudinal section, because the material is pushed from one (advancing) side to the other (retreating) side by the pin (Fig. 1.31). A welded joint has a nugget, a thermo-mechanically affected zone (TMAZ), a heat affected zone (HAZ) and a base metal (BM).

The advantages of friction stir welding are: single-pass welds of very deep penetration (up to 75 mm thick aluminium alloys (Martin 2006)), no consumables, possibility to weld pieces made of dissimilar materials, cost benefits. The drawbacks of the method are: exit hole, large downwards force of the tool, less flexibility compared to arc welding.

Welding of 1–30 mm thick aluminium alloys has become widely spread in shipbuilding, plane-, rocket-, car-making and construction, as well as for joining magnesium, copper and other ductile alloys.

The main parameters of the welding conditions are the following: welding (travel) speed v, tool rotation speed ω, axial (plunge) force F_z and longitudinal force F_x, tilt angle of the tool α, shape and dimensions of the tool (Fig. 1.30). Conditions of friction, depending on the material of the tool and welded material, and the yield strength of the base metal at the deformation temperature are considered too.

The heat source in friction stir welding is a very complicated object of the control system. Differently from many welding processes, for instance, arc welding, heat generation in friction stir welding is not directly controllable and depends on the process itself. It can be controlled indirectly by selecting welding parameters. The mechanisms of heat generation are determined by friction and plastic deformation, which, in turn, depend on the conditions of the tool-workpiece contact and the field of the material flow speed (Schmidt 2010).

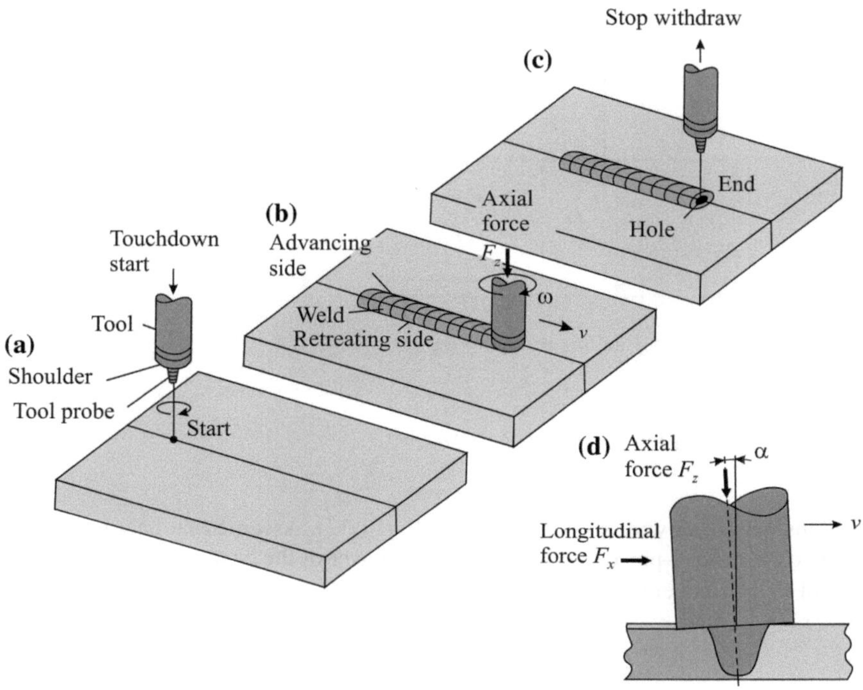

Fig. 1.30 Schematic diagram of the friction stir welding process: start (**a**), welding (**b**), end (**c**), longitudinal section (**d**) (Threadgill et al. 2009)

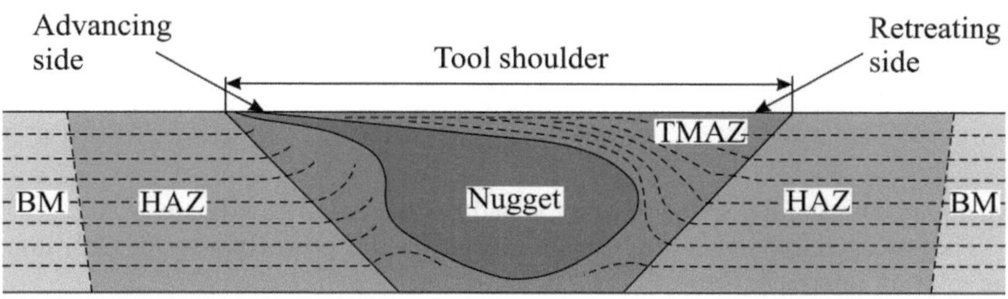

Fig. 1.31 Cross section of a friction stir weld showing different zones of the weld

In order to evaluate heat distribution, a phenomenological model is used. It is based on the balance of forces acting on the tool-workpiece interface. If the force F_z, directed along the tool axis, (Fig. 1.30) is known, pressure and friction force can be found. The torque, applied to the tool, almost completely converts into heat and can be easily measured (Schmidt 2010).

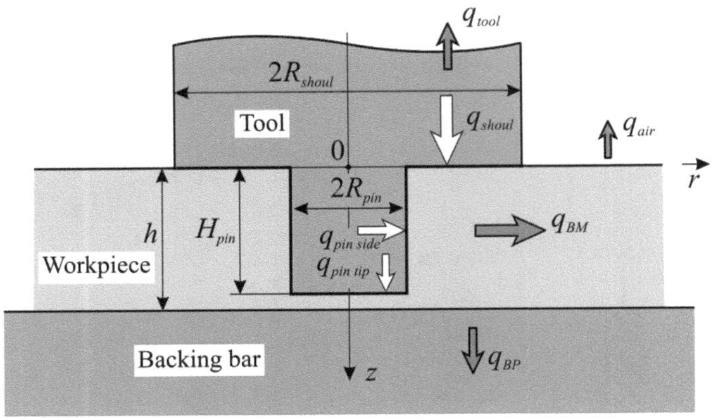

Fig. 1.32 Heat flow pathways in friction stir welding

Consider a tool with a cylindrical pin (Fig. 1.32). Assume that heat is only produced on the tool—piece interface. The total power of the heat source represents the sum of its components:

$$q = q_{shoul} + q_{pin\,side} + q_{pin\,tip}, \tag{1.12.1}$$

where q_{shoul} is the source power on the shoulder-surface interface, $q_{pin\,side}$ is the source power on the lateral side of the pin-workpiece interface, $q_{pin\,tip}$ is the source power on the pin tip—workpiece interface.

Assume that shear stresses τ_{cont} are constant along the entire interface. Then the local density of heat power on this interface is equal to

$$q_2 = \tau_{cont}\omega r, \tag{1.12.2}$$

where ω is the angular speed of rotation and r is the radial coordinate.

The power components are determined by integrating power density (1.12.2) along the relevant interface:

$$q_{shoul} = \int_{R_{pin}}^{R_{shoul}} \int_{0}^{2\pi} (\tau_{cont}\omega r)\, r d\theta dr = \frac{2}{3}\pi\omega\tau_{cont}\left(R_{shoul}^3 - R_{pin}^3\right); \tag{1.12.3}$$

$$q_{pin\,side} = \int_{0}^{H_{pin}} \int_{0}^{2\pi} (\tau_{cont}\omega r)\, r d\theta dz = 2\pi\omega\tau_{cont} R_{pin}^2 H_{pin}; \tag{1.12.4}$$

$$q_{pintip} = \int_{0}^{R_{pin}} \int_{0}^{2\pi} (\tau_{cont}\omega r)\, r d\theta dr = \frac{2}{3}\pi\omega\tau_{cont} R_{pin}^3. \tag{1.12.5}$$

By summing up the Eqs. (1.12.3)–(1.12.5), the total power of the source is obtained:

$$q = \frac{2}{3}\pi\omega\tau_{cont}\left(R_{shoul}^3 + 3R_{pin}^2 H_{pin}\right). \quad (1.12.6)$$

If the pin has a conical shape, the equation will be a bit more complicated (Schmidt 2010).

Contribution of the three components in total power is different. For example, during welding a 3 mm thick aluminium alloy with yield strength equal to 20 MPa at an elevated temperature and by using a cylinder-shaped tool ($R_{shoul} = 9$ mm, $R_{pin} = 3$ mm, $H_{pin} = 3$ mm) 1419 W of heat is generated. Herewith, the share of the shoulder q_{shoul} is 87%, that of the lateral side $q_{pin\ side}$ is 10%, and that of the pin tip $q_{pin\ tip}$ is 3% of the generated heat (Schmidt 2010).

In order to calculate the temperature fields, it is necessary to know the power distribution of the source. Separate out the stress τ_{cont} from Eq. (1.12.6) and substitute it in (1.12.2) in order to obtain plane power density (Schmidt 2010):

$$q_2 = \frac{3qr}{2\pi(R_{shoul}^3 + 3R_{pin}^2 H_{pin})}. \quad (1.12.7)$$

Until now it was presumed that the metal of the workpiece does not move on the interface (there is external sliding friction and no plastic deformations). It is, to a larger degree, true for the shoulder-workpiece contact, and to a smaller extent for the pin-workpiece contact, where metal moves together with the pin tip. The contribution of the sliding friction and viscosity (plastic deformations) in the heat is marked with a relation between the local velocity of the surface layer of the workpiece and the local velocity of the tool (Schmidt 2010):

$$\delta = \frac{v_{layer}}{\omega r}, \quad (1.12.8)$$

where v_{layer} is the speed of the welded metal on the tool-workpiece interface. The condition $\delta = 0$ corresponds to pure sliding, condition $\delta = 1$—to full sticking of the workpiece material with the tool, condition $0 < \delta < 1$—to both sliding and sticking.

Allowing for the sliding and sticking phenomena, the power density of the heat source (1.12.2) can be written in a form of the total (Schmidt 2010)

$$q_2 = q_{2\,fric} + q_{2\,plast}; \quad (1.12.9)$$

$$q_{2fric} = (1 - \delta)\tau_{fric}\omega r; \quad (1.12.10)$$

$$q_{2\,plast} = \delta\tau_{plast}\omega r. \quad (1.12.11)$$

where τ_{fric} is the frictional shear stress, τ_{plast} is the yield strength.

The frictional stress is determined by the friction coefficient μ and normal pressure p at the contact interface:

$$\tau_{fric} = \mu\, p = \mu \frac{F_z}{\pi R_{shoul}^2}. \quad (1.12.12)$$

Coefficient μ is within a range of 0–0.6 (mostly $\mu = 0.3$) and depends on many factors. It can be determined experimentally by measuring the plunge force F_z and torque M (Schmidt 2010).

The yield strength of the material depends on the local temperature, $\tau_{plast} = \tau_{plast}(T)$. It falls down to about zero in case the temperature increased up to the solidus temperature T_S. The rate of falling $\partial \tau_{plast}(T)/\partial T$ is governed by the nature of the material.

In reality, contact conditions, as a rule, are unknown and so some simplified hypotheses are accepted: coefficient δ is constant, stresses on an interface are limited by the yield strength $\tau_{plast,}$ depending on the local temperature. The latter hypothesis allows considering a self-regulating process. Shear stresses on the interface are a determinant factor of heat generation. With a growing temperature the yield strength falls down, so the temperature is principally limited by the solidus temperature of the welded material.

Similarly to other welding methods, the effective power q is only part of the gross power q_{gross}, transferred by the tool and determined by torque M:

$$q = q_{gross} - q_{tool} = \frac{q_{gross} - q_{tool}}{q_{gross}} q_{gross} = \eta_h q_{gross};$$
$$q_{gross} = \omega M, \quad (1.12.13)$$

where the efficiency η_h considers the heat flowing to the tool, $\eta_h = 0.7$–0.9 (Schmidt 2010) (Fig. 1.32).

By substituting (1.12.13) into (1.12.7), we obtain the effective power distribution of the source

$$q_2 = \frac{3\eta_h q_{gross} r}{2\pi \left(R_{shoul}^3 + 3R_{pin}^2 H_{pin} \right)}. \quad (1.12.14)$$

If the effect of the lateral side of the tool is neglected, the formula (1.12.14) is simplified:

$$q_2 = \frac{3\eta_h q_{gross} r}{2\pi R_{shoul}^3}. \quad (1.12.15)$$

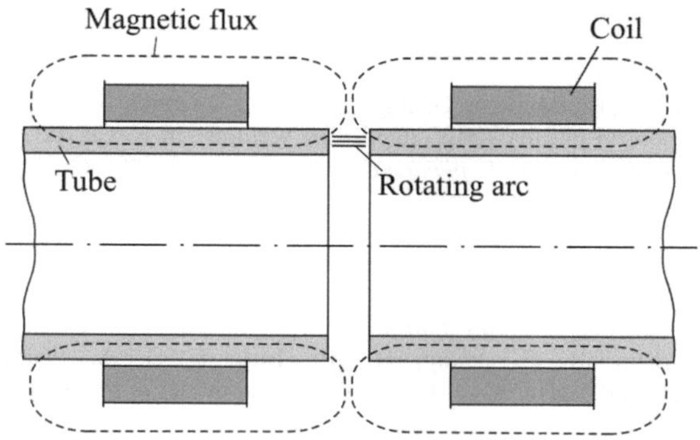

Fig. 1.33 Basic schematic of magnetically impelled arc welding (Gelman 1970)

Heat flows from the workpieces to the backing plate q_{BP} depend on the thermal conductivity and thickness of the backing plate, and to the air q_{air}—on the state of the workpiece surface and environment.

There are other more complex models for calculating the source power which consider many other factors (Mishra and Mahoney 2007; Lohwasser and Chen 2010; Mishra and Ma 2005).

1.13 Magnetically Impelled Arc Welding

In magnetically impelled arc welding, an electric arc of direct current rotates between pipe ends in the magnetic field, created by electro-magnetic coils (Fig. 1.33). The pipe and the coil are positioned coaxially, so that the magnetic fluxes will be directed towards each other and the radial component of the magnetic field will be produced. The arc current crosses the radial component of the field at a right angle, thus creating a force, which rotates the arc along the perimeter of the pipes. That is the very reason why this method is also called rotating arc welding. After sufficient heating of the edges of the pipes, an axial force is applied, which is accompanied by fin (flash) formation.

The heating time depends on the diameter of pipes and wall thickness: 15–20 s for pipes 30 × 1.5 mm, 120–150 s for pipes 325 × 9 mm (Gelman 1970).

In the steady state the arc moves at a high speed achieving 100 m s^{-1}, and ensures that the pipe edges should be uniformly heated. So, a rotating arc can be seen as a plane uniformly distributed source positioned perpendicularly to the axis of the pipes and having the power density

$$q_2 = \frac{\eta_h U I}{A}, \tag{1.13.1}$$

where A is the cross-sectional area. The arc power can vary in time.

References

Alaluss, K., Buerkner, G., Nguyen-Chung, T., Gehde, M., & Mennig, G. (2010). Simulation of weld pool in plasma—MIG deposition welding. In H. Cerjak & N. Enzinger (Eds.), *Mathematical modelling of weld phenomena* (Vol. 9, pp. 93–109). Graz: Verlag der Technischen Universitaet Graz.

Belousov, Yu. V. (2002). Evaluation of concentration of surface heat source with normally distributed heat power. *Welding Production, 8*, 8–12 (in Russian).

Bosworth, M. R. (1991) Effective heat input in pulsed current gas metal arc welding with solid wire electrodes. *Welding Journal, 5*, 111-s–117-s.

Carlson, B. E., Wang, H. -P., Turichin, G. A., Valdaitseva, Y. A., Ivanov, S. Yu., & Karkhin, V. A. (2013). Mathematical model of plasma jet for plasma arc brazing. In C. Sommitsch & N. Enzinger (Eds.), *Mathematical modelling of weld phenomena* (Vol. 10, pp. 737–751). Graz: Verlag der Technischen Universitaet Graz.

Cho, W. -I., Na, S. -Y., Cho, M. -H., & Lee, J. -S. (2010). A transient investigation of laser–arc hybrid welding by numerical simulation. In H. Cerjak & N. Enzinger (Eds.), *Mathematical modelling of weld phenomena* (Vol. 9, pp. 57–63). Graz: Verlag der Technischen Universitaet Graz.

Choo, R. T. C., Szekely, J., & Westhoff, R. C. (1990). Modelling of high-current arcs with emphasis on free surface phenomena in the weld pool. *Welding Journal, 9*, 346-s–361-s.

Christensen, N., Davies, V. L., & Gjermundsen, K. (1965). Distribution of temperatures in arc welding. *British Welding Journal*, 54–75.

Doan, G. E., & Lorentz, R. E. (1941). Crater formation and the force of the electric welding arc in various atmospheres. *Welding Journal, 20*, 103-s–108-s.

Dresvin, S. V. (Ed.). (1972). *Physics and techniques of low-temperature plasma* (351 pp.). Moscow: Atomizdat (in Russian).

DuPont, J. N., & Marder, A. R. (1995). Thermal efficiency of arc welding processes. *Welding Journal, 12*, 406-s–416-s.

Eagar, T. W., & Tsai, N. -S. (1983). Temperature fields produced by traveling distributed heat sources. *Welding Journal, 12*, 346-s–355-s.

Evans, D. M., Huang, D., McClure, J. C., & Nunes, A. C. (1998). Arc efficiency of plasma arc welding. *Welding Journal, 2*, 53-s–58-s.

Farmer, A. J. D., Haddad, G. N., & Cram, L. E. (1986). Temperature determinations in a free-burning arc: III measurements with molten anodes. *Journal of Physics D: Applied Physics, 19*, 1723–1730.

Finkelnburg, W., & Maecker, H. (1961). *Electric arcs and thermal plasma* (370 pp.). Moscow: Foreign Literature Publishing (in Russian).

Frolov (Ed.). (1988). *Theory of welding processes* (559 pp.). Moscow: Vysshaya Shkola (in Russian).

Fuerschbach, P. W. (1995). A dimensionless parameter model for arc welding processes/trends in welding research. In *Proceedings of the 4th International Conference* (pp. 493–497), 5–8 June 1995, Gatlinburg, Tennessee.

Gage, R. M. (1959). Principles of the modern arc torch. *Welding Journal, 38*(10), 959–962.

Galler, M., Ernst, W., Vallant, R., & Enzinger, N. (2010). Simulation based determination of the electrical contact resistance during resistance spot welding. In H. Cerjak & N. Enzinger (Eds.), *Mathematical modelling of weld phenomena* (Vol. 9, pp. 883–900). Graz: Verlag der Technischen Universitaet Graz.

Gelman, A. S. (1970). *Principles of resistance welding* (312 pp.). Moscow: Mashinostroenie (in Russian).

Gick, A. E. F., Quigley, M. B. C., & Richards, P. H. (1973). The use of electrostatic probes to measure the temperature profiles of welding arcs. *Journal of Physics D: Applied Physics, 6*, 1941–1949.

Giedt, W. H., Tallerico, L. N., & Feurschbach, P. W. (1989). GTA welding efficiency: Calorimetric and temperature field measurements. *Welding Journal, 1*, 28-s–32-s.

Glickstein, S. S. (1981). Basic studies of the arc welding process. In *Trends in welding research in the United States. Proceedings of a Conference* (pp. 3–51).
Glickstein, S. S., & Friedmann, E. (1983). Temperature transients in gas tungsten arc weldments. *Welding Review*, *62*(5), 72–73.
Grigoryants, A. G. (1994). *Basics of laser material processing* (313 pp.). Taylor and Francis Inc.
Haddad, G. N., & Farmer, A. Y. D. (1984). Temperature determinations in a free–burning arc. I: experimental techniques and results in argon. *Journal of Physics D*, *17*, 1189–1196.
Haddad, G. N., Farmer, A. Y. D., Kovitya, P., & Cram, L. E. (1985). Physical processes in gas—tungsten arcs. *IIW Doc*. 212-627-85.
Haelsig, A., Pehle, S., Kusch, M., & Mayr, P. (2017). Reducing potential errors in the calculation of cooling rates for typical arc welding processes. *Welding in the World*, *61*, 745–754.
Hertel, M., Fuessel, U., Schnick, M., Reisgen, U., Mokrov, O., Zabirov, A., & Spille-Kohoff, A. (2013). Numerical simulation of arc and metal transfer in gas metal arc welding. In C. Sommitsch & N. Enzinger (Eds.), *Mathematical modelling of weld phenomena* (Vol. 10, pp. 67–81). Graz: Verlag der Technischen Universitaet Graz.
Hiraoka, K., Shiwaku, T., & Ohji, T. (1997). Determining temperature distributions of gas tungsten arc (TIG) plasma by spectroscopic methods. *Welding International*, *11*(9), 688–696.
Hsu, K. C., Etemadi, K., & Pfender, E. (1983). Study of the free-burning high intensity argon arc. *Journal of Applied Physics*, *54*, 1293–1301.
Ishchenko, A. Ya., Podielnikov, S. V., & Poklyatsky, A. G. (2007). Friction stir welding of aluminium alloys (Review). *The Paton Welding Journal*, *11*, 25–30.
Jackson, C. E. (1960). The science of arc welding. *Welding Journal*, *39*, 129-s–140-s, 177-s–190-s, 225-s–230-s.
Katsaounis, A. (1993). Heat flow and arc efficiency at high pressures in argon and helium tungsten arcs. *Welding Journal*, *9*, 447-s–454-s.
Key, J. F., Chan, J. W., & McIlwain, M. E. (1983). Process parameter influence on arc temperature distribution. *Welding Journal*, *62*, 179-s–184-s. IIW Doc. 212-549-83.
Kobayashi, M., & Suga, T. (1979). A method for the spectral temperature measurement of a welding arc. In W. Lucas (Ed.), *Arc physics and weld pool behaviour* (pp. 25–37). Cambridge: The Welding Institute.
Kochergin, K. A. (1987). *Resistance welding* (240 pp.). Leningrad: Mashinostroenie (in Russian).
Kopayev, B. V., Rybachuk, A. M., & Lebedev, V. A. (2006). On selection of empirical formulae for arc pressure distribution. *Welding Production*, *4*, 3–8 (in Russian).
Kou, S., & Le, Y. (1984). Heat flow during the autogeneous GTA welding of aluminum alloy pipes. *Metallurgical Transactions A*, *15A*(6), 1165–1171.
Kovitya, P., & Lowke, J. J. (1982). Two-dimensional calculations in welding arcs in argon. *IIW Doc*. 212-534-82.
Kovitya, P., & Lowke, J. J. (1985). Two-dimensional analysis of free-burning arcs in argon. *Journal of Physics D*, *18*, 53–70.
Kudinov, V. V., & Ivanov, V. M. (1981). *Plasma refractory coating* (192 pp.). Moscow: Mashinostroenie (in Russian).
Lancaster, J. F. (Ed.). (1986). *The physics of welding* (2nd ed., 340 pp.). Oxford: Pergamon Press.
Lancaster, J. F. (1987). The physics of fusion welding part 1: The electric arc in welding. In *IEEE Proceedings*, *134*, Pt. B(5), 233–254.
Lebedev, V. K., Chernenko, I. A., & Vill, V. I. (Eds.). (1987). *Friction welding. Handbook* (236 pp.). Leningrad: Mashinostroenie (in Russian).
Lee, S.-Y., & Na, S.-J. (1996). A numerical analysis of a stationary gas tungsten welding arc considering various electrode angles. *Welding Journal*, *9*, 269-s–279-s.
Leskov, G. I. (1970). *Electric welding arc* (335 pp.). Moscow: Mashinostroenie (in Russian).
Lindgren, L.-E. (2007). *Computational welding mechanics. Thermomechanical and microstructural simulations* (248 pp.). Cambridge: Woodhead Publishing Ltd.

Lohwasser, D., & Chen, Z. (Eds.). (2010). *Friction stir welding: From basics to applications* (424 pp.). Oxford: Woodhead Publishing.

Lopota, V. A., Turichin, G. A., Valdaytseva, E. A., Malkin, P. E., & Gumenyuk, A. V. (2006). Computer system for modelling of electron beam and laser welding. *Automatic Welding, 2*, 18–21 (in Russian).

Lowke, J. J., & Tanaka, M. (2006). LTE—Diffusion approximation for arc calculations. *Journal of Physics D: Applied Physics, 39*, 3634–3643.

Lu, M., & Kou, S. (1988). Power and current distributions in gas tungsten arcs. *Welding Journal, 2*, 29-s–34-s.

Makhnenko, V. I., & Kravtsov, T. G. (1976). *Thermal processes in mechanized deposition on circular cylinder-shaped workpieces* (159 pp.). Kiev: Naukova Dumka (in Russian).

Martin, J. (2006, Jan/Feb). Pushing the boundaries—Friction stir goes deeper than before. *TWI Connect, 1*.

Matsunawa, A., & Nishiguchi, M. (1979). The cathode mechanism in free burning arcs with refractory electrodes: Probe measurement in low pressure arcs and the mechanism of a cathode plasma ball. In W. Lucas (Ed.), *Arc physics and weld pool behaviour* (pp. 67–77). Cambridge: The Welding Institute.

Messler, R. W. Jr. (1999). *Principles of welding: Processes, physics, chemistry, and metallurgy* (662 pp.). New York: Wiley.

Metcalfe, J. C., & Quingley, M. B. C. (1975). Heat transfer in plasma-arc welding. *Welding Journal, 54*(3), 99-s–103-s.

Mishra, R. S., & Ma, Z. Y. (2005). Friction stir welding and processing. *Reports: A Review Journal. Materials Science and Engineering R, 50*, 1–78.

Mishra, R. S., & Mahoney, M. W. (Eds.). (2007). *Friction stir welding and processing* (352 pp.). Materials Park, Ohio: ASM International.

Mochizuki, M., Tanaka, M., & Okano, S. (2010). Distortion analysis by combining arc plasma process with weld mechanics. In H. Cerjak & N. Enzinger (Eds.), *Mathematical modelling of weld phenomena* (Vol. 9, pp. 551–578). Graz: Verlag der Technischen Universitaet Graz.

Murphy, A. B., Tanaka, M., Yamamoto, K., Tashiro, S., Sato, T., & Lowke, J. J. (2009). Modelling of thermal plasmas for arc welding: The role of the shielding gas properties and of metal vapour. *Journal of Physics D: Applied Physics, 42*, 1–20.

Murphy, A. B., & Thomas, D. G. (2017). Prediction of arc, weld pool and weld properties with a desktop computer model of metal-inert-gas welding. *Welding in the World, 61*, 623–633.

Nerovny, V. M. (Ed.). (2016). *Theory of welding processes* (2nd ed., 702 pp.). Moscow: MVTU Publishing (in Russian).

Nestor, O. H. (1962). Heat intensity and current density distributions at the anode of high-current, inert gas. *Journal of Applied Physics, 33*(5), 1638–1648.

Niles, R. W., & Jackson, C. E. (1975). Weld thermal efficiency of the GTAW process. *Welding Journal, 1*, 25 s–32 s.

Nomura, K., Ogino, Y., Murakami, K., & Hirata, Y. (2010). Features of magnetic controlled TIG arc plasma—Modelling and experiment. In H. Cerjak & N. Enzinger (Eds.), *Mathematical modelling of weld phenomena* (Vol. 9, pp. 83–91). Graz: Technischen Universitaet Graz.

Olsen, H. N. (1963). The electric arc as a light source for quantitative spectroscopy. *Journal of Quantitative Spectroscopy and Radiative Transfer, 3*, 305–333.

Olshansky, N. A. (Ed.). (1978). *Welding in engineering industry. 1. Handbook* (504 pp.). Moscow: Mashinostroenie (in Russian).

Petrov, G. L., & Tumarev, A. S. (1977). *Theory of welding processes* (2nd ed., 392 pp.). Moscow: Vysshaya Shkola Publishing (in Russian).

Petrunichev, V. A. (1960). Thermal and mechanical effect of high-power arc on weld pool. In N. N. Rykalin (Ed.), *Processes of melting of base metal during welding* (pp. 117–166). Moscow: Publishing House of the Academy of Sciences of the USSR (in Russian).

Prokhorov, N. N. (1976). *Physical processes in metals during welding. 2 Stresses, deformations and phase transformations* (600 pp.). Moscow: Metallurgiya (in Russian).

Radaj, D. (1992). *Heat effects of welding. Temperature field, residual stress, distortion* (348 pp.). Berlin: Springer.
Rykalin, N. N. (1951). *Calculation of heat flow in welding* (Z. Paley & C. M. Adams, Jr. Trans.) (337 pp.). Moscow.
Rykalin, N. N. (1957). *Berechnung der Waermevorgaenge beim Schweissen* (326 pp.). Berlin: VEB Verlag Technik (in German).
Rykalin, N. N. (1974). Energy sources for welding. *Welding in the World*, *12*(9/10), 227–248.
Rykalin, N. N., & Kulagin, I. D. (1953). Thermal parameters of the welding arc. In V. P. Nikitin (Ed.), *Thermal processes in welding* (pp. 10–58). Moscow: Publication of the USSR Academy of Sciences (in Russian).
Rykalin, N. N., & Shorshorov, M. H. (1953). Heating of thin metal sheets and massive workpieces with gas flame torch. In V. P. Nikitin (Ed.), *Thermal processes in welding* (pp. 89–111). Moscow: Publication of the USSR Academy of Sciences (in Russian).
Rykalin, N., Uglov, A., & Kokora, A. (1978). *Laser machining and welding* (312 pp.). Moscow: Mir Publishers.
Rykalin, N., Uglov, A., Zuev, I., & Kokora, A. (1988). *Laser and electron beam material processing: Handbook* (591 pp.). Moscow: Mir Publishers.
Sandvik (1977). *Welding handbook* (136 pp.). Sandviken: Sandvik Publication.
Schmidt, H. N. B. (2010). Modelling thermal properties in friction stir welding. In D. Lohwasser, Z. Chen (Eds.), *Friction stir welding. From basics to applications* (pp. 277–313). Oxford: Woodhead Publishing.
Schnick, M., Fussel, U., Hertel, M., Spille-Kohoff, A., & Murphy, A. B. (2010). Effects of metal vapour on the arc behaviour in GMA welding. In H. Cerjak & N. Enzinger (Eds.), *Mathematical modelling of weld phenomena* (Vol. 9, pp. 43–56). Graz: Verlag der Technischer Universitaet Graz.
Seyffarth, P., & Krivtsun, I. (2002). *Laser-arc processes and their applications in welding and material treatment* (200 pp.). Boca Raton: CRC Press.
Smaars, E. A., & Acinger, K. (1968). Material transport and temperature distribution in arc between melting aluminium electrodes. *IIW Doc*. 212-162-68.
Smartt, H. B., Stewart, J. A., & Einerson, C. J. (1986). Heat transfer in gas tungsten arc welding. *ASM Metals/Materials Technology Series, 8511–011*. Metals Park, Ohio, 1–14.
Sosnin, N. A., Yermakov, S. A., & Topolyansky, P. A. (2008). *Plasma technologies* (406 pp.). St. Petersburg: Polytechnic University Publishing (in Russian).
Sudnik, V. A., & Ivanov, A. V. (1998). Mathematical model for heat source in gas metal arc welding. Part 1. Normal process. *Welding Production, 9*, 3–9 (in Russian).
Sudnik, V. A., & Rybakov, A. S. (1992). Calculation and experimental models of the moving arc of a non-consumable electrode in argon. *Welding International*, *6*(4), 301–303.
Sudnik, V. A., Rybakov, A. S., & Zaytsev, O. I. (2005). Mathematical and computer software TIGSIM for analysis of arc welding with a non-consumable electrode in argon. In V. A. Sudnik (Ed.), *Proceedings of International Conference on Computer Technologies in Joining of Materials* (pp. 128–145). Tula: Tula State University Publishing (in Russian).
Sudnik, V. A., & Yerofeyev, V. A. (1988). *Computer methods for research of welding processes* (94 pp.). Tula: Publishing House of the Technical University (in Russian).
Szekely, J. (1989). Transport phenomena in welds with emphasis on free surface phenomena. In *Proceedings of 2nd International Conference on Trends in Welding Research* (pp. 3–11).
Thomas, W. M., Nicholas, E. D., Needhamm, J. C., Murch, M. G., Temple-Smith, P., & Dawes, C. J. (1991). *Improvements relating to friction welding*. European Patent Specification 0 615 480 B1 1991.
Threadgill, P. L., Leonard, A. J., Shercliff, H. R., & Withers, P. J. (2009). Friction stir welding of aluminum alloys. *International Materials Reviews*, *54*(2), 49–93.
Tikhodeyev, G. M. (1961). *Energetic properties of electric welding arc* (254 pp.). Moscow: Publishing House of the USSR Academy of Sciences (in Russian).

Tsai, N. S., & Eagar, T. W. (1985). Distribution of the heat and current fluxes in gas tungsten arcs. *Metallurgical Transactions B, 16B*(12), 841–846.

Tsarkov, A. V., & Orlik, G. V. (2001). Determination of concentration factor of welding arc in tungsten arc welding. *Welding Production, 6*, 3–5 (in Russian).

Turichin, G., Valdaitseva, E., Pozdeeva, E., Dilthey, U., & Gumeniuk, A. (2008). Theoretical investigation of dynamic behaviour of molten pool in laser and hybrid welding with deep penetration. *The Paton Welding Journal, 7*, 11–15.

Turichin, G. A., Valdaytseva, E. A., Karkhin, V. A., Wang, H.-P., & Carlson, B. E. (2013). Modelling of plasma jet temperature field with slope incident on the surface with plasma and hybrid processing materials. In *Proceedings of the 7th International Scientific and Technical Conference on Beam Technologies and Laser Application* (pp. 18–21) September 2013. St. Petersburg, Russia. St. Petersburg: St. Petersburg State Polytechnic University Publishing (pp. 52–64).

Ushio, M., & Matsuda, F. (1982). Mathematical modeling of heat transfer of welding arc (Part 1). *IIW Doc*. 212-528-82.

Vill, V. I. (1962). *Friction welding of metals* (114 pp.). American Welding Society.

Watkins, A. D., Smartt, H. B., & Einerson, C. Y. (1990). Heat transfer in gas metal arc welding. In *Proceedings of 3rd Conference on Recent Trends in Welding Science and Technology. Metals Park* (pp. 19–23). Ohio: ASM International.

Wendelstorf, J., Decker, I., & Wohlfahrt, H. (1997). TIG and plasma arc modelling: A survey. In H. Cerjak (Ed.), *Mathematical modelling of weld phenomena* (Vol. 3, pp. 848–897). London: The Institute of Materials.

Yamauchi, N., & Taka, T. (1979). TIG arc welding with hollow tungsten electrodes. *IIW Doc*. 212-452-79.

Yampolsky, V. M. (1972). Investigation of features of vacuum arc discharge with hollow cathode of welding type. *Transactions of Institutes of Higher Education. Engineering, 7*, 67–68 (in Russian).

Yerofeyev, V. A., & Maslennikov, A. V. (2005). Physical-mathematical model for multi-pass arc welding process/transactions of Tula State University. In *Computer Technologies in Joining Materials, 3* (pp. 246–255). Tula: Tula State Technical University Publishing (in Russian).

Yushchenko, K. A., Chervyakov, N. O., & Kalina, P. P. (2006). Energy characteristics of low-amperage arcs. *The Paton Welding Journal, 4*, 17–21.

Zaehr, J., Schnick, M., Fuessel, U., Lohse, M., & Sende, M. (2010). Numerical investigations of process gases and their influence on TIG—Welding. In H. Cerjak & N. Enzinger (Eds.), *Mathematical modelling of weld phenomena* (Vol. 3, pp. 111–126). Graz: Technischen Universitaet Graz.

Zhu, P., Lowke, J. J., Morrow, R., & Haidar, J. (1995). Prediction of anode temperatures of free burning arcs. *Journal of Physics D: Applied Physics, 28*, 1369–a1376.

Chapter 2
Thermophysical Properties of Metals

2.1 Overview

In the analysis of thermal processes in welding the main properties of the metal are considered: thermal conductivity λ (W m^{-1} K^{-1}), specific heat capacity c (J g^{-1} K^{-1}) and density ρ (g m^{-3}) and additional ones: thermal diffusivity a (m^2 s^{-1}) and specific enthalpy (specific heat content) H (J g^{-1}),

$$a = \frac{\lambda}{c\rho}, \ H(T) = \int_{T_0}^{T} c(T)dt. \tag{2.1.1}$$

Thermal conductivity is one of the types of heat transfer from more heated parts of the body to less heated ones, which results in temperature equalisation. The coefficient λ is numerically equal to the amount of heat transferred through one unit of surface in a unit of time, with the temperature gradient being equal to 1 K m^{-1}. Its amount depends on the chemical composition and temperature.

The thermal conductivity of steel between room temperature and solidus temperature is a function of temperature T and the total volume proportion of alloying elements V:

$$\lambda = \lambda(T, V); \quad V = \sum_{i=1}^{N} m_i \frac{\rho_{\text{Fe}}}{\rho_i}, \tag{2.1.2}$$

where N is the amount of alloying elements, m_i is the mass proportion, ρ_{Fe} is the density of iron, ρ_i is the density of the ith alloying element (Radaj 2003).

The specific heat capacity depends on the chemical composition of the substance and temperature (Fig. 2.1). In case of phase transformations (change of crystal lattice, melting and solidification), the enthalpy changes sharply due to latent heat (more precisely, the Gibbs energy of phase transitions) and, consequently, so does the

V. A. Karkhin, *Thermal Processes in Welding*, Engineering Materials,
https://doi.org/10.1007/978-981-13-5965-1_2

specific heat capacity, which can be seen in the curves $c(T)$. In aluminium alloys the share of the latent heat of melting in enthalpy can be 40% at liquidus temperature.

It is convenient to use the mean heat capacity in the temperature interval $T_1 - T_2$ for calculations:

$$c = \frac{1}{T_2 - T_1} \int_{T_1}^{T_2} c(T)dT = \frac{H(T_2) - H(T_1)}{T_2 - T_1} \quad (2.1.3)$$

If the heat capacity is referred in the definition to a body volume unit, rather than to a mass unit, then the *volumetric heat capacity* $c\rho$ (J m^{-3} K^{-1}) is obtained.

Thermal diffusivity is a physical quantity which characterises the rate of temperature equalisation of the substance in non-equilibrium heat processes. The thermal diffusivity a depends on the nature of the substance and temperature.

Specific enthalpy (specific heat content) is a thermodynamic potential which characterises the state of the system in thermodynamic equilibrium. Enthalpy in welding (isobaric) processes is determined by formula (2.1.1).

In technical calculations either temperature $T_0 = 273$ K (0 °C) or 298 K (25 °C) is used. Enthalpy depends on the chemical composition and microstructural state of the substance.

Note that the relation between the heat capacity $c(T)$ and enthalpy $H(T)$ can be established by singling out the enthalpy of phase transformations $H^{tr}(T)$ from the latter:

$$H(T) = \int_{T_0}^{T} c(T)dT + H^{tr}(T); \quad H^{tr} = f(T)L^{tr}, \quad (2.1.4)$$

where $f(T)$ is the fraction of the new phase, L^{tr} is transformation heat. This technique is usually used to consider only latent heat of melting since the heat of melting is much bigger than the heat of phase transformation in solid state. For example, for steel, containing 0.44% C and 0.71% Mn, the transformation heat of bainite is equal to 20.9 J g^{-1}, that of martensite –75.3 J g^{-1} (Inoue et al. 1990), which is one order of magnitude less than the melting heat (Fig. 2.1).

The relation between the enthalpy $H(T)$ and heat capacity $c(T)$ is single valued. It is predominantly enthalpy which is used in modern algorithms for solving heat conduction problems by numerical methods, since it is a continuous function of temperature, while the heat capacity $c(T)$ can have finite breaks.

Density is the mass ratio of body element to its volume. In case of a homogeneous body, its density is constant along the entire volume V and the body mass $m = \rho V$. Density depends little on temperature. In phase transformations of a material in the solid state, when the crystal lattice restructures, a density jump is observed.

It should be emphasised that the published values of thermal conductivity and heat capacity of a material often have a wide scatter. The scatter of experimental

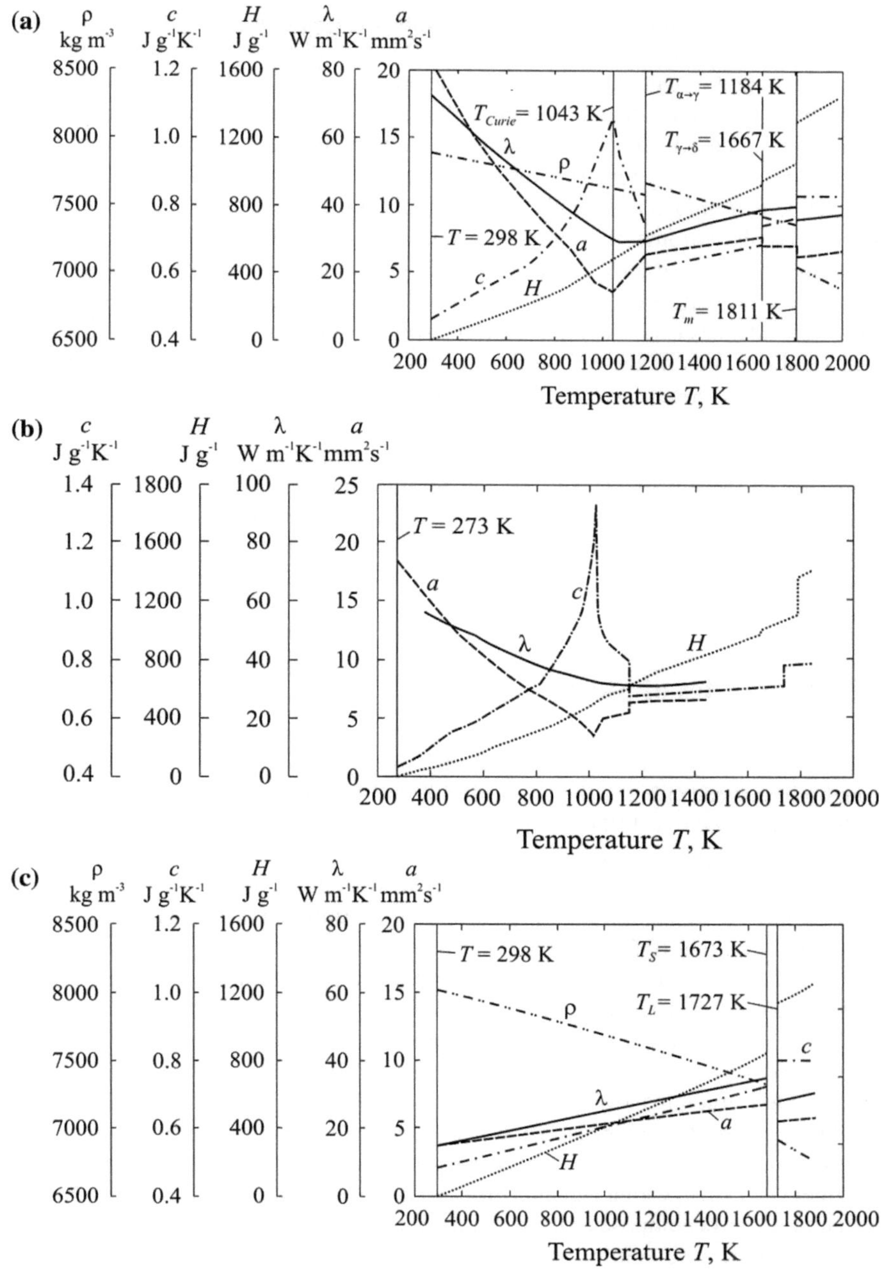

Fig. 2.1 The temperature dependence of density ρ, specific heat capacity c, enthalpy H, thermal conductivity λ and thermal diffusivity a for pure iron (**a**) (Mills 2002), low-carbon steel (0.1% C) (**b**) (Rykalin 1947, 1951, 1952, 1957) and 304 austenitic stainless steel (19.0% Cr, 9.5% Ni) (**c**) (Mills 2002)

data depends on the methodology used to define the properties (for example, the way how the changed microstructure of a material is considered in the experiment). The properties taken in calculations are also determined by a mathematical model. For example, if the model does not include the liquid metal convection of the welding pool, then increased values of thermal conductivity are introduced at a temperature exceeding the melting temperature (effective thermal conductivity) and, thus, the convective heat transfer is roughly considered. The values of heat capacity depend on how the phase transformation heat is taken into account.

Further on some graphs are presented showing temperature dependences of thermophysical properties (ρ, c, H, λ and a) of some alloys used in welded structures. Such dependences serve as input data for solving the heat conduction problem by numerical methods. If the problem is solved by analytical methods, the properties are adopted as constant, averaged in the temperature interval, where the accuracy of calculation is especially important.

Note that in the figures of this chapters, the temperature dependences of alloy properties in the mushy state in the temperature interval $T_L - T_S$ are not shown. Within a rigorous approach the properties depend on the temperature dependence of phase fractions $f(T)$ (see Sect. 9.4). In approximation the function $f(T)$ can be defined as linear one.

2.2 Properties of Steels

The main component of all steels is iron. Two crystal lattices are known for pure iron: body-centred cubic (α-Fe up to 1184 K and δ-Fe with 1667–1811 K) and face-centred cubic (γ-Fe at 1184–1667 K) (Fig. 2.1a). During transition through critical points at the temperature scale ($T_{\alpha\rightarrow\gamma}$, $T_{\gamma\rightarrow\delta}$, melting temperature T_m and the Curie point T_{Curie}) breaks in iron properties are observed (Fig. 2.1a).

Depending on the carbon content steel is differentiated as low carbon (<0.25% C), medium carbon (0.25–0.6% C) and high carbon (>0.6% C). In terms of alloy addition (i.e. the total content of alloying elements) there are low alloy (<2.5%), medium alloy (2.5–10%) and high alloy (>10%) steels.

Carbon in iron shifts the critical points and affects the thermophysical properties (Fig. 2.1b).

In welded structures low-carbon, low-alloy and high-alloy steels are mostly used. The temperature dependences of the phase state are close for the first two groups: during heating α-Fe (ferrite-pearlite) transforms in γ-Fe (austenite) in the temperature range A_{C1}–A_{C3}, which depends little on the heating rate, whereas during cooling austenite γ-Fe decomposes at the temperature largely dependent on the cooling rate. The higher the cooling rate is the lower the temperature of decomposition is. The products of austenite decomposition can be bainite, martensite and other microstructural phases. Thus, within a rigorous approach, the temperature dependences of properties are multifunctions, since these properties are also dependent on the entire previous stage of the thermal cycle $T(t)$.

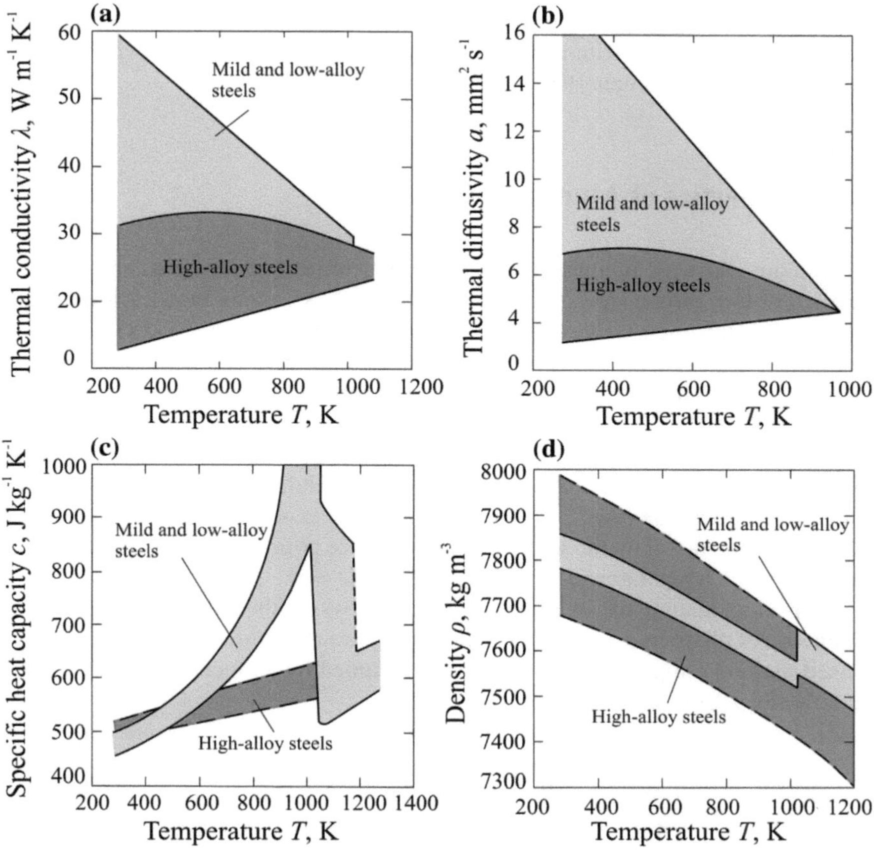

Fig. 2.2 Thermal conductivity λ (**a**), thermal diffusivity a (**b**), specific heat capacity c (**c**) and density ρ (**d**) of steels dependent on temperature (Radaj 1992, 2003; Richter 1973)

High-alloy steels, containing high concentration of the so-called austenitisers (Ni, Mn, C, N), can have a stable austenite structure at room temperature and lower. A typical example is stainless austenitic steels, containing 18–20% Cr and 9–10% Ni (such steels as Kh18N10T and 304). Figure 2.1c shows the temperature dependences of thermophysical properties of steel 304 (0.08% C, 19.0% Cr, 0.3% Cu, 2% Mn, 9.5% Ni).

There are very many steel grades, different in the chemical composition, microstructure and, consequently, in properties. Figure 2.2 shows value bands of the properties of low-alloy and low-carbon steels and high-alloy austenitic steels at the equilibrium state. In the temperature range A_{C1}–A_{C3} the properties of low-alloy and low-carbon steels change dramatically, which is caused by phase transformation α-Fe → γ-Fe. At a high temperature, the bandwidth reduces considerably.

The temperature dependencies of thermophysical properties of many steels are published (Grong 1994; Kim 1975; Kuzminov 1974; Grigoryev and Meylikhov 1991; Radaj 1992, 2003; Richter 1973).

2.3 Properties of Aluminium Alloys

Aluminium alloys are lightweight and at the same time their strength can exceed the strength of low-carbon steels. In welding Al (>99% Al, series 1xxx), Al–Cu (series 2xxx), Al–Mn (series 3xxx), Al–Si (series 4xxx), Al–Mg (series 5xxx), Al–Mg–Si (series 6xxx), Al–Zn (series 7xxx) and other aluminium alloys are used. One commercial alloy in each series, applied in welded structures, is discussed as examples below (Table 2.1).

The thermophysical properties of aluminium alloys (density ρ, specific heat capacity c, enthalpy H, thermal conductivity λ and thermal diffusivity a) depend on their chemical composition (Fig. 2.3). The accuracy of data is determined by experimental and computational methodologies. The reference point of enthalpy is accepted as enthalpy at the temperature $T = 298$ K, H (298 K) $= 0$.

A specific feature of aluminium and its alloys is the relatively high energy (enthalpy) of phase transformations. Aluminium transits from solid to liquid state at a fixed melting temperature T_m and is accompanied by an increase in heat content by 60% (Fig. 2.3a).

Alloys in the temperature range of solidus T_S – liquidus T_L are in solid-liquid state. The fraction of solid phase f_S in the interval $T_S - T_L$ depends both on the chemical composition of an alloy and on the heating and cooling rates (Fig. 2.4). Alloy properties in this interval $P(T)$ are recommended to be estimated according to a simple rule (Mills 2002):

$$P(T) = f_S(T)P(T_S) + (1 - f_S)P(T_L). \tag{2.3.1}$$

As the function $f_S(T)$ is non-linear in the general case (Fig. 2.3), the properties P depend non-linearly on the temperature T. If there is no data on the dependence $f_S(T)$, it is usually accepted as linear:

Table 2.1 The chemical composition of aluminium alloys, %

Alloy	Cr	Cu	Fe	Mg	Mn	Si	Ti	Zn
2024	0.10	4.4	0.50	1.5	0.6	0.50	0.15	0.25
3004		0.2	0.43	1.0	1.0	0.14		0.25
5182	0.15	0.1	0.35	4.5	0.3	0.2		0.25
6061	0.4	0.3	0.7	1.0	0.15	0.6	0.15	0.25
7075	0.2	1.6	0.50	2.5	0.30	0.4	0.2	5.6

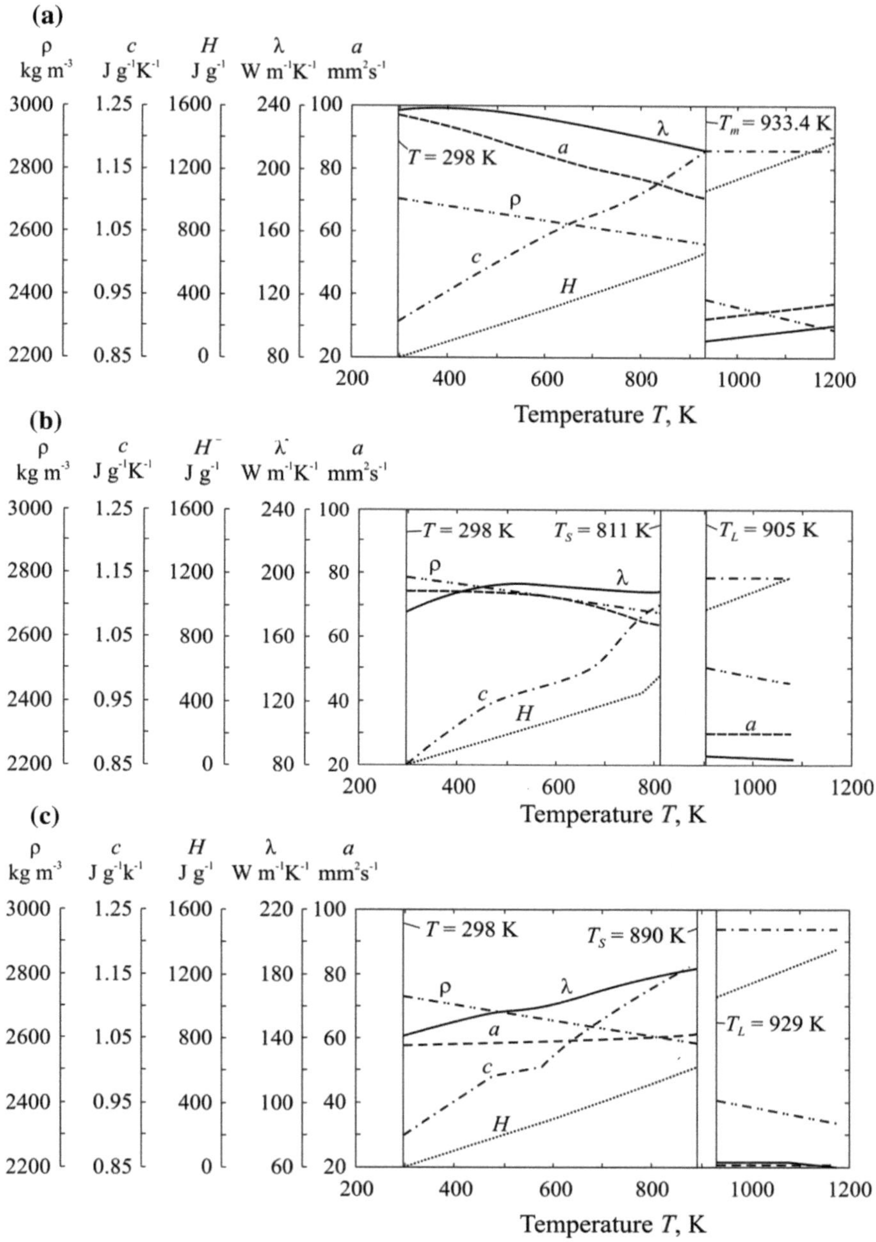

Fig. 2.3 Temperature-dependent density ρ, specific heat capacity c, enthalpy H, thermal conductivity λ and thermal diffusivity a: pure aluminium (**a**), 2024 aluminium alloy (**b**), 3004 aluminium alloy (**c**), 5182 aluminium alloy (**d**), 6061 aluminium alloy (**e**) and 7075 aluminium alloy (**f**) (Mills 2002)

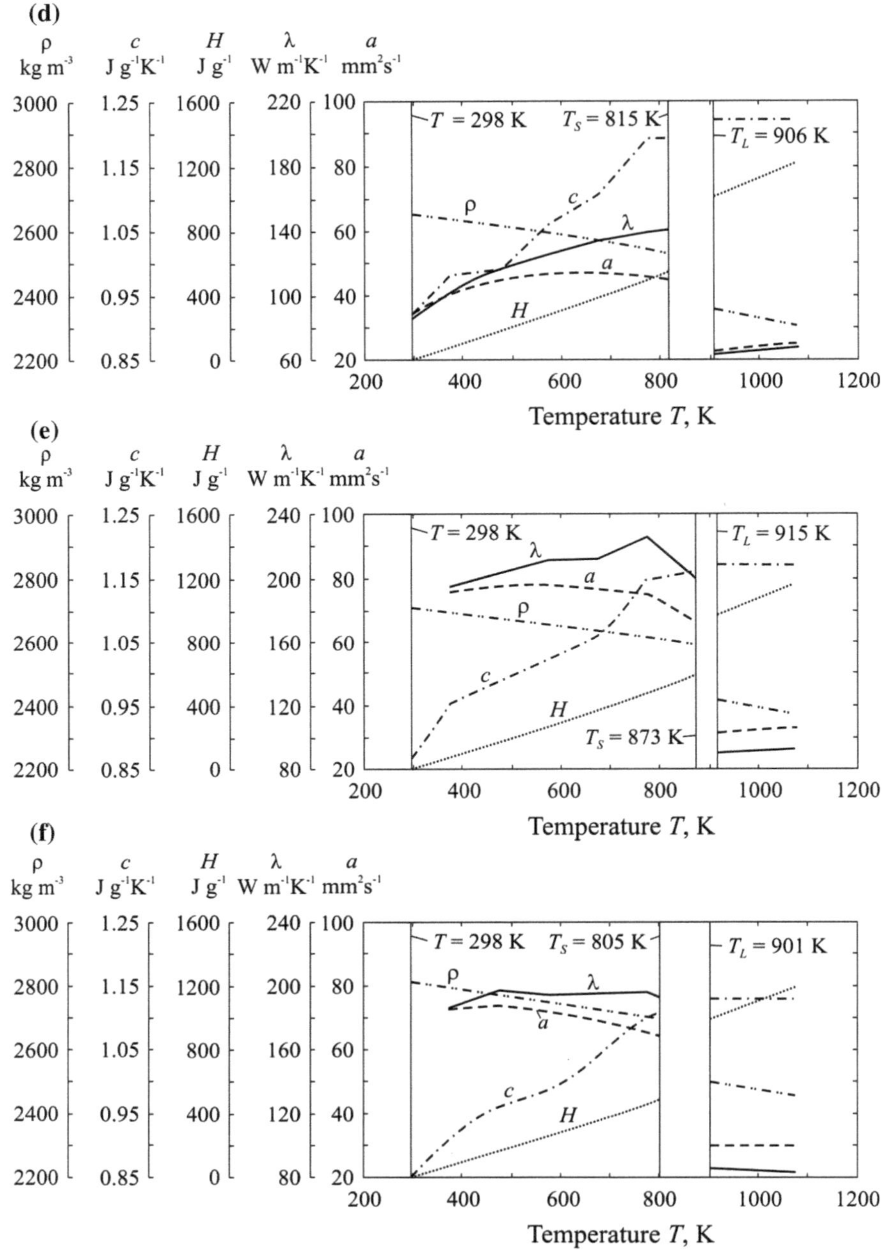

Fig. 2.3 (continued)

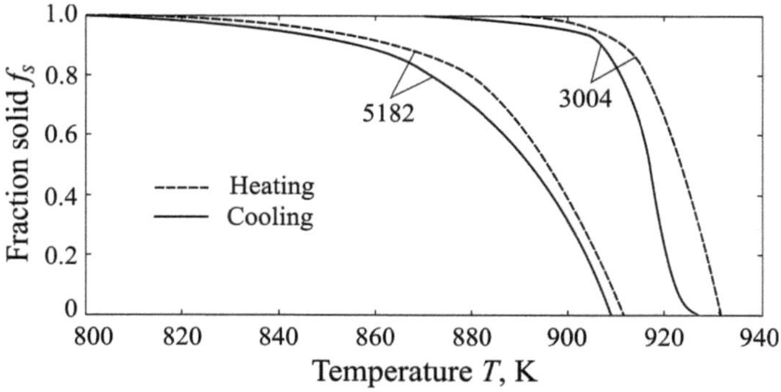

Fig. 2.4 Fraction solid of aluminium alloys 3004 and 5182 at cooling rate 0.17 K s^{-1} as a function of temperature (Mills 2002)

$$f_S(T) = \frac{T_L - T}{T_L - T_S}, \; T_S \le T \le T_L. \tag{2.3.2}$$

The thermophysical properties of aluminium alloys are published in literature (Kuzminov 1974; Mills 2002; Grigoryev and Meylikhov 1991).

2.4 Properties of Titanium Alloys

Lightweightness, high strength from cryogenic to relatively high temperatures and excellent corrosion resistance make titanium alloys applicable as structural materials in many fields, in particular, in aviation and other transport engineering industries.

The thermophysical properties of pure titanium are given in Fig. 2.5a. With $T_{tr} =$ 1155 K phase transformation occurs in solid state, with $T_m = 1941$ K the metal melts. Titanium exists in the form of two allotropic modifications: below the temperature $T_{tr} = 1155$ K it is a stable α-phase with a hexagonal lattice, whereas above T_{tr} it is a β-phase with a body-centred cubic lattice. Admixtures and alloying additions can considerably change the α/β transformation temperature.

The high strength alloy Ti–6Al–4V (5.5–6.7% Al, 0.03% Fe, 3.5–4.5% V, 0.0125% H, 0.25% O2+N2) refers to α + β alloys and its composition is similar to alloy VT6 (5.5–7.0% Al, 4.2–6.0% V). The temperature of $\alpha \to \beta$ transformation $T_{\alpha\to\beta} = 1268 \pm 5$ K (Mills 2002).

The thermophysical properties of titanium alloys are discussed in literature (Mills 2002; Grigoryev and Meylikhov 1991).

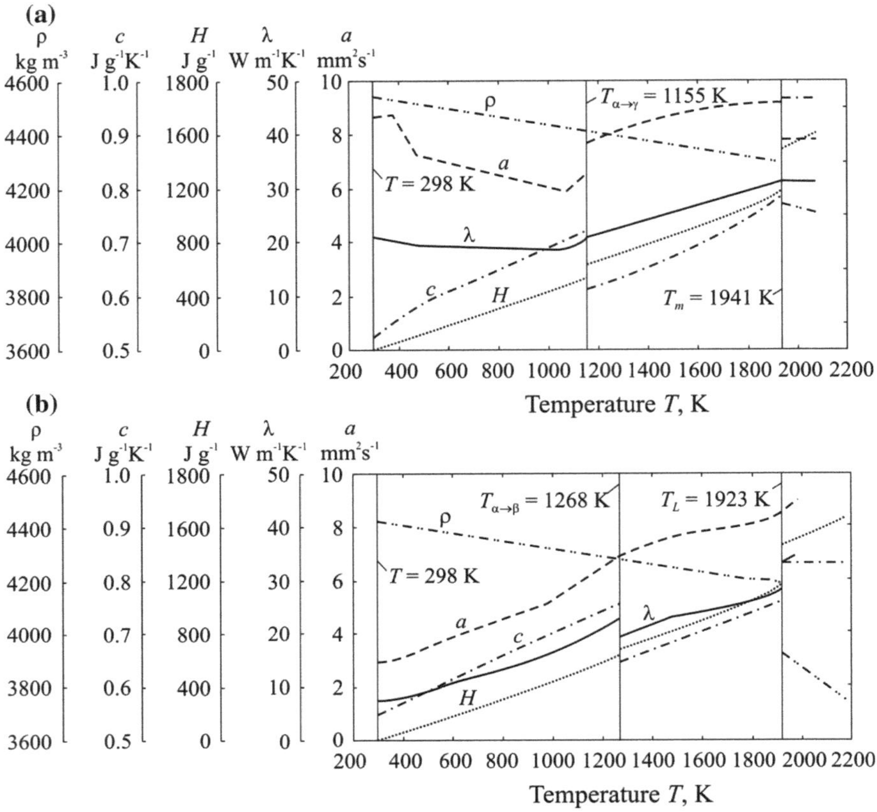

Fig. 2.5 Temperature-dependent density ρ, specific heat capacity c, enthalpy H, thermal conductivity λ and thermal diffusivity a of pure titanium (**a**) and titanium alloy Ti–6Al–4V (5.5–6.7% Al, 3.5–4.5% V) (**b**) (Mills 2002)

2.5 Properties of Magnesium

Magnesium alloys are the most lightweight metallic structural material (its density fluctuates in the range of 1360–2000 kg m^{-3}). Thanks to their low density, high specific strength (the ultimate strength to density ratio), magnesium alloys are applied in automotive, aerospace and other industry sectors.

The thermophysical properties of pure magnesium depending on temperature are shown in Fig. 2.6. The properties of magnesium alloys are presented in the reference books (Mills 2002; Grigoryev and Meylikhov 1991).

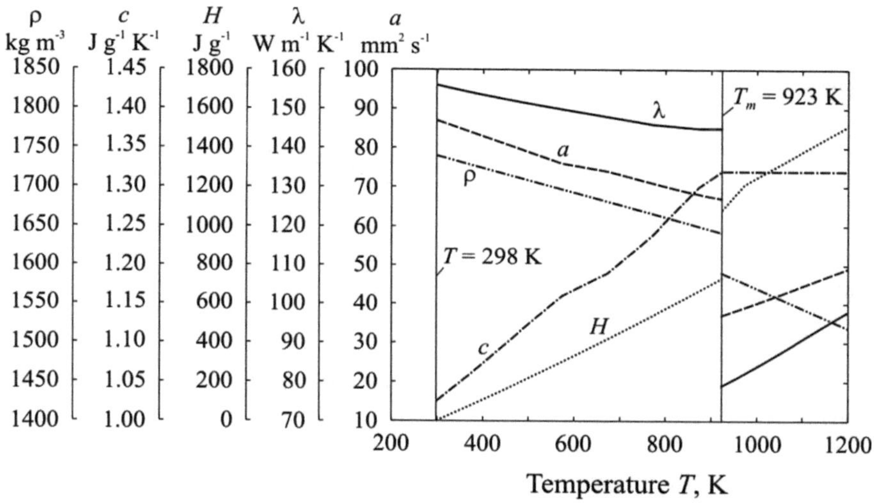

Fig. 2.6 Temperature-dependent density ρ, specific heat capacity c, enthalpy H, thermal conductivity λ and thermal diffusivity a of pure magnesium (Mills 2002)

2.6 Properties of Copper Alloys

Copper alloys are produced by alloying copper with alloying elements and are subdivided into brasses (from 0 to 50% Zn), various types of bronzes and copper-nickel alloys.

Pure copper has a face-centred cubic lattice. The thermal conductivity is very high (Fig. 2.7a).

There are aluminium, tin, beryllium and other bronzes. The thermophysical properties of aluminium bronze, type Br. AZhN10-4-4 (9.7% Al, 80.5% Cu, 4.6% Fe, 0.64% Mn, 4.6% Ni) are presented in Fig. 2.7b. The solid phase fraction in the interval $T_S - T_L$ depends, the same as for aluminium alloys, on the heating and cooling rates.

The properties of copper alloys are published in the reference book (Mills 2002; Grigoryev and Meylikhov 1991).

2.7 Properties of Nickel

The capacity of nickel to dissolve many other metals and remain ductile has resulted in many nickel alloys. High-temperature alloys of nimonic type work at a temperature of 1100–1300 K.

Under normal conditions, pure nickel exists as a β-modification, having a face-centred cubic lattice. The thermophysical properties of pure nickel are shown in

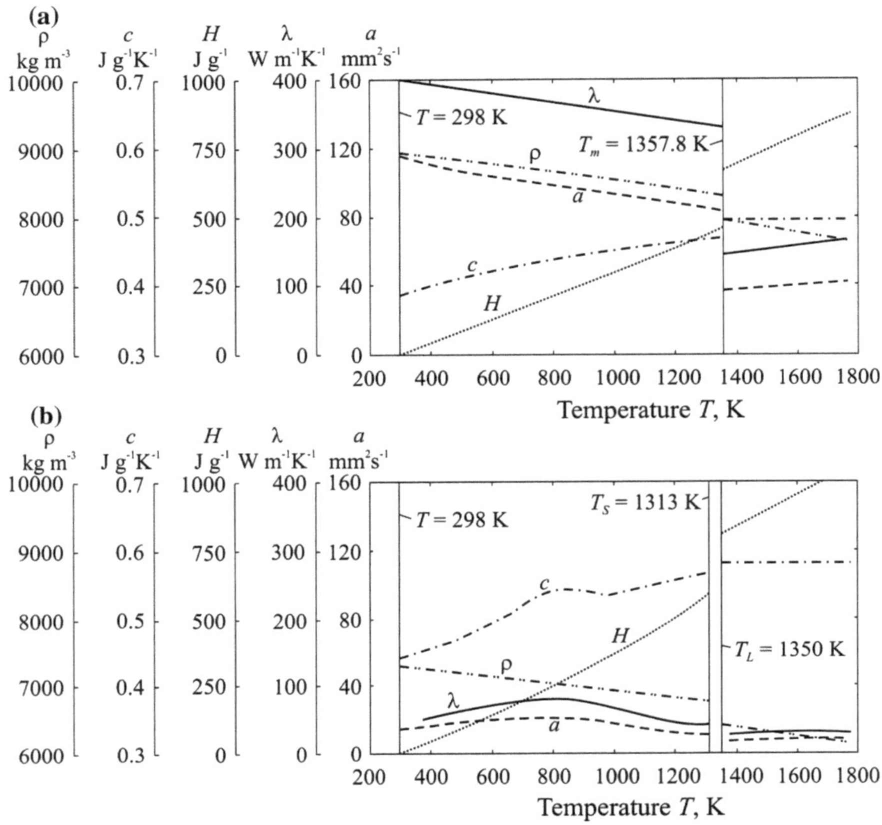

Fig. 2.7 Temperature-dependent density ρ, specific heat capacity c, enthalpy H, thermal conductivity λ and thermal diffusivity a of pure copper (**a**) and aluminium bronze (9.7% Al, 80.5% Cu, 4.6% Fe, 4.6% Ni) (**b**) (Mills 2002)

Fig. 2.8. The properties of nickel alloys are published in the reference book (Grigoryev and Meylikhov 1991).

If analytical methods (functional analysis methods) are used to solve heat conduction problems, the constant values of thermophysical properties of the material have to be used. Average values are recommended for use as input data (Table 2.2).

The thermophysical properties of alloys make a dominating impact on thermal processes in welding. As shown below (Sect. 5.1.3.4), under the same welding conditions and with the same body geometry, the shape and size of the weld pool can vary considerably. Moreover, within the same alloy, a variation of properties (λ, c, ρ) can cause significant changes of the peak temperature, cooling time and size of the weld pool (Michailov et al. 2005, 2012; Schwenk et al. 2007; Sudnik et al. 1998).

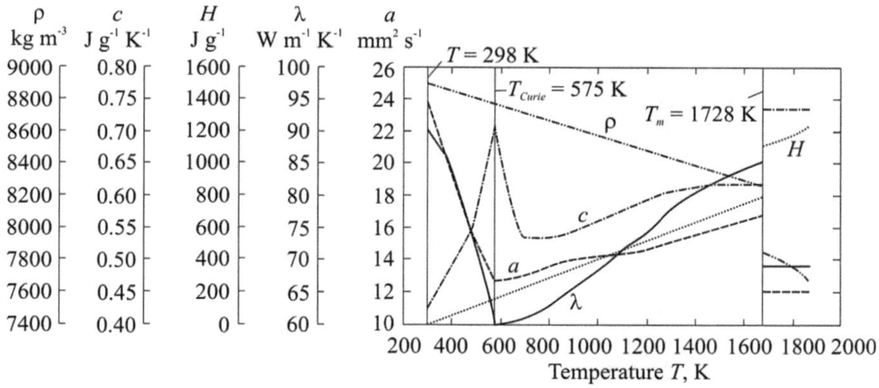

Fig. 2.8 Temperature-dependent density ρ, specific heat capacity c, enthalpy H, thermal conductivity λ and thermal diffusivity a of pure nickel (Mills 2002)

Table 2.2 Physical properties for metals and alloys (Grong 1994)

Material	Thermal conductivity λ (W mm^{-1} K^{-1})	Thermal diffusivity a (mm^2 c^{-1})	Volumetric heat capacity $c\rho$ (J mm^{-3} K^{-1})	Melting temperature T_m (K)	Enthalpy H^a (J mm^{-3})	Latent heat of melting L (J mm^{-3})
Carbon steel	0.040	8	0.005	1793	7.50	2.0
Low alloy steel	0.025	5	0.005	1793	7.50	2.0
High alloy steel	0.020	4	0.005	1773	7.40	2.0
Titanium alloys	0.030	10	0.003	1923	4.89	1.4
Aluminium (>99% Al)	0.230	85	0.0027	933	1.73	0.8
Al–Mg–Si alloys	0.167	62	0.0027	925	1.71	0.8
Al–Mg alloys	0.149	55	0.0027	923	1.70	0.8

[a]Latent heat of melting L is not included. Enthalpy $H = 0$ at $T_0 = 293$ K

References

Grigoryev, I. S., & Meylikhov, E. Z. (Eds.). (1991). *Physical values: Handbook* (1232 pp.). Moscow: Energoatomizdat (in Russian).

Grong, O. (1994). *Metallurgical modelling of welding* (581 pp.). London: The Institute of Materials.

Inoue, K., Ohmura, E., Ohata, N., & Takamachi Y. (1990). Analysis of heat flow coupled with structural changes in laser transformation hardening process. *Transactions of Japan Welding Research Institute, 19*(2), 89–97.

Kim, C. S. (1975). *Thermophysical properties of stainless steels* (28 pp.). Rep. N ANL-75-75. Argonne, III: Argonne National Laboratory.

Kuzminov, S. A. (1974). *Welding deformations of ship structures* (286 pp.). Leningrad: Sudostroenie (in Russian).

Michailov, V., Doynov, N., & Ossenbrink, R. (2012). *Sensibilitaetanalyse der thermomechanischen FE-Schweisssimulation* (137 pp.). Aachen: Shaker Verlag (in German).

Michailov, V., Stadtaus, M., Dilger, K., & Rethmeier, M. (2005). Numerical MIG-welding simulation of magnesium alloy. In H. Cerjak, H. K. D. H. Bhadeshia, & E. Kozenschnik (Eds.), *Mathematical modelling of weld phenomena* (Vol. 7, pp. 77–89). Graz: Verlag der Technischen Universitaet Graz.

Mills, K. C. (2002). *Recommended values of thermophysical properties for selected commercial alloys* (244 pp.). Cambridge: Woodhead Publishing.

Radaj, D. (1992). *Heat effects of welding. Temperature field, residual stress, distortion* (348 pp.). Berlin: Springer.

Radaj, D. (2003). *Welding residual stresses and distortion. Calculation and measurement* (397 pp.). Duesseldorf: DVS-Verlag.

Richter, F. (1973). Die wichtingsten physikalischen Eingenschaften von 52 Eisenwerkstoffen. In *Stahleisen – Sonderberichte Heft* (Vol. 8, 32 pp.). Duesseldorf: Verlag Stahleisen (in German).

Rykalin, N. N. (1947). *Thermal principles in welding. Part 1. Heat flow in arc welding* (271 pp.). Moscow: Publication of the USSR Academy of Sciences (in Russian).

Rykalin, N. N. (1951). *Calculation of heat flow in welding* (Z. Paley & C. M. Adams, Jr., Trans.) (337 pp.). Moscow.

Rykalin, N. N. (1952). *Die Waermegrundlagen des Schweissvorganges* (265 pp.). Berlin: Verlag Technik (in German).

Rykalin, N. N. (1957). *Berechnung der Waermevorgaenge beim Schweissen* (326 pp.). Berlin: VEB Verlag Technik (in German).

Schwenk, C., Rethmeier, M., Dilger, K., & Michailov, V. (2007). Sensitivity analysis of welding simulation depending on material properties value variation. In H. Cerjak, H. K. D. H. Bhadeshia, & E. Kozenschnik (Eds.), *Mathematical modelling of weld phenomena* (Vol. 8, pp. 1107–1128). Graz: Verlag der Technischen Universitaet Graz.

Sudnik, W., Radaj, D., & Erofeew, W. (1998). Validation of computerized simulation of welding processes. In H. Cerjak (Ed.), *Mathematical modelling of weld phenomena* (Vol. 4, pp. 477–493). London: IOM Communications.

Chapter 3
Physical Fundamentals of Heat Conduction During Welding

3.1 Principal Definitions

Temperature is a physical quantity that characterizes the degree of body heat. When the system is in the state of thermodynamic equilibrium, the temperatures of all the bodies forming the system are the same.

A temperature field is a set of temperature values T at all points of the body at a given time:

$$T = T(x, y, z, t)\,\text{at}\,t = \text{const}, \tag{3.1.1}$$

where x, y, z are the spatial coordinates in the Cartesian system; t is the time. In a cylindrical coordinate system $T = T(r, \theta, z, t)$, where r is a radius, θ is the angle, z is the axis. If the temperature at any point of the body does not change over time, the field is called stationary, otherwise it is non-stationary.

A temperature field (3.1.1) is volumetric (three-dimensional). It can also be plane (two-dimensional), $T = T(x, y, t)$, or linear (one-dimensional), $T = T(x, t)$. Figure 3.1 shows a plane field $T(x, y, \infty)$ in a plate after a prolonged action of the moving source.

An isotherm is a geometric place of field points that have the same temperature (Fig. 3.1c). Isotherms cannot intersect, because one point of the body there cannot have two different temperature values simultaneously. Isotherms cannot have boundaries within the body. They either close inside the body, or end on its surface.

The temperature does not change along the isotherm. The greatest temperature difference per unit length occurs in the direction of the normal to the isotherm and it is characterised by a *temperature gradient* (K m^{-1}):

$$\text{grad}\,T = \partial T/\partial n. \tag{3.1.2}$$

The temperature gradient at point P is a vector directed along the normal to the isotherm in the direction of temperature increase (see Fig. 3.1c).

V. A. Karkhin, *Thermal Processes in Welding*, Engineering Materials,
https://doi.org/10.1007/978-981-13-5965-1_3

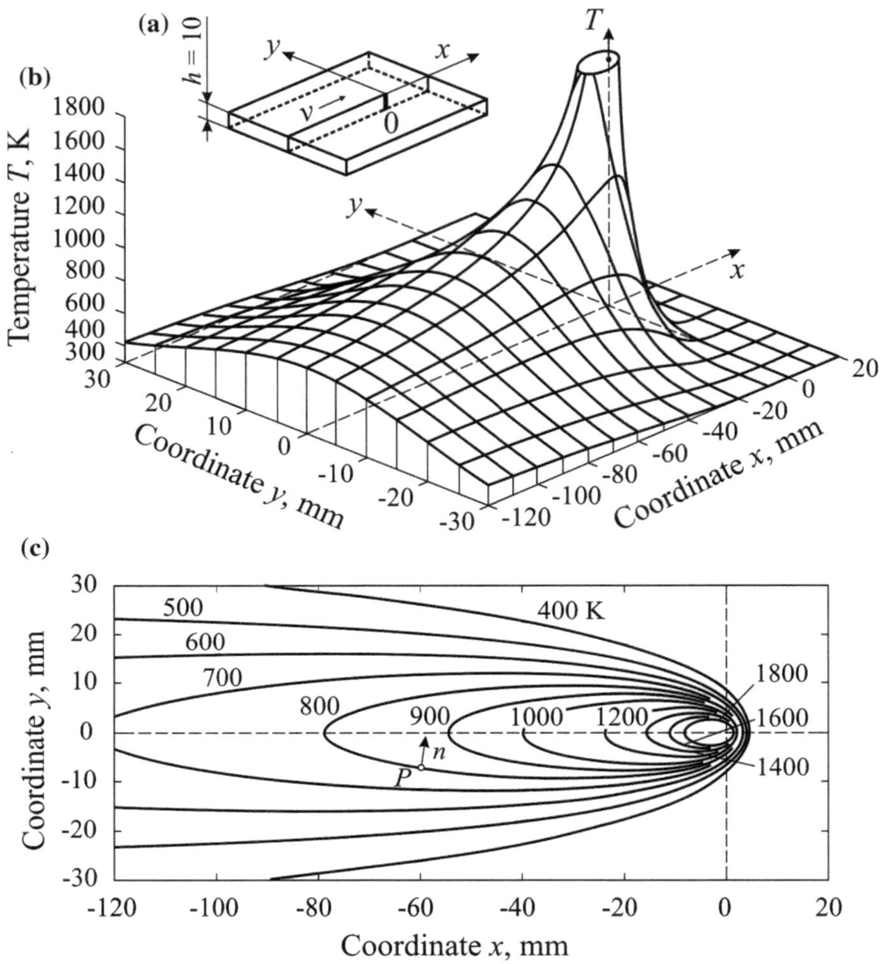

Fig. 3.1 The temperature field around a moving line source in an infinite plate 10 mm thick with the heat-impermeable surface (power $q = 5000$ W, speed $v = 5$ mm s^{-1}, thermal conductivity $\lambda = 0.04$ W mm^{-1} K^{-1}, thermal diffusivity $a = 8$ mm^2 s^{-1}, initial temperature $T_0 = 300$ K): reference frame (**a**), temperature distribution (**b**) and isotherms (**c**)

The cooling (heating) rate of the point x, y, z (K s^{-1}) at time t is determined by the derivative $\partial T/\partial t(x, y, z, t)$.

The temperature-time relationship of a fixed point of the body is called a *thermal cycle* and is represented as a curve with one or more maxima (for example, in multi-pass welding).

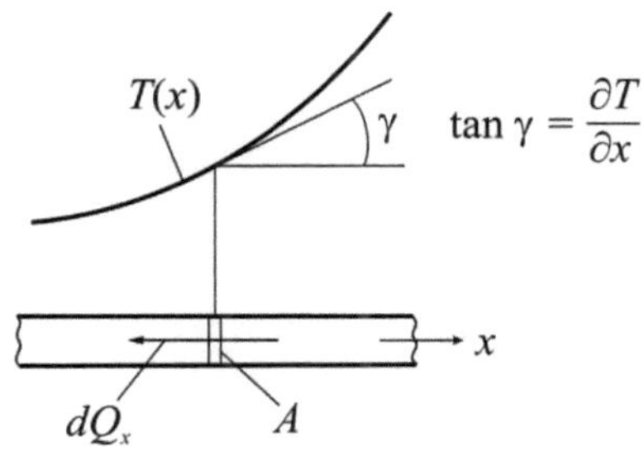

Fig. 3.2 Temperature distribution in a rod

3.2 Fourier's Law of Heat Conduction

Heat in a solid body is conducted from places with a higher temperature to places with a lower temperature by means of molecular heat conduction.

Consider heat transfer in a rod with a non-uniform temperature distribution $T(x)$ (Fig. 3.2). The quantity of heat dQ_x (J), flowing through the cross-sectional area A of the rod in the time dt, is proportional to the temperature gradient $\partial T/\partial x$ (Fourier's hypothesis):

$$dQ_x = -\lambda \frac{\partial T}{\partial x} A dt, \tag{3.2.1}$$

where λ is the thermal conductivity, W m^{-1} K^{-1}. The minus sign indicates that heat is transferred in the direction of temperature decrease.

The heat flow density (W m^{-2}) is determined by the formula

$$q_{2x} = \frac{dQ_x}{Adt} = -\lambda \frac{\partial T}{\partial x} \tag{3.2.2}$$

and equals the amount of heat flowing through the cross section of 1 m^2 per 1 s.

In the general three-dimensional case, Fourier's law of heat conduction for an isotropic body has the form

$$q_2 = -\lambda \frac{\partial T}{\partial n} = -\lambda \operatorname{grad} T. \tag{3.2.3}$$

This equation is the main hypothesis of the mathematical theory of heat conduction.

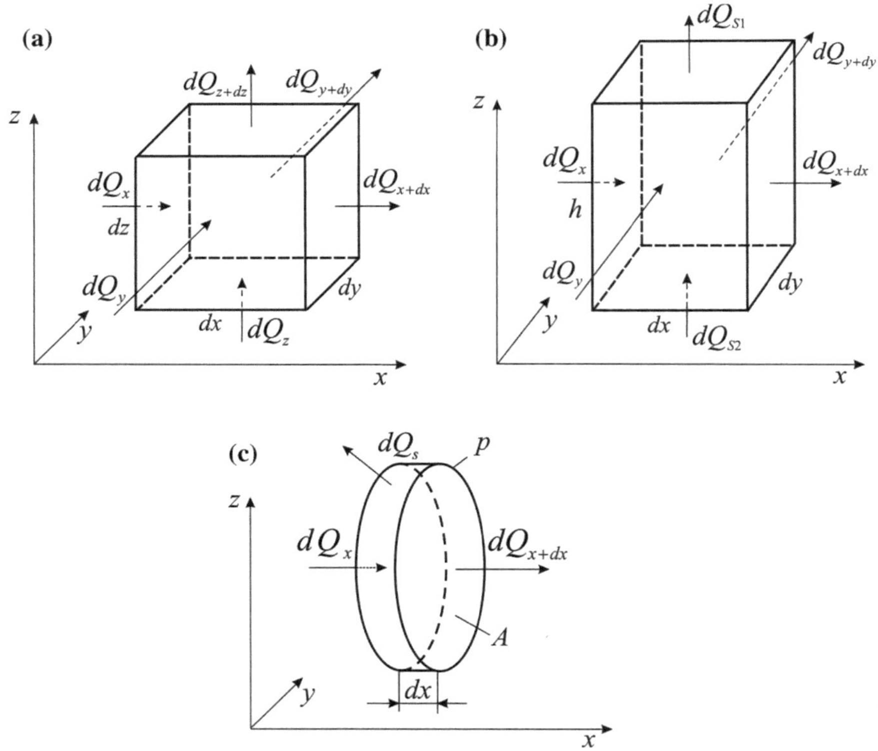

Fig. 3.3 Models for derivation of the differential equation for heat conduction: solid body (**a**), plate of thickness h (**b**) and a rod of cross-sectional area A and perimeter p (**c**)

3.3 Differential Equation for Heat Conduction

The physical meaning of the differential equation for heat conduction implies that it connects the spatial distribution of temperature with the temperature change in time. The derivation is based on the law of conservation of energy and Fourier's law of heat conduction.

Consider an elementary cuboid with edges dx, dy, dz in an orthotropic body (Fig. 3.3a), i.e. we assume for generality that the thermal conductivity along the axes λ_x, λ_y, λ_z can be different. Let there be a heat source (sink) inside with a volumetric power density (volume-specific power) q_3, W m^{-3}. In welding, for example, Joule heat may be such a source.

According to Fourier's law of heat conduction (3.2.1), the amount of heat applied to the surface of the cuboid $dydz$ along the x-axis in the time dt,

$$dQ_x = -\lambda_x \frac{\partial T}{\partial x} dydzdt. \tag{3.3.1}$$

The heat dQ_{x+dx}, extracted through the opposite face, can be determined by expanding it into a Taylor series and confining to the first two terms:

$$dQ_{x+dx} = dQ_x + \frac{\partial(dQ_x)}{\partial x}dx = dQ_x + \frac{\partial}{\partial x}\left(-\lambda_x \frac{\partial T}{\partial x}\right)dxdydzdt. \tag{3.3.2}$$

The change in the amount of heat in the volume $dV = dxdydz$, caused by the thermal conductivity along the x-axis in the time dt, is equal to

$$dQ_x - dQ_{x+dx} = \frac{\partial}{\partial x}\left(\lambda_x \frac{\partial T}{\partial x}\right)dVdt. \tag{3.3.3}$$

Similarly, the differences in heat flows along the y- and z-axes are taken into account:

$$\begin{aligned} dQ_y - dQ_{y+dy} &= \frac{\partial}{\partial y}\left(\lambda_y \frac{\partial T}{\partial y}\right)dVdt; \\ dQ_z - dQ_{z+dz} &= \frac{\partial}{\partial z}\left(\lambda_z \frac{\partial T}{\partial z}\right)dVdt. \end{aligned} \tag{3.3.4}$$

The amount of heat generated (absorbed) by the internal heat source (sink) in the parallelepiped in the time dt, is equal to

$$dQ_v = q_3 dVdt. \tag{3.3.5}$$

The accumulated heat in the volume dV in the time dt is connected with the change in its temperature

$$dQ = \frac{\partial(\rho H)}{\partial t}dVdt \tag{3.3.6}$$

or

$$dQ = c\rho dV \frac{\partial T}{\partial t}dt, \tag{3.3.7}$$

where H is the enthalpy, J g^{-1}, $c\rho$ is the volume-specific heat capacity, J m^{-3} K^{-1}.

By equating the accumulated heat (3.3.7) to the heat due to the thermal conductivity (3.3.3), (3.3.4) and action of internal sources (3.3.5), and dividing all by $dVdt$, we obtain the differential heat equation for an orthotropic body in a body-fixed coordinate system:

$$\frac{\partial(\rho H)}{\partial t} = \frac{\partial}{\partial x}\left(\lambda_x \frac{\partial T}{\partial x}\right) + \frac{\partial}{\partial y}\left(\lambda_y \frac{\partial T}{\partial y}\right) + \frac{\partial}{\partial z}\left(\lambda_z \frac{\partial T}{\partial z}\right) + q_3 \tag{3.3.8}$$

or

$$c\rho\frac{\partial T}{\partial t} = \frac{\partial}{\partial x}\left(\lambda_x\frac{\partial T}{\partial x}\right) + \frac{\partial}{\partial y}\left(\lambda_y\frac{\partial T}{\partial y}\right) + \frac{\partial}{\partial z}\left(\lambda_z\frac{\partial T}{\partial z}\right) + q_3. \tag{3.3.9}$$

Equations (3.3.8) and (3.3.9) are equivalent.

Repeat the derivation of the heat conduction equation for a thin plate with the thickness h allowing for the heat transfer from both surfaces. Consider an elementary cuboid whose volume $dV = dxdyh$ (Fig. 3.3b). We assume that the heat flows along the z-axis through upper and lower surfaces of the plate dQ_{S1} and dQ_{S2} are proportional to the temperature difference between surface T and the temperature of the medium T_∞ (Newton's law):

$$\begin{aligned} dQ_{S1} &= \alpha_1(T - T_\infty)dxdydt; \\ dQ_{S2} &= \alpha_2(T - T_\infty)dxdydt, \end{aligned} \tag{3.3.10}$$

where α_1 and α_2 are the coefficients of surface heat transfer for upper and lower surfaces, respectively, W m^{-2} K^{-1}. We also assume that the temperature is instantaneously equalized along the thickness (the thermal conductivity coefficient λ_z is large and thickness h is small), so the field remains two-dimensional, $T = T(x, y, t)$.

By equating the accumulated heat (3.3.7) to the heat due to the thermal conductivity (3.3.3) and (3.3.4), the actions of surface sinks (3.3.10) and internal sources (3.3.5), we obtain a differential heat conduction equation for the orthotropic plate with surface heat transfer in a fixed coordinate system:

$$c\rho\frac{\partial T}{\partial t} = \frac{\partial}{\partial x}\left(\lambda_x\frac{\partial T}{\partial x}\right) + \frac{\partial}{\partial y}\left(\lambda_y\frac{\partial T}{\partial y}\right) - \frac{\alpha_1 + \alpha_2}{h}(T - T_\infty) + q_3. \tag{3.3.11}$$

Now consider a rod with a cross-sectional area A and perimeter p and its elementary volume (Fig. 3.3c). The heat flow dQ_S through the surface pdx is equal to

$$dQ_S = \alpha(T - T_\infty)pdxdt. \tag{3.3.12}$$

If we ignore the non-uniformity of temperature over the cross section (which is valid if the coefficients λ_y and λ_z are large and area A is small), then from the energy conservation law taking into account (3.3.3), (3.3.5) and (3.3.12) we obtain a differential heat conduction equation for a rod with surface heat transfer in a fixed coordinate system:

$$c\rho\frac{\partial T}{\partial t} = \frac{\partial}{\partial x}\left(\lambda\frac{\partial T}{\partial x}\right) - \frac{\alpha p}{A}(T - T_\infty) + q_3. \tag{3.3.13}$$

If the heat transfer coefficient α is not constant along perimeter p, then the product αp in (3.3.13) should be replaced by an integral $\int_p \alpha dp$.

It should be noted that welding is always accompanied by elastic and plastic deformation of material, which is related to the absorption and release of energy.

According to the law of thermodynamics, this energy must be taken into account in the general energy balance in the form of the following additional term on the right side of the heat equation (Boley and Weiner 1960; Radaj 1992; Ziegler 1983):

$$-\frac{E\alpha T}{1-2\nu}\frac{\partial \varepsilon_{ij}^{e}}{\partial t}+\xi S_{ij}\frac{\partial \varepsilon_{ij}^{p}}{\partial t}, \tag{3.3.14}$$

where E, α and ν are the elastic modulus, the coefficient of thermal expansion, and the Poisson's ratio of the material, respectively. Here, the deviatoric stress S, the elastic strain ε^{e} and the plastic (inelastic) strain ε^{p} are written in a tensor form. The coefficient $\xi \leq 1$ takes into account the fact that not all inelastic deformation is dissipated in heat. It can be partially spent on microstructure change. However, as shown in paper (Argyris et al. 1982), in welding problems the thermal processes are almost completely determined by the input of energy from outside and the mechanical component (3.3.14) can be ignored. So, in the following we consider the temperature problem as uncoupled, i.e. not related to the mechanical problem. Friction welding is an exception because the mechanical component of the generated heat can be dominant.

By using the concepts of divergence and gradient, Eqs. (3.3.8), (3.3.9), (3.3.11) and (3.3.13) can be written in another form

$$\frac{\partial(\rho H)}{\partial t}=\operatorname{div}(\lambda \operatorname{grad} T)+c\rho b(T-T_{\infty})+q_3 \tag{3.3.15}$$

or

$$c\rho\frac{\partial T}{\partial t}=\operatorname{div}(\lambda \operatorname{grad} T)+c\rho b(T-T_{\infty})+q_3, \tag{3.3.16}$$

where b is the coefficient of heat loss:

$$b=\begin{cases}0 & \text{for the massive body;}\\ (\alpha_1+\alpha_2)/(c\rho h) & \text{for the plate;}\\ \alpha p/(c\rho A) & \text{for the rod.}\end{cases} \tag{3.3.17}$$

In Eqs. (3.3.9), (3.3.11) and (3.3.13) the material characteristics $c\rho$, λ_x, λ_y, λ_z, as well as coefficient α and the power of the internal source q_3 can generally be the functions of temperature T, i.e. these equations are non-linear.

If the thermal conductivity of the material is constant, the heat conduction equations can be simplified. For example, Eq. (3.3.9) for a homogeneous orthotropic solid takes the form:

$$\frac{\partial T}{\partial t}=a_x\frac{\partial^2 T}{\partial x^2}+a_y\frac{\partial^2 T}{\partial y^2}+a_z\frac{\partial^2 T}{\partial z^2}+\frac{q_3}{c\rho}, \tag{3.3.18}$$

where a_x, a_y, a_z are the thermal diffusivity in the direction of the x-, y- and z-axes:

$$a_x = \lambda_x/(c\rho);\ a_y = \lambda_y/(c\rho); a_z = \lambda_z/(c\rho). \tag{3.3.19}$$

The orthotropy of properties is typical for composites, single crystals, layered and fibrous materials; it is insignificant for structural materials used in welded structures.

The solution of the heat conduction Eq. (3.3.16) is considerably simplified if we assume that the body is isotropic, the thermal conductivity λ, volume-specific heat capacity $c\rho$ and coefficient of heat transfer α are constant, the intensity of the distributed sources q_3 is a linear function of temperature, and the ambient temperature T_∞ is constant. Then Eqs. (3.3.9), (3.3.11) and (3.3.13) for the infinite body, the plate and the rod, respectively, will be as follows:

$$\frac{\partial T}{\partial t} = a\left(\frac{\partial^2 T}{\partial x^2} + \frac{\partial^2 T}{\partial y^2} + \frac{\partial^2 T}{\partial z^2}\right) + \frac{q_3}{c\rho} = a\nabla^2 T + \frac{q_3}{c\rho}; \tag{3.3.20}$$

$$\frac{\partial T}{\partial t} = a\left(\frac{\partial^2 T}{\partial x^2} + \frac{\partial^2 T}{\partial y^2}\right) - bT + \frac{q_3}{c\rho}; \tag{3.3.21}$$

$$\frac{\partial T}{\partial t} = a\frac{\partial^2 T}{\partial x^2} - bT + \frac{q_3}{c\rho}, \tag{3.3.22}$$

where ∇^2 is the Laplace operator. These equations are parabolic linear partial differential equations of the second order. It is convenient to write them in a general form:

$$\frac{\partial T}{\partial t} = a\nabla^2 T - bT + \frac{q_3}{c\rho}, \tag{3.3.23}$$

In literature this equation is also found in tensor notation:

$$\dot{T} = aT_{i,j} - bT + \frac{q_3}{c\rho}. \tag{3.3.24}$$

Equations (3.3.20) and (3.3.21) can be written in other coordinate systems. For example, taking into account the relationship between Cartesian x, y, z and cylindrical r, θ, z coordinates $x = rcos\theta$, $y = rsin\theta$, $z = z$, Eq. (3.3.20) can be written in cylindrical coordinates as follows

$$\frac{\partial T}{\partial t} = a\left(\frac{\partial^2 T}{\partial r^2} + \frac{1}{r}\frac{\partial T}{\partial r} + \frac{1}{r^2}\frac{\partial^2 T}{\partial \theta^2} + \frac{\partial^2 T}{\partial z^2}\right) + \frac{q_3}{c\rho}. \tag{3.3.25}$$

For the stationary thermal processes ($\partial T/\partial t = 0$) the Eq. (3.3.20) takes the form

$$\nabla^2 T + \frac{q_3}{\lambda} = 0 \tag{3.3.26}$$

and is called the Poisson's equation, and for $q_3 = 0$ (no sources)

$$\nabla^2 T = 0 \tag{3.3.27}$$

is called the Laplace's equation. Equations (3.3.26) and (3.3.27) are partial differential equations of the elliptic type.

All previously obtained equations describe the processes of heat transfer in a coordinate system fixed in reference to the solid. The heat processes in welding are described most simply if the equations are referred to moving coordinates associated with a moving heat source.

Let the moving coordinate system x, y, z move along the axis x_0 of the fixed coordinate system x_0, y_0, z_0 for time t with a constant velocity v, i.e. $x = x_0 - vt$, $y = y_0$, $z = z_0$. From here

$$\frac{\partial T}{\partial x_0} = \frac{\partial T}{\partial x}\frac{\partial x}{\partial x_0} = \frac{\partial T}{\partial x}; \quad \frac{\partial^2 T}{\partial x_0^2} = \frac{\partial^2 T}{\partial x^2}. \tag{3.3.28}$$

The relation between $\partial T/\partial t$ in fixed coordinates $(\partial T/\partial t)_{x_0, y_0, z_0}$ and moving coordinates $(\partial T/\partial t)_{x, y, z}$ is as follows:

$$\left(\frac{\partial T}{\partial t}\right)_{x_0, y_0, z_0} = \left(\frac{\partial T}{\partial t}\right)_{x, y, z} + \frac{\partial T}{\partial x}\frac{\partial x}{\partial t} = \left(\frac{\partial T}{\partial t}\right)_{x, y, z} - v\frac{\partial T}{\partial x}. \tag{3.3.29}$$

By substituting (3.3.28) and (3.3.29) in (3.3.20), we obtain a heat conduction equation in the moving coordinate system x, y, z:

$$\frac{\partial T}{\partial t} = a\nabla^2 T + v\frac{\partial T}{\partial x} + \frac{q_3}{c\rho}. \tag{3.3.30}$$

If $\partial T/\partial t = 0$, the temperature field is called *quasi-stationary* (it does not change in moving coordinates, but varies in fixed coordinates). If we assume that the internal sources (sinks) are proportional to the temperature, $q_3 = -c\rho bT$, then the quasi-stationary field is described by the equation.

$$\nabla^2 T + \frac{v}{a}\frac{\partial T}{\partial x} - \frac{b}{a}T = 0. \tag{3.3.31}$$

The substitution

$$T(x, y, z) = U(x, y, z)\exp\left(-\frac{vx}{2a}\right) \tag{3.3.32}$$

leads to an equation of the elliptic type

$$\nabla^2 U - \left(\frac{v^2}{4a^2} + \frac{b}{a}\right)U = 0, \tag{3.3.33}$$

which is symmetric with respect to the variables and is solved more simply. The asymmetry of the temperature field is taken into account due to the factor $\exp(-vx/(2a))$ in (3.3.32). This factor is present in all cases of moving sources (Sect. 5.1.3).

Fourier's equation of heat conduction (3.3.9) is derived with an implicit assumption that the heat propagation velocity is infinitely high. This assumption is valid for moderate intensity processes, including almost all welding processes. Consideration of the finite velocity of heat propagation leads to a differential heat conduction equation of the hyperbolic type (Luikov 1968), and we do not consider it here.

3.4 Boundary Conditions

The differential equation of heat conduction has an infinite aggregate of solutions. For example, if the function $T(x, y, z, t)$ satisfies the equation, then the function $T(x, y, z, t)$ + const also does. In order to choose a solution out of the aggregate which characterises a certain practical case, geometrical conditions (shape and size of the body), thermophysical parameters (a, $c\rho$, q_3) and boundary conditions must be set.

Boundary conditions are understood as a set of initial and boundary conditions.

The initial condition is determined by setting the temperature distribution law inside the body at the initial time ($t = 0$):

$$T(x, y, z, 0) = f(x, y, z). \tag{3.4.1}$$

In welding it is common to assume a uniform distribution of temperature with $t = 0$ $T(x, y, z, 0) = T_0 = \text{const}$ and value T_0 as the temperature zero reference point.

Boundary conditions indicate the interaction of the body surface (boundary) with the ambient medium. It is common to use the so-called boundary conditions of the first, second, third and fourth kinds out of the entire diversity in the calculation of processes in welding.

The boundary condition of the first kind (Dirichlet type) implies prescribing the temperature distribution along the body surface S at any time:

$$T_S = \varphi(x, y, z, t), \quad x, y, z \in S. \tag{3.4.2}$$

A special case is an isothermal boundary condition, when $\varphi(x, y, z, t) = \text{const}$. Let us give it a graphic interpretation (Fig. 3.4a). The heat flow is proportional to the magnitude of the temperature gradient, which is numerically equal to the tangent of the line slope to the temperature distribution curve along normal n to surface S, i.e. $(\partial T/\partial n)_S = \tan\gamma$. According to the fixed value T_S ($T_S = \text{const}$), distributions T are found and so are the line slopes ($\gamma = \text{var}$) and the amount of heat lost by the body surface.

In welding practice the condition of the first kind is a rare thing. It is used, for example, to describe the contact of the body with massive heat sink clamps.

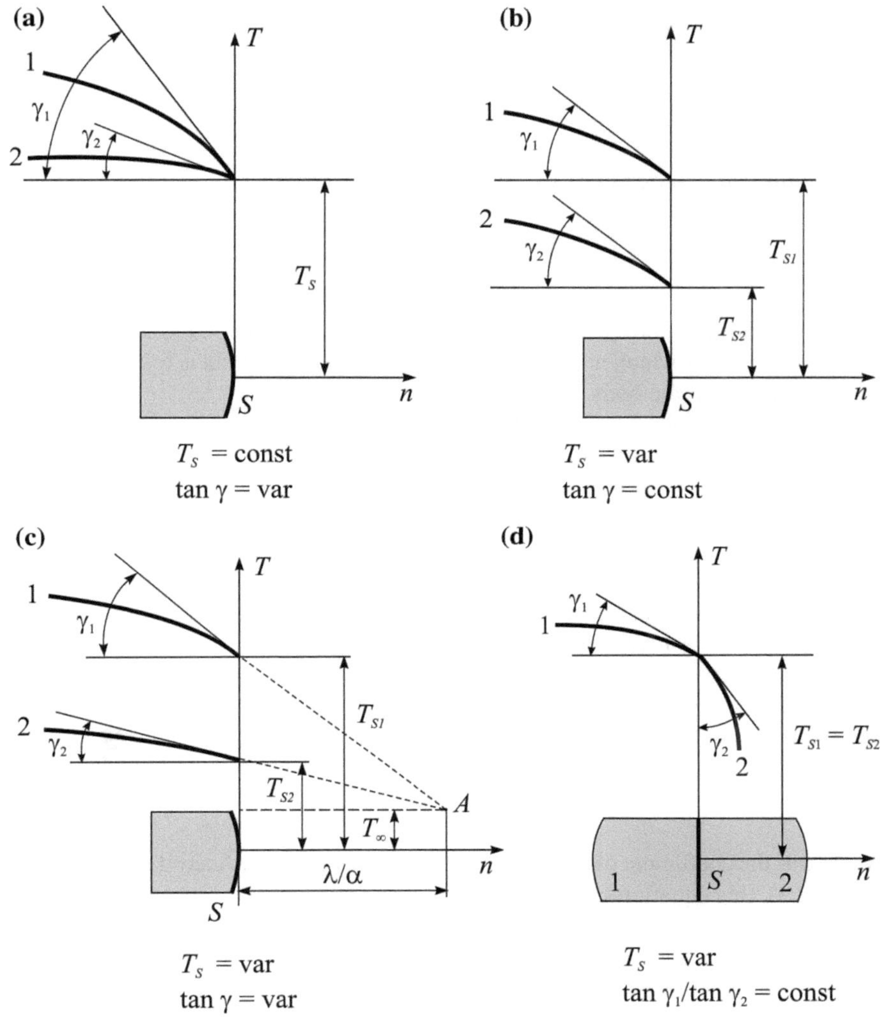

Fig. 3.4 Boundary conditions: prescribed surface temperature (condition of the first kind) (**a**), prescribed flux across the surface (condition of the second kind) (**b**), heat transfer at the surface (condition of the third kind) (**c**) and the continuous flux over the surface of separation of two media (condition of the fourth kind) (**d**) (Luikov 1968)

The boundary condition of the second kind (Neumann type) implies prescribing the density of heat flow for every boundary point of body S as a time function:

$$q_{2S} = \psi(x, y, z, t), \quad x, y, z \in S. \tag{3.4.3}$$

According to Fourier's law (3.2.3), this equation can be written differently:

$$-\lambda\left(\frac{\partial T}{\partial n}\right)_S = \psi(x, y, z, t), \; x, y, z \in S. \tag{3.4.4}$$

The geometrical interpretation of the condition of the second kind is shown in Fig. 3.4b. It is $\tan\gamma$ which is prescribed, and the temperature field is found, including the temperature of the body surface T_S.

The adiabatic boundary is a special case of the condition of the second kind and reflects the condition of the so-called heat insulation

$$q_{2S} = 0, \quad (\partial T/\partial n)_S = 0. \tag{3.4.5}$$

When heat processes are calculated in the welding of bounded bodies, the last condition is used if heat transfer to the ambient medium is insignificant in comparison to the heat flows inside the body.

The boundary condition of the third kind implies prescribing the flux across the surface to be proportional to the temperature difference between the body surface T_S and the ambient medium T_∞ according to Newton's law:

$$q_{2S} = \alpha(T_S - T_\infty), \tag{3.4.6}$$

where α is the coefficient of surface heat transfer, W m^{-2} K^{-1}. According to Fourier's law (3.2.3), this equation can be written differently:

$$-\lambda\left(\frac{\partial T}{\partial n}\right)_S = \alpha(T_S - T_\infty). \tag{3.4.7}$$

The geometrical interpretation of the boundary condition is shown in Fig. 3.4c. Although T_S and $\tan\gamma$ are variables, the condition (3.4.7) requires that all tangent lines to the temperature curve should go through point A, which is at a distance λ/α from the boundary of the body.

Note, that from the condition of the third kind (3.4.7), it is possible to obtain an isothermal condition ($T_S = T_\infty$ with $\alpha \to \infty$) and an adiabatic condition ($(\partial T/\partial n)_S = 0$ with $\alpha \to 0$), as limiting cases.

Heat is transferred from the surface of bodies through convection and radiation.

With convective heat transfer, heat is carried away by a gas or liquid, which are in motion due to the different density of inequally heated zones or under the action of

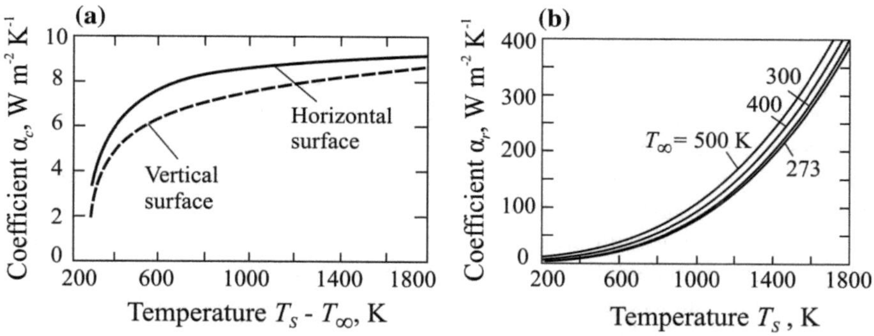

Fig. 3.5 The temperature-dependent coefficient of convective heat transfer α_c in air (**a**) (Argyris et al. 1985; Radaj 2003) and the coefficient of radiative heat transfer α_r with emissivity $\varepsilon = 1$ (**b**)

external forces (for example, gas blowing of the body). The density of the heat flow is approximately described by Newton's law.

$$q_{2S\,c} = \alpha_c(T_S - T_\infty), \tag{3.4.8}$$

where α_c is the coefficient of convective heat transfer (Fig. 3.5a). It depends on the shape, size and spatial position of the body surface, properties of the ambient medium and the body surface, difference of $T_S - T_\infty$ and other factors. With an increase in temperature, the coefficient α_c first grows fast and then slower, tending to limiting values (Fig. 3.5a).

Cooling of the body through radiative heat transfer is described by the Stefan–Boltzmann's law, according to which the density of the heat flow is

$$q_{2S\,r} = \varepsilon C_0(T_S^4 - T_\infty^4) = \alpha_r(T_S - T_\infty). \tag{3.4.9}$$

Hence

$$\alpha_r = \varepsilon C_0(T_S + T_\infty)(T_S^2 + T_\infty^2), \tag{3.4.10}$$

where ε is the emissivity of the surface, $0 < \varepsilon < 1$; C_0 is the Stefan-Boltzmann constant, $C_0 = 5.67\ 10^{-8}$ W m^{-2} K^{-4}. It is seen that the coefficient α_r is proportional to coefficient ε and grows with a temperature increase in conformity with the cubic parabola law (Fig. 3.5b). Magnitude ε depends on the condition of the surface. For polished metal surfaces coefficient ε varies within the range of 0.2–0.4, for oxidized and rough surfaces of steels—within the range of 0.6–0.95 (Carslaw and Jaeger 1973; Rykalin 1951, 1957), for aluminium—within the range of 0.05–0.2 (Frolov 1970).

The total heat transfer (3.4.6) can be represented as the sum of convective (3.4.8) and radiative (3.4.9) components:

$$q_{2S} = q_{2Sc} + q_{2Sr} = \alpha(T_S - T_\infty), \tag{3.4.11}$$

where $\alpha = \alpha_c + \alpha_r$ is the coefficient of the total surface heat transfer.

In case of natural heat transfer of the welded body surface to the ambient air at the surface temperature up to 600 K most of the heat is removed by convection. At a higher temperature, radiative heat transfer prevails, which constitutes about 80% of the total heat transfer at 1100 K. The recommended values of coefficient α for calculating the temperature fields in plates during fusion welding are presented in Table 3.1. The presented data is obtained with a distance from the joint axis to the heat-removing clamps equal to $(1.5 - 1.8)h$, where h is the plate thickness. Z is the distance from the midplane of the plate to the axes of channels with a cooling liquid.

The boundary condition of the fourth kind implies prescribing a perfect heat contact of two dissimilar bodies: continuous temperature ($T_{S1} = T_{S2}$) and continuous heat flow ($q_{2S1} = q_{2S2}$). Herewith $\tan\gamma_1/\tan\gamma_2 = \lambda_1/\lambda_2 = \text{const}$ (Fig. 3.4d).

Boundary conditions can be of a mixed kind, when in various areas of the boundary various conditions are prescribed. They can be very complicated, for example, in case of multi-pass narrow-gap welding (Gabriel et al. 2007).

A differential equation jointly with boundary conditions fully determines the problem, i.e. if geometrical, physical, initial and boundary conditions are known, the function of temperature distribution can be found at any instant of time $T(x, y, z, t)$. Moreover, it can be proved that this function is the unique solution to this problem.

3.5 Accounting for Variability of Thermal Properties

The methods of functional analysis are mostly used to solve linear problems of heat conduction when all thermophysical properties of the material do not depend on temperature. If the properties do not depend on coordinates, but depend on temperature, in particular cases the non-linear Eqs. (3.3.9), (3.3.11) and (3.3.13) can be solved by analytical methods, having been linearized first (Carslaw and Jaeger 1973). As exemplified by Eq. (3.3.9), the thermal diffusivity a is constant, and the thermal conductivity λ and the volume-specific heat capacity $c\rho$ are a non-linear function of temperature:

$$\lambda(T) = \lambda_0(1 + mT);\ c\rho(T) = c\rho_0(1 + mT). \tag{3.5.1}$$

Experiments prove that $\lambda(T)$ and $c\rho(T)$ grow or decrease within a certain range, so, for many materials, magnitude a can be considered constant, $a = \lambda/(c\rho) = \text{const}$.

Let us introduce a Kirchhoff's variable (Carslaw and Jaeger 1973; Nedoseka 1988; Ravi Vishnu and Easterling 1993):

Table 3.1 The coefficient of heat transfer under various conditions of heat removal (Kazimirov et al. 1968)

Magnitude, characterizing the efficiency of heat transfer	Heat-removing cooled clamps						Heat-removing uncooled platen	Transfer to air
	Copper		Aluminium		Steel		Steel	
	The surface of the plate is etched		There is a layer of oil between pressure bars and the plate	The surface of the plate and pressure bars is etched	The surfaces are degreased		The surfaces of the plate are cleaned and degreased	
	Z = 10°mm	Z = 30 mm	Z = 10 mm	Z = 30 mm	Z = 10 mm	Z = 30 mm		
Contour contact surface, % or pressure, kPa	80–90	80–90	80–90	80–90	80–90	80–90	50	
Coefficient of heat transfer α, $W\ m^{-2}\ K^{-1}$	4600–5450	3560–4190	3350–4190	2930–4190	1680–2100	293–419	84–168	13–25

$$\Theta = \frac{1}{\lambda_0}\int_{T_0}^{T} \lambda(T)dT = T - T_0 + 0.5m(T^2 - T_0^2). \tag{3.5.2}$$

From (3.5.2) it follows that

$$\begin{aligned}
&\frac{\partial\Theta}{\partial t} = (1+mT)\frac{\partial T}{\partial t}; \quad \frac{\partial\Theta}{\partial x} = (1+mT)\frac{\partial T}{\partial x};\\
&\frac{\partial\Theta}{\partial y} = (1+mT)\frac{\partial T}{\partial y};\\
&\frac{\partial\Theta}{\partial z} = (1+mT)\frac{\partial T}{\partial z}.
\end{aligned} \tag{3.5.3}$$

By substituting (3.5.3) in (3.3.9), we obtain

$$\frac{\partial\Theta}{\partial t} = a\left(\frac{\partial^2\Theta}{\partial x^2} + \frac{\partial^2\Theta}{\partial y^2} + \frac{\partial^2\Theta}{\partial z^2}\right) + \frac{q_3}{c\rho}, \tag{3.5.4}$$

which coincides in form with Eq. (3.3.20), if q_3 does not depend on T. Note that if q_3 depends on temperature, then q_3 must be expressed through a new variable Θ. By solving the linearized equation of heat conduction (3.5.4) for Θ, the temperature distribution can be obtained from (3.5.2):

$$T = \left(\sqrt{(1+mT_0)^2 + 2m\Theta} - 1\right)/m. \tag{3.5.5}$$

Other linearisation methods of a non-linear equation of heat conduction are described in books (Carslaw and Jaeger 1973; Luikov 1968).

3.6 Accounting for the Latent Heat of Phase Transformations

During welding heating and cooling, phase transformations occur in the body. They are accompanied by additional heat consumption and generation (see Sect. 2.1). If the metal is pure, phase transformation occurs at a fixed temperature and the relevant problem is called the problem of Stefan (Meyermanov 1986). In alloys, transformations occur within a temperature interval. This effect takes place, for example, during phase transformation α-Fe $\rightarrow$ γ-Fe and during melting. Consider the heat effect of phase transformations in solving heat conduction problems (Karkhin et al. 2004, 2005).

Since the inequation $H\partial\rho/\partial T \ll \rho\partial H/\partial T$ is true for welding processes, Eq. (3.3.15) in the fixed system of coordinates can be simplified

$$\rho \frac{\partial H}{\partial t} = \operatorname{div}(\lambda \operatorname{grad} T) + c\rho b(T - T_\infty) + q_3. \tag{3.6.1}$$

This equation is input if solved by a numerical method allowing for the dependence of the heat capacity c on temperature T, and, consequently, on time t (see Sect. 4.3).

Equation (3.6.1) can be integrated by an analytical method with an assumption that the properties (λ, c, ρ, α) do not depend on temperature. Let us show it by an example of a source which has been moving for quite a while along the x-axis with a constant velocity v. Then Eq. (3.6.1) in the moving system of coordinates, taking into account (3.3.29) and the condition $\partial H/\partial t = 0$, takes the form:

$$\lambda \nabla^2 T + v\rho \frac{\partial H}{\partial x} + c\rho b(T - T_\infty) + q_3 = 0. \tag{3.6.2}$$

Separate out the heat of phase transformations ΔH from enthalpy.

$$H(T) = c(T - T_0) + \Delta H(T). \tag{3.6.3}$$

By substituting (3.6.3) in (3.6.2), we obtain

$$\nabla^2 T + \frac{v}{a}\frac{\partial T}{\partial x} - \frac{b}{a}(T - T_\infty) + \frac{q_3}{\lambda} + \frac{v}{ac}\frac{\partial(\Delta H)}{\partial x} = 0. \tag{3.6.4}$$

If $T_0 = T_\infty = 0$ this equation coincides with (3.3.31), except for the last two terms, considering both the real volume source q_3, and the heat ΔH. Here, in principle, any deviation from linearity $H(T)$ with $c = \text{const}$ can be understood as ΔH. However, this deviation is significant in case of phase transitions (Figs. 2.1–2.8).

The heat of phase transformation is distributed unevenly (Fig. 3.6). However, the total power of heat sources (in the cooling zone) and sinks (in the heating zone) is equal to zero, i.e. the heat of phase transformations does not affect the heat balance.

The Eq. (3.6.4) can be solved by the well-known methods of functional analysis, as shown in Sect. 9.4.

3.7 Models for Heat Sources

A common feature of all welding methods is local heating of the workpiece. For this various concentrated heat sources are used with the density of heat flow up to 10^{10} W m^{-2} and even above (see Chap. 1).

Usually when calculating heat processes, all welding sources (electric arc, gas flame, electron and laser beams, electroslag pool and so on) are idealised. Idealised sources can be classified in terms of the duration of heat generation into *instantaneous* (heat generation time $t_w = 0$) (Fig. 3.7a) and *continuous* ($t_w > 0$) sources (Fig. 3.7b). In practice, a constant power source is often applied, $q(t) = \text{const}$. Periodic sources are becoming widely used (Fig. 3.7c).

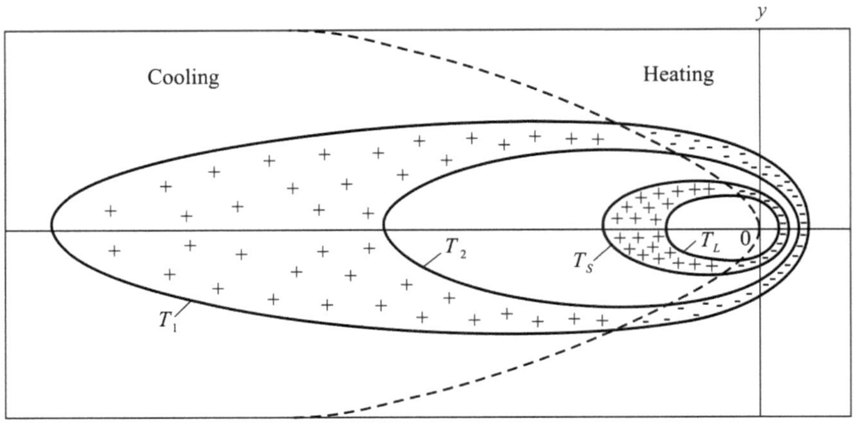

Fig. 3.6 Diagram of sinks (−) and sources (+) of the heat of melting and solidification in mushy zone (between isotherms T_S and T_L) and solid-state phase transformation zone (between isotherms T_1 and T_2)

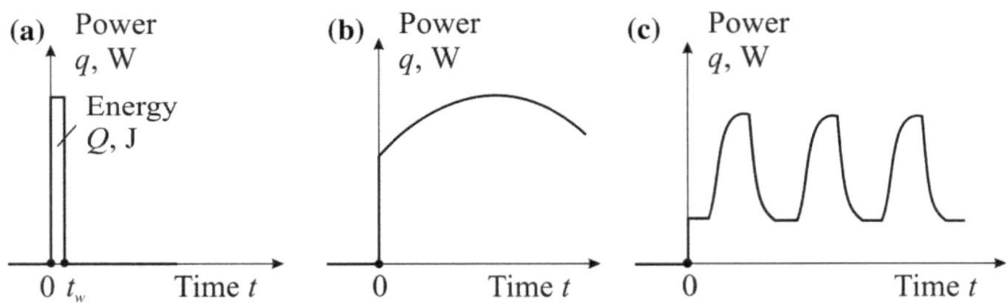

Fig. 3.7 Idealised temporal distributions of the heat source: the instantaneous heat source ($t_w \to 0$) (**a**), the continuous heat source (**b**) and the periodic (pulsed power) source (**c**)

According to the spatial distribution, idealised sources are divided into *point* (dim = 0), *line* (dim = 1), *plane* (dim = 2) and *volumetric* (dim = 3) sources (Fig. 3.8). Indexes $i = 0; 1; 2; 3$ of heat Q_i (J m^{-i}) and power q_i (W m^{-i}) will be used below to show the dimensionality of the source.

In turn, power and energy distribution along each s-axis ($s = x, y, z$) can be diverse (Fig. 3.9).

Continuous sources can be *stationary* (travel velocity $v = 0$) and *moving* ($v \neq 0$). In turn, *rapidly moving heat sources* are distinguished out of moving sources and introduced in a calculation model when it is possible to ignore the heat flows along the axis of source movement, which considerably simplifies the calculation (reduces the problem dimensionality by one).

In fusion welding, surface sources are used as heat sources. In many cases the surface of the weld pool is deformed insignificantly, so the welding source on the body surface can be considered as a plane one.

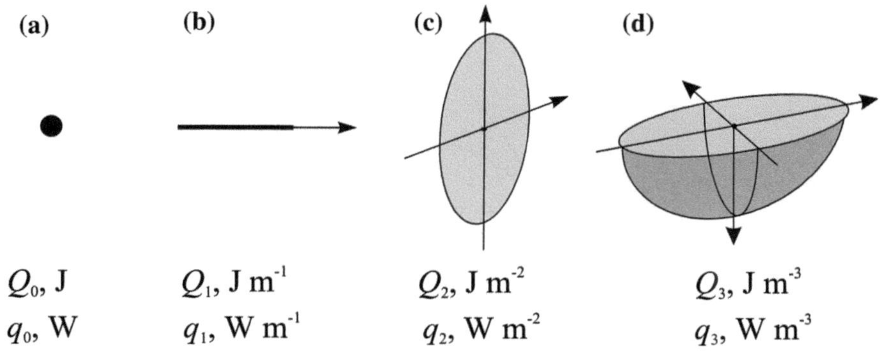

Fig. 3.8 Idealised spatial distributions of the point heat source (**a**), line heat source (**b**), plane heat source (**c**) and volume heat source (**d**)

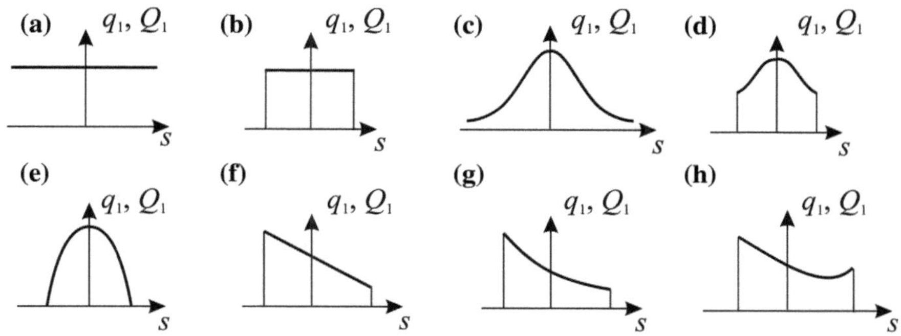

Fig. 3.9 Distributions of power density q_1 and energy density Q_1 for a line source along the s-axis: unbounded uniform (**a**); bounded uniform (**b**); unbounded normal (**c**); bounded normal (**d**); bounded sinusoidal (**e**); bounded linear (**f**); bounded exponential (**g**) and bounded parabolic (**h**)

Numerous experiments show (Chap. 1), that the majority of welding sources (electric arc, gas flame, electron beam, plasma jet, laser beam, etc.) have an axial symmetry and the radial distribution of the heat flow density is satisfactorily described by the Gaussian curve (normal distribution law):

$$q_2(r) = q_{2\max} \exp(-Cr^2), \tag{3.7.1}$$

where $q_{2\max}$ is the density of heat flow in the heat centre (Fig. 1.5) is the concentration factor of normal distribution heat source.

The heat spot radius r_e and the concentration factor of the normally distributed heat source C characterise the sharpness of the normal distribution curve (Fig. 3.10). The smaller is r_e (the bigger is C), the quicker the power density decreases with distance away from the centre of the source. If $r_e = 0$ $(C = \infty)$ the plane source turns into a point source on the body surface.

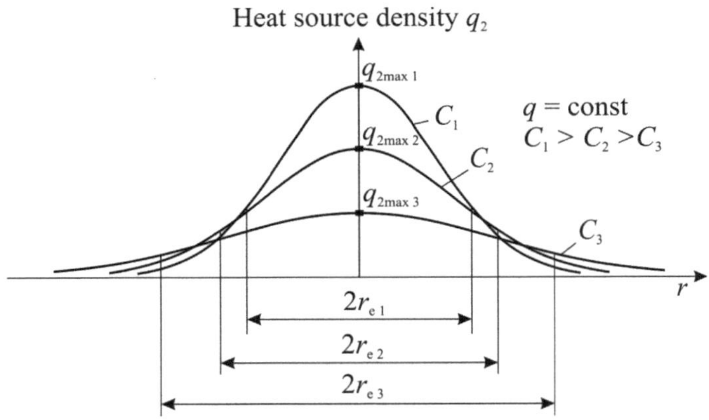

Fig. 3.10 Heat source density q_2 over radial distance from centre r, normally distributed sources, equal value $q = \text{const}$, different concentration factors C

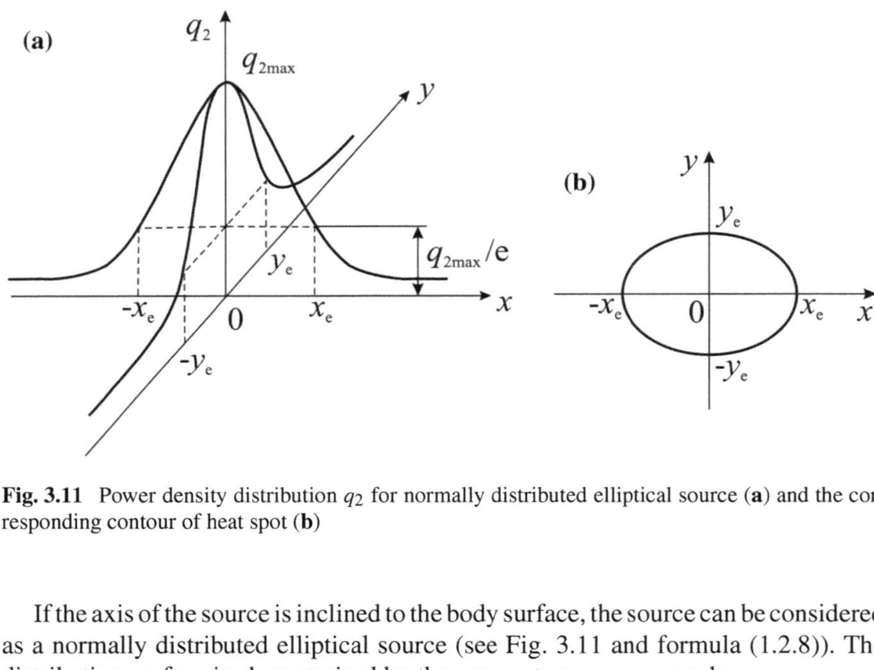

Fig. 3.11 Power density distribution q_2 for normally distributed elliptical source (**a**) and the corresponding contour of heat spot (**b**)

If the axis of the source is inclined to the body surface, the source can be considered as a normally distributed elliptical source (see Fig. 3.11 and formula (1.2.8)). The distribution surface is characterised by the parameters $q_{2\max}$, x_e and y_e.

In case the normally distributed elliptical source travels rapidly along the x-axis (Fig. 3.12), the heat flow seems to be concentrated in the direction of movement and its distribution has to be considered only in the transverse direction y (a model of a *normally distributed line source*). For this we project the surface source on the y-axis. Then, allowing for (1.2.8) we obtain (Rykalin 1951, 1957):

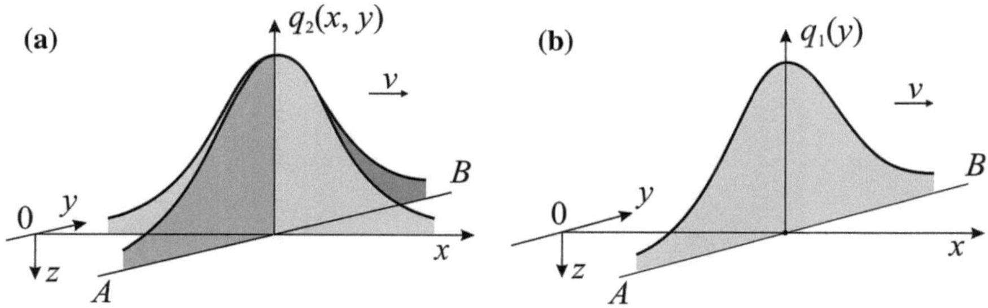

Fig. 3.12 Heating the surface by a rapidly moving normally distributed elliptical source (**a**) and its equivalent normally distributed line source (**b**)

$$q_1(y) = \int_{-\infty}^{\infty} q_2(x, y)dx = \sqrt{\frac{C_y}{\pi}} q \exp(-C_y y^2) = q_{1\max} \exp(-C_y y^2);$$

$$q_{1\max} = \sqrt{C_y/\pi} q, \tag{3.7.2}$$

where $q_{1\max}$ is the maximum line power density of a normal line source (Fig. 3.12).

We have looked into idealised heat sources acting on the plane of a heated body. In reality, due to the mechanical pressure on the weld pool, the heat source (arc, plasma jet, beam, etc.) can go deep in the body and thus become volume distributed (Fig. 3.8d).

Complex physical processes occur in the weld pool (interaction of the heat source and liquid metal, convection and evaporation of liquid, phase transformations, etc.). These processes are complicated to calculate. However, their effects can be approximately considered by adding equivalent virtual volume sources (see Chap. 11). Let us mention here the main types of such sources (normally, exponentially and parabolically distributed ones) (Karkhin et al. 2006, 2007a, b, c, 2010a, b) (Fig. 3.13).

The normal power distribution of the source f along the axis s (where $s = x, y, z$) is given by an equation

$$f(s) = \frac{2}{\pi^{1/2}} \frac{1}{s_e[\Phi(s''/s_e) - \Phi(s'/s_e)]} \exp\left(-\left(\frac{s}{s_e}\right)^2\right); \tag{3.7.3}$$

$$\Phi(u) = \frac{2}{\sqrt{\pi}} \int_0^u \exp(-t^2)dt, \tag{3.7.4}$$

where $\Phi(u)$ is a probability integral (the Gauss error function) (Fig. 3.14), s_e is a distance from the origin of coordinates to the point, where the value of the function f is e times lower than at the origin of coordinates (Figs. 1.5 and 3.12b). Here the s-axis represents one of the x-, y-, z-axes. The factor of function f is chosen so as to make the area under the curve equal 1. So such a source can be considered as unit.

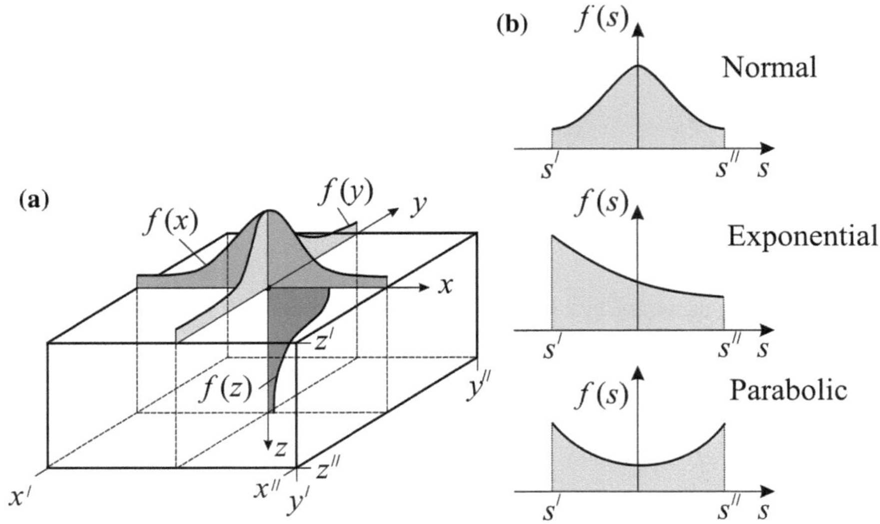

Fig. 3.13 Bounded volume source (**a**) and bounded volumetric unit power distribution along the axes: $s = x, y, z$ (**b**) (Karkhin et al. 2010a)

The exponential power distribution of source f along the s-axis (where $s = x, y, z$) is given by the equation

$$f(s) = \frac{1}{s_e} \frac{1}{\exp(-s'/s_e) - \exp(-s''/s_e)} \exp\left(-\frac{s}{s_e}\right). \tag{3.7.5}$$

The parabolic power distribution of source f along the s-axis (where $s = x, y, z$) is given by the equation.

$$f(s) = \frac{1}{a_0(s'' - s') + \frac{1}{2}a_1(s''^2 - s'^2) + \frac{1}{3}a_2(s''^3 - s'^3)}(a_0 + a_1 s + a_2 s^2). \tag{3.7.6}$$

Virtually any function of power distribution can be approximated by a piecewise parabolic function to the prescribed accuracy.

Out of the stipulated unit sources (3.7.3)–(3.7.6) it is possible to obtain various combinations of power distribution for a volume source:

$$q_3(x, y, z) = q\, f(x) f(y) f(z), \tag{3.7.7}$$

where q is the effective power of the real source. Some combinations will be reviewed in Sect. 5.2.

It is natural that the balance of energy in the entire body should be retained:

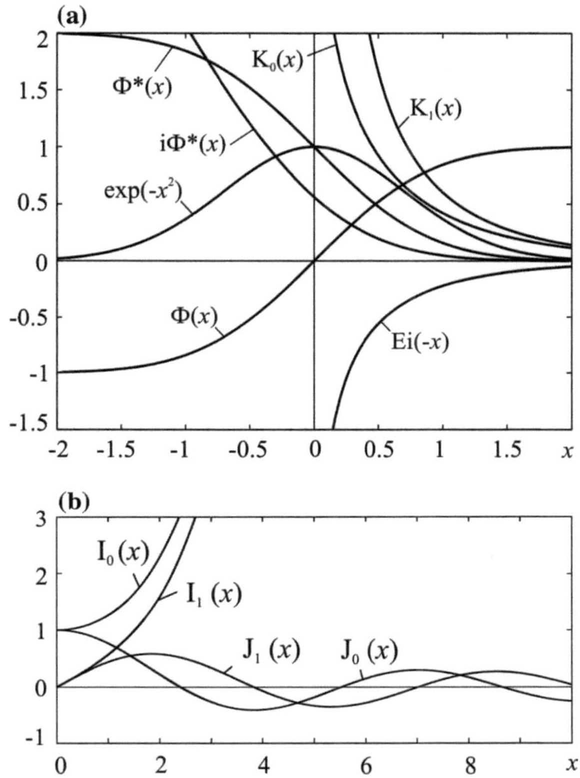

Fig. 3.14 Functions: normal distribution $\exp(-x^2)$, error function $\Phi(x)$ (3.7.4), complementary error function $\Phi^*(x)$ (4.2.16), exponential integral $\mathrm{Ei}(-x)$ (5.1.32), modified Bessel function of the second kind of order n (Macdonald function) $\mathrm{K_n}(x)$ (5.2.116), integral error function $\mathrm{i}\Phi^*(x)$ (5.1.37) (**a**), modified Bessel function of the first kind of order n $\mathrm{I_n}(x)$ (5.2.115) and Bessel function of the first kind of order n $\mathrm{J_n}(x)$ (4.2.44) (**b**)

$$\int_{-\infty}^{\infty}\int_{-\infty}^{\infty}\int_{-\infty}^{\infty} q_3(x, y, z)dxdydz = q\int_{x'}^{x''} f(x)dx\int_{y'}^{y''} f(y)dy\int_{z'}^{z''} f(z)dz = q. \quad (3.7.8)$$

Note that $q, f(x), f(y)$ and $f(z)$ can be functions of the time t.

The presentation of the power distribution of a volume source (3.7.7) is a generalisation of the previously described sources, by taking the relevant function f in the form of δ—function, line, Gaussian curve, exponent, etc. Let us show it as an example of a popular double-ellipsoidal model by Goldak et al. (1984), Goldak and Akhlaghi (2005). This model represents a combination of unbounded normal distributions along the y- and z-axes, as well as along x in front of the centre of the source (index f) and along x behind the source (index r). Moreover, the parameters of every distribution $\left(x_{0.05f} = x_{ef}\sqrt{3},\ x_{0.05r} = x_{er}\sqrt{3},\ y_{0.05} = y_e\sqrt{3},\ z_{0.05} = z_e\sqrt{3}\right)$ are different (Fig. 3.15). The power distribution of the source in front of and behind centre O is described by the formulas:

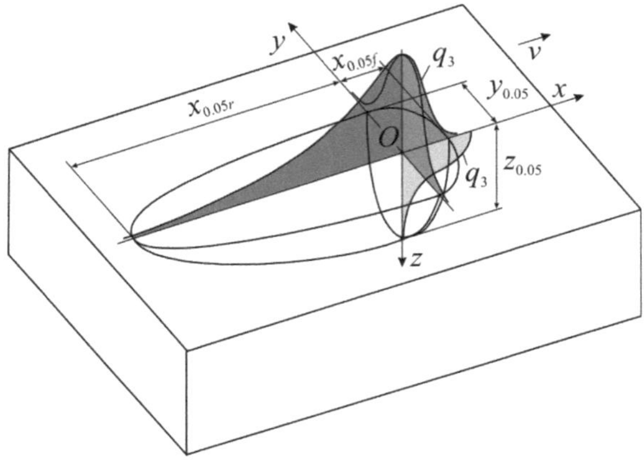

Fig. 3.15 Normally distributed double-ellipsoidal volume heat source with characteristic parameters (Goldak et al. 1984; Goldak and Akhlaghi 2005)

$$q_{3f}(x, y, z) = f_f \frac{6\sqrt{3}q}{\pi^{3/2} x_{0.05f} y_{0.05} z_{0.005}} \times \exp\left(-3\left[\left(\frac{x}{x_{0.05f}}\right)^2 + \left(\frac{y}{y_{0.05}}\right)^2 + \left(\frac{z}{z_{0.05}}\right)^2\right]\right); \qquad (3.7.9)$$

$$q_{3r}(x, y, z) = f_r \frac{6\sqrt{3}q}{\pi^{3/2} x_{0.05r} y_{0.05} z_{0.005}} \times \exp\left(-3\left[\left(\frac{x}{x_{0.05r}}\right)^2 + \left(\frac{y}{y_{0.05}}\right)^2 + \left(\frac{z}{z_{0.05}}\right)^2\right]\right), \qquad (3.7.10)$$

where factors $f_f + f_r = 2$. These equations follow from (3.7.3) and (3.7.7). The advantage of this model is its simplicity, its drawback is physical inaccuracies (for example, an input of energy is assumed inside the body out of the weld pool). An egg-configuration heat source model is developed (Yadaiah and Bag 2014).

Generally, the heat source distributions and their combinations depend on the welding method. For example, a 3D cylindrical volume heat source (functions $f(x)$ and $f(y)$ are normal, function $f(z)$ is constant) was proposed for arc welding (Artinov et al. 2018; Bachorski et al. 1999). A 3D conical Gaussian heat source (functions $f(x)$ and $f(y)$ are normal, function $f(z)$ is linear) was used in simulation of laser welding of butt and T-joints (Bakir et al. 2016; Balasubramanian et al. 2008; Shanmugam et al. 2010). An overview of different heat source distributions and their combinations is given in Chukkan et al. (2015).

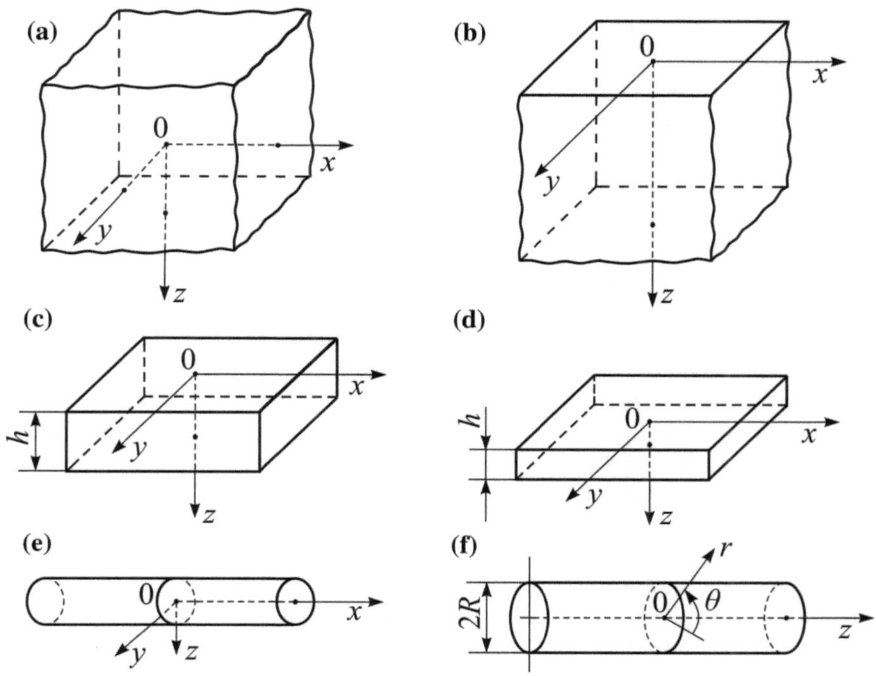

Fig. 3.16 Basic geometrical bodies for temperature calculation by functional-analytical methods: infinite solid (**a**), semi-infinite solid (**b**), infinite slab (**c**), infinite plate (**d**), infinite rod (**e**) and infinite cylinder (**f**) (Karkhin 1990)

3.8 Schematisation of Heated Bodies

Consider the main models of heated bodies. Consideration of the actual shape of welded workpieces can significantly complicate the solution of temperature problems. Therefore, the real shape is idealised in heat calculations and conditionally reduced to one of the following basic models (Fig. 3.16). Here the schematised bodies are characterised not only by linear sizes but also by the dimension of the temperature field.

An infinite body (solid) takes the area $-\infty < x < \infty$, $-\infty < y < \infty$, $-\infty < z < \infty$, (Fig. 3.16a). The heat flow in the solid is spatial, the temperature field is three-dimensional, $T = T(x, y, z, t)$.

A semi-infinite body (solid) takes a space on one side of the plane in the infinite solid $-\infty < x < \infty$, $-\infty < y < \infty$, $0 \le z < \infty$ (Fig. 3.16b). The heat flow is spatial, the temperature field is three- dimensional, $T = T(x, y, z, t)$.

An infinite slab occupies an interspace between two parallel planes in the infinite solid $-\infty < x < \infty$, $-\infty < y < \infty$, $0 \le z \le h$ (Fig. 3.16c). The heat flow is spatial, the temperature field is three-dimensional, $T = T(x, y, z, t)$.

An infinite plate is only different from a slab because the heat flow is plane, the temperature field is two-dimensional, $T = T(x, y, t)$ and $\partial T/\partial z = 0$ (Fig. 3.16d).

An infinite rod is a body of an arbitrary section in the plane $y0z$, $-\infty < x < \infty$ (Fig. 3.16e). The heat flow is linear, the temperature field is one-dimensional, $T = T(x, t)$ and $\partial T/\partial y = \partial T/\partial z = 0$.

An infinite cylinder occupies the area $0 \leq r \leq R$, $0 \leq \theta < 2\pi$, $-\infty < z < \infty$ (Fig. 3.16e). The heat flow is spatial, the temperature field is three-dimensional, $T = T(r, \theta, z, t)$.

It should be emphasised that the presented schematisation of heated bodies is conventional. For example, a semi-infinite body can be considered as a slab of an infinite thickness; a plate—as a slab of a small thickness, where the temperature difference along the thickness can be ignored; a rod—as a cylinder without temperature difference over the section. Moreover, various body models can correspond to the same thick-wall welded workpiece: an infinite solid—when root passes are carried out in two-sided multi-pass narrow-gap welding (heat flows are in all directions); a semi-infinite solid—when the final pass is made on the surface (heat flows are directed deep into metal); a slab—in electron beam welding with incomplete penetration, but complete heating of the workpiece (the heat insulation of the lower surface of the workpiece makes an impact) and a plate—in electron beam welding with complete penetration in the workpiece (heat flows are directed virtually only across the axis of the electron beam). Only understanding the physical model helps to select the relevant calculation model correctly.

In the following, we will often use mainly five combinations of models for heat sources and heated bodies: a point source in a semi-infinite solid or on its surface, a point source in an infinite slab or on its surface, a line source in an infinite plate, a plane source in an infinite rod and a point source in an infinite circular cylinder, since these are the main in constructing analytical solutions to diverse problems of heat conduction. Note that with the assumed definitions some combinations of sources and bodies are unacceptable. For example, there can be no point source in a plate or rod, since the plate implies uniform distribution of the source across thickness, while the rod—over the cross section.

References

Argyris, J. H., Szimmat, J., & Willam, K. J. (1982). Computational aspects of welding stress analysis. *Computer Methods in Applied Mechanics and Engineering, 33*, 635–666.

Argyris, J. H., Szimmat, J., & Willam, K. J. (1985). Finite element analysis of arc welding processes. *Numerical Methods in Heat Transfer, 3*, 1–34.

Artinov, A., Bachmann, M., & Rethmeier, M. (2018). Equivalent heat source approach in a 3D transient heat transfer simulation of full-penetration high power laser beam welding of thick metal plates. *International Journal of Heat and Mass Transfer, 122*, 1003–1013.

Bachorski, A., Painter, M. J., Smailes, A. J., & Wahab, M. A. (1999). Finite-element prediction of distortion during gas metal arc welding using the shrinkage volume approach. *Journal of Materials Processing Technology, 92*, 405–409.

Bakir, N., Gumenyuk, A., & Rethmeier, M. (2016). Numerical simulation of solidification crack formation during laser beam welding of austenitic stainless steel under external load. *Welding in the World, 60*(5), 1001–1008.

Balasubramanian, K. R., Shanmugam, N. S., Buvanashekaran, G., & Sankaranarayanasamy, K. (2008). Numerical and experimental investigation of laser beam welding of AISI 304 stainless steel sheet. *Advances in Production Engineering and Management, 3*(2), 93–105.

Boley, B. A., & Weiner, J. H. (1960). *Theory of thermal stresses* (586 pp.). Wiley.

Carslaw, H. S., & Jaeger, J. C. (1973). *Conduction of heat in solids* (510 pp.). London: Oxford University Press.

Chukkan, J. R., Vasudevan, M., Muthukumaran, S., Kumar, R. R., & Chandrasekhar, N. (2015). Simulation of laser butt welding of AISI 316L stainless steel sheet using various heat sources and experimental validation. *Journal of Materials Processing Technology, 219*, 48–59.

Frolov, V. V. (1970). *Theoretical principles of welding* (592 pp.). Moscow: Vysshaya Shkola (in Russian).

Gabriel, F., Ayrault, D., Fontes, A., Roatta, J. L., & Raynaud, M. (2007). Global method for estimation of heat source parameters dedicated to narrow gap GTA welding. In H. Cerjak, H. K. D. H. Bhadeshia & E. Kozenschnik (Eds.), *Mathematical modelling of weld phenomena* (Vol. 8, pp. 891–906). Graz: Verlag der Technischen Universitaet Graz.

Goldak, J., Chakravarti, A., & Bibby, M. (1984). A new finite element model for welding heat sources. *Metallurgical Transactions B, 15B*(6), 299–305.

Goldak, J. A., & Akhlaghi, M. (2005). *Computational welding mechanics* (321 pp.). Springer.

Karkhin, V. A. (1990). *Principles of heat transfer in weldments* (100 pp.). Leningrad: Leningrad State Technical University Publishing (in Russian).

Karkhin, V. A., Ilin, A. S., Ploshikhin, V. V., & Prikhodovsky, A. A. (2004). Effect of heat of aluminium alloy melting on shape and sizes of weld pool. *Paton Welding Journal*, (1), 2–7.

Karkhin, V. A., Ivanov, S. Yu., Homich, P. N., & Rajamaki, P. (2010). Analysis of solute macro- and microsegregation in fusion welding. In H. Cerjak & N. Enzinger (Eds.), *Mathematical modelling of weld phenomena* (Vol. 9, pp. 155–175). Graz: Verlag der Technischen Universitaet Graz.

Karkhin, V. A., Khomich, P. N., & Michailov, V. G. (2006). Prediction of microstructure and mechanical properties of weld metal with consideration for real weld geometry. In W. Lucas & V. I. Makhnenko (Eds.), *Proceedings of Joint International Conference "Computer Technology in Welding and Manufacturing (16th International Conference) and Information Technologies in Welding and Related Processes (3rd International Conference)"*, Kiev (pp. 162–166).

Karkhin, V. A., Homich, P. H., & Michailov, V. G. (2007a). Analytical-experimental technique for calculating the temperature fields in laser welding. In *11th NOLAMP Conference in Laser Processing of Materials*, Lappeenranta, Finland (pp. 21–27).

Karkhin, V. A., Khomich, P. N., & Michailov, V. G. (2007b). Models for volume heat sources and functional-analytical technique for calculating the temperature fields in butt welding . In H. Cerjak, H. K. D. H. Bhadeshia & E. Kozeschnik (Eds.), *Mathematical modelling of weld phenomena* (Vol. 8, pp. 819–834). Graz: Verlag der Technischen Universitaet Graz.

Karkhin, V. A., Khomich, P. N., Ossenbrink, R., & Michailov, V. G. (2007c). Calculation-experimental method for the determination of the temperature field in laser welding. *Welding International, 21*(5), 391–395.

Karkhin, V. A., Pittner, A., Schwenk, C., & Rethmeier, M. (2010a). Heat source models in simulation of heat flow in fusion welding. In *Proceedings of 5th International Conference "Mathematical Modelling and Information Technologies in Welding and Related Processes"* (pp. 56–60). Crimea, 25–28 May 2010.

Karkhin, V. A., Homich, P. N., & Ivanov, S. Yu. (2010b). Heat source models for prediction of temperature fields in fusion welding. In *Proceedings of the Tula State University. Technical Sciences, Part 1* (Vol. 4, pp. 241–254). Tula: Tula State University Publishing (in Russian).

Karkhin, V. A., Pesch, H. J., Ilin, A. S., Prikhodovsky, A. A., Plochikhine, V. V., Makhutin, M. V., & Zoch, H.-W. (2005). Effects of latent heat of fusion on thermal processes during welding.

In H. Cerjak, H. K. D. H. Bhadeshia, & E. Kozeschnik (Eds.), *Mathematical modelling of weld phenomena* (Vol. 7, pp. 39–62). Graz: Verlag der Technischen Universitaet Graz.
Kazimirov, A. A., Nedoseka, A. Ya., Lobanov, A. M., & Radchenko I. S. (1968). *Calculation of temperature fields in fusion welding of plates* (832 pp.). Kiev: Naukova dumka (in Russian).
Luikov, A. (1968). *Analytical heat conduction theory* (702 pp.). Elsevier.
Meyermanov, A. M. (1986). *Stefan problem* (239 pp.). Novosibirsk: Nauka (in Russian).
Nedoseka, A. Ya. (1988). *Fundamentals of calculating welded structures* (263 pp.). Kiev: Vishcha Shkola Publishing (in Russian).
Radaj, D. (1992). *Heat effects of welding. Temperature field, residual stress, distortion* (348 pp.). Berlin: Springer.
Radaj, D. (2003). *Welding residual stresses and distortion. Calculation and measurement* (397 pp.). Duesseldorf: DVS.
Ravi Vishnu, P., & Easterling, K. E. (1993). Phenomenological modelling of heat flow and microstructural changes in pulsed GTA welds in a quenched and tempered steel. In H. Cerjak & K. Easterling (Eds.), *Mathematical modelling of weld phenomena* (pp. 241–299). London: The Institute of Materials.
Rykalin, N. N. (1951). *Calculation of heat flow in welding.* (337 pp.) (Z. Paley & C. M. Adams, Jr., Trans.) . Moscow.
Rykalin, N. N. (1957). *Berechnung der Waermevorgaenge beim Schweissen* (326 pp.). Berlin: VEB Verlag Technik (in German).
Shanmugam, N. S., Buvanashekaran, G., Sankaranarayanasamy, K., & Kumar, S. R. (2010). A transient finite element simulation of the temperature and bead profiles of T-joint laser welds. *Materials and Design, 31*(9), 4528–4542.
Yadaiah, N., & Bag, S. (2014). Development of egg-configuration heat source model in numerical simulation of autogenous fusion welding process. *International Journal of Thermal Sciences, 86*, 125–138.
Ziegler, H. (1983). *An introduction to thermomechanics* (2nd ed.) (355 pp.). Amsterdam: North–Holland.

Chapter 4
Methods for Solving the Problems of Heat Conduction in Welding

4.1 Classification of Methods for Solving the Problems of Heat Conduction in Welding

The existing methods for solving the problems of heat conduction in welding are easy to classify according to the form of the solution results (Fig. 4.1). Functional-analytical methods lead to solving problems using formulas, whereas numerical ones use the numerical values of the function (temperature) in some fixed values of the argument.

Analytical methods are developed mostly to solve linear problems of the theory of heat conduction. In reference to the conditions of welding, the following methods are

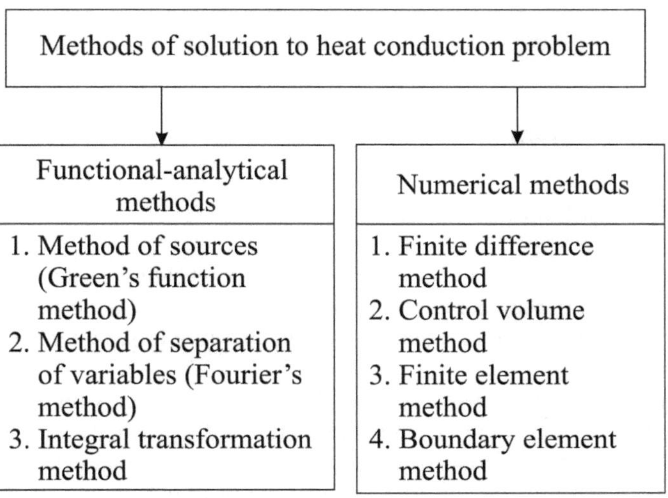

Fig. 4.1 Classification of the main methods of solution to heat conduction problem in welding

V. A. Karkhin, *Thermal Processes in Welding*, Engineering Materials,
https://doi.org/10.1007/978-981-13-5965-1_4

used: the method of sources (Green's functions), the method of separation of variables (Fourier's method), and the method of integral transformations. Non-linear problems are solved only in exceptional cases and only under considerable assumptions.

Numerical methods allow us to solve complex problems which cannot be solved through functional-analytical methods, for example, for bodies of complex shape having regard to the temperature dependence of the material properties. The finite difference method (FDM) and the finite element method (FEM) are the most commonly used methods in the theory of heat conduction in welding.

In comparison to the numerical methods, analytical methods make it possible to obtain more demonstrative solutions and analyse the factors impacting the solution results in an easier and faster way. In order to analyse functional-analytical solutions, numerical methods often have to be applied, for example, to calculate infinite series and definite integrals.

4.2 Functional-Analytical Methods for Calculation of Thermal Processes in Welding

In this section, the following prerequisites are assumed (if not stipulated otherwise):

- heat is distributed only according to Fourier's law of heat conduction;
- microstructural and phase transformations occur without heat generation or heat absorption;
- the thermal conductivity λ, volume-specific heat capacity $c\rho$ and the coefficient of surface heat transfer α do not depend on temperature;
- the initial temperature T_0 and ambient temperature T_∞ are constant and equal to each other, $T_0 = T_\infty$.

4.2.1 Source Method

4.2.1.1 Basis of Source Method

A specific feature of heat conduction problems in welding is that in an area of high temperature, the temperature field is affected mostly by the particularities of heat input (specific nature of welding sources) and to a lesser degree by boundary conditions.

The source method proposed by Lord Kelvin (called Green's function method in mathematical physics) is preferred due to its illustrative capacity and its simplicity when considering the features of welding heat sources; at the same time it has mathematical rigor. All this has governed its popularity in calculating the temperature fields in welded structures.

The physical essence of the source method implies that any process of heat dissipation in a solid can be represented as a total of temperature equalisation processes from a set of elementary heat sources distributed both in space and in time. It should be emphasised that the superposition principle, which is a part of the source method, is only applicable if the boundary value problem is linear (in the heat conduction equation and boundary conditions, $c\rho$, λ, a, q_{2S}, α do not depend on temperature and the power density $q_3(T)$ is a linear function of temperature T). In this case the spatial-temporal temperature fields generated by individual sources do not affect each other and can be combined, forming a field of mutual action of a number of sources. The heat from this source dissipates according to the law inherent to it, determined by a heat conduction equation and boundary conditions and not dependent on the initial distribution and action of other sources.

An instantaneous point source in a homogeneous infinite solid is assumed as an elementary source, of which any source, distributed in space and time, is composed. This combination of the source and solid represents a special principle interest. The following problem statement corresponds to this combination.

1. The equation of conduction of heat in Cartesian coordinates:

$$\frac{\partial T}{\partial t} = a\left(\frac{\partial^2 T}{\partial x^2} + \frac{\partial^2 T}{\partial y^2} + \frac{\partial^2 T}{\partial z^2}\right). \tag{4.2.1}$$

2. Initial condition. The elementary volume $dv = dxdydz$ is placed at the origin. At the initial moment $t = 0$, the heat quantity Q_0 (J) is introduced. Then the distribution of the initial temperature is expressed as

$$T(x, y, z, 0) = \begin{cases} T_0 + \frac{Q_0}{c\rho\, dv}, & \text{if } x, y, z \in dv; \\ T_0, & \text{if } x, y, z \notin dv. \end{cases} \tag{4.2.2}$$

3. Boundary conditions. Far away from the source, the temperature does not change (boundary conditions of the first kind):

$$T(\pm\infty, y, z, t) = T(x, \pm\infty, z, t) = T(x, y, \pm\infty, t) = T_0. \tag{4.2.3}$$

The following formula is a solution to this problem:

$$T(R, t) - T_0 = \frac{Q_0}{c\rho(4\pi a t)^{3/2}} \exp\left(-\frac{R^2}{4at}\right), \tag{4.2.4}$$

where $R^2 = x^2 + y^2 + z^2$ is volumetric radius vector squared (the squared distance from the point x, y, z to the heat source). It can be easily verified that the solution (4.2.4) satisfies the differential equation of conduction of heat (4.2.1), and initial and boundary conditions (4.2.2) and (4.2.3), while an increase in the heat content of the infinite solid

$$\int_0^{\infty} c\rho[T(R,t) - T_0]4\pi R^2 dR = Q_0 = \text{const} \tag{4.2.5}$$

remains constant throughout the entire period of temperature equalisation. The solution (4.2.4) is called *fundamental*.

If the instantaneous point source is located at point ξ, η, ζ, rather than in the centre of the coordinates and acts at moment τ, then the temperature at any moment t ($t > \tau$) is determined by the formula

$$T(x,y,z,t) - T_0 = \frac{Q_0}{c\rho[4\pi a(t-\tau)]^{3/2}} \times \exp\left(-\frac{(x-\xi)^2 + (y-\eta)^2 + (z-\zeta)^2}{4a(t-\tau)}\right), \tag{4.2.6}$$

where $t - \tau$ is the time of heat dissipation. It can be seen that the increment in temperature tends to zero, when $t \to \tau$ at all points except for one (ξ, η, ζ), where it becomes infinitely large.

The analytical solutions are also known for the instantaneous point source in an infinite anisotropic solid and in an infinite composite solid with an ideal contact of two semi-infinite bodies, whose properties (λ, $c\rho$, a) differ (Carslaw and Jaeger 1973).

Let us present a continuous source of power q_0 (W) as a sequence of elementary instantaneous sources. If defined according to (4.2.6) the increment of temperature at moment t due to the elementary source $dQ_0 = q_0(\tau)d\tau$ (J), operating and being at point (ξ, η, ζ) at moment τ, is:

$$dT(x,y,z,t) = \frac{q_0(\tau)d\tau}{c\rho[4\pi a(t-\tau)]^{3/2}} \times \exp\left(-\frac{[x-\xi(\tau)]^2 + [y-\eta(\tau)]^2 + [z-\zeta(\tau)]^2}{4a(t-\tau)}\right). \tag{4.2.7}$$

Then the temperature due to any source distributed in space and time can be obtained by summing all elementary sources:

$$T(x,y,z,t) - T_0 = \int_{-\infty}^{\infty}\int_{-\infty}^{\infty}\int_{-\infty}^{\infty}\int_0^t \frac{q_3(\xi,\eta,\zeta,\tau)}{c\rho[4\pi a(t-\tau)]^{3/2}} \times \exp\left(-\frac{[x-\xi(\tau)]^2 + [y-\eta(\tau)]^2 + [z-\zeta(\tau)]^2}{4a(t-\tau)}\right) d\xi d\eta d\zeta d\tau, \tag{4.2.8}$$

where q_3 is the volumetric density of the source power, W m^{-3}. This function is a solution to Eq. (3.3.20) in a homogeneous infinite solid. Most of the analytical solutions to heat conduction problems presented in this book are special forms of Eq. (4.2.8).

The following five combinations of bodies and instantaneous concentrated sources are selected as the main ones:

(1) an instantaneous point source in a semi-infinite solid and infinite wedge (Figs. 3.8a and 3.16b);
(2) an instantaneous point source in an infinite slab (Figs. 3.8a and 3.16c);
(3) an instantaneous line source uniformly distributed across the thickness in an infinite plate (Figs. 3.8b and 3.16d);
(4) an instantaneous plane source uniformly distributed over the cross-section in an infinite rod (Figs. 3.8c and 3.16e);
(5) an instantaneous point source in an infinite circular solid cylinder (Figs. 3.8a and 3.16f).

These combinations are the main ones when solutions are constructed for any sources distributed in space and time in the relevant bodies.

Combination 2 is usually not pointed out as a main one. It is pointed out here due to its exceptional practical value. It can be considered as a generalisation of combinations 1 (the thickness of the slab is infinitely increased) and 3 (the thickness is extremely reduced).

4.2.1.2 Instantaneous Point Source in a Semi-infinite Body and an Infinite Wedge

An infinite solid can be divided into $2n$ wedges by planes crossing on the y-axis (Fig. 4.2b). The angle of every wedge at the vertex is equal to $\beta = \pi/n$, where $n = 1, 2, \ldots$. The semi-infinite solid can be considered a wedge with a vertex angle $\beta = \pi$ $(n = 1)$.

When looking into a semi-infinite solid with an instantaneous point source inside, it is first assumed that its surface is heat-insulated (adiabatic), i.e. the coefficient of heat transfer $\alpha = 0$, and then the effect of heat transfer from the boundary plane is evaluated. The source is considered to act at the moment $t = 0$. The origin of coordinates is placed at point O on the surface, and point A_1 with the instantaneous source Q_0 is placed at the depth ζ (Fig. 4.2a). The boundary condition $\partial T/\partial z(x, y, 0, t) = 0$ is carried out using the *method of images.*

The essence of the method of images is to expand the bounded solid along the relative coordinates towards infinity and to select additional sources in the expanded area in such a way, that the boundary conditions on the surface of the bounded solid are satisfied. In the case under consideration, the semi-infinite solid expands in the area $z < 0$ and the instantaneous source Q_0 is introduced at the point $A_{-1}(\xi, \eta, -\zeta)$, which is a mirror image of the point $A_1(\xi, \eta, \zeta)$ on the plane $z = 0$. The algebraic sum of fluxes q_{2z} from both sources is equal to zero on the plane of symmetry $z = 0$. That is, at any moment t, the boundary conditions $\partial T/\partial z(x, y, 0, t) = 0$ are met.

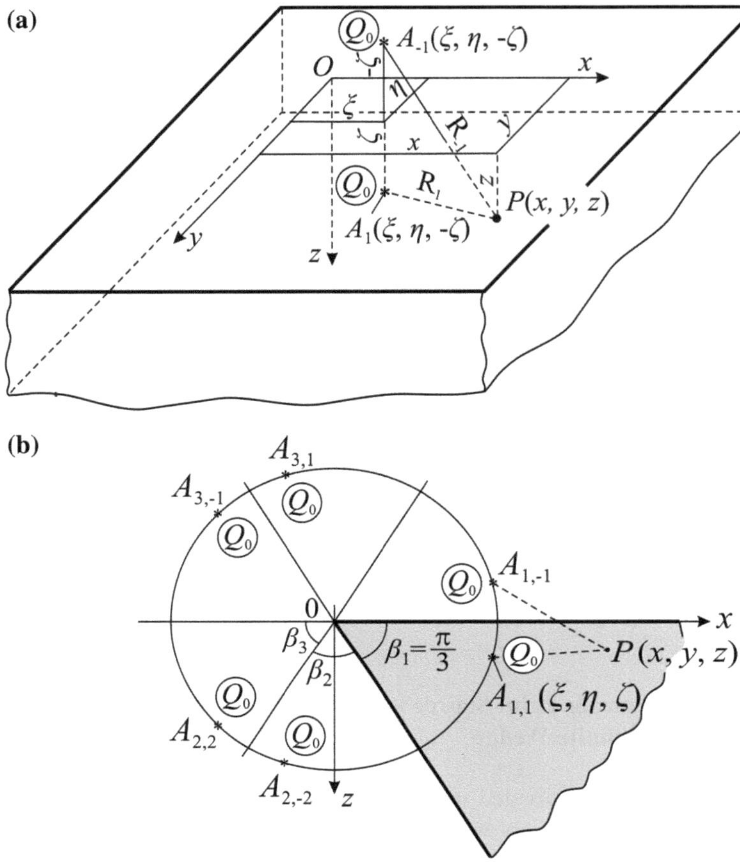

Fig. 4.2 The reflection of a concentrated source Q_0 in point A_1 of a semi-infinite solid $z > 0$ in the impermeable plane $z = 0$ (**a**) and the reflection of a concentrated source Q_0 in point $A_{1,1}$ of an infinite wedge in the impermeable planes $\beta_i = \text{const}$ ($i = 1, \ldots, n$; $n = 3$) (**b**)

The process of heat transfer in the area $z > 0$ is obtained by superimposing thermal processes (4.2.6) in the infinite solid from the instantaneous sources at points A_1 and A_{-1}:

$$T(x, y, z, t) - T_0 = \frac{Q_0}{c\rho(4\pi at)^{3/2}}\left[\exp\left(-\frac{R_1^2}{4at}\right) + \exp\left(-\frac{R_{-1}^2}{4at}\right)\right], \tag{4.2.9}$$

where $R_1^2 = (x-\xi)^2 + (y-\eta)^2 + (z-\zeta)^2$; $R_{-1}^2 = (x-\xi)^2 + (y-\eta)^2 + (z+\zeta)^2$, or in a more compact notation

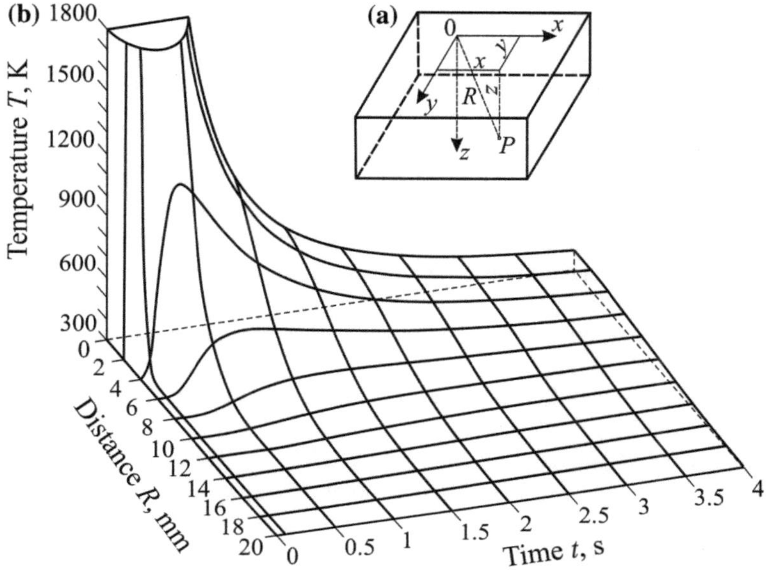

Fig. 4.3 The process of heat diffusion from an instantaneous point source on a semi-infinite solid ($Q = 2000$ J, $c\rho = 0.005$ J mm^{-3} K^{-1}, $a = 8$ mm^2 s^{-1}, $T_0 = 300$ K): model of point source (**a**) and the temperature distribution over radial distance from centre R and time t (**b**)

$$T(x, y, z, t) - T_0 = \frac{Q_0}{c\rho(4\pi at)^{3/2}} \sum_{j=-1,1} \exp\left(-\frac{R_j^2}{4at}\right), \tag{4.2.10}$$

where $R_j^2 = (x - \xi)^2 + (y - \eta)^2 + (z - j\zeta)^2$.

If the source is located on the surface of the semi-infinite solid at the origin ($\xi = \eta = \zeta = 0$), then

$$T(R, t) - T_0 = \frac{2Q_0}{c\rho(4\pi at)^{3/2}} \exp\left(-\frac{R^2}{4at}\right), \tag{4.2.11}$$

where $R^2 = x^2 + y^2 + z^2$ is the radius vector squared (Fig. 4.3a).

Let us analyse formula (4.2.11). The isothermal surfaces T = const represent spherical surfaces. Relation $T(R)$ when t = const is described by the law of normal distribution (Fig. 4.3b). Relation $T(t)$ when R = const is a thermal cycle of the point with the coordinate R. At the first moment ($t \to 0$), the temperature differs from the initial temperature only at point $R = 0$ [the temperature field is described by the Dirac delta function (1.2.6)]. Furthermore, the temperature at the point of origin drops according to the hyperbolic law $t^{-3/2}$. The temperature at all points tends to initial T_0 when $t \to \infty$. The farther the point of observation P lies from the source, the lower the peak temperature will be, and the later it will be reached (Fig. 4.3b). The projective boundary of the whole surface $T(R, t)$ on the plane T, R forms the radial distribution curve of the peak temperature $T_{\max}(R)$. It allows the size of the fusion zone from condition $T_{\max}(R) = T_m$ to be determined, where T_m is the melting temperature. It can also be seen from (4.2.11) that the temperature at any point is proportional to the amount of introduced heat Q_0 and is two times higher than by the same conditions in an infinite solid [see formula (4.2.4)]. The material properties $c\rho$ and a affect the temperature field. The temperature is inversely proportional to the volumetric heat capacity $c\rho$. The thermal diffusivity a appears only in the complex at, i.e., the increment of a accelerates the process of heat transfer, while its decrement causes deceleration. The dimensionless criterion at/R^2 is called the Fourier number Fo.

Let us consider another extreme case, when the surface heat transfer coefficient is infinitely great ($\alpha = \infty$) and consequently, the surface temperature remains equal to the ambient temperature T_0 at all times (the surface is isothermal, which corresponds to boundary conditions of the first kind).

An instantaneous source is given at point $A_1(\xi, \eta, \zeta)$. The solid is extended to the area $z < 0$ and the sink $-Q_0$ is placed at point $A_{-1}(\xi, \eta, -\zeta)$, which is the mirror image of point A (ξ, η, ζ) in the plane $z = 0$ (Fig. 4.2a). The temperature field is expressed by the algebraic sum

$$T(x, y, z, t) - T_o = \frac{Q_0}{c\rho(4\pi at)^{3/2}} \sum_{j=-1,1} j \exp\left(-\frac{R_j^2}{4at}\right). \quad (4.2.12)$$

It is evident that this expression satisfies the boundary condition $T(x, y, 0, t) = T_0$.

Let us allow now for the surface heat transfer. Suppose there is a heat transfer in the boundary plane $z = 0$ of a semi-infinite solid $z > 0$ to the surrounding medium with the temperature $T_\infty = T_0$ according to Newton's law [condition of the third kind (3.4.7)]:

$$\lambda \frac{\partial T}{\partial z} = \alpha(T - T_\infty). \quad (4.2.13)$$

The solution to the problem with the boundary condition (4.2.13) can be represented in the form (Carslaw and Jaeger 1973)

$$T(x, y, z, t) - T_0 = [T(x, y, z, t)|_{\alpha=0} - T_0] F_\alpha(z, t), \quad (4.2.14)$$

where

$$T(x, y, z, t)|_{\alpha=0} - T_o = \frac{Q_0}{c\rho(4\pi at)^{3/2}} \left[\exp\left(-\frac{R_1^2}{4at}\right) + \exp\left(-\frac{R_{-1}^2}{4at}\right)\right]; \quad (4.2.15)$$

$$F_\alpha(z, t) = 1 - 2\sqrt{\pi}\frac{\alpha}{\lambda}\sqrt{at}\left[1 + \exp\left(\frac{z\zeta}{at}\right)\right]^{-1} \exp(u^2)\, \Phi^*(u);$$

$$u = \frac{\alpha}{\lambda}\sqrt{at} + \frac{z}{2\sqrt{at}};$$

$$\Phi^*(u) = 1 - \Phi(u) = 1 - \frac{2}{\sqrt{\pi}}\int_0^u \exp(-t^2)dt. \tag{4.2.16}$$

The function $\Phi^*(u)$ is called the complementary error function (Fig. 3.14).

The first term between the square brackets in (4.2.14) expresses the temperature field without considering heat transfer [see (4.2.9)], while function F_α takes into account the influence of surface heat transfer on the process of heat distribution and depends on time, the source coordinates, and the coordinates of the point of observation. At $\alpha = 0$, the function $F_\alpha = 1$ and is less than 1, the bigger α/λ is. For large values of parameter u (coefficient α), the following expansion can be used (Luikov 1968):

$$\exp(u^2)\Phi^*(u) = \frac{1}{\sqrt{\pi}}\left(\frac{1}{u} - \frac{1}{2u^3} + \frac{1\cdot 3}{2^2u^5} - \frac{1\cdot 3\cdot 5}{2^3u^7} + \cdots\right). \tag{4.2.17}$$

It is easily proved that in the extreme case when $\alpha \to \infty$, formula (4.2.14) is transformed into (4.2.12).

If the source is located on the surface at the origin ($\xi = \eta = \zeta = 0$), then formula (4.2.14) is simplified

$$T(x, y, z, t) - T_o = \frac{2Q_0}{c\rho(4\pi at)^{3/2}}\exp\left(-\frac{x^2 + y^2 + z^2}{4at}\right)F_\alpha(z, t), \tag{4.2.18}$$

where

$$F_\alpha(z, t) = 1 - \sqrt{\pi}\frac{\alpha}{\lambda}\sqrt{at}\exp(u^2)\Phi^*(u); \quad u = \frac{\alpha}{\lambda}\sqrt{at} + \frac{z}{2\sqrt{at}}. \tag{4.2.19}$$

It is evident that function F_α depends only on the dimensionless heat transfer coefficient $\alpha/\lambda\sqrt{at}$ and the dimensionless coordinate $z/\sqrt{at}$. This dependence is shown in Fig. 4.4. It follows that the significant effect ($F_\alpha < 0.9$) of heat transfer only happens when $\alpha/\lambda\sqrt{at} > 0.05$. This condition corresponds to the very intensive forced cooling of the surface. Therefore, the heat transfer of the surface of semi-infinite solids is not considered below. If necessary, Eqs. (4.2.14) and (4.2.18) can be used as fundamental for constructing more complex computational schemes (instantaneous, distributed, moving, and other sources).

Let us now consider an infinite wedge with heat-insulated (adiabatic) surfaces and with an instantaneous source at point $A_{1,1}$ (Fig. 4.2b). The boundary condition of the wedge is met using the method of images. To do this, sources are additionally introduced into the infinite solid so that the planes of the wedge, forming the angle $\beta = \pi/n$, are planes of symmetry. The figure shows the wedge at $n = 3$. Then the

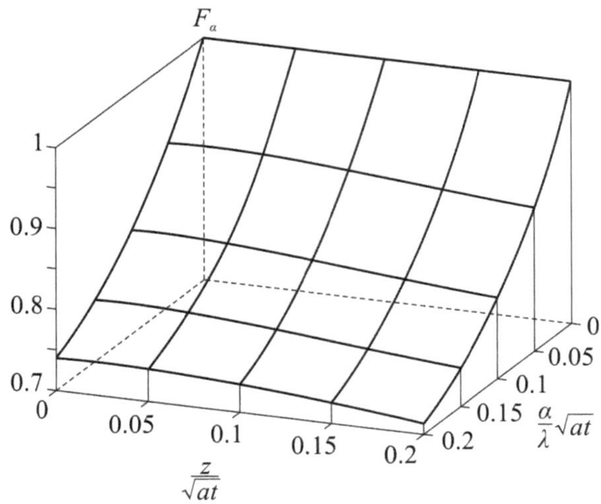

Fig. 4.4 Influence function of dimensionless heat transfer coefficient $\alpha/\lambda\sqrt{at}$ and dimensionless coordinate $z/\sqrt{at}$ for temperature in a semi-infinite solid

increment of temperature at points x, y, z of the wedge are determined by the sum of the temperature increments due to all sources:

$$T(x, y, z, t) - T_0 = \sum_{i=1}^{n} \sum_{j=-1,1} (T_{i,j}(x, y, z, t) - T_0), \tag{4.2.20}$$

where $T_{i,j}$ is determined by using formula (4.2.6), taking into account the source coordinates at points $A_{i,j}$ $\left(\xi_{i,j}, \eta, \zeta_{i,j}\right)$.

In the case where angle $\beta \neq \pi/n$ and a heat transfer takes place on the surface of the wedge, the solution to the problem becomes more complicated (Carslaw 1945; Carslaw and Jaeger 1973).

4.2.1.3 Instantaneous Point Source in an Infinite Slab

First it is assumed that the boundaries of an infinite slab (plane layer) are adiabatic ($\partial T/\partial z = 0$ for $z = 0$ and $z = h$, Fig. 4.5). This assumption is valid if the slab thickness is large and no additional forced cooling is used, as the heat fluxes in metal are much greater than the fluxes through the surface of the solid.

Let us find the temperature field using the method of images and having taken the scheme of an instantaneous point source in an infinite solid as the basis [Eq. (4.2.6)]. The infinite slab is expanded along z to infinity (to an infinite solid). On a vertical line passing through point A with a source, an infinite sequence of sources is placed so that the planes $z = 0$ and $z = h$ are planes of symmetry (Fig. 4.5). Then the temperature at any point P of the infinite slab due to the instantaneous source Q_0 is defined by summing all ith source pairs in the infinite solid:

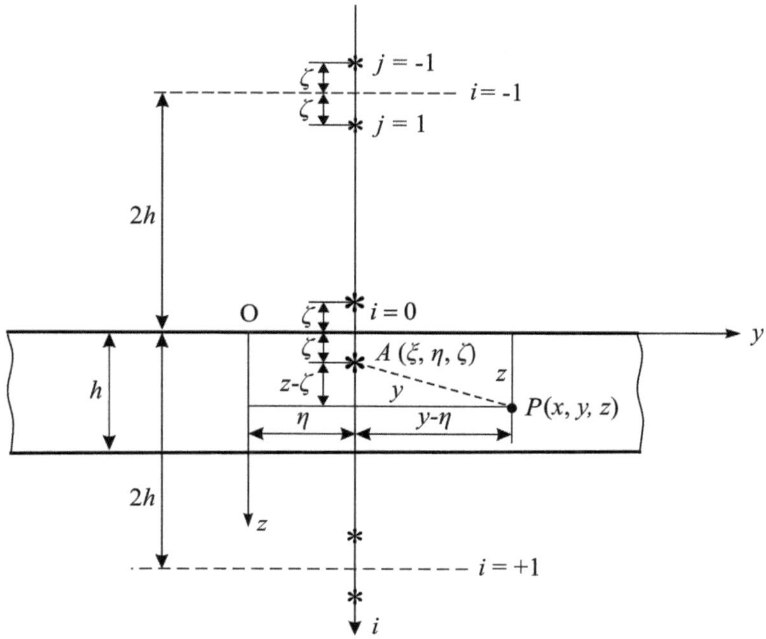

Fig. 4.5 Model of a point source in the infinite slab with heat-insulated surface

$$T(r, z, t) - T_0 = \frac{Q_0}{c\rho(4\pi at)^{3/2}} \exp\left(-\frac{r^2}{4at}\right) \times \sum_{i=-\infty}^{\infty} \sum_{j=-1,1} \exp\left(-\frac{(z - j\varsigma - 2ih)^2}{4at}\right), \tag{4.2.21}$$

where $r = \sqrt{(x-\xi)^2 + (y-\eta)^2}$ is a plane radius vector.

If the source is located on the surface of a slab at the origin ($\xi = \eta = \varsigma = 0$), then from (4.2.21) we obtain the following

$$T(R, z, t) - T_0 = \frac{2Q_0}{c\rho(4\pi at)^{3/2}} \exp\left(-\frac{R^2}{4at}\right) F_1(z, t), \tag{4.2.22}$$

where

$$R^2 = x^2 + y^2 + z^2; F_1(z, t) = \sum_{i=-\infty}^{\infty} \exp\left(-\frac{ih(ih - z)}{at}\right) \tag{4.2.23}$$

or in dimensionless form

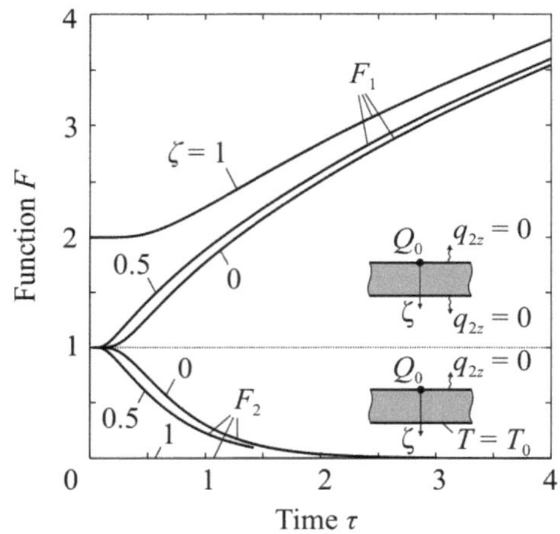

Fig. 4.6 Influence function of dimensionless time τ and boundary conditions at upper ($\zeta = 0$), middle ($\zeta = 0.5$) and lower ($\zeta = 1$) planes of a slab

$$F_1(\zeta, \tau) = \sum_{i=-\infty}^{\infty} \exp\left(-\frac{i(i-\zeta)}{\tau}\right); \quad \zeta = \frac{z}{h}; \; \tau = \frac{at}{h^2}. \tag{4.2.24}$$

Function F_1 shows the degree of temperature increase in the slab in comparison to a semi-infinite solid, since it is always true that $F_1 > 1$ (Fig. 4.6).

The series in formulas (4.2.21), (4.2.23) and (4.2.24) always converge, and the faster they do, the greater the slab thickness h is and the smaller the time t is. At $h \to \infty$, the formulas (4.2.21) and (4.2.22) change to (4.2.9) and (4.2.11) for a semi-infinite solid. In the interval of slow convergence, these series can be expanded into Fourier cosine series as sums of even functions using the well-known formula (Rykalin 1947, 1952):

$$\sum_{i=-\infty}^{\infty} f(u + 2ih) = \frac{1}{h}\int_0^{\infty} f(u)du + \frac{2}{h}\sum_{i=1}^{\infty} \cos\frac{\pi i u}{h}\int_0^{\infty} f(u)\cos\frac{\pi i u}{h}du. \tag{4.2.25}$$

Then

$$\sum_{i=-\infty}^{\infty} \exp\left(-\frac{(u-2ih)^2}{4at}\right) = \frac{\sqrt{\pi a t}}{h}\left[1 + 2\sum_{i=1}^{\infty} \cos\frac{\pi i u}{h} \exp\left(-i^2\pi^2\frac{at}{h^2}\right)\right]. \tag{4.2.26}$$

In this sum, the main variables u and t have been separated and the rate of convergence of the series has been increased.

Substituting (4.2.26) into (4.2.21), the following for a point source in a slab is obtained

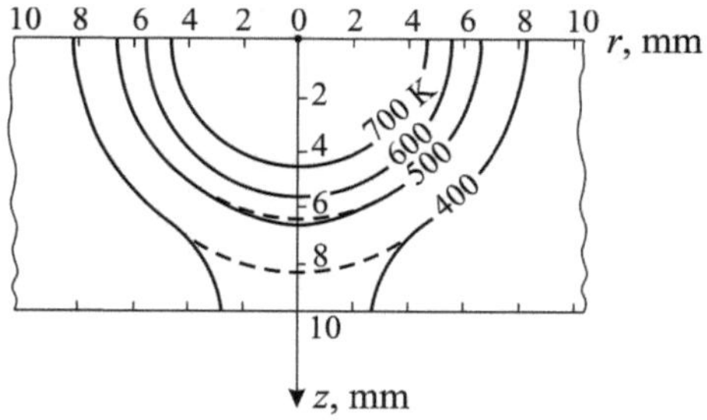

Fig. 4.7 Isotherms in cross-section of a 10 mm thick slab (solid lines) and a semi-infinite solid (broken lines) in 1 s after heat generation by instantaneous point source on the surface ($Q = 2000$ J, $c\rho = 0.005$ J mm^{-3} K^{-1}, $a = 8$ mm^2 s^{-1}, $T_0 = 300$ K)

$$T(r, z, t) - T_0 = \frac{Q_0/h}{4\pi\lambda t}\exp\left(-\frac{r^2}{4at}\right) \times \left[1 + 2\sum_{i=1}^{\infty}\cos\frac{\pi i z}{h}\cos\frac{\pi i \zeta}{h}\exp\left(-i^2\pi^2\frac{at}{h^2}\right)\right] \qquad (4.2.27)$$

and for a point source on the surface of a slab at the origin ($\xi = \eta = \zeta = 0$)

$$T(r, z, t) - T_0 = \frac{Q_0/h}{4\pi\lambda t}\exp\left(-\frac{r^2}{4at}\right) \times \left[1 + 2\sum_{i=1}^{\infty}\cos\frac{\pi i z}{h}\exp\left(-i^2\pi^2\frac{at}{h^2}\right)\right]. \qquad (4.2.28)$$

On the right side of Eqs. (4.2.27) and (4.2.28), the term before the square brackets represents, as it will be shown in Sect. 4.2.1.4, the temperature caused by a line source in a plate, uniformly distributed across the thickness with no surface heat transfer; it is dependent on the plane radius vector r. The term in the square brackets considers a deviation of the actual temperature distribution across the thickness from uniform distribution; it is dependent on coordinate z.

Formulas (4.2.21) and (4.2.27) are equivalent but in the case of distributed sources during integration with respect to ζ, formula (4.2.27) appears to be more convenient.

As can be seen from Fig. 4.7, in contrast to a semi-infinite solid, the isotherms in a slab are not spherical surfaces, whereby the maximum deviations are observed at the lower surface and the deviations increase with time.

Now the solution to the same problem is given but taking into account the heat transfer from the upper and the lower surfaces:

$$\lambda\frac{\partial T}{\partial z} = \alpha_1(T - T_0) \quad \text{at } z = 0;$$

$$\lambda \frac{\partial T}{\partial z} = -\alpha_2 (T - T_0) \quad \text{at}\, z = h. \tag{4.2.29}$$

The solution to this problem has the form (Carslaw and Jaeger 1973):

$$T(x, y, z, t) - T_0 = \frac{Q_0/h}{c\rho(4\pi at)} \exp\left(-\frac{(x-\xi)^2 + (y-\eta)^2}{4at}\right) \times \sum_{n=1}^{\infty} A_n B_n(\zeta) B_n(z) \exp\left(-\mu_n^2 \frac{at}{h^2}\right), \tag{4.2.30}$$

where

$$A_n = \frac{2}{\left(\mu_n^2 + \mathrm{Bi}_1^2\right)\left(1 + \frac{\mathrm{Bi}_2}{\mu_n^2 + \mathrm{Bi}_2^2}\right) + \mathrm{Bi}_1};$$

$$B_n(u) = \mu_n \cos\left(\mu_n \frac{u}{h}\right) + \mathrm{Bi}_1 \sin\left(\mu_n \frac{u}{h}\right).$$

Here μ_n is the positive roots of the equation:

$$\cot(\mu) = \frac{\mu^2 - \mathrm{Bi}_1 \mathrm{Bi}_2}{\mu(\mathrm{Bi}_1 + \mathrm{Bi}_2)}, \tag{4.2.31}$$

where Bi_1 and Bi_2 are Biot numbers:

$$\mathrm{Bi}_1 = \frac{\alpha_1 h}{\lambda}, \quad \mathrm{Bi}_2 = \frac{\alpha_2 h}{\lambda}. \tag{4.2.32}$$

It can be noted that the right part of Eq. (4.2.30) up to the summation sign describes the temperature field caused by the instantaneous uniformly distributed line source in a plate with thickness h without heat transfer (see Sect. 4.2.1.4). The expression under the summation sign takes into account the non-uniform temperature distribution across the thickness of the slab.

For some values of Bi_1 and Bi_2, the roots μ_n are shown in Fig. 4.8. It is worth noting that solution (4.2.27) cannot be obtained by setting the coefficients of heat transfer $\alpha_1 = \alpha_2 = 0$ in formula (4.2.30), since there are no roots of Eq. (4.2.31) in this case.

Of practical interest is the extreme case when the source is located on the heat-insulated upper surface $z = 0$ ($\alpha_1 = 0$), and the lower surface $z = h$ is maintained at a constant temperature ($T(h, t) = T_0$). The solution to this problem can be easily obtained by the method of images (method of fictitious sources and sinks):

$$T(R, z, t) - T_0 = \frac{2Q}{c\rho(4\pi at)^{3/2}} \exp\left(-\frac{R^2}{4at}\right) F_2(z, t), \tag{4.2.33}$$

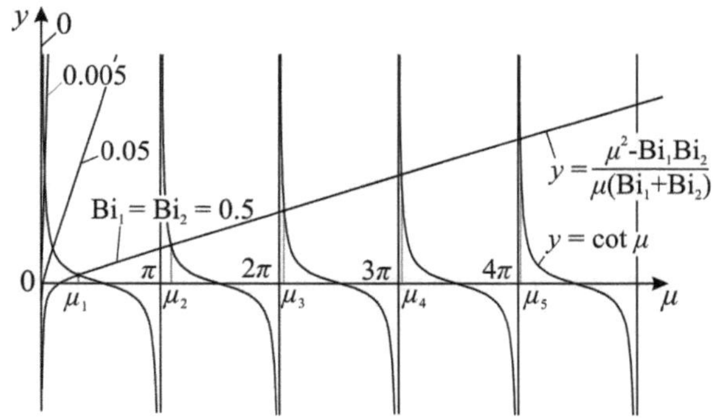

Fig. 4.8 Determination of roots of characteristic Eq. (4.2.31)

where

$$F_2(z, t) = \sum_{i=-\infty}^{\infty} (-1)^i \exp\left(-\frac{ih(ih - z)}{at}\right) \tag{4.2.34}$$

or in dimensionless form

$$F_2(\zeta, \tau) = \sum_{i=-\infty}^{\infty} (-1)^i \exp\left(-\frac{i(i - \zeta)}{\tau}\right). \tag{4.2.35}$$

The structure of formulas (4.2.22)–(4.2.24) and (4.2.33)–(4.2.35) is the same. Function $F_2(\zeta, \tau)$ is shown in Fig. 4.6.

4.2.1.4 Instantaneous Line Source in an Infinite Plate

Suppose that along the vertical z-axis of an infinite solid (see Fig. 3.16a) elementary instantaneous point sources with a linear heat density (heat input per unit length) Q_1, J m^{-1} are uniformly distributed. This type of source is called an *instantaneous line source*. Then the temperature field can be obtained by integrating all the elementary sources:

$$\begin{aligned} T(r, t) - T_0 &= \int_{-\infty}^{\infty} \frac{Q_1}{c\rho(4\pi at)^{3/2}} \exp\left(-\frac{x^2 + y^2 + z^2}{4at}\right) dz \\ &= \frac{Q_1}{c\rho(4\pi at)} \exp\left(-\frac{r^2}{4at}\right), \end{aligned} \tag{4.2.36}$$

where $r = \sqrt{x^2 + y^2}$ is a plane radius vector (distance to the z-axis).

If a thin layer with thickness h is cut out perpendicular to the z-axis using two parallel planes, then the scheme for an instantaneous line source in an infinite plate is obtained:

$$T(r,t) - T_0 = \frac{Q/h}{c\rho(4\pi at)} \exp\left(-\frac{r^2}{4at}\right), \tag{4.2.37}$$

where Q is the amount of heat in a plate of thickness h, J. The temperature field $T(x, y, t)$ is two-dimensional.

It was assumed that there were no heat fluxes across the planes, $q_{2z} = 0$. Under strict consideration of surface heat transfer, the line source should be represented in the form of elementary instantaneous sources and the formula for point sources in a slab (4.2.30) should be used. However, the temperature field in this case becomes volumetric, $T = T(x, y, z, t)$, and therefore this case is examined in the section on distributed sources in a slab (Sect. 5.2). Here it is assumed that the plate is thin with no difference in temperature across the thickness ($\partial T/\partial z = 0$), and the temperature field is plane, $T = T(x, y, t)$. At the same time, the heat transfer from both surfaces of the plate will be approximated.

Let it be supposed that the heat transfer occurs according to Newton's law (3.4.6), but the temperature across the thickness is instantaneously equalized (the field stays two-dimensional), which is tantamount to the existence of a volume sink in a plate with an adiabatic surface (W m^{-3}),

$$q_3 = -\frac{\alpha_1 T}{h} - \frac{\alpha_2 T}{h} = -\frac{\alpha_1 + \alpha_2}{h} T, \tag{4.2.38}$$

where α_1 and α_2 are the coefficients of surface heat transfer on the first and second sides. Then the solution to Eq. (3.3.21) for a two-dimensional case is obtained by taking into account (4.2.38) and using the substitution $T(x, y, t) = U(x, y, t)\exp(-bt)$ (Rykalin 1951, 1957):

$$T(r,t) - T_0 = \frac{Q/h}{c\rho(4\pi at)} \exp\left(-\frac{r^2}{4at} - bt\right), \tag{4.2.39}$$

where $b = (\alpha_1 + \alpha_2)/(c\rho h)$ is the coefficient of heat loss for a plate, s^{-1}. It can be seen that the influence of surface heat transfer is more significant, the thinner the plate is.

The radial distributions of temperature $T(r)$ at $t = \text{const}$ and the thermal cycles $T(t)$ at $r = \text{const}$ in a plate (Fig. 4.9) are qualitatively similar to those in a semi-infinite solid (Fig. 4.3).

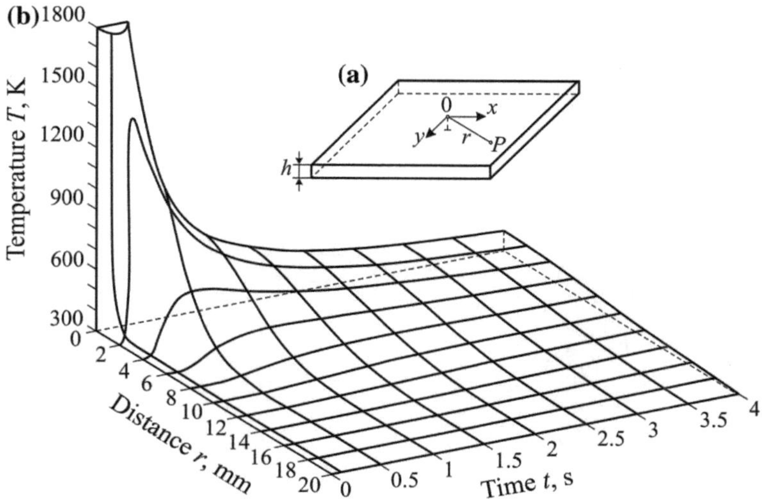

Fig. 4.9 The process of heat diffusion from an instantaneous line source in a infinite plate ($h = 5$ mm, $Q = 1000$ J, $c\rho = 0.005$ J mm^{-3} K^{-1}, $a = 8$ mm^2 s^{-1}, $\alpha_1 = \alpha_2 = 100$ W m^{-2} K^{-1}, $T_0 = 300$ K): model of a line source (**a**) and the temperature distribution over radial distance r and time t (**b**)

4.2.1.5 Instantaneous Plane Source in an Infinite Rod

Suppose elementary instantaneous point sources with a plane density of Q_2, J m^{-2} are evenly distributed in an infinite solid along the entire plane $x = 0$ (see Fig. 3.16a). This source is called an *instantaneous plane source*. Then the temperature field can be obtained by integrating all elementary sources:

$$T(x,t) - T_0 = \int_{-\infty}^{\infty}\int_{-\infty}^{\infty} \frac{Q_2}{c\rho(4\pi at)^{3/2}} \exp\left(-\frac{x^2 + y^2 + z^2}{4at}\right) dydz$$
$$= \frac{Q_2}{c\rho(4\pi at)^{1/2}} \exp\left(-\frac{x^2}{4at}\right). \tag{4.2.40}$$

If a cylinder with cross-sectional area A and perimeter p is cut out perpendicular to plane $x = 0$, then the scheme of an instantaneous plane source in an infinite rod is obtained:

$$T(x,t) - T_0 = \frac{Q/A}{c\rho(4\pi at)^{1/2}} \exp\left(-\frac{x^2}{4at}\right), \tag{4.2.41}$$

where Q is the amount of released heat in the rod, J.

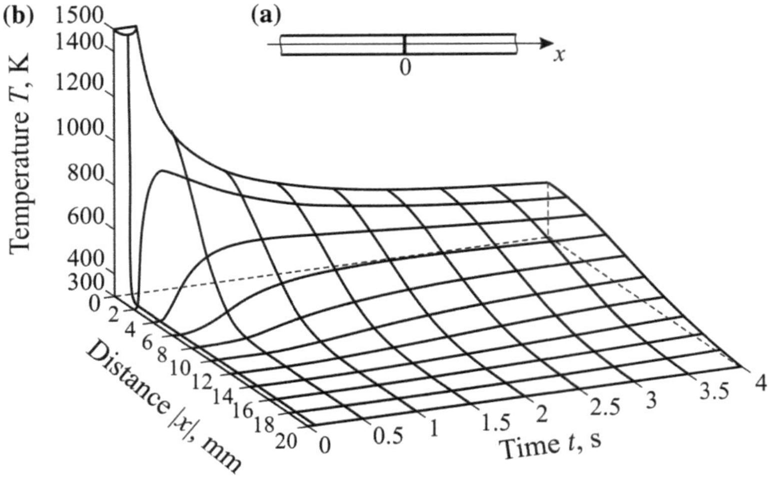

Fig. 4.10 The process of heat diffusion from an instantaneous plane source in a round rod 10 mm in diameter ($Q = 2000$ J, $c\rho = 0.005$ J mm^{-3} K^{-1}, $a = 8$ mm^2 s^{-1}, $\alpha = 100$ W m^{-2} K^{-1}, $T_0 = 300$ K): model of a plane source (**a**) and the temperature distribution over axial distance $|x|$ and time t (**b**)

If, like in the case of a plate, the heat transfer from the surface of the rod is considered but the non-uniformity of the temperature over the section is disregarded, then we obtain the following:

$$T(x,t) - T_0 = \frac{Q/A}{c\rho(4\pi at)^{1/2}} \exp\left(-\frac{x^2}{4at} - bt\right), \tag{4.2.42}$$

where $b = \alpha p/(c\rho A)$ is the coefficient of heat loss for the rod, s^{-1}.

The axial temperature distribution $T(x)$ at $t =$ const and thermal cycles $T(t)$ at $|x| =$ const in a rod (Fig. 4.10) are qualitatively similar to those in a semi-infinite solid (Fig. 4.3) and plate (Fig. 4.9).

4.2.1.6 Instantaneous Point Source in an Infinite Solid Cylinder

Suppose heat, Q_0, J, is released at point A (ρ, θ', ζ) of an infinite solid cylinder at moment τ (Fig. 4.11a). On the surface of the cylinder $r = R$, a heat transfer to the medium, with temperature $T_\infty = T_0$, obeys Newton's law (3.4.7). Then the temperature at any point P (r, θ, z) at any moment $t > \tau$ is described by the following expression (Carslaw and Jaeger 1973; Makhnenko and Kravtsov 1976):

$$T(r,\theta,z,t) - T_0 = \frac{Q_0/(\pi R^2)}{c\rho[4\pi a(t-\tau)]^{1/2}} \exp\left(-\frac{(z-\zeta)^2}{4a(t-\tau)}\right) \sum_{n=0}^{\infty}{}' \cos(n(\theta-\theta'))$$

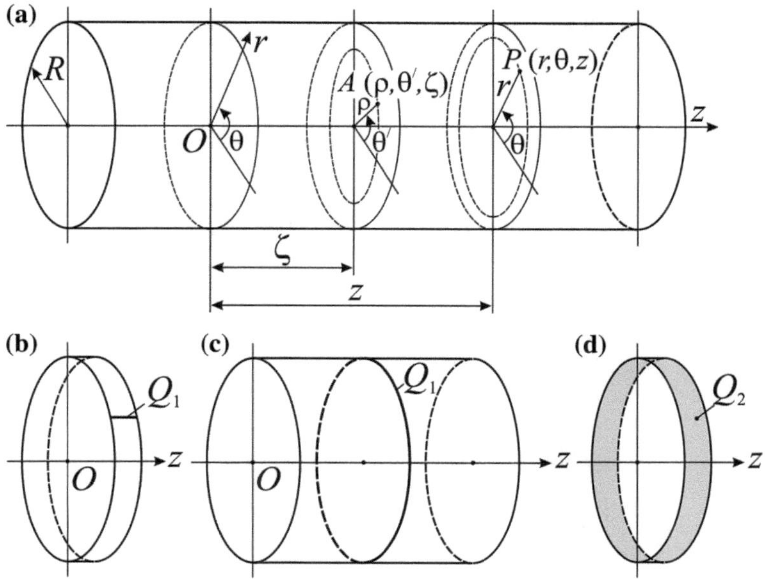

Fig. 4.11 Schematic representation of instantaneous heat sources: a point source in a cylinder (**a**); a line source on a disk (**b**); a ring source on a cylinder (**c**); an area source on a disk (**d**)

$$\times \sum_{m=1}^{\infty} A_{n,m} \mathrm{J_n}\left(\mu_{n,m}\frac{r}{R}\right)\mathrm{J_n}\left(\mu_{n,m}\frac{\rho}{R}\right)/\mathrm{J_n}(\mu_{n,m})\exp\left(-\mu_{n,m}^2\frac{a(t-\tau)}{R^2}\right);$$

$$A_{n,m} = \frac{\mu_{n,m}^2}{(\mathrm{Bi}^2+\mu_{n,m}^2-n^2)\mathrm{J_n}(\mu_{n,m})}; \quad \mathrm{Bi} = \frac{\alpha R}{\lambda}; \tag{4.2.43}$$

$$\mathrm{J_n}(u) = \sum_{m=0}^{\infty}\frac{(-1)^m}{m!\,\Gamma(n+m+1)}\left(\frac{u}{2}\right)^{n+2m}, \tag{4.2.44}$$

where $\mathrm{J_n}$ is the Bessel function of a real argument of the first kind of order n [alternating function with variable amplitude and variable period, depending on the order n, Fig. 3.14; there are tables and representations in the form of an integral (Abramowitz and Stegun 1965; Janke et al. 1960)]; Γ is the gamma function; Bi is the Biot number, characterizing the surface heat transfer; the primed sum symbol sign indicates that all terms of the series $n \neq 0$ are multiplied by two; $\mu_{n,m}$ are the positive roots of a characteristic equation:

$$\mu\frac{d\mathrm{J_n}(\mu)}{d\mu} + \mathrm{Bi}\,\mathrm{J_n}(\mu) = 0. \tag{4.2.45}$$

The Equation (4.2.43) satisfies the equation of heat conduction in cylindrical coordinates r, θ, z (3.3.25).

If the source is located on the surface of a cylinder ($\rho = R$), then formula (4.2.43) is simplified:

$$T(r,\theta,z,t) - T_0 = \frac{Q_0/\left(\pi R^2\right)}{c\rho[4\pi a(t-\tau)]^{1/2}} \exp\left(-\frac{(z-\zeta)^2}{4a(t-\tau)}\right) \times \sum_{n=0}^{\infty}{}' \cos(n(\theta-\theta')) \sum_{m=1}^{\infty} A_{n,m} \mathrm{J_n}\left(\mu_{n,m}\frac{r}{R}\right) \exp\left(-\mu_{n,m}^2 \frac{a(t-\tau)}{R^2}\right). \tag{4.2.46}$$

If the surface of the cylinder is heat-impermeable ($\alpha = 0$ and $\mathrm{Bi} = 0$), then the solution to the problem has the form:

$$T(r,\theta,z,t) - T_0 = \frac{(Q_0/\pi R^2)}{c\rho[4\pi a(t-\tau)]^{1/2}} \exp\left(-\frac{(z-\zeta)^2}{4a(t-\tau)}\right) \times \sum_{n=0}^{\infty}{}' \cos(n(\theta-\theta')) \sum_{m=1}^{\infty} A_{n,m} \mathrm{J_n}\left(\mu_{n,m}\frac{r}{R}\right) \exp\left(-\mu_{n,m}^2 \frac{a(t-\tau)}{R^2}\right); \tag{4.2.47}$$

$$A_{n,m} = \frac{\mu_{n,m}^2}{(\mu_{n,m} - n^2)\mathrm{J_n}(\mu_{n,m})},$$

where $\mu_{n,m}$ are the positive roots of the equation:

$$\frac{d\mathrm{J_n}(\mu)}{d\mu} = 0. \tag{4.2.48}$$

At $n = 0$ and $m = 1$ ($\mu_{n,m} = 1$) the value is $A_{n,m} = 1$.

It can be noted that the temperature field due to an instantaneous plane source in an infinite rod with no surface heat transfer is described in formulas (4.2.43), (4.2.46) and (4.2.47) by the expression in front of the summation sign [cf. (4.2.41)]. The deviation from uniform temperature distribution over the cross-section of a cylinder and the effect of surface heat transfer are taken into account by a factor under the summation sign. As time t increases, the sum tends to 1.

In Fig. 4.12, the thermal cycles of the points located in a plane of action of a surface instantaneous point source are shown. The temperature and time are dimensionless. It can be seen that, as in the case of a point source on the surface of a semi-infinite solid (Fig. 4.3), the peak temperature decreases while moving from the source. If Fourier number is $\mathrm{Fo} > 2$, the temperature is uniformly distributed over the cross-section. The higher the thermal diffusivity of the material a is and the smaller the radius R is, the faster the temperature is equalised.

The solutions obtained for a solid cylinder can be easily generalised for the case of a wedge of finite width with the vertex angle $\beta = \pi/n$ ($n = 1, 2, \ldots$), infinitely long and with heat-impermeable planes. To do this, the method of images should be used in the same way as in the case of an infinite wedge [Eq. (4.2.20) and Fig. 4.2b].

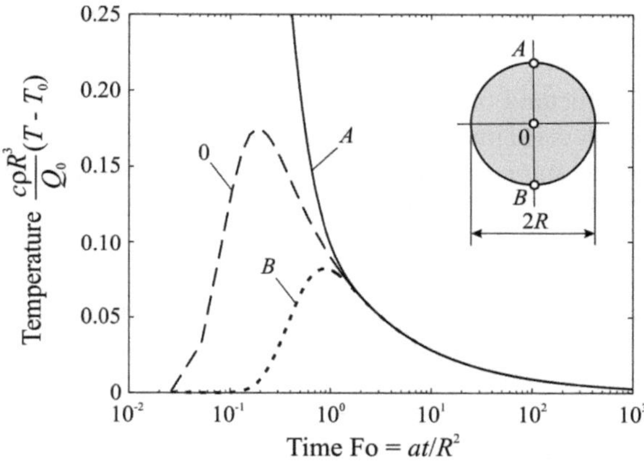

Fig. 4.12 Thermal cycles of the points in cross-section of an instantaneous point source acting in point *A* at time $t = 0$ on an infinite solid cylinder

It can be noted that there are known solutions to the problem of heat conduction for an instantaneous point source of unit energy (Green's function) in any region bounded by coordinate surfaces of a cylindrical coordinate system (Carslaw and Jaeger 1973). For example, Green's function for a hallow cylinder was used to analyze the thermal processes when surfacing cylindrical solids with a strip electrode (Yazovskikh and Belenky 2011).

It can be seen by comparing formulas (4.2.6), (4.2.37) and (4.2.41) that the temperature distributions in an infinite solid, plate and rod with no surface heat transfer are similar in structure. However close to the source in a plate does the temperature fall more slowly (according to law t^{-1}), than in an infinite solid (according to law $t^{-3/2}$), but faster than in a rod (according to law $t^{-1/2}$). For convenience, all three formulas can be written in a general form:

$$T(x_1, \ldots, x_n, t) = \frac{Q_{3-n}}{c\rho(4\pi at)^{n/2}} \exp\left(-\frac{\sum_{i=1}^{n} x_i^2}{4at}\right), \tag{4.2.49}$$

where n is the dimensionality of space ($n = 1, 2, 3$).

In a slab, the law of temperature drop is more complicated: at first, the temperature falls like in an infinite solid (according to law $t^{-3/2}$), and in the end, like in a plate (according to law t^{-1}). In a solid cylinder, the temperature first falls like in an infinite solid (according to law $t^{-3/2}$), and in the end, like in a rod (according to law $t^{-1/2}$).

The formulas presented (4.2.9), (4.2.14), (4.2.21), (4.2.27), (4.2.30), (4.2.39), (4.2.42), (4.2.43) and (4.2.47) are fundamental solutions to the problem of heat conduction (Green's functions) for a homogeneous semi-infinite solid, slab, plate, rod, and cylinder. They are used below to construct the solution for sources distributed in time and space.

4.2.2 Method of Separation of Variables

The method of separation of variables has found wide application in solving linear problems of mathematical physics, including problems of heat conduction in welding. In literature, this method is often found under the name Fourier's method.

The main idea of the method of separation of variables implies that the general solution to the problem of heat conduction, dependent on three independent spatial variables *x, y, z* and time *t*, is sought by forming several unknown functions, each of which depends only on one independent variable.

In general, the solution sequence of the equation of conduction of heat

$$\partial T(x, y, z, t)/\partial t = a\nabla^2 T(x, y, z, t) \tag{4.2.50}$$

is the following:

1. The sought solution is represented in the form of four functions

$$\partial T(x, y, z, t) = X(x)Y(y)Z(z)\theta(t). \tag{4.2.51}$$

   This expression is substituted in (4.2.50), as a result of which four ordinary differential equations are obtained.
2. Each of the equations is solved with their boundary conditions.
3. A particular solution is chosen according to (4.2.51) with the unknown coefficients; all particular solutions are summed. As a result, a general solution to a boundary value problem is obtained and contains unknown coefficients.
4. Unknown coefficients from the boundary conditions are defined.
   It can be noted that the linearity of the equation of heat conduction and the linearity of the boundary conditions in the original problem are essential to the applicability of the method.

Let us consider the method of separation of variables in a simple example of a heat conduction problem for an unbounded rod with impermeable surface (Kartashov 2001; Luikov 1968):

$$\frac{\partial T(x, t)}{\partial t} = a\frac{\partial^2 T(x, t)}{\partial x^2}, \quad -\infty < x < +\infty, \; t > 0; \tag{4.2.52}$$

$$T(x, 0) = T_0(x), -\infty < x < +\infty. \tag{4.2.53}$$

Here the initial temperature $T_0(x)$ is a continuous and bounded function.

First, let us seek the solution of Eq. (4.2.52) in the form of a product of two functions

$$T(x, t) = \theta(t)\psi(x), \tag{4.2.54}$$

where θ depends only on time t and ψ depends only on coordinate x. Substituting (4.2.54) in (4.2.52), we obtain the following

$$\theta'(t)\psi(x) = a\theta(t)\psi''(x), \tag{4.2.55}$$

where the primes denote derivatives. Dividing this equality by $\theta(t)\psi(x)$ gives:

$$\frac{\theta'(t)}{\theta(t)} = a\frac{\psi''(x)}{\psi(x)}. \tag{4.2.56}$$

Equality is only possible in the case of the right and left sides being equal to some negative constant $-ak^2$:

$$\frac{\theta'(t)}{\theta(t)} = -ak^2; \quad a\frac{\psi''(x)}{\psi(x)} = -ak^2. \tag{4.2.57}$$

The last equations are rewritten:

$$\theta'(t) + ak^2\theta(t) = 0; \tag{4.2.58}$$

$$\psi''(x) + k^2\psi(x) = 0. \tag{4.2.59}$$

By integrating these ordinary differential equations, we obtain the following

$$\theta(t) = A\ \exp(-ak^2 t); \tag{4.2.60}$$

$$\psi(x) = B\ \cos(kx) + C\ \sin(kx), \tag{4.2.61}$$

where A, B and C are arbitrary constants.

Accordingly, the partial solution of Eq. (4.2.52) has the form:

$$T(x,t) = [AB\ \cos(kx) + AC\ \sin(kx)]\exp(-ak^2 t) \tag{4.2.62}$$

or

$$T(x,t) = [a_0\cos(kx) + b_0\sin(kx)]\exp\left(-ak^2 t\right), \tag{4.2.63}$$

where $a_0 = AB$, $b_0 = AC$. Here k, A, B, C and, consequently, a_0 and b_0 are arbitrary constants. Constant k is determined from the boundary conditions, and constants a_0 and b_0 are determined from the initial ones; they assume quite defined values depending on the conditions of the problem.

Function (4.2.63) at any fixed k satisfies Eq. (4.2.52) and constants a_0 and b_0 can be chosen for each value k. This means that a_0 and b_0 can be arbitrary functions of k, so that finally the following family of partial equations (4.2.52) is obtained:

$$T(x, t) = [a_0(k)\cos(kx) + b_0(k)\sin(kx)]\exp\left(-ak^2t\right). \tag{4.2.64}$$

Since the boundary conditions are missing from the problem, parameter k remains arbitrary and can assume all values from $-\infty$ to $+\infty$. Thereby, the first part of the method of separation of variables is completed.

The second part of the method of separation of variables is to superpose the partial solutions obtained, that is, to integrate with respect to variable k in (4.2.64):

$$T(x, t) = \int_{-\infty}^{+\infty} [a_0(k)\cos(kx) + b_0(k)\sin(kx)]\exp\left(-ak^2t\right)dk. \tag{4.2.65}$$

The unknown functions $a_0(k)$ and $b_0(k)$ are determined below in such a way that the initial condition (4.2.53) is fulfilled, that is

$$T(x, 0) = T_0(x) = \int_{-\infty}^{+\infty} [a_0(k)\cos(kx) + b_0(k)\sin(kx)]\,dk. \tag{4.2.66}$$

The resulting equality denotes the expansion of the function $T_0(x)$ in a Fourier's integral. The coefficients of this expansion are found using the formulas:

$$a_0(k) = \frac{1}{2\pi}\int_{-\infty}^{+\infty} T_0(\xi)\cos(k\xi)d\xi; \tag{4.2.67}$$

$$b_0(k) = \frac{1}{2\pi}\int_{-\infty}^{+\infty} T_0(\xi)\sin(k\xi)d\xi. \tag{4.2.68}$$

Substituting (4.2.67) and (4.2.68) in (4.2.65), the following is found

$$\begin{aligned} T(x, t) &= \frac{1}{2\pi}\int_{-\infty}^{\infty} \exp(-ak^2t)dk \int_{-\infty}^{\infty} T_0(\xi)\cos(k(\xi - x))d\xi \\ &= \frac{1}{\pi}\int_{0}^{\infty} \exp(-ak^2t)dk \int_{-\infty}^{\infty} T_0(\xi)\cos(k(\xi - x))d\xi. \end{aligned} \tag{4.2.69}$$

Changing the order of integration, it is written

$$T(x, t) = \frac{1}{\pi}\int\int_{-\infty}^{+\infty} T_0(\xi)d\xi \int_{0}^{+\infty} \exp\left(-ak^2t\right)\cos(k(\xi - x))dk. \tag{4.2.70}$$

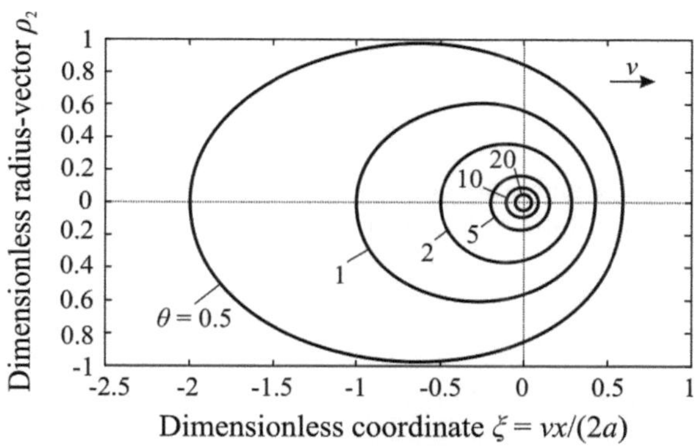

Fig. 4.13 Dimensionless temperature field θ (ξ, ρ_2) around a point source moving at speed v relative to the medium (Wilson 1904)

Considering the integral (Prudnikov et al. 1986)

$$\int_0^{\infty} \exp(-ax^2)\cos(bx)dx = \sqrt{\frac{\pi}{4a}}\exp\left(-\frac{b^2}{4a}\right) \tag{4.2.71}$$

finally, the solution to the problem in question is reached:

$$T(x,t) = \frac{1}{\sqrt{4\pi at}}\int_{-\infty}^{+\infty} T_0(\xi)\exp\left(-\frac{(x-\xi)^2}{4at}\right)d\xi. \tag{4.2.72}$$

A direct verification shows that Green's function

$$G(x,\xi,t) = \frac{1}{\sqrt{4\pi at}}\exp\left(-\frac{(x-\xi)^2}{4at}\right) \tag{4.2.73}$$

satisfies Eq. (4.2.52). Function (4.2.73) is called a fundamental solution to the heat conduction Eq. (4.2.52). It corresponds to the problem solution for an infinite heat-impermeable rod with an instantaneous plane source of unit capacity, $Q_2/(c\rho) = 1$ K m [cf. formula (4.2.41)].

Now let us demonstrate the method of separation of variables with an example of a point source that is moving in an unbounded medium rectilinearly along the x-axis at the constant speed v relative to the medium with an initial zero temperature, $T_0 = 0$ (Fig. 4.13).

If the source moves long enough (theoretically $t \to \infty$), then a quasi-stationary (steady) state ($\partial T/\partial t = 0$) is established in the coordinate system related to the source. This state is described by a differential equation similar to (3.3.31):

$$\nabla^2 T + \frac{v}{a}\frac{\partial T}{\partial x} = 0. \tag{4.2.74}$$

Let us seek for a solution in the form of a product of two functions:

$$T = A\ \exp(-\alpha x)U(R), \tag{4.2.75}$$

where α and A are constants, $R = \left(x^2 + y^2 + z^2\right)^{1/2}$ is a volumetric radius vector.

Substituting (4.2.75) in (4.2.74), we get

$$\nabla^2 U + \left(-2a + \frac{v}{a}\right)\frac{\partial U}{\partial x} + \left(a^2 - \frac{v}{a}a\right)U = 0. \tag{4.2.76}$$

Let $\alpha = v/(2a)$. Equation (4.2.76) simplifies:

$$\nabla^2 U(R) - \alpha^2 U(R) = 0 \tag{4.2.77}$$

or taking into account spherical symmetry:

$$\frac{\partial^2 U}{\partial R^2} + \frac{2}{R}\frac{\partial U}{\partial R} - \alpha^2 U = 0. \tag{4.2.78}$$

The complete solution to this equation is an expression:

$$U(R) = \frac{1}{R}\left[C\ \exp(\alpha R) + B\ \exp(-\alpha R)\right], \tag{4.2.79}$$

where C and B are constants. At $R \to \infty$, function $U(R)$ should tend to 0, from which $C = 0$. Then the solution (4.2.75) of Eq. (4.2.74) takes the form:

$$T(x, R) = A\frac{1}{R}\exp\left(-\frac{v(x + R)}{2a}\right). \tag{4.2.80}$$

If radius R is very small, then $T = A/R$ and $dT/dR = -A/R^2$. Based on the law of conservation of energy and Fourier's law for heat conduction, the thermal flux over the surface of a sphere with the centre at a point source is equal to the power of the source:

$$-4\pi R^2 \lambda \frac{dT}{dR} = q \tag{4.2.81}$$

from which follows $A = q/(4\pi\lambda)$.

Thereby, the general solution

$$T(x, R) = \frac{q}{4\pi\lambda R}\exp\left(-\frac{v(x + R)}{2a}\right) \tag{4.2.82}$$

satisfies the differential equation (4.2.74) and boundary conditions ($T \to 0$ at $R \to \infty$).

The solution (4.2.82) can be represented in dimensionless form:

$$\theta(\xi, \rho_2) = \frac{1}{\sqrt{\xi^2 + \rho_2^2}} \exp\left(-\xi - \sqrt{\xi^2 + \rho_2^2}\right), \tag{4.2.83}$$

where

$$\theta = \frac{8\pi\lambda a}{qv}; \quad \xi = \frac{vx}{2a}, \ \rho_2^2 = \left(\frac{vy}{2a}\right)^2 + \left(\frac{vz}{2a}\right)^2.$$

In Fig. 4.13 a steady-state dimensionless temperature field is given around a point source located in the centre of the coordinate system ($\xi = \rho_2 = 0$) and moving relative to the infinite medium along the longitudinal x-axis at a constant speed v. The temperature field is analysed later in Sect. 5.1.3.1.

It can be noted that the solution (4.2.82) was derived by H. A. Wilson more than 100 years ago (Wilson 1904). He also solved problems under stationary conditions using the method of separation of variables and the source method (Lord Kelvin's method) for the following main combinations of a moving source and a solid: a point source in an infinite solid, a line source in a plate, and a plane source in a rod (Wilson 1904). Later, similar results were published by other researchers (Roberts 1923; Rosenthal 1935, 1941; Rykalin 1938a, b, 1941).

The method of separation of variables continues to develop. There is extensive literature on the method of the theory of conduction of heat in solids (Belyaev and Ryadno 1982; Berezovsky 2006; Carslaw and Jaeger 1973; Kartashov 2001; Luikov 1968).

4.2.3 Method of Integral Transformations

Over the last decades, operational methods have been used in solving a variety of heat conduction problems in welding. The Laplace integral transformation has been used the most often. The idea of this method is that the function (original) T itself is not studied, but the transform of the function (image) $\overline{T}$ is instead. This transformation occurs by multiplying the function T by an exponential function and integrating it over time t within certain limits (Bateman and Erdelyi 1954; Belyaev and Ryadno 1982; Boley and Weiner 1960; Carslaw and Jaeger 1973; Luikov 1968), as a result of which, time t is excluded:

$$\overline{T}(x, p) = \int_0^\infty T(x, t)\, \mathrm{e}^{-pt} dt = \mathrm{L}[T(x, t)]. \tag{4.2.84}$$

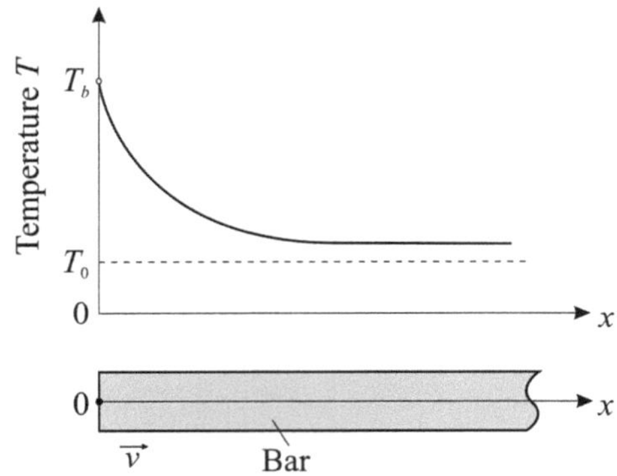

Fig. 4.14 Schematic of flashing the bar

Here $\exp(-pt)$ is the kernel of integral transformation. Variable p can also be a complex number.

If the problem is solved for images, then the original T is found from the image $\overline{T}$ (inverse transformation) in general by using the inversion formula (Carslaw and Jaeger 1973; Luikov 1968):

$$T(x, t) = \mathrm{L}^{-1}\left[\overline{T}(x, p)\right] = \frac{1}{2\pi i} \int\limits_{\sigma - i\infty}^{\sigma + i\infty} \overline{T}(x, p) \mathrm{e}^{pt} dp. \tag{4.2.85}$$

Integration occurs in a complex plane p along the straight line $\sigma = \text{const}$, parallel to the imaginary axis. The integration technique is detailed in special manuals on the theory of functions of a complex variable. In the vast majority of cases, the inverse transformation can be carried out without resorting to integration, but instead, by using tables (Bateman and Erdelyi 1954; Luikov 1968).

The method of integral transformations has its physical substantiation. Any integral transformation, taken in regard to time or spatial coordinates, is a certain averaging of the temperature from the physical point of view. Naturally, this averaging should be performed in accordance with both the nature of the process and shape of the solid (the form of the differential equation) and with the boundary conditions. In this case, the solution for the image of function $\overline{T}$ is of independent interest, as a transformation of this kind physically represents a transition from analysing real temperature values to averaged values.

The technique of applying the Laplace transformation will be shown in an example of flash welding of two rods (Fig. 1.28b).

The following assumptions are made (Fig. 4.14): the properties of a homogeneous rod material (a, $c\rho$, ρ_e), the flash rate (movement of the butt) of rod v, the current density j, and the temperature of the butt T_b are constant, the rod is infinitely long, and the temperature field is one-dimensional.

The formulation of the problem of heat conduction in a coordinate system moving along with the butt relative to the rod in the direction of the x-axis is the following (Fig. 4.14).

(1) The equation of heat conduction in a moving (relative to the rod) coordinate system:

$$\frac{\partial T}{\partial t} = a\frac{\partial^2 T}{\partial x^2} + v\frac{\partial T}{\partial x} + \frac{\rho_e j^2}{c\rho}, \tag{4.2.86}$$

where ρ_e is the specific electrical resistance.

(2) Initial condition ($t = 0$):

$$T(x, 0) = T_0 = \text{const.} \tag{4.2.87}$$

(3) Boundary conditions:

$$T(0, t) = T_b; \tag{4.2.88}$$

$$T(\infty, t) = T_0. \tag{4.2.89}$$

The Laplace transformation L is applied relative to variable t. The equation of heat conduction (4.2.86) takes the form:

$$\mathrm{L}\left[\frac{\partial T}{\partial t}\right] = a\mathrm{L}\left[\frac{\partial^2 T}{\partial x^2}\right] + v\mathrm{L}\left[\frac{\partial T}{\partial x}\right] + \frac{\rho_e j^2}{c\rho}\mathrm{L}[1]$$

or

$$p\overline{T}(x, p) - T_0 = a\frac{d^2\overline{T}(x, p)}{dx^2} + v\frac{d\overline{T}(x, p)}{dx} + \frac{\rho_e j^2}{c\rho}\cdot\frac{1}{p}. \tag{4.2.90}$$

This is a linear ordinary differential equation of the second order (the function $\overline{T}$ depends on one independent variable x). The left side of the equation is obtained by integrating by parts:

$$\int\limits_0^\infty \frac{\partial T}{\partial t}\mathrm{e}^{-pt}dt = T\mathrm{e}^{-pt}|_0^\infty + p\int\limits_0^\infty T\mathrm{e}^{-pt}dt = -T_0 + p\overline{T} \tag{4.2.91}$$

The right side is obtained with the following formulas

$$\int\limits_0^\infty \frac{\partial^n T}{\partial x^n}\mathrm{e}^{-pt}dt = \frac{\partial^n}{\partial x^n}\int\limits_0^\infty T\mathrm{e}^{-pt}dt; \tag{4.2.92}$$

$$\int_0^{\infty} e^{-pt} dt = \frac{1}{p}. \tag{4.2.93}$$

Boundary conditions (4.2.88) and (4.2.89) take the form:

$$\overline{T}(0, p) = \frac{T_b}{p}; \tag{4.2.94}$$

$$\overline{T}(\infty, p) = \frac{T_0}{p}. \tag{4.2.95}$$

Problems (4.2.90), (4.2.94) and (4.2.95) have the solution (Carslaw and Jaeger 1973):

$$\overline{T}(x, p) = \left(\frac{T_b - T_0}{p} - \frac{\rho_e j^2}{c\rho \cdot p^2}\right) \exp\left(-\frac{vx}{2a} - x\left[\frac{v^2}{4a^2} + \frac{p}{a}\right]^{1/2}\right) + \left(\frac{\rho_e j^2}{c\rho} + av\right)\frac{1}{p^2} + \frac{T_0}{p}. \tag{4.2.96}$$

Using the table of Laplace transforms, the original is obtained (Carslaw and Jaeger 1973; Karkhin et al. 2007, 2008):

$$T(x,t) - T_0 = \frac{\rho_e j^2}{c\rho} t + \frac{1}{2}(T_в - T_0)\left[\Phi^*\left(\frac{x+vt}{\sqrt{4at}}\right) + \exp\left(-\frac{vx}{a}\right)\Phi^*\left(\frac{x-vt}{\sqrt{4at}}\right)\right] + \frac{\rho_e j^2}{2vc\rho}\left[(x-vt)\exp\left(-\frac{vx}{a}\right)\Phi^*\left(\frac{x-vt}{\sqrt{4at}}\right) - (x+vt)\Phi^*\left(\frac{x+vt}{\sqrt{4at}}\right)\right]. \tag{4.2.97}$$

This formula is used later on when analysing thermal cycles in flash welding (Sect. 6.2).

A one-sided Laplace transformation was examined in relation to time variable t. There is another series of integral transformations, the kernel of which includes the functions of time variable t or spatial variables x, y, z: Laplace-Carson, Fourier, Hankel, and others. Which method of transformation to use is determined by the length of the solid (unbounded, semi-bounded, bounded), by the heterogeneity of the solid (composite or uniform), by the initial conditions (if the initial temperature is a constant or depends on the coordinates), by the type of boundary conditions, by the coordinate system (Cartesian, cylindrical, polar). The method of integral transformations is detailed in special literature (Belyaev and Ryadno 1982; Carslaw and Jaeger 1973; Luikov 1968, etc .).

It is worthy of note that the main stages of integral transformations are standard, which facilitates the solution of problems. However, the method is devoid of the vivid physical interpretation which is inherent in the source method. Special advantages of integral transformations are revealed when solving systems of partial differential equations, describing, for instance, the heating of a welded solid of dissimilar materials. By using the method of integral transformations, the solutions to the following welding problems were obtained:

1. a moving line source between two dissimilar plates (Makhnenko 1967);
2. a cylindrical source moving in an infinite plate with a constant surface temperature (keyhole technique) (Laplace transformation) (Carslaw and Jaeger 1973);
3. a plane source moving in an infinite rod with a temperature given in time (friction welding, Laplace transformation) (Wichelhaus 1975);
4. an instantaneous plane source in a given cross-section of a composite infinite rod (Carslaw and Jaeger 1973);
5. an instantaneous plane source in a given cross-section of a composite rod of finite length (Carslaw and Jaeger 1973);
6. an instantaneous plane source between similar bounded rods joined at both ends with semi-bounded rods of another material (Luikov 1968);
7. an instantaneous point source in a solid made up of two different semi-infinite solids with no contact resistance between them (Carslaw and Jaeger 1973);
8. an instantaneous line source on the boundary of two dissimilar plates (Kiselev et al. 1981).

Solutions 4, 5 and 6 will be used in the section on rapidly moving line sources in a plate (Sect. 5.1.4).

The method of integral transformations is also effective when solving problems of mass transfer in welded joints (Karkhin and Marx 1994).

It can be noted that there are also other analytical methods of solving problems of heat conduction (thermal potentials, Duhamel, conformal mapping, and so on) (Boley and Weiner 1960; Kartashov 2001); however, they are practically not applied to solving problems in welding.

4.3 Numerical Methods of Calculating Thermal Processes in Welding

The main disadvantage of the analytical methods examined (method of sources, separation of variables, integral transformations) is the required temperature independence of thermophysical characteristics ($a, c\rho, \lambda$ and α), as well as the constrained simplification of the geometric shape of a welded solid. The main advantage of these methods is the relative simplicity in solving any problem at any source distributed in space and time. The structure of analytical solutions indicates the dimensionless criteria (e.g., Fourier, Peclet, Biot numbers, etc.), which allows the results to be presented in a convenient, generalised form. If the solution cannot be successfully obtained in closed form (the integrals are not expressed as known functions), then the integrals can be quantified according to the well-known quadrature formulas of Simpson, Gauss, Chebyshev, and others (corresponding subroutines are included in the mathematical software of all modern computers). The computation of convergent series, which are obtained by using the method of images for considering the boundedness of solids, presents no fundamental difficulty either. As a rule, the algorithms and computational programs are much simpler when applying analytical methods than when applying numerical methods.

Considering the heterogeneity and complex geometric shape of a welded structure, the temperature dependence of the material properties, the complex non-linear boundary conditions, the distribution of welding sources in space and time, and other factors, the problem of conduction of heat can be solved by numerical methods only. This is the main advantage of numerical methods and has led to the intensive implementation of them in the modelling of welding processes. It can be argued that, at present, the accuracy of the calculation results of thermal processes in welding is limited not by the error of the numerical methods but by the incompleteness of the initial physical model (the lack of knowledge of the laws of heat input and heat transfer with the medium, inaccuracy of material properties, etc.).

Of all the numerical methods in welding practice, the finite difference method and the finite element method have become the most widespread. Less used are the boundary element method and the control volume method. The physical approach to solving a non-stationary temperature problem with these methods is generic: the whole heating and cooling period is divided into separate stages, and the temperature is consistently traced in separate discrete zones of the workpiece. Let us examine these numerical methods.

4.3.1 Finite Difference Method

The finite difference method (FDM) is the replacement of the sought continuous function T by its aggregate value, defined on a discrete set of points in a given region. These points are called nodes, and a set of nodes connected to each other is a mesh.

There is extensive mathematical literature on FDM in problems of heat conduction (Belyaev and Ryadno 1982; Godunov and Ryabenky 1973; Kalitkin 1978; Samarskii 1987, 2001; Samarskii and Vabishchevich 1995a, b, 2000; Tikhonov and Samarskii 1990). In it, the method is examined from mathematical points of view (accuracy, stability of the solution, etc.). Only the fundamentals of the method as applied to the thermal processes in welding are described below.

FDM is first considered in the simple example of linear heat distribution in a rod, one end of which is maintained at the ambient temperature (which is taken as the origin, $T_0 = 0$), while the rest of the surface is heat-impermeable (Fig. 4.15). This case occurs, for example, in the pressure welding of two similar rods, the ends of

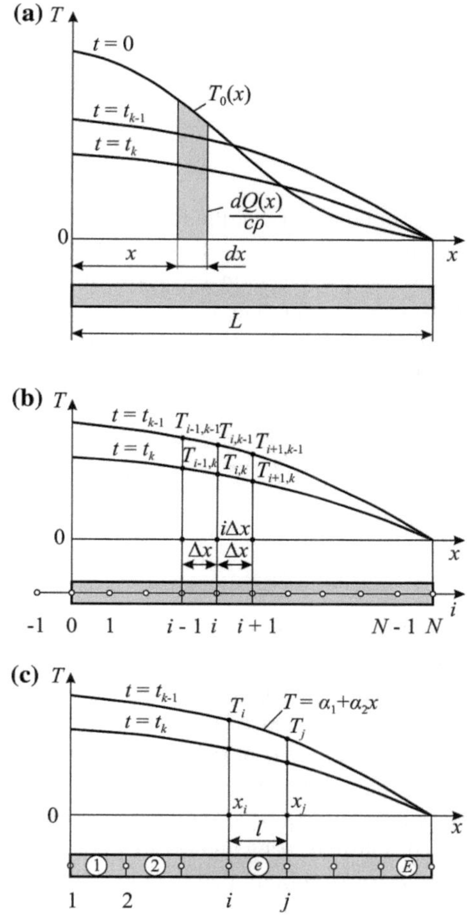

Fig. 4.15 Temperature distribution in a rod at instants of time t_{k-1} and t_k: Green's function method (**a**); finite difference method (**b**) and finite element method (**c**)

which are rigidly fixed in cooled copper clamps. The statement of a one-dimensional boundary value problem has the following form:

1. Heat conduction equation

$$c\rho \frac{\partial T}{\partial t} = \frac{\partial}{\partial x}\left(\lambda \frac{\partial T}{\partial x}\right), \tag{4.3.1}$$

or at $\lambda = \text{const}$

$$\frac{\partial T}{\partial t} = a \frac{\partial^2 T}{\partial x^2}. \tag{4.3.2}$$

2. Initial condition ($t = 0$):

$$T(x,0) = T_0(x). \tag{4.3.3}$$

3. Boundary conditions:

$$\frac{\partial T(0,t)}{\partial x} = 0;\ T(L,t) = 0. \tag{4.3.4}$$

It can be noted that the formulated linear problem (4.3.2)–(4.3.4) is solved using the analytical source method. To do this, the initial temperature distribution $T_0(x)$ should be presented in the form of elementary instantaneous plane heat sources $dQ_2(x) = c\rho T_0(x)dx$ (J m^{-2}), and the temperature increments induced by them should be summed using formula (4.2.41) with consideration for the boundary conditions (4.3.4), using the method of images:

$$T(x,t) = \int_0^L \frac{T_0(\xi)}{(4\pi at)^{1/2}} \sum_{i=-\infty}^{\infty} \sum_{j=-1,1} \left\{ \exp\left(-\frac{(x - j\xi + 4iL)^2}{4at} \right) \right. \\ \left. - \exp\left(-\frac{[x - j\xi + (4i+2)L]^2}{4at} \right) d\xi \right. \tag{4.3.5}$$

In Fig. 4.15a smooth curves are given, describing the temperature distribution in a rod at the successive moments of time t_{k-1} and t_k.

The same problem will now be solved using the FDM. The method is based on the replacement of derivatives by their approximate values expressed through differences in value at individual points—nodes of a mesh at individual instances of time (Fig. 4.15b). As a result of these transformations, differential equation (4.3.2) is replaced by equivalent relations in finite differences.

The entire heating and cooling period is divided into separate stages. Suppose for the sake of simplicity that the mesh is uniform (the grid step Δx is constant) and the time interval between stages $k-1$ and k is equal to Δt, that is, $\Delta t = t_k - t_{k-1}$. Then for node i with coordinate $x = i\Delta x$ at moment $t = t_{k-1}$ the following differential relations are obtained

$$\frac{\partial T}{\partial t} \approx \frac{T_{i,k} - T_{i,k-1}}{\Delta t};\ \frac{\partial T}{\partial x} \approx \frac{T_{i+1,k-1} - T_{i,k-1}}{\Delta x}, \tag{4.3.6}$$

the same derivative to the left of the node

$$\begin{aligned} \frac{\partial T}{\partial x} &\approx \frac{T_{i,k-1} - T_{i-1,k-1}}{\Delta x}; \\ \frac{\partial^2 T}{\partial x^2} &= \frac{\partial}{\partial x}\left(\frac{\partial T}{\partial x}\right) \approx \frac{1}{\Delta x}\left(\frac{T_{i+1,k-1} - T_{i,k-1}}{\Delta x} - \frac{T_{i,k-1} - T_{i-1,k-1}}{\Delta x}\right) \\ &= \frac{T_{i-1,k-1} - 2T_{i,k-1} + T_{i+1,k-1}}{\Delta x^2}. \end{aligned} \tag{4.3.7}$$

Here the subscripts indicate the numbers of the grid nodes and the moments of time. The derivatives with respect to x are referred to the previous stage, that is, to moment t_{k-1}. By substituting (4.3.6) and (4.3.7) into (4.3.2) and solving it for $T_{i,\,k}$, we obtain the following

$$T_{i,k} = T_{i,k-1} + \frac{a\Delta t}{\Delta x^2}\left(T_{i-1,k-1} - 2T_{i,k-1} + T_{i+1,k-1}\right). \tag{4.3.8}$$

If the mesh is not regular (step Δx is not constant), then formula (4.3.8) becomes more complicated, and the accuracy of approximation of the given equation decreases.

Analogously, formulas can also be obtained in the case of a non-linear equation (4.3.1) when the properties of welded joint $c\rho$ and λ are not constant. Then the second derivative is approximated with the expression

$$\frac{\partial}{\partial x}\left(\lambda\frac{\partial T}{\partial x}\right) \approx \frac{1}{\Delta x^2}\left[\lambda_{i+1/2}\left(T_{i+1,k-1} - T_{i,k-1}\right) - \lambda_{i-1/2}\left(T_{i,k-1} - T_{i-1,k-1}\right)\right], \tag{4.3.9}$$

where $\lambda_{i+1/2}$, $\lambda_{i-1/2}$ are the average current values of thermal conductivity between nodes i and $i + 1$, i and $i - 1$, respectively. The average values in the simplest case can be defined as a half-sum of thermal conductivity of the medium at neighboring nodes. With a large difference in thermal conductivity at neighboring nodes, it is more correct to take the harmonic mean (Sudnik and Yerofeyev 1986, 1988) or the integral mean (Kozdoba 1976). Values $c\rho$ are taken depending on the current temperature at the node. The phase transformation temperature during the melting and solidification of the alloy can be considered, taking the increased value of the specific heat capacity c at the solidus-liquidus temperature interval, that is, considering that at any temperature $c = dH/dT$, where H is specific enthalpy (J g^{-1}), including the heat of phase and allotropic transformations.

Boundary conditions (4.3.4) in finite differentials are easily expressed. The temperature on the right side of all stages is assumed to equal zero: $T_{N,k-1} = T(L,t) = 0$. On the left end of the mesh, an additional node is introduced (node $i = -1$ in Fig. 4.15b), which allows the zero value of a derivative to be set for all k by satisfying the condition $T_{-1,k-1} = T_{0,k-1}$ (more precisely the symmetry condition on the adiabatic boundary $T_{-1,k-1} = T_{1,k-1}$).

In this way, the temperature $T_{i,k}$ at any node i ($i = 0, 1, \ldots, N - 1$) at the current moment t_k can be found using the formula (4.3.8), knowing the temperature only for this node and two neighboring nodes at the previous moment t_{k-1} (Fig. 4.16a). Starting from moment $t = \Delta t$, when $k = 1$ and $T_{i,k-1} = T_0(i\Delta x)$ according to (4.3.3), the process of constructing a solution can be continued arbitrarily far if the time interval satisfies the stability condition of the solution:

$$\Delta t \le \Delta x^2/(2a). \tag{4.3.10}$$

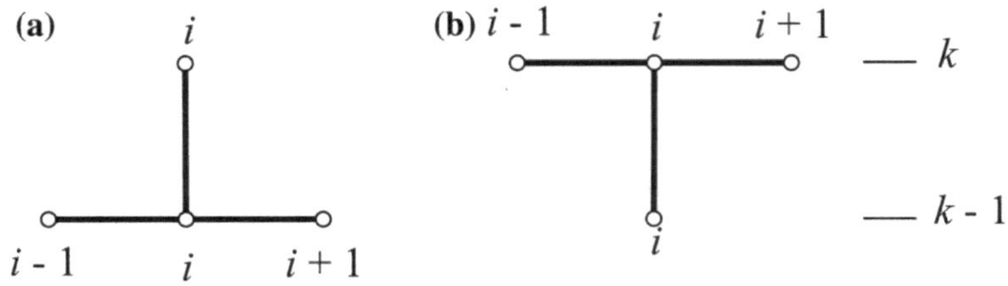

Fig. 4.16 Patterns for unsteady heat conduction: explicit scheme (**a**) and implicit scheme (**b**)

If step Δx and coefficient a are variable, then the minimum $\Delta x^2/a$ should be assumed. The physical meaning of the condition (4.3.10) is that the temperature in a node at a small time interval Δt is affected only by neighboring nodes.

In fundamentally the same way, the formulas for calculating a two-dimensional and three-dimensional temperature fields in a welded joint of any shape can be obtained, taking into account the temperature dependence of the thermophysical properties of the metal of a heterogeneous welded joint, boundary conditions of another kind, additional heat sources and sinks, and other factors. For example, in Fig. 4.17a, a two-dimensional rectangular mesh with steps Δx and Δy is shown on the left side of a cross-section of a symmetrical butt joint. The formula for calculating the temperature at node i, j at the current moment k can be obtained analogously to formula (4.3.8), introducing an additional derivative with respect to y:

$$T_{i,j,k} = T_{i,j,k-1} + \frac{a\Delta t}{\Delta x^2}\left(T_{i-1,j,k-1} - 2T_{i,j,k-1} + T_{i+1,j,k-1}\right) + \frac{a\Delta t}{\Delta y^2}\left(T_{i,j-1,k-1} - 2T_{i,j,k-1} + T_{i,j+1,k-1}\right). \tag{4.3.11}$$

The stability condition of the solution to a two-dimensional problem has the form

$$\Delta t \leq \frac{1}{2a(\Delta x^{-2} + \Delta y^{-2})}. \tag{4.3.12}$$

In this way, the surface $T(x, y)$ at any moment is approximated by an aggregate of points $T_{i,j}$ (Fig. 4.17b).

A three-dimensional problem of heat conduction is solved similarly.

A so called explicit scheme was given here. Its disadvantage is the restriction on time interval Δt according to the stability condition (4.3.10) and (4.3.12). An implicit scheme, which is always stable, can be used as well and a large time interval can be taken (restricted only by physical considerations). With an implicit scheme, the derivatives with respect to the coordinates in (3.3.9), (3.3.20), (4.3.1) or (4.3.2) should be calculated according to the temperature values at the nodes of the mesh at the current moment k instead of the previous moment $k - 1$ (see Fig. 4.16b). For example, for a one-dimensional case analogous to (4.3.7) the following can be written

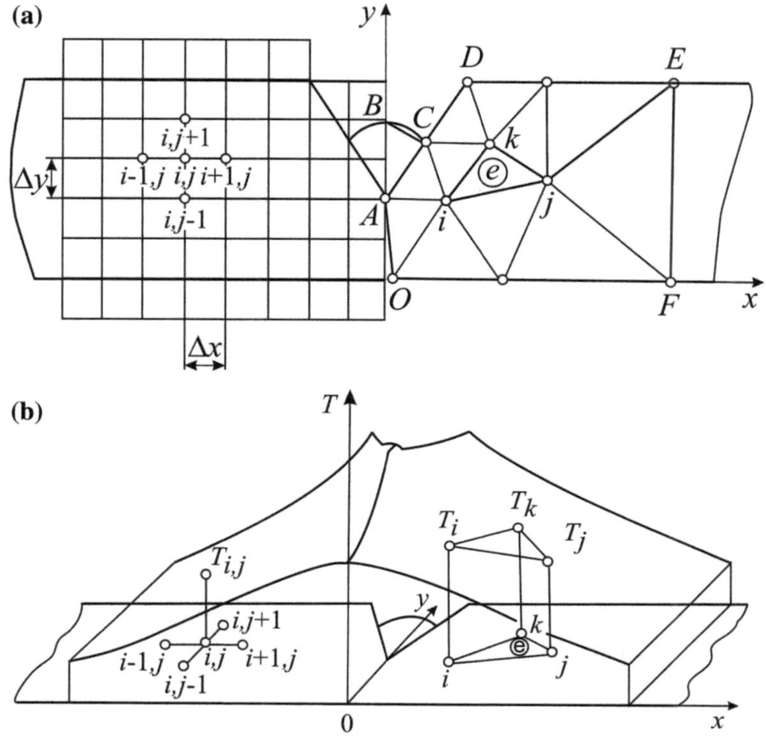

Fig. 4.17 Example of an finite difference mesh (at the left) and a finite element mesh (at the right) for the cross-section of a butt welded joint (**a**) and modelling the temperature field (**b**)

$$\frac{\partial^2 T}{\partial x^2} \approx \frac{T_{i-1,k} - 2T_{i,k} + T_{i+1,k}}{\Delta x^2}, \tag{4.3.13}$$

as a result of which the equation for node i is obtained

$$T_{i,k} = T_{i,k-1} + \frac{a\Delta t}{\Delta x^2}\left(T_{i-1,k} - 2T_{i,k} + T_{i+1,k}\right), \tag{4.3.14}$$

from which

$$-T_{i-1,k} + \left(\frac{\Delta x^2}{a\Delta t} + 2\right)T_{i,k} - T_{i+1,k} = \frac{\Delta x^2}{a\Delta t}T_{i,k-1}. \tag{4.3.15}$$

An equation of type (4.3.15) is formed for every i from N of nodes ($i = 0, 1, \ldots, N - 1$), the temperature of which is unknown at moment k. As a result, a system of linear equations is obtained, which can conveniently be written down in matrix notation

$$[A]\{T\}_k = \{B\}, \tag{4.3.16}$$

where matrix $[A]$, column vector of the temperature at the nodes $\{T\}_k$, and inhomogeneous vector $\{B\}$ have the form

$$[A] = \begin{bmatrix} & & \vdots & & \\ & & \vdots & 0 & \\ & & -1 & & \\ \dots & -1 & \frac{\Delta x^2}{a\Delta t}+2 & -1 & \dots \\ & & -1 & & \\ & & \vdots & & \\ & 0 & \vdots & & \end{bmatrix} \begin{matrix} \\ \\ \\ i\text{th row;} \\ \\ \\ \\ \end{matrix} \quad (i\text{th column}) \tag{4.3.17}$$

$$\{T\}_k = \begin{Bmatrix} T_{0,k} \\ \vdots \\ T_{i-1,k} \\ T_{i,k} \\ T_{i+1,k} \\ \vdots \\ T_{N-1,k} \end{Bmatrix} \; i\text{th row;} \quad \{B\} = \begin{Bmatrix} \vdots \\ \frac{\Delta x^2}{a\Delta t} T_{i,k-1} \\ \vdots \end{Bmatrix} \; i\text{th row.}$$

It is easy to see that by directly multiplying the ith row of matrix $[A]$ by column $\{T\}_k$, one Eq. (4.3.15) is obtained. It can be seen that matrix $[A]$ is symmetric and banded; only on three diagonals do the elements of the matrix differ from zero. Therefore, solving the system using computers, for example the Gauss elimination method, presents no fundamental complexity, even if the mesh contains thousands of nodes.

For a two-dimensional (or three-dimensional) temperature field, the equations have a structure analogous to (4.3.15), only on the left side of the equations for five (or seven) non-zero coefficients at $T_{i,j,k}$, $T_{i-1,j,k}$, $T_{i+1,j,k}$, $T_{i,j-1,k}$, $T_{i,j+1,k}$ (see mesh template in Fig. 4.17a on the left). The bandwidth of matrix $[A]$ is greater than five (or seven) and is defined by the mesh size and a numbering system for nodes.

In the same way, a system of differential equations for a boundary value problem in cylindrical coordinates (3.3.25) can be constructed.

The advantage of the explicit system is the simplicity of its algorithm, as it uses a simple cycle of mesh nodes with a small number of arithmetic operations at each step. The disadvantage of it is the stability condition (4.3.10) or (4.3.12), which requires the small time step Δt, even if the rate of temperature change $\partial T/\partial t$ is small. The advantage of the implicit scheme is the unconditional stability. The disadvantage is the relatively complex algorithm which requires solving a system of equations of high rank, itself requiring the use of computers with a large enough memory. Combining the advantages of the explicit (volume of computations proportional to the number of mesh nodes) and implicit (unconditional stability) schemes has led to the development of economical calculation schemes, from which the alternating direction method is the best for solving two-dimensional, non-stationary problems. For three-dimensional problems, the splitting method (locally one-dimensional scheme) is best (Kalitkin 1978; Samarskii 1987, 2001).

The accuracy of a numerical solution is higher, the smaller intervals Δx, Δy and Δt are; in other words, the more precise the equation of heat conduction is approximated. It is recommended to assume the minimal values Δx and Δy in the areas of high gradients of the temperature field (near the weld), and Δt in the period of a high rate of temperature change. For example, in the case of beam welding methods, a mesh with a step of less than 1 mm near the heating source and with a time step of no more than 0.01 s is recommended.

The method of finite differences can be used to solve problems in a moving coordinate system for a quasi-stationary thermal state (Makhnenko et al. 1969; Pavelic et al. 1969). In the case of a plate, the difference scheme, approximating (3.3.31), has the form (Karkhin 1990)

$$\left(-\frac{2}{\Delta x^2}-\frac{2}{\Delta y^2}+\frac{v}{a\Delta x}-\frac{b}{a}\right)T_{i,j}+\left(\frac{1}{\Delta x^2}-\frac{v}{a\Delta x}\right)T_{i-1,j}$$
$$+\frac{1}{\Delta x^2}T_{i+1,j}+\frac{1}{\Delta y^2}T_{i,j-1}+\frac{1}{\Delta y^2}T_{i,j+1}=0. \tag{4.3.18}$$

As an example, let us look at the thermal processes in an infinite plate near a moving weld pool (Makhnenko et al. 1969; Makhnenko and Kravtsov 1976). The description of heat transfer in Fourier's law does not show any noticeable errors at a sufficient distance from the weld pool but leads to large discrepancies with experimental data near the mushy zone, since the convective flows of liquid metal and other factors are not considered. The experimental-calculation method is convenient when the experimental shape and size of the weld pool with isothermal interface ($T = T_L$) are given as input data. Then the problem can be formulated as the following: find the solution to Eq. (3.3.31) in the finite rectangular area *ABCDEFGA* (Fig. 4.18) with boundary conditions of the first and second kinds, whereby the temperature on boundary *ABC* is $T = T_L$, on *GA* and *CD* $\partial T/\partial y = 0$ (since the *x*-axis is a symmetrical axis of the temperature field), and on *DEFG* the temperature is calculated according to formula (5.1.81) for a moving line source in a plate, considering that the source is located at point *O*. By approximating area *ABCDEFGA* with a finite difference mesh and forming equations of type (4.3.18) for all unknown temperature values at the mesh nodes, while considering boundary conditions, a system of algebraic equations is obtained. By solving this system, the temperature at every node

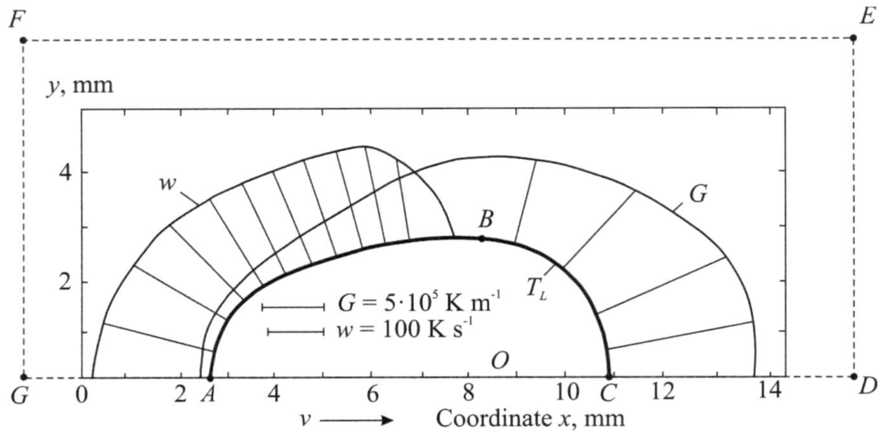

Fig. 4.18 Distributions of temperature gradient G and cooling rate w along the weld pool contour ABC in plate (Makhnenko et al. 1969)

is found and, therefore, the cooling rate, temperature gradients, and other characteristics of the temperature field as well. In Fig. 4.18, the dimensions and shape of a weld pool contour on a 4.5 mm thick titanium alloy plate are given, as are the calculated curves of the temperature gradient and cooling rate distributions along the pool contour. The alloy properties are: $a = 3$ mm^2 s^{-1}, $\lambda = 0.02$ W mm^{-1} K^{-1}, $T_m = 1913$ K. Welding conditions: $I = 130$ A, $U = 14$ V, $v = 2.22$ mm s^{-1}, $\eta_h = 0.7$ (Makhnenko et al. 1969).

From the above, it follows that the main advantage of the FDM is its simplicity. The main disadvantages of the method are the poor approximation of the curvilinear region by a rectangular mesh and the need for constant grid steps, otherwise the calculation scheme becomes more complicated and the inaccuracy of the solution increases.

The control volume method is close to the FDM in its mathematical idea (Patankar 1980).

4.3.2 Finite Element Method

The finite element method (FEM) has been widely applied in solving complex problems in a variety of scientific and technical fields. The FEM is devoid of the main disadvantages of the finite difference method, despite being much more complicated and requiring more powerful computers. The method has been used since the 1970s in researching thermal processes in welding (Krutz and Segerling 1978; Negoda and Karkhin 1977; Nickell and Hibbitt 1975).

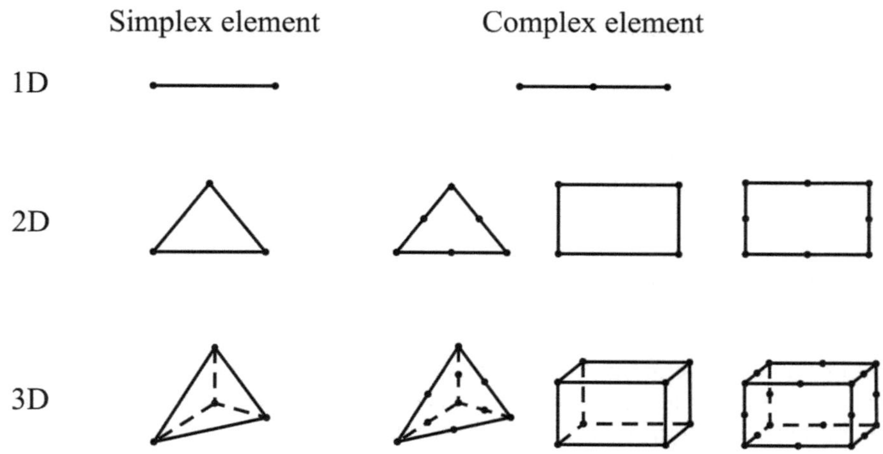

Fig. 4.19 Examples of one-, two- and three-dimensional finite elements

Extensive literature has been published about the FEM (Gallagher 1975; Lindgren 2001a, b, c, 2007; Ottosen and Petersson 1992; Rao 1999; Segerlind 1984; Strang and Fix 1973; Zienkiewicz 1971; Zienkiewicz and Morgan 1983; Zienkiewicz and Taylor 2000). However, it mainly examines the method from a mathematical point of view, which makes it difficult for welding engineers to master. Therefore, all stages of formulating and solving problems by the FEM related to welding are briefly laid out below.

The main idea of the method is that any continuous function (for example temperature) can be approximated using piecewise continuous functions defined on a finite number of subdomains called elements.

Let us illustrate the main idea with an example of temperature distribution in a rod (Fig. 4.15). The rod is divided into individual elements 1, 2, ..., *e*, ..., *E*, bounded by two neighbouring nodes 1, 2, ..., *i*, *j*, ... (Fig. 4.15c). Within any *e*th element, let us approximate the temperature distribution using a straight line, whereby points T_i and T_j uniquely determine this straight line, that is, at any moment $t = t_{k-1}$, curve $T(x)$ will represent a piecewise linear function.

In a two-dimensional case, the solid can be divided into plane finite simplex elements in the shape of a triangle, which are connected to each other by three nodes (Fig. 4.19 and right side of Fig. 4.17a). The temperature distribution within the element is now depicted by a plane (Fig. 4.17b). Therefore, at any moment, surface $T(x, y)$ will be approximated by an assembly of piecewise plane surfaces. Naturally, in order to best approximate surface $T(x, y)$ in a high gradient area (near the weld), it is necessary to assume the density of elements at its largest (the smallest dimensions of the elements).

When constructing a discrete model for a one-dimensional region, a segment of three nodes can be taken, for a two-dimensional area, rectangular element of four or eight nodes, and for a three-dimensional solid, tetrahedrons and cuboids (Fig. 4.19).

However, only the simplest elements are used below: one-dimensional elements with two nodes and triangular elements with three nodes. These elements are theoretically simple, and any one-dimensional and two-dimensional bodies can be discretised by them.

Let us return to the example of a one-dimensional temperature field in a rod (see Fig. 4.15). In general, the heat distribution $T(x)$ (the temperature values at nodes T_1, T_2, …, T_i, T_j, …) is unknown. Our task is to find it, and do so in a way that the sequence of values $T_1, T_2, \ldots$ is best approximated to curve $T(x)$, which satisfies the one-dimensional equation of heat conduction. This best possible approximation can be achieved by varying all temperature values at the nodes to minimize a certain functional which is uniquely associated with the differential equation of heat conduction. Let us explain the term functional. In mathematics, a function is a relation between argument x (a set of numbers) and output $f(x)$ (a set of numbers). The term functional refers to mapping from a space of functions into a field of numbers. For example, in the case of a function of two variables the following integral is a functional:

$$\mathrm{X}[f(x, y)] = \int\limits_S F\left(x, y, f(x, y), \frac{\partial f}{\partial x}, \frac{\partial f}{\partial y}\right) ds. \tag{4.3.19}$$

Thus, the sequence of defining the temperature field for the FEM is the following:

(1) formulate the problem of heat conduction, that is, define the equation of heat conduction, and initial and boundary conditions;
(2) choose a functional with the property where the function, at which it becomes minimal, satisfies both the initial differential equation of heat conduction and the boundary conditions;
(3) divide the studied region into elements (discretise the region) and choose functions that approximate the sought temperature field within each element;
(4) express the functional through the temperature values at the nodes of elements;
(5) differentiate the functional with respect to every unknown temperature value at the nodes and equate the derivatives to zero;
(6) solve the obtained system of equations for the unknown temperature values at the nodes.

The main part of the section will be concerned with all stages of solving this variational problem.

An important aspect of the FEM is the construction of interpolation functions, which approximate the sought temperature field within each element. For this function, let us take a linear polynomial. It is expressed in terms of values at the nodal points.

For a one-dimensional element (see Fig. 4.15c), function T has the form

$$T = \alpha_1 + \alpha_2 x. \tag{4.3.20}$$

Coefficients α_1 and α_2 are defined by conditions at nodal points i and j:

$$T = T_i \text{ at } x = x_i;$$
$$T = T_j \text{ at } x = x_j. \tag{4.3.21}$$

Setting of these conditions into the formula (4.3.20) leads to a system of equations

$$T_i = \alpha_1 + \alpha_2 x_i;$$
$$T_j = \alpha_1 + \alpha_2 x_j, \tag{4.3.22}$$

by solving this, we obtain the following

$$\alpha_1 = \frac{T_i x_j - T_j x_i}{l}; \quad \alpha_2 = \frac{T_j - T_i}{l}, \tag{4.3.23}$$

where $l = x_j - x_i$.

Substituting values α_1 and α_2 into Eq. (4.3.20), we have

$$T = \frac{T_i x_j - T_j x_i}{l} + \frac{T_j - T_i}{l} x, \tag{4.3.24}$$

or in another form

$$T = \frac{x_j - x}{l} T_i + \frac{x - x_i}{l} T_j. \tag{4.3.25}$$

This equation can be written in matrix notation

$$T = N_i T_i + N_j T_j = [N]\{T\}, \tag{4.3.26}$$

where $[N] = \left[N_i N_j\right]$ is a row vector; functions N_i and N_j are called shape functions

$$N_i = 1/l\left(x_j - x\right); \quad N_j = 1/l(x - x_i). \tag{4.3.27}$$

It is characteristic of a shape function to be equal to 1 at the node, whose number coincides with the subscript of the function, and 0 at all other nodes.

The interpolation function is analogously defined in the form (4.3.26) for a two-dimensional element. The nodes are numbered counterclockwise in the sequence i, j, k (see Fig. 4.17). The interpolation polynomial is

$$T = \alpha_1 + \alpha_2 x + \alpha_3 y \tag{4.3.28}$$

with the nodal conditions

$$T = T_i \text{ at } x = x_i, y = y_i;$$
$$T = T_j \text{ at } x = x_j, y = y_j;$$

$$T = T_k \text{ at } x = x_k,\ y = y_k.$$

These nodal conditions lead to a system of equations

$$\begin{aligned} T_i &= \alpha_1 + \alpha_2 x_i + \alpha_3 y_i; \\ T_j &= \alpha_1 + \alpha_2 x_j + \alpha_3 y_j; \\ T_k &= \alpha_1 + \alpha_2 x_k + \alpha_3 y_k, \end{aligned} \tag{4.3.29}$$

the solution of which gives

$$\begin{aligned} \alpha_1 &= \frac{1}{2A}\left[\left(x_j y_k - x_k y_j\right)T_i + \left(x_k y_i - x_i y_k\right)T_j + \left(x_i y_j - x_j y_i\right)T_k\right]; \\ \alpha_2 &= \frac{1}{2A}\left[\left(y_j - y_k\right)T_i + \left(y_k - y_i\right)T_j + \left(y_i - y_j\right)T_k\right]; \\ \alpha_3 &= \frac{1}{2A}\left[\left(x_k - x_j\right)T_i + \left(x_i - x_k\right)T_j + \left(x_j - x_i\right)T_k\right], \end{aligned} \tag{4.3.30}$$

where A is the area of triangle i, j, k; it is related to the determinant of the system

$$2\,A = \begin{vmatrix} 1 & x_i & y_i \\ 1 & x_j & y_j \\ 1 & x_k & y_k \end{vmatrix}. \tag{4.3.31}$$

By substituting (4.3.30) into (4.3.28) and bringing the result to the form (4.3.26), we obtain the following

$$T = N_i T_i + N_j T_j + N_k T_k = [N]\{T\}, \tag{4.3.32}$$

where three shape functions have the form

$$\begin{aligned} N_i &= \frac{1}{2A}(a_i + b_i x + c_i y) \text{ where } \begin{cases} a_i = x_j y_k - x_k y_j, \\ b_i = y_j - y_k, \\ c_i = x_k - x_j; \end{cases} \\ N_j &= \frac{1}{2A}\left(a_j + b_j x + c_j y\right) \text{ where } \begin{cases} a_j = x_k y_i - x_i y_k, \\ b_j = y_k - y_i, \\ c_j = x_i - x_k; \end{cases} \\ N_k &= \frac{1}{2A}(a_k + b_k x + c_k y) \text{ where } \begin{cases} a_k = x_i y_j - x_j y_i, \\ b_k = y_i - y_j, \\ c_k = x_j - x_i. \end{cases} \end{aligned} \tag{4.3.33}$$

Let us formulate a two-dimensional problem of heat conduction in differential form with sufficient consideration of welding conditions.

1. The non-stationary equation of heat conduction has the form

$$c\rho\frac{\partial T}{\partial t}=\frac{\partial}{\partial x}\left(\lambda_x\frac{\partial T}{\partial x}\right)+\frac{\partial}{\partial y}\left(\lambda_y\frac{\partial T}{\partial y}\right)+q_3, \tag{4.3.34}$$

where λ_x and λ_y are the thermal conductivities in the x- and y-directions; q_3 is the volumetric density of the heat source (sink) power inside the solid, W m^{-3}; $c\rho$, λ_x, λ_y and q_3 can be functions of x, y and t, that is, the solid can be heterogeneous, and its properties depend on temperature. The equation is applied both to isotropic ($\lambda_x=\lambda_y$) and anisotropic ($\lambda_x\neq\lambda_y$) solids.
2. Initial condition ($t=0$):

$$T(x,y,0)=T_0(x,y). \tag{4.3.35}$$

3. Boundary conditions:

(a) on a part of the boundary surface S_1, conditions of the second and third kind are given

$$\lambda_x\frac{\partial T}{\partial x}l_x+\lambda_y\frac{\partial T}{\partial y}l_y-q_2+\alpha(T-T_\infty)=0; \tag{4.3.36}$$

(b) on a part of the boundary surface S_2, conditions of the first kind are given

$$T=T_S; \tag{4.3.37}$$

the union of S_1 and S_2 forms the full boundary S. Here, l_x, l_y are the direction cosines of the outward normal; q_2 is the heat flux density on surface S_1 (positive if heat is introduced), W m^{-2}; α is the coefficient of surface heat transfer; T_∞ is the ambient temperature. If both quantities q_2 and α are equal to zero on boundary S_1, then the equality (4.3.36) is reduced to the equation

$$\lambda_x\frac{\partial T}{\partial x}l_x+\lambda_y\frac{\partial T}{\partial y}l_y=0, \tag{4.3.38}$$

which reflects the lack of heat transfer over boundary S_1. In the case of an isotropic solid, the last equation can be written in the form

$$\partial T/\partial n=0, \tag{4.3.39}$$

where n is the outward normal to the surface.

From a practical point of view, the problem (4.3.34)–(4.3.37) is sufficiently complete to describe thermal processes in welding. For example, by using the volume source q_3 the heat input due to molten filler metal, the released heat during the passage of an electric current, the heat of melting and solidification, etc., can all be modelled. When examining half of the symmetrical cross-section of a butt joint (see Fig. 4.17a), the boundary condition (4.3.36) can be used to describe heat transfer ($q_2 = 0, \alpha > 0$) on surfaces *BCDE* and *OF*, and the thermal insulation ($q_2 = 0, \alpha = 0$) of boundary *OAB*. At the conditional boundary *EF*, sufficiently far from weld *ABCA*, the condition (4.3.37) can be given, by calculating the temperature using analytical formulas, for example, formula (4.2.42); the input quantities $q_2, q_3, \alpha, T_\infty, T_S$ and the surface can change in the welding process of heating and cooling. Although a welded joint is considered isotropic as a rule, the differences between thermal conductivity λ_x and λ_y are defined below.

Equation (4.3.34), along with boundary conditions (4.3.35)–(4.3.37), uniquely determines the problem. It serves as a starting point in solving the problem by the FDM. The FEM is based on a variational approach. In calculus of variations it is established that the solution to (4.3.34) with boundary conditions (4.3.36) is equivalent to finding the minimum of the functional

$$\mathrm{X} = \int\limits_V \left[\frac{1}{2}\lambda_x\left(\frac{\partial T}{\partial x}\right)^2 + \frac{1}{2}\lambda_y\left(\frac{\partial T}{\partial y}\right)^2 - T\left(q_3 - c\rho\frac{\partial T}{\partial t}\right)\right]dv + \int\limits_S \left[-q_2 T + \frac{1}{2}\alpha(T - T_\infty)^2\right]ds. \tag{4.3.40}$$

Let us show this. Present the functional in the form

$$\mathrm{X} = \int\limits_V F(x, y, T, \partial T/\partial x, \partial T/\partial y)dv + \int\limits_S f(x, y, T)ds, \tag{4.3.41}$$

where F and f are the integrands in (4.3.40):

$$F = \frac{1}{2}\lambda_x\left(\frac{\partial T}{\partial x}\right)^2 + \frac{1}{2}\lambda_y\left(\frac{\partial T}{\partial y}\right)^2 - T\left(q_3 - c\rho\frac{\partial T}{\partial t}\right);$$
$$f = -q_2 T + 1/2\,\alpha(T - T_\infty)^2. \tag{4.3.42}$$

The infinitesimal changes in $F(x, y)$ and $f(x, y)$ correspond to a variation of the functional

$$\delta\mathrm{X} = \int\limits_V \left[\frac{\partial F}{\partial T}\delta T + \frac{\partial F}{\partial(\partial T/\partial x)}\delta(\partial T/\partial x) + \frac{\partial F}{\partial(\partial T/\partial y)}\delta(\partial T/\partial y)\right]dv + \int\limits_S \left(\frac{\partial f}{\partial T}\delta T\right)ds. \tag{4.3.43}$$

Since $\delta(\partial T/\partial x) = \partial/\partial x(\delta T)$ and $\delta(\partial T/\partial y) = \partial/\partial y(\delta T)$, then

$$\delta X = \int_V \left[\frac{\partial F}{\partial T}\delta T + \frac{\partial F}{\partial(\partial T/\partial x)}\frac{\partial}{\partial x}(\delta T) + \frac{\partial F}{\partial(\partial T/\partial y)}\frac{\partial}{\partial y}(\delta T)\right] dv + \int_S \left(\frac{\partial f}{\partial T}\delta T\right) ds. \tag{4.3.44}$$

Let us integrate by parts the second term in the first integral:

$$\int_V \frac{\partial F}{\partial(\partial T/\partial x)}\frac{\partial}{\partial x}(\delta T) dv = \int_V \frac{\partial}{\partial x}\left[\frac{\partial F}{\partial(\partial T/\partial x)}\delta T\right] dv - \int_V \frac{\partial}{\partial x}\left[\frac{\partial F}{\partial(\partial T/\partial x)}\right]\delta T dv \tag{4.3.45}$$

and apply the Gauss's theorem

$$\int_V \frac{\partial F}{\partial(\partial T/\partial x)}\frac{\partial}{\partial x}(\delta T) dv = \int_S \frac{\partial F}{\partial(\partial T/\partial x)}\delta T l_x ds - \int_V \frac{\partial}{\partial x}\left[\frac{\partial F}{\partial(\partial T/\partial x)}\right]\delta T dv. \tag{4.3.46}$$

By analogously transforming the third term in the first integral (4.3.44) and by combining the results of integration, we obtain the following

$$\begin{aligned}\delta X = &\int_V \left\{\frac{\partial F}{\partial T} - \frac{\partial}{\partial x}\left[\frac{\partial F}{\partial(\partial T/\partial x)}\right] - \frac{\partial}{\partial y}\left[\frac{\partial F}{\partial(\partial T/\partial y)}\right]\right\}\delta T dv \\ &+ \int_S \left\{\frac{\partial F}{\partial(\partial T/\partial x)} l_x + \frac{\partial F}{\partial(\partial T/\partial y)} l_y + \frac{\partial f}{\partial T}\right\}\delta T ds.\end{aligned} \tag{4.3.47}$$

Functional X takes a minimum (stationary) value if its variation is equal to zero, $\delta X = 0$, which in turn happens only when the expressions in curly brackets in both integrals (4.3.47) reduce to zero:

$$\frac{\partial F}{\partial T} - \frac{\partial}{\partial x}\left[\frac{\partial F}{\partial(\partial T/\partial x)}\right] - \frac{\partial}{\partial y}\left[\frac{\partial F}{\partial(\partial T/\partial y)}\right] = 0; \tag{4.3.48}$$

$$\frac{\partial F}{\partial(\partial T/\partial x)} l_x + \frac{\partial F}{\partial(\partial T/\partial y)} l_y + \frac{\partial f}{\partial T} = 0. \tag{4.3.49}$$

Each summond is found separately by differentiating the integrand functions (4.3.42):

$$\begin{aligned}\frac{\partial F}{\partial T} &= c\rho\frac{\partial T}{\partial t} - q_3; \\ \frac{\partial F}{\partial(\partial T/\partial x)} &= \lambda_x\frac{\partial T}{\partial x}; \quad \frac{\partial F}{\partial(\partial T/\partial y)} = \lambda_y\frac{\partial T}{\partial y};\end{aligned}$$

$$\frac{\partial f}{\partial T} = -q_2 + \alpha(T - T_\infty). \tag{4.3.50}$$

The substitution of these expressions in (4.3.48) leads to (4.3.34), and in (4.3.49) to (4.3.36).

So, function T, resulting in the minimal value of the functional (4.3.40), should satisfy the differential equation of heat conduction (4.3.34) and boundary conditions (4.3.36). Initial (4.3.35) and boundary conditions of the first kind (4.3.37) are considered later.

Let us minimise the functional (4.3.40) using a set of approximating functions, each of which is defined on a separate element and expressed through values at nodes $\{T\}$. Since the nodal values determine the quantity of the functional, the minimisation should also be performed using these values. To this end, the integrals (4.3.40) are divided between individual elements, and the temperature distribution $T^{(e)}$ within each element is expressed by its values at nodes $\{T\}$:

$$\mathrm{X} = \sum_{e=1}^{E} \chi^{(e)}, \tag{4.3.51}$$

where E is the total number of elements, and $\chi^{(e)}$ is the contribution of each element in X:

$$\begin{aligned}\chi^{(e)} = &\int\limits_{V^{(e)}} \left[\frac{1}{2}\lambda_x^{(e)}\left(\frac{\partial T^{(e)}}{\partial x}\right)^2 + \frac{1}{2}\lambda_y^{(e)}\left(\frac{\partial T^{(e)}}{\partial y}\right)^2\right] dv \\ &- \int\limits_{V^{(e)}} T^{(e)}\left[q_3^{(e)} - c\rho^{(e)}\frac{\partial T^{(e)}}{\partial t}\right] dv \\ &- \int\limits_{S^{(e)}} q_2^{(e)} T^{(e)} ds + \int\limits_{S^{(e)}} \frac{1}{2}\alpha^{(e)}\left[\left(T^{(e)}\right)^2 - 2T^{(e)}T_\infty + T_\infty^2\right] ds.\end{aligned} \tag{4.3.52}$$

Here, the superscript (e) indicates that the quantity belongs to the given element e. It can be seen that the properties of elements can differ by λ_x, λ_y, $c\rho$ and α, that is, a welded solid can be heterogeneous. The superscript (e) of these coefficients, as well as of q_2 and q_3, are omitted below for ease of notation.

Let us write Eq. (4.3.52) in matrix notation. To do this, the following matrix and the gradient vector for the element are introduced

$$\left[D^{(e)}\right] = \begin{bmatrix} \lambda_x & 0 \\ 0 & \lambda_y \end{bmatrix} \tag{4.3.53}$$

$$\left\{g^{(e)}\right\} = \begin{Bmatrix} \frac{\partial T^{(e)}}{\partial x} \\ \frac{\partial T^{(e)}}{\partial y} \end{Bmatrix}. \tag{4.3.54}$$

By substituting (4.3.32) into (4.3.54), we have

$$\{g^{(e)}\} = \begin{bmatrix} \frac{\partial N_i^{(e)}}{\partial x} & \frac{\partial N_j^{(e)}}{\partial x} & \frac{\partial N_k^{(e)}}{\partial x} \\ \frac{\partial N_i^{(e)}}{\partial y} & \frac{\partial N_j^{(e)}}{\partial y} & \frac{\partial N_k^{(e)}}{\partial y} \end{bmatrix} \begin{Bmatrix} T_i \\ T_j \\ T_k \end{Bmatrix} = [B^{(e)}]\{T\}, \tag{4.3.55}$$

where the matrix $[B^{(e)}]$ is obtained by differentiating shape functions (4.3.33)

$$B^{(e)} = \frac{1}{2A}\begin{bmatrix} b_i & b_j & b_k \\ c_i & c_j & c_k \end{bmatrix}. \tag{4.3.56}$$

Now the integrals (4.3.52) can be written in matrix notation taking into account (4.3.32), (4.3.53) and (4.3.54)

$$\begin{aligned} \chi^{(e)} = & \int\limits_{V^{(e)}} 1/2\{T\}^T\left[B^{(e)}\right]^T\left[D^{(e)}\right]\left[B^{(e)}\right]\{T\}dv - \int\limits_{V^{(e)}} q_3\left[N^{(e)}\right]\{T\}dv \\ & + \int\limits_{V^{(e)}} c\rho\left[N^{(e)}\right]\{T\}\left[N^{(e)}\right]\frac{\partial\{T\}}{\partial t}dv - \int\limits_{S^{(e)}} q_2\left[N^{(e)}\right]\{T\}ds \\ & + \int\limits_{S^{(e)}} 1/2\alpha\{T\}^T\left[N^{(e)}\right]^T\left[N^{(e)}\right]\{T\}ds - \int\limits_{S^{(e)}} \alpha T_\infty\left[N^{(e)}\right]\{T\}ds \\ & + \int\limits_{S^{(e)}} 1/2\alpha T_\infty^2 ds, \end{aligned} \tag{4.3.57}$$

since according to matrix algebra

$$\begin{aligned} & \lambda_x\left(\frac{\partial T^{(e)}}{\partial x}\right)^2 + \lambda_y\left(\frac{\partial T^{(e)}}{\partial y}\right)^2 = \{g^{(e)}\}^T\left[D^{(e)}\right]\{g^{(e)}\} \\ & = \left(\left[B^{(e)}\right]\{T\}\right)^T\left[D^{(e)}\right]\left(\left[B^{(e)}\right]\{T\}\right) = \{T\}^T\left[B^{(e)}\right]^T\left[D^{(e)}\right]\left[B^{(e)}\right]\{T\} \end{aligned} \tag{4.3.58}$$

and $[N^{(e)}]$ is only a coordinate function and does not depend on time:

$$\frac{\partial T^{(e)}}{\partial t} = \frac{\partial}{\partial t}\left(\left[N^{(e)}\right]\{T\}\right) = \left[N^{(e)}\right]\frac{\partial\{T\}}{\partial t}. \tag{4.3.59}$$

Superscript T signifies a transpose of the matrix, that is, the operation when the matrix row forms a column and the column forms a row.

So, the functional has been expressed through the temperature at the nodes. The minimisation of functional X with respect to the temperature at the nodes leads to the equation

$$\frac{\partial X}{\partial\{T\}} = \begin{Bmatrix} \frac{\partial X}{\partial T_1} \\ \frac{\partial X}{\partial T_2} \\ \cdot \\ \cdot \\ \cdot \end{Bmatrix} = \frac{\partial}{\partial\{T\}} \sum_{e=1}^{E} \chi^{(e)} = \sum_{e=1}^{E} \frac{\partial \chi^{(e)}}{\partial\{T\}} = 0. \tag{4.3.60}$$

The partial derivatives $\frac{\partial \chi^{(e)}}{\partial\{T\}}$ are found by differentiating the expression (4.3.57):

$$\begin{aligned} \frac{\partial \chi^{(e)}}{\partial\{T\}} &= \int\limits_{V^{(e)}} \left[B^{(e)}\right]^T \left[D^{(e)}\right]\left[B^{(e)}\right]\{T\}dv - \int\limits_{V^{(e)}} q_3\left[N^{(e)}\right]^T dv \\ &+ \left(\int\limits_{V^{(e)}} c\rho\left[N^{(e)}\right]^T \left[N^{(e)}\right]dv \right) \frac{\partial\{T\}}{\partial t} - \int\limits_{S^{(e)}} q_2\left[N^{(e)}\right]^T ds \\ &+ \int\limits_{S^{(e)}} \alpha\left[N^{(e)}\right]^T \left[N^{(e)}\right]\{T\}ds - \int\limits_{S^{(e)}} \alpha T_\infty\left[N^{(e)}\right]^T ds + 0, \end{aligned} \tag{4.3.61}$$

since in accordance with the rules of differentiation of matrix products

$$\frac{\partial}{\partial\{T\}}\left(\left[N^{(e)}\right]\{T\}\right) = \frac{\partial}{\partial\{T\}}\left(N_i^{(e)}T_i + N_j^{(e)}T_j + N_k^{(e)}T_k\right) = \begin{Bmatrix} N_i^{(e)} \\ N_j^{(e)} \\ N_k^{(e)} \end{Bmatrix} = \left[N^{(e)}\right]^T, \tag{4.3.62}$$

$$\frac{\partial}{\partial\{T\}}\left(\{T\}^T[M]\{T\}\right) = 2[M]\{T\}, \tag{4.3.63}$$

where $[M]$ is the matrix or matrix product, independent of $\{T\}$.

The sum of integrals (4.3.61) is written in a more convenient matrix notation

$$\frac{\partial \chi^{(e)}}{\partial\{T\}} = \left[c^{(e)}\right]\frac{\partial\{T\}}{\partial t} + \left[k^{(e)}\right]\{T\} - \left\{f^{(e)}\right\}, \tag{4.3.64}$$

where $[c^{(e)}]$ is the heat capacity matrix; $[k^{(e)}]$ is the matrix of thermal conductivity of the element; $\{f^{(e)}\}$ is the heat load vector:

$$\left[c^{(e)}\right] = \int\limits_{V^{(e)}} c\rho\left[N^{(e)}\right]^T \left[N^{(e)}\right]dv;$$

$$\left[k^{(e)}\right] = \int\limits_{V^{(e)}} \left[B^{(e)}\right]^T \left[D^{(e)}\right]\left[B^{(e)}\right]dv + \int\limits_{S^{(e)}} \alpha\left[N^{(e)}\right]^T \left[N^{(e)}\right]ds;$$

$$\{f^{(e)}\} = \int\limits_{V^{(e)}} q_3\left[N^{(e)}\right]^T dv + \int\limits_{S^{(e)}} q_2\left[N^{(e)}\right]^T ds + \int\limits_{S^{(e)}} \alpha T_\infty\left[N^{(e)}\right]^T ds. \tag{4.3.65}$$

Let us calculate all these matrices for a triangular element:

$$\left[c^{(e)}\right] = \int\limits_{V^{(e)}} c\rho \begin{bmatrix} N_iN_i & N_iN_j & N_iN_k \\ N_jN_i & N_jN_j & N_jN_k \\ N_kN_i & N_kN_j & N_kN_k \end{bmatrix} dv = \frac{c\rho V}{12}\begin{bmatrix} 2 & 1 & 1 \\ 1 & 2 & 1 \\ 1 & 1 & 2 \end{bmatrix}; \tag{4.3.66}$$

$$\left[k^{(e)}\right] = \int\limits_{V^e} \frac{1}{2A}\begin{bmatrix} b_i & c_i \\ b_j & c_j \\ b_k & c_k \end{bmatrix}\begin{bmatrix} \lambda_x & 0 \\ 0 & \lambda_y \end{bmatrix}\frac{1}{2A}\begin{bmatrix} b_i & b_j & b_k \\ c_i & c_j & c_k \end{bmatrix} dv$$

$$+ \int\limits_{S^{(e)}} \alpha \begin{bmatrix} N_iN_i & N_iN_j & N_iN_k \\ N_jN_i & N_jN_j & N_jN_k \\ N_kN_i & N_kN_j & N_kN_k \end{bmatrix} ds$$

$$= \frac{\lambda_x h}{4A}\begin{bmatrix} b_ib_i & b_ib_j & b_ib_k \\ b_jb_i & b_jb_j & b_jb_k \\ b_kb_i & b_kb_j & b_kb_k \end{bmatrix} + \frac{\lambda_y h}{4A}\begin{bmatrix} c_ic_i & c_ic_j & c_ic_k \\ c_jc_i & c_jc_j & c_jc_k \\ c_kc_i & c_kc_j & c_kc_k \end{bmatrix}$$

$$+ \frac{2\alpha A}{12}\begin{bmatrix} 2 & 1 & 1 \\ 1 & 2 & 1 \\ 1 & 1 & 2 \end{bmatrix} + \frac{\alpha_{ij}L_{ij}h}{6}\begin{bmatrix} 2 & 1 & 0 \\ 1 & 2 & 0 \\ 0 & 0 & 0 \end{bmatrix} + \frac{\alpha_{jk}L_{jk}h}{6}\begin{bmatrix} 0 & 0 & 0 \\ 0 & 2 & 1 \\ 0 & 1 & 2 \end{bmatrix}$$

$$+ \frac{\alpha_{ki}L_{ki}h}{6}\begin{bmatrix} 2 & 0 & 1 \\ 0 & 0 & 0 \\ 1 & 0 & 2 \end{bmatrix}; \tag{4.3.67}$$

$$\left\{f^{(e)}\right\} = \int\limits_{V^{(e)}} q_3 \begin{bmatrix} N_i \\ N_j \\ N_k \end{bmatrix} dv + \int\limits_{S^{(e)}} q_2 \begin{bmatrix} N_i \\ N_j \\ N_k \end{bmatrix} ds + \int\limits_{S^{(e)}} \alpha T_\infty \begin{bmatrix} N_i \\ N_j \\ N_k \end{bmatrix} ds$$

$$= \frac{q_3 V}{3}\begin{Bmatrix} 1 \\ 1 \\ 1 \end{Bmatrix} + \frac{q_2 A}{3}\begin{Bmatrix} 1 \\ 1 \\ 1 \end{Bmatrix} + \frac{q_{2ij}L_{ij}h}{2}\begin{Bmatrix} 1 \\ 1 \\ 0 \end{Bmatrix} + \frac{q_{2jk}L_{jk}h}{2}\begin{Bmatrix} 0 \\ 1 \\ 1 \end{Bmatrix} + \frac{q_{2ki}L_{ki}h}{2}\begin{Bmatrix} 1 \\ 0 \\ 1 \end{Bmatrix}$$

$$+ \frac{2\alpha T_\infty A}{3}\begin{Bmatrix} 1 \\ 1 \\ 1 \end{Bmatrix} + \frac{\alpha_{ij}T_\infty L_{ij}h}{2}\begin{Bmatrix} 1 \\ 1 \\ 0 \end{Bmatrix} + \frac{\alpha_{jk}T_\infty L_{jk}h}{2}\begin{Bmatrix} 0 \\ 1 \\ 1 \end{Bmatrix} + \frac{\alpha_{ki}T_\infty L_{ki}h}{2}\begin{Bmatrix} 1 \\ 0 \\ 1 \end{Bmatrix}. \tag{4.3.68}$$

Here $V = Ah$ is the volume of the triangular element. The subscripts at length L, q_2 and α signify the sides of the triangle to which these quantities pertain; q_2 and α without subscripts refer to facial surfaces (2 is kept at α, if the heat transfer from both sides is identical, otherwise 2α should be replaced with the sum of coefficients on both surfaces). In deriving formulas (4.3.66)–(4.3.68), the following integral values were used (Segerlind 1984):

$$\int_A N_l N_m dA = \begin{cases} \frac{A}{6} \text{ at } l = m, \\ \frac{A}{12} \text{ at } l \neq m; \end{cases}$$

$$\int_{L_{ij}} N_l^2 dL = \begin{cases} \frac{L_{ij}}{3} \text{ at } l = i \text{ or } l = j, \\ 0 \text{ at } l \neq i \text{ or } l \neq j; \end{cases}$$

$$\int_{L_{ij}} N_l N_m dL = \begin{cases} \frac{L_{ij}}{6} \text{ at } l = i \text{ or } l = j \text{ and } m = i \text{ or } m = j, \\ 0 \text{ at } l \neq i \text{ or } l \neq j \text{ and } m \neq i \text{ or } m \neq j; \end{cases}$$

$$\int_A N_l dA = A/3 \text{ at any } l;$$

$$\int_{L_{ij}} N_l dL = \begin{cases} \frac{L_{ij}}{2} \text{ at } l = i \text{ or } l = j, \\ 0 \text{ at } l \neq i \text{ or } l \neq j. \end{cases} \tag{4.3.69}$$

Substituting expression (4.3.64) into (4.3.60), a final system of differential equations for the temperature at the nodal points of the whole assembly of finite elements is obtained

$$\frac{\partial X}{\partial \{T\}} = \sum_{e=1}^{E} \left(\left[c^{(e)} \right] \frac{\partial \{T\}}{\partial t} + \left[k^{(e)} \right] \{T\} - \left\{ f^{(e)} \right\} \right) = 0 \tag{4.3.70}$$

or

$$[C] \frac{\partial \{T\}}{\partial t} + [K]\{T\} = \{F\}, \tag{4.3.71}$$

where

$$[C] = \sum_{e=1}^{E} \left[c^{(e)} \right]; [K] = \sum_{e=1}^{E} \left[k^{(e)} \right]; \{F\} = \sum_{e=1}^{E} \left\{ f^{(e)} \right\}. \tag{4.3.72}$$

The terms of Eq. (4.3.71) have the dimension of power (W).

In order to obtain the values $\{T\}$ in time, the linear differential equation (4.3.71) must be solved. There are different pocedures for solution; the simplest, FDM, is chosen (see Sect. 4.3.1).

Equation (4.3.71) is considered in the middle point $t_{k-1/2}$ of the time interval (t_{k-1}, t_k) with the length of the interval Δt. Analogous to expression (4.3.6), we obtain the following

$$\left.\begin{aligned}\left(\frac{\partial\{T\}}{\partial t}\right)_{k-1/2} &= \frac{\{T\}_k - \{T\}_{k-1}}{\Delta t};\\ \{T\}_{k-1/2} &= \frac{\{T\}_k + \{T\}_{k-1}}{2};\\ \{F\}_{k-1/2} &= \frac{\{F\}_k + \{F\}_{k-1}}{2}.\end{aligned}\right\} \tag{4.3.73}$$

Substituting expressions (4.3.73) into the differential equation (4.3.71) gives

$$\frac{1}{\Delta t}[C]\{T\}_k - \frac{1}{\Delta t}[C]\{T\}_{k-1} + \frac{1}{2}[K]\{T\}_k + \frac{1}{2}[K]\{T\}_{k-1} = \frac{1}{2}\{F\}_k + \frac{1}{2}\{F\}_{k-1} \tag{4.3.74}$$

or

$$\left[\overline{K}\right]\{T\}_k = \left\{\overline{F}\right\}, \tag{4.3.75}$$

where

$$\left.\begin{aligned}\left[\overline{K}\right] &= [K] + \tfrac{2}{\Delta t}[C];\\ \left\{\overline{F}\right\} &= \left(\tfrac{2}{\Delta t}[C] - [K]\right)\{T\}_{k-1} + \{F\}_k + \{F\}_{k-1}.\end{aligned}\right\} \tag{4.3.76}$$

The resulting Eq. (4.3.75) is fundamental. Considering the temperature at the nodes at the previous moment t_{k-1} is known, the temperature at the nodes at the current moment t_k can be obtained using this equation. At the first stage ($k = 1$), the initial conditions (4.3.35) are taken into account. In this case, the temperature at the nodes $\{T\}_0$ is taken as the temperature at the previous stage ($k - 1 = 0$) according to given function $T_0(x, y)$.

Until now, one of the boundary conditions of the problem has not been considered, namely: condition (4.3.37). Let us take into account the known temperature values at the boundary nodes at this stage of the solution, modifying the matrix $\left[\overline{K}\right]$ and vector $\left\{\overline{F}\right\}$ in Eq. (4.3.75). For example, if the temperature at a node with number n is given by $T_n = a$, then the row n in $\left[\overline{K}\right]$, $\{T\}$, $\left\{\overline{F}\right\}$ and column n in $\left[\overline{K}\right]$ are eliminated, and the product $\overline{k}_{mn} \times a (m = 1, 2, \ldots)$, where $\overline{k_{mn}}$ is the element of the matrix $\left[\overline{K}\right]$, is subtracted from every element $\overline{F_m}$ of the inhomogeneous vector $\left\{\overline{F}\right\}$. After this kind of modification, matrix $\left[\overline{K}\right]$ remains symmetric and banded. The final formulated system of Eq. (4.3.75) can be solved by any known method, for example, by the Gaussian elimination. It is necessary to solve this system at each stage of time tracking.

Thus, all stages of solving two-dimentional problem of heat conduction by the FEM in welding have been examined. A general flowchart of the programme compiled on the basis of the above algorithm is shown in Fig. 4.20.

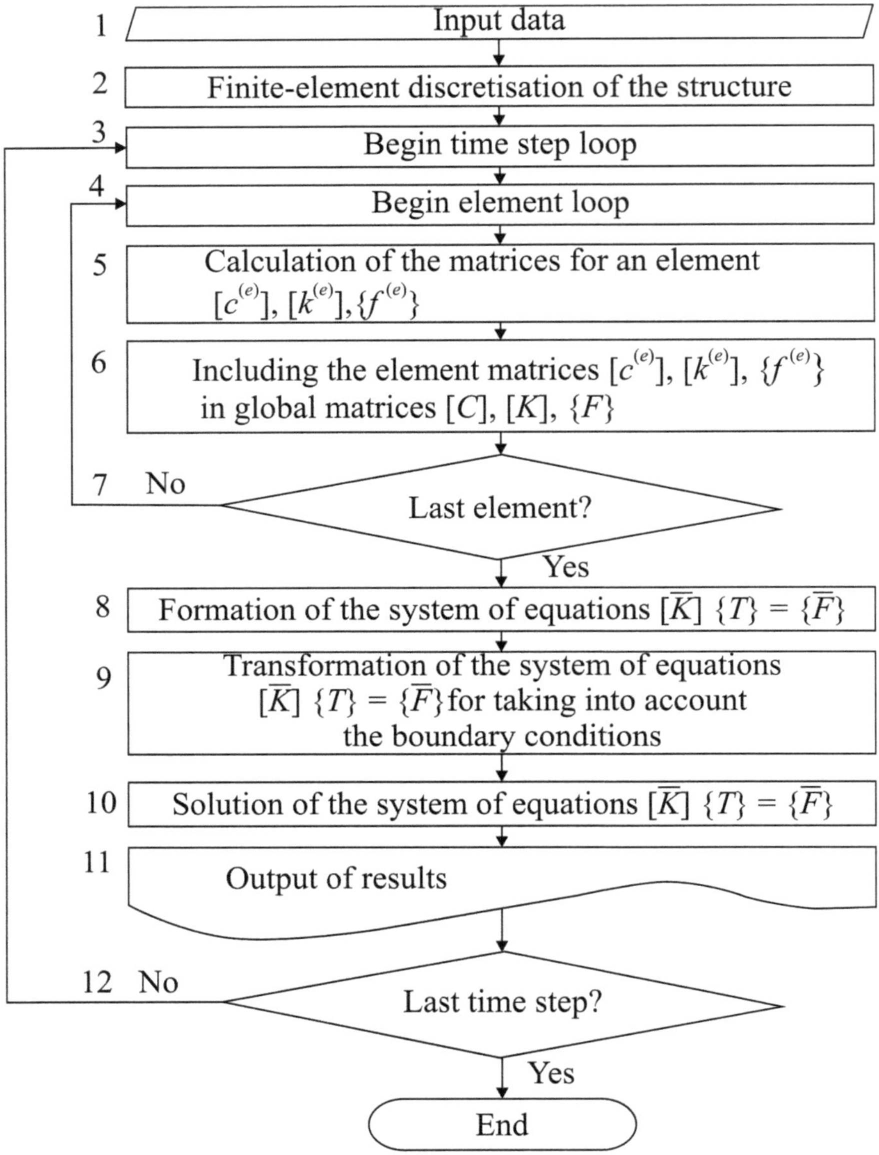

Fig. 4.20 Flowchart for an incremental finite element programme (Karkhin 1990)

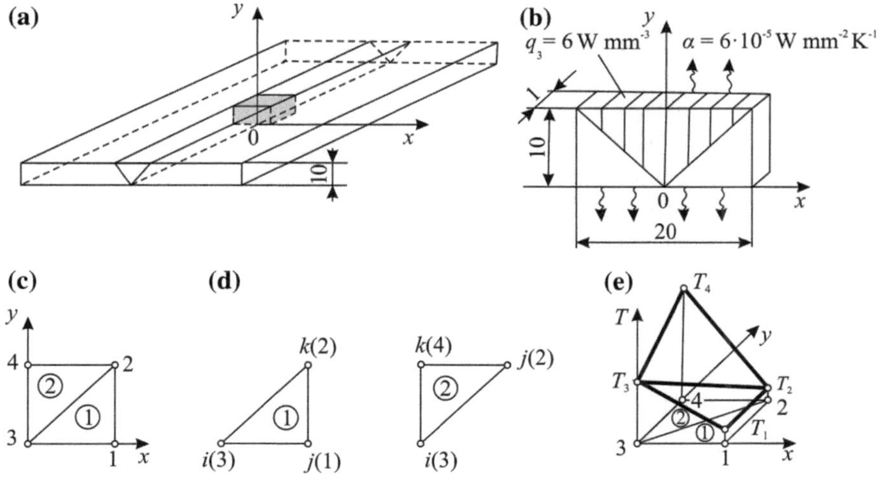

Fig. 4.21 Example of calculating the temperature in welding of a plate (Karkhin 1990): plate (**a**); separated strip across the weld (**b**); subdivision of the strip into finite elements (**c**); finite elements with nodal numbers (**d**) and temperature distribution in the finite elements (**e**)

After an abundance of formulas and mathematical calculations, it is best to consider all stages of solving the problem using a simple example which imitates welding conditions in a very simplified manner.

Example 4.3.1 Suppose a 10 mm plate is being welded (Fig. 4.21a). The material of the plate is homogeneous; thermal conductivity is $\lambda_x = \lambda_y = \lambda = 0.04$ W mm^{-1} K^{-1}; volume-specific heat capacity is $c\rho = 0.0048$ J mm^{-3} K^{-1}. We select a 1 mm thick 20 mm × 10 mm transverse strip from the central zone of the plate (Fig. 4.21b). Suppose in the upper half of the strip, which imitates a single-run weld and is shaded in the figure, a volume heat source with a power of $q_3 = 6$ W mm^{-3} is active. We assume the front and back surfaces of the strip to be heat-impermeable ($\alpha = 0$), on the upper and lower surfaces there is a heat transfer to the medium with coefficient $\alpha = 6 \times 10^{-5}$ W mm^{-2} K^{-1}, and on the side, a temperature is given according to function (4.2.42), which simulates an ideal contact of the selected strip along the side with the remaining part of the plate. There are no additional surface sources along the entire surface ($q_2 = 0$). The initial temperature is $T_0 = 0$ K. The ambient temperature is $T_\infty = 0$ K (taken as the origin). It is required to determine the temperature field in a strip 1 s after the start of the action of the heat source.

All operations are numbered in sequence according to the numbering of blocks in Fig. 4.20.

1. In the condition of the problem, initial data is formulated (geometry of a solid, material properties, heating mode, initial and boundary conditions).
2. The solid is subdivided into elements. Since the solid and the temperature field are symmetric relative to plane $x = 0$, only the right side is considered, assuming

the plane $x = 0$ as an adiabatic boundary ($\alpha = 0$). For the ease of calculation, this side is divided into only two triangular elements with four nodes, but in such a way that the difference between the numbers of the nodes of the element is minimal (Fig. 4.21c). The element thickness is $h^{(1)} = h^{(2)} = 1$ mm; the area $A^{(1)} = A^{(2)} = 50$ mm^2; the volume $V^{(1)} = V^{(2)} = 50$ mm^3. The nodes of each element are numbered counterclockwise with the letters i, j, k, whereby the first node i can be chosen arbitrarily (Fig. 4.21d).

3. The entire heating period given is represented in the form of only one interval, $\Delta t = 1$ s.
4. Sequentially, all matrices are calculated for elements 1 and 2.
5. The heat capacity matrix for element 1 is calculated according to (4.3.66):

$$\left[c^{(1)}\right] = \begin{bmatrix} c_{ii} & c_{ij} & c_{ik} \\ c_{ji} & c_{jj} & c_{jk} \\ c_{ki} & c_{kj} & c_{kk} \end{bmatrix} = \frac{c\rho V}{12}\begin{bmatrix} 2 & 1 & 1 \\ 1 & 2 & 1 \\ 1 & 1 & 2 \end{bmatrix}$$
$$= \frac{0.0048 \cdot 50}{12}\begin{bmatrix} 2 & 1 & 1 \\ 1 & 2 & 1 \\ 1 & 1 & 2 \end{bmatrix} = \begin{bmatrix} 0.04 & 0.02 & 0.02 \\ 0.02 & 0.04 & 0.02 \\ 0.02 & 0.02 & 0.04 \end{bmatrix}. \tag{4.3.77}$$

Here, the numbering system of nodes i, j, k can be understood as a local system.

The thermal conductivity matrix $[k^{(1)}]$ at condition $\lambda_x = \lambda_y = \lambda$ and $\alpha_{jk} = \alpha_{ki} = 0$ according to formula (4.3.67) equals

$$\left[k^{(1)}\right] = \begin{bmatrix} k_{ii} & k_{ij} & k_{ik} \\ k_{ji} & k_{jj} & k_{jk} \\ k_{ki} & k_{kj} & k_{kk} \end{bmatrix} = \frac{\lambda h}{4A}\begin{bmatrix} b_ib_i + c_ic_i & b_ib_j + c_ic_j & b_ib_k + c_ic_k \\ b_jb_i + c_jc_i & b_jb_j + c_jc_j & b_jb_k + c_jc_k \\ b_kb_i + c_kc_i & b_kb_j + c_kc_j & b_kb_k + c_kc_k \end{bmatrix}$$
$$+ \frac{\alpha_{ij}L_{ij}h}{6}\begin{bmatrix} 2 & 1 & 0 \\ 1 & 2 & 0 \\ 0 & 0 & 0 \end{bmatrix}. \tag{4.3.78}$$

The constants b and c are calculated according to formulas (4.3.33), and for element 1 (Fig. 4.21d) equal

$$\begin{aligned} &b_i = y_j - y_k = 0 - 10 = -10; \; c_i = x_k - x_j = 10 - 10 = 0; \\ &b_j = y_k - y_i = 10 - 0 = 10; \quad c_j = x_i - x_k = 0 - 10 = -10; \\ &b_k = y_i - y_j = 0 - 0 = 0; \qquad c_k = x_j - x_i = 10 - 0 = 10. \end{aligned} \tag{4.3.79}$$

The substitution of the resulting numerical values b and c and the length of the lower side of triangle L_{ij} in (4.3.78) gives

$$\left[k^{(1)}\right] = \frac{0.04 \cdot 1}{4 \cdot 50}\begin{bmatrix} 100 + 0 & -100 - 0 & -0 + 0 \\ -100 - 0 & 100 + 100 & 0 - 100 \\ -0 - 0 & 0 - 100 & 0 + 100 \end{bmatrix} + \frac{6 \cdot 10^{-5} \cdot 10 \cdot 1}{6}\begin{bmatrix} 2 & 1 & 0 \\ 1 & 2 & 0 \\ 0 & 0 & 0 \end{bmatrix}$$

$$= \begin{bmatrix} 0.02 & -0.02 & 0 \\ -0.02 & 0.04 & -0.02 \\ 0 & -0.02 & 0.02 \end{bmatrix} + \begin{bmatrix} 0.0002 & 0.0001 & 0 \\ 0.0001 & 0.0002 & 0 \\ 0 & 0 & 0 \end{bmatrix}$$

$$= \begin{bmatrix} 0.0202 & -0.0199 & 0 \\ -0.0199 & 0.0402 & -0.02 \\ 0 & -0.02 & 0.02 \end{bmatrix}. \quad (4.3.80)$$

It can be seen that the matrix is symmetric.

The thermal load vector at moments $k - 1 = 0$ and $k = 1$ for element 1 is determined by formula (4.3.68)

$$\{f^{(1)}\}_0 = \{f^{(1)}\}_1 = \begin{Bmatrix} f_i^{(1)} \\ f_j^{(1)} \\ f_k^{(1)} \end{Bmatrix} = \begin{Bmatrix} 0 \\ 0 \\ 0 \end{Bmatrix}, \quad (4.3.81)$$

since for element 1 $q_3 = 0$, $q_2 = 0$ and $T_\infty = 0$.

All matrices for element 2 are determined analogously:

$$[c^{(2)}] = [c^{(1)}] = \begin{bmatrix} 0.04 & 0.02 & 0.02 \\ 0.02 & 0.04 & 0.02 \\ 0.02 & 0.02 & 0.04 \end{bmatrix},$$

$$b_i = 0, b_j = 10, b_k = -10, c_i = -10, c_j = 0, c_k = 10, L_{jk} = 10, \quad (4.3.82)$$

$$[k^{(2)}] = \frac{0.04 \cdot 1}{4 \cdot 50} \begin{bmatrix} 0 + 100 & 0 - 0 & -0 - 100 \\ 0 - 0 & 100 + 0 & -100 + 0 \\ -0 - 100 & -100 + 0 & 100 + 100 \end{bmatrix}$$

$$+ \frac{6 \cdot 10^{-5} \cdot 10 \cdot 1}{6} \begin{bmatrix} 0 & 0 & 0 \\ 0 & 2 & 1 \\ 0 & 1 & 2 \end{bmatrix} = \begin{bmatrix} 0.02 & 0 & -0.02 \\ 0 & 0.0202 & -0.0199 \\ -0.02 & -0.0199 & 0.0402 \end{bmatrix}, \quad (4.3.83)$$

$$\{f^{(2)}\}_0 = \{f^{(2)}\}_1 = \frac{6 \cdot 50}{3} \begin{Bmatrix} 1 \\ 1 \\ 1 \end{Bmatrix} = \begin{Bmatrix} 100 \\ 100 \\ 100 \end{Bmatrix}, \quad (4.3.84)$$

since $q_3 = \text{const}$.

6. The global matrices for the system of elements $[C]$, $[K]$ and $\{F\}$ are compiled by summing the corresponding element matrices $[c^{(e)}]$, $[k^{(e)}]$ and $\{f^{(e)}\}$. This procedure is considered in detail when calculating $[C]$.

The size of matrix $[C]$ is determined by the total number of nodes:

$$[C]=\begin{bmatrix} c_{11} & c_{12} & c_{13} & c_{14} \\ c_{21} & c_{22} & c_{23} & c_{24} \\ c_{31} & c_{32} & c_{33} & c_{34} \\ c_{41} & c_{42} & c_{43} & c_{44} \end{bmatrix}. \tag{4.3.85}$$

Every element c_{mn} of matrix $[C]$ is the sum of corresponding elements of matrix $[c^{(e)}]$ according to (4.3.72). In the case under consideration, the matrices $[c^{(1)}]$ and $[c^{(2)}]$ in the numbering system of nodes i, j, k are known. Inclusion of $[c^{(1)}]$ and $[c^{(2)}]$ into $[C]$ in the global numbering system of nodes can occur with the help of relations $i=3, j=1, k=2$ for element 1, and $i=3, j=2, k=4$ for element 2 (see Fig. 4.21d). Thus, we obtain the following:

$$[C]=\sum_{e=1}^{2}\left[c^{(e)}\right]=\left[c^{(1)}\right]+\left[c^{(2)}\right]=\begin{bmatrix} c_{jj}^{(1)}+0 & c_{jk}^{(1)}+0 & c_{ji}^{(1)}+0 & 0+0 \\ c_{kj}^{(1)}+0 & c_{kk}^{(1)}+c_{jj}^{(2)} & c_{ki}^{(1)}+c_{ji}^{(2)} & 0+c_{jk}^{(2)} \\ c_{ij}^{(1)}+0 & c_{ik}^{(1)}+c_{ij}^{(2)} & c_{ii}^{(1)}+c_{ii}^{(2)} & 0+c_{ik}^{(2)} \\ 0+0 & 0+c_{kj}^{(2)} & 0+c_{ki}^{(2)} & 0+c_{kk}^{(2)} \end{bmatrix}$$

$$=\begin{bmatrix} 0.04+0 & 0.02+0 & 0.02+0 & 0 \\ 0.02+0 & 0.04+0.04 & 0.02+0.02 & 0+0.02 \\ 0.02+0 & 0.02+0.02 & 0.04+0.04 & 0+0.02 \\ 0 & 0+0.02 & 0+0.02 & 0+0.04 \end{bmatrix}$$

$$=\begin{bmatrix} 0.04 & 0.02 & 0.02 & 0 \\ 0.02 & 0.08 & 0.04 & 0.02 \\ 0.02 & 0.04 & 0.08 & 0.02 \\ 0 & 0.02 & 0.02 & 0.04 \end{bmatrix}. \tag{4.3.86}$$

It can be seen that matrix $[C]$ is symmetric and banded; beyond the bandwidth of four diagonals the elements of the matrix equal zero. The bandwidth is determined by the maximum difference in the node numbers for an element.

Analogously, let us include the matrices $[k^{(e)}]$ and $\{f^{(e)}\}$ into the global matrices, as a result of which we obtain the following

$$[K]=\sum_{e=1}^{2}\left[k^{(e)}\right]=\begin{bmatrix} 0.0402+0 & -0.02+0 & -0.0199+0 & 0+0 \\ -0.02+0 & 0.02+0.0202 & 0+0 & 0-0.0199 \\ -0.0199+0 & 0+0 & 0.0202+0.02 & 0-0.02 \\ 0+0 & 0-0.0199 & 0-0.02 & 0+0.0402 \end{bmatrix}$$

$$=\begin{bmatrix} 0.0402 & -0.02 & -0.0199 & 0 \\ -0.02 & 0.0402 & 0 & -0.0199 \\ -0.0199 & 0 & 0.0402 & -0.02 \\ 0 & 0.0199 & -0.02 & +0.0402 \end{bmatrix}; \tag{4.3.87}$$

$$\{F\}_0 = \begin{Bmatrix} F_1 \\ F_2 \\ F_3 \\ F_4 \end{Bmatrix}_0 = \sum_{e=1}^{2} \left\{ f^{(e)} \right\}_0 = \begin{Bmatrix} f_j^{(1)} + 0 \\ f_k^{(1)} + f_i^{(2)} \\ f_i^{(1)} + f_i^{(2)} \\ 0 + f_k^{(2)} \end{Bmatrix}_0$$

$$= \begin{Bmatrix} 0+0 \\ 0+100 \\ 0+100 \\ 0+100 \end{Bmatrix} = \begin{Bmatrix} 0 \\ 100 \\ 100 \\ 100 \end{Bmatrix}, \ \{F\}_1 = \{F\}_0. \tag{4.3.88}$$

7. The global matrices $[C]$, $[K]$ and $[F]$ are formed taking into account all (two) elements.
8. With the help of the calculated matrices $[C]$, $[K]$, $\{F\}_0$ and $\{F\}_1$, according to formulas (4.3.76), a system of Eq. (4.3.75) is finally formed:

$$\left[\overline{K}\right]\{T\}_1 = \left\{\overline{F}\right\}, \tag{4.3.89}$$

where

$$\left[\overline{K}\right] = \begin{bmatrix} 0.0402 & -0.02 & -0.0199 & 0 \\ -0.02 & 0.0402 & 0 & -0.0199 \\ -0.0199 & 0 & 0.0402 & -0.02 \\ 0 & -0.0199 & -0.02 & 0.0402 \end{bmatrix}$$

$$+ \frac{2}{1} \begin{bmatrix} 0.04 & 0.02 & 0.02 & 0 \\ 0.02 & 0.08 & 0.04 & 0.02 \\ 0.02 & 0.04 & 0.08 & 0.02 \\ 0 & 0.02 & 0.02 & 0.04 \end{bmatrix} = \begin{bmatrix} 0.1202 & 0.02 & 0.0201 & 0 \\ 0.02 & 0.2002 & 0.08 & 0.0201 \\ 0.0201 & 0.08 & 0.2002 & 0.02 \\ 0 & 0.0201 & 0.02 & 0.1202 \end{bmatrix}; \tag{4.3.90}$$

$$\left\{\overline{F}\right\} = \begin{Bmatrix} 0 \\ 0 \\ 0 \\ 0 \end{Bmatrix} + \begin{Bmatrix} 0 \\ 100 \\ 100 \\ 100 \end{Bmatrix} + \begin{Bmatrix} 0 \\ 100 \\ 100 \\ 100 \end{Bmatrix} = \begin{Bmatrix} 0 \\ 200 \\ 200 \\ 200 \end{Bmatrix}. \tag{4.3.91}$$

It can be noticed that the matrix $\left[\overline{K}\right]$ is a symmetric, banded, positive definite matrix. Positive definiteness signifies that all elements on the main diagonal are positive. The listed properties of the matrix make it ideal to use in computer technology; the solution of such a system of equations is existent and unique.

9. Let us present the system of Eq. (4.3.89) for unknowns T_1, T_2, T_3, T_4 in expanded form:

$$\left.\begin{aligned}&0.12T_1+0.02T_2+0.0201T_3=0\\&0.02T_1+0.2002T_2+0.08T_3+0.0201T_4=200\\&0.0201T_1+0.08T_2+0.2002T_3+0.02T_4=200\\&0.0201T_2+0.02T_3+0.1202T_4=200\end{aligned}\right\} \tag{4.3.92}$$

In fact, the temperature at nodes 1 and 2 is given according to the condition of the problem. According to Eq. (4.2.42), it equals 61 K at $Q = 2 \times 300$ J and $p = 2$ mm. The system of equations is transformed, taking into account these boundary conditions. To do this, let us set $T_1 = 61$ and $T_2 = 61$ into the system (4.3.92), eliminate the first and second equations, and exclude the first and second columns of the remaining equations, rearranging them to the right side of the system. In the end, a transformed system of equations for unknowns T_3 and T_4 is obtained:

$$\begin{aligned}&0.2002T_3+0.02T_4=193.8;\\&0.02T_3+0.1202T_4=198.8.\end{aligned} \tag{4.3.93}$$

10. The solution to the transformed system of equations is $T_3 = 817$ K, $T_4 = 1518$ K.

The temperature distribution in the selected strip can be constructed with the known temperature values at the nodes (Fig. 4.21e). Based on the possibility of a manual calculation, a cross-section was very roughly divided into elements which, naturally, led to very rough results. Such a class of two-dimensional problems requires discretisation into several hundreds and even thousands of finite elements, while the time interval Δt should be assumed as equal to fractions of a second (Goldak et al. 2011).

Schemes of concentrated heat sources are very popular among welding researchers. Let us examine a triangular element of thickness h with a concentrated heat source $q_1 = q/h$ (W m^{-1}) at point x_0, y_0 (Fig. 4.22a). The power distribution of such a source is described using the δ-function (1.2.6):

$$q_3(x, y) = q/h\delta(x - x_0)\delta(y - y_0). \tag{4.2.94}$$

Then the first integral in (4.3.68) considering the property of the δ-function (1.2.7) is

$$\int\limits_{V^{(e)}} q_3\left[N^{(e)}\right]^T dv = \int\limits_{A^{(e)}} q\begin{Bmatrix}N_i\\N_j\\N_k\end{Bmatrix}\delta(x - x_0)\delta(y - y_0)dxdy = q\begin{Bmatrix}N_i(x_0, y_0)\\N_j(x_0, y_0)\\N_k(x_0, y_0)\end{Bmatrix}. \tag{4.3.95}$$

The rest of the terms in Eqs. (4.3.66)–(4.3.68) remain unchanged.

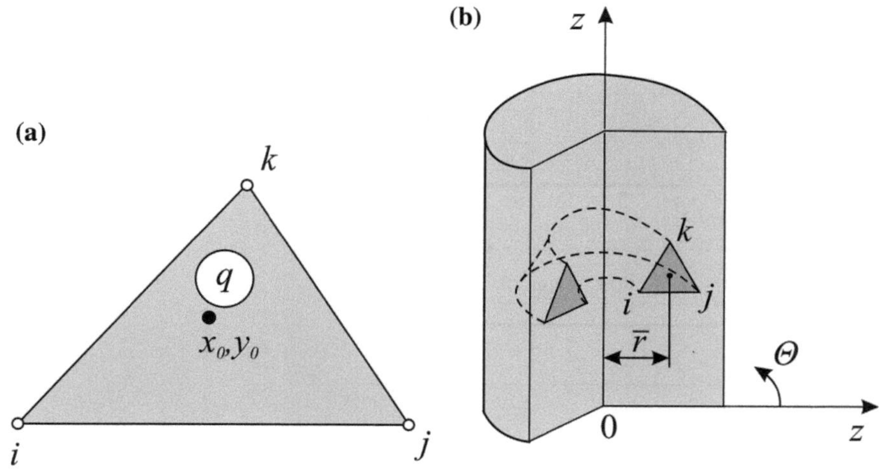

Fig. 4.22 Concentrated heat source in the finite element (**a**) and a two-dimensional axisymmetric (toroidal) element of an axisymmetric solid (**b**)

Let us briefly consider the particularities of calculating the temperature field in bodies of revolution, which make up a large class of welded structures (pipes, cylinders, shells, etc.; Fig. 4.22b). If the material properties, distribution of volumetric sources q_3, initial and boundary conditions do not depend on the circumferential direction (azimuth angle θ), then the temperature field is axisymmetric.

The calculation of the axisymmetric temperature field can be reduced to the already analysed plane problem if x is understood as r, and y as z, and if the thickness of the element is taken as a variable and dependent on radius r, $h = 2\pi r$. In this case, the integration of formulas (4.3.66)–(4.3.68) when calculating the matrices $[c^{(e)}]$, $[k^{(e)}]$ and $\{f^{(e)}\}$ becomes more complicated (Segerlind 1984), insomuch as thickness h is not factored outside the integral sign. If the area of the elements is sufficiently small, then the element thickness can be considered constant and equal to its thickness at the centre of the element with the coordinate $\overline{r} = \left(r_i + r_j + r_k\right)/3$ (Fig. 4.22b). With this approximate approach, the form of formulas (4.3.66)–(4.3.66) is retained if thickness h is understood as quantity $2\pi\overline{r}$.

The accuracy of the FEM is estimated in the example of the manual arc welding of 5 mm rectangular low-carbon steel plates (Fig. 4.23a) (Negoda and Karkhin 1977). Welding conditions: $q = 3270$ W, $v = 4$ mm s^{-1}, the concentration factor of the normally distributed circular source $C = 1.8 \times 10^{-4}$ m^{-2}. The problem was solved by dividing half of the joint into 84 triangular elements. The time step was $\Delta t =$ 0.5 s. The temperature was also calculated using formula (5.2.110) for a moving normally distributed source in a bounded plate. As can be seen in Fig. 4.23b, the deviation of an approximate numerical solution of a linear problem from the exact analytical solution, when λ, a, $c\rho$ and α are assumed as constants, is insignificant (<3%) even with a relatively small number of elements and a large time step Δt,

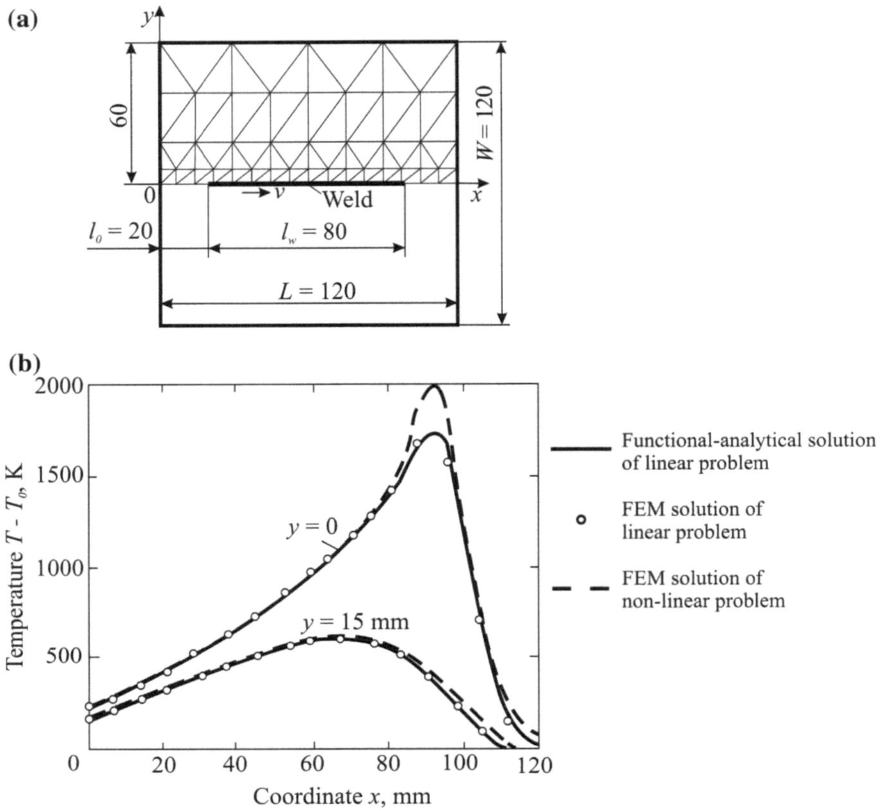

Fig. 4.23 Finite element discretisation of a rectangular plate (symmetry half) for numerical analysis (**a**) and the temperature distribution curves at distance y from the centric weld line at the end of welding ($t_w = 20$ s) (**b**) (Negoda and Karkhin 1977)

which characterises the high accuracy of the FEM. The solution can differ if the problem is formulated as non-linear, when the temperature dependence of λ, $c\rho$ and α is taken into account.

Computer programs for solving problems in welding have been developed on the basis of the FEM (Goldak et al. 2001; Goldak and Akhlaghi 2005; Karkhin 1995; Kurkin et al. 2002; Lindgren 2007; Makarov et al. 2002). Algorithms used in commercial programs, for example, in Ansys, Marc, Abaqus, and Sysweld, are based on FEM. Initially, the codes Sysweld were focused on problems in welding because it allows many (though not all) particularities of the welding process to be considered. When solving three-dimensional welding problems of heat conduction with a present-day computer, tens of thousands and more finite complex elements are used. The required accuracy of the solution to the problem gives the number and types of finite elements and is practically determined both by the accuracy of the input data and by the further use of the calculation results (Goldak et al. 2007).

4.3.3 *Boundary Element Method*

Physical, mechanical, and chemical processes proceed in the solid metal around the moving molten weld pool during the fusion welding. These processes affect considerably the local microstructure and mechanical properties of the weld metal and heat affected zone. In order to analyse these processes, data on the properties of the temperature field (peak temperature and temperature gradient, heating and cooling rates, etc.) is required. The direct measurement of these parameters involves complex experiments. Application of merely computational methods leads to the great difficulties with the description of the heat source and workpiece interaction and heat transfer in the molten weld pool (Kumar et al. 2005).

The inverse modelling of thermal processes based on available information about the weld, heat affected zone, thermal cycle in individual points, etc. is an efficient strategy. If the molten pool geometry is known, the temperature field in the solid part of the workpiece can be determined using the boundary element method (BEM). Using this method when solving such problems is really successful for its special feature, namely, the decrease in the problem dimensionality and simple allowance for large weldments (Banerjee and Butterfield 1981; Brebbia et al. 1984; Hang and Okada 1993; Karkhin et al. 2016; Karkhin et al. 2017a, b; Wrobel and Brebbia 1992).

Let us consider the BEM with the following assumptions: the object under study is a solid part of the homogeneous infinite plate of constant thickness h; thermophysical properties (thermal conductivity λ, thermal diffusivity a, heat capacity per unit volume $c\rho$; upper and lower surface heat transfer coefficients α_1 and α_2) are temperature-independent; the weld pool moves along the x-axis with the constant speed v (Fig. 4.24); coordinates and temperature of the pool boundary Γ are known and constant (temperature field is steady in the moving reference frame); the heat energy is introduced only through the weld pool; initial and ambient temperatures T_0 are constant. Then the two-dimensional heat conduction equation for a thin plate with constant temperature across its thickness in the moving coordinate system becomes similar to (3.3.31):

$$a\left(\frac{\partial^2 T}{\partial x^2} + \frac{\partial^2 T}{\partial y^2}\right) + v\frac{\partial T}{\partial x} - \frac{\alpha_1 + \alpha_2}{\lambda h}(T - T_0) = 0. \tag{4.3.96}$$

The boundary conditions for the solid (Fig. 4.24a):

$$T(x, y) = T_S, \text{ if } x \text{ and } y \text{ belong to the boundary of the pool } \Gamma; \tag{4.3.97}$$

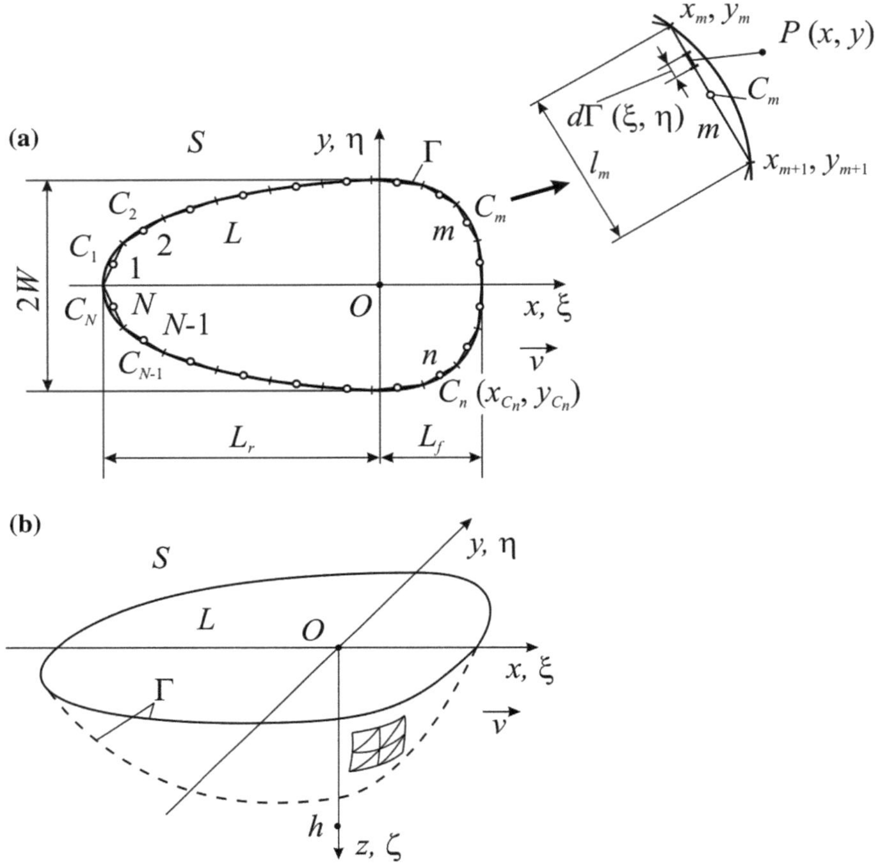

Fig. 4.24 Schematic of the moving weld pool interface Γ division into boundary elements: two-dimensional problem (**a**), three-dimensional problem (division of pool interface fragment into triangular boundary elements is shown) (**b**)

$$T(\pm\infty, y) = T(x, \pm\infty) = T_0, \tag{4.3.98}$$

where T_S is the solidus temperature.

Three-dimensional heat conduction Eq. (3.3.30) for a thick plate (slab) without inner sources and sinks ($q_3 = 0$) takes the form:

$$a\left(\frac{\partial^2 T}{\partial x^2} + \frac{\partial^2 T}{\partial y^2} + \frac{\partial^2 T}{\partial z^2}\right) + v\frac{\partial T}{\partial x} = 0. \tag{4.3.99}$$

The boundary conditions are as follows (Fig. 4.24b):

$$T(x, y, z) = T_S, \text{ if } x, y, \text{ and } z \text{ belong to the pool interface } \Gamma; \tag{4.3.100}$$

$$T(\pm\infty, y, z) = T(x, \pm\infty, z) = T_0; \tag{4.3.101}$$

$$\lambda \frac{\partial T}{\partial z}(x, y, 0) = \alpha_1 (T - T_0) \quad \text{heat transfer at an upper surface } z = 0; \tag{4.3.102}$$

$$\lambda \frac{\partial T}{\partial z}(x, y, h) = -\alpha_2 (T - T_0) \quad \text{heat transfer at a lower surface } z = h. \tag{4.3.103}$$

We shall first consider the two-dimensional problem (4.3.96)–(4.3.98) exemplified by the thin plate welding using a moving source of heat (Fig. 4.24a). A closed Γ-boundary with the set solidus temperature T_S separates the solid part of the weldment S from the liquid and mushy parts L. We shall use the BEM for the homogeneous domain S (Banerjee and Butterfield 1981) and expand the domain S up to infinite domain $S + L$ made of the same material. Then we introduce the fictitious heat sources with unknown linear power density (source power per unit length) $q_1(\xi, \eta)$ into the Γ-curve within the infinite domain $S + L$ using the ξ and η coordinates for the corresponding sources on the Γ-curve. The origins of x, y and ξ, η systems are congruous (Fig. 4.24a). Let us apply the method of sources (see Sect. 4.2.1.1). Now a temperature increase at any point can be defined by the cumulative effect of all the fictitious heat sources along the curve $\Gamma(\xi, \eta)$ (Karkhin et al. 2016, 2017a, b):

$$T(x, y) - T_0 = \int_\Gamma q_1(\xi, \eta) G(x, y, \xi, \eta) d\Gamma(\xi, \eta), \tag{4.3.104}$$

where $G(x, y, \xi, \eta)$ is the fundamental solution (Green's function) to the basic differential equation (4.3.96) for the solid homogeneous infinite plate:

$$\begin{aligned} G(x, y, \xi, \eta) = \frac{1}{2\pi\lambda h} \exp\Big(-\frac{v}{2a}(x - \xi)\Big) \\ \times \mathrm{K}_0\Big(\mu \frac{v}{2a}[(x - \xi)^2 + (y - \eta)^2]^{1/2}\Big); \\ \mu = [1 + 4a(\alpha_1 + \alpha_2)/(c\rho h v^2)]^{1/2}, \end{aligned} \tag{4.3.105}$$

where K_0 is the modified Bessel function of the first kind of order zero (Fig. 3.14a), μ is the coefficient of the surface heat transfer influence. Green's function conforms to the temperature at point x, y due to a line source of unit power acting at point ξ, η (Sect. 5.1.3.4).

To determine unknown function $q_1(\xi, \eta)$ in (4.3.104), the temperature on the Γ-curve at the infinite domain $S + L$ is to be equal to T_S at boundary Γ of the domain S:

$$\frac{1}{2\pi\lambda h} \int_\Gamma q_1(\xi, \eta) \exp\Big(\frac{v}{2a}(x - \xi)\Big)$$

$$\times \mathrm{K}_0\Big(\mu\frac{v}{2a}[(x-\xi)^2+(y-\eta)^2]^{1/2}\Big)d\Gamma(\xi,\eta) = T_S(\xi,\eta) - T_0 \tag{4.3.106}$$

Local T_S temperature varies in some cases, e.g. if base and weld metals are dissimilar or there is a dependence of liquidus temperature on local concentration of liquating elements non-uniformly distributed along the pool tail boundary.

Actually, it is impossible to find the exact solution to integral Eq. (4.3.106) for function q_1, therefore, approximate methods are to be used. Hereby, the closed curve can be approximated by a piecewise polynomial curve (a continuous chain of boundary elements), while the linear power density of each element—by a polynomial distribution. Let us consider a simple computation scheme: a pool boundary is divided into N rectilinear boundary elements (Fig. 4.24a) and linear power density of the fictitious source q_{1m} along each element is constant:

$$q_{1m} = \frac{q_m}{\ell_m};\ \sum_{m=1}^{N} q_m = q, \tag{4.3.107}$$

where q_m is the source power in element m, l_m is the mth element length, q is the total power of all elements (or effective power of the welding source if to neglect the pool surface heat transfer).

The position of each mth element is described by the equation of the straight line which connects terminal points x_m, y_m and x_{m+1}, y_{m+1} (Fig. 4.3a):

$$y = k_m x + b_m;\ k_m = \frac{y_{m+1} - y_m}{x_{m+1} - x_m};\ b_m = y_m - \frac{y_{m+1} - y_m}{x_{m+1} - x_m}x_m. \tag{4.3.108}$$

Therefore, we replace unknown continuous function $q_1(\xi,\eta)$ in Eq. (4.3.106) by a piecewise-constant function (a series of unknown function values $q_{11}, q_{12}, \ldots, q_{1N}$). These unknown values are found from the condition that the total temperature at the midpoint (node) of each element C_n ($n = 1, \ldots, N$) due to each source q_{1m} ($m = 1, \ldots, N$) is equal to the prescribed temperature T_{Sn}. As a result, we have the system of N linear equations with respect to unknown values q_{1n}. Each equation n takes the form:

$$\begin{aligned} &\frac{1}{2\pi\lambda h}\sum_{m=1}^{N} q_{1m}\int_{\ell_m} \exp\Big(-\frac{v}{2a}(x_{C_n} - \xi_m)\Big) \\ &\times \mathrm{K}_0\Big(\mu\frac{v}{2a}[(x_{C_n} - \xi_m)^2 + (y_{C_n} - \eta_m)^2]^{1/2}\Big)dl_m(\xi_m,\eta_m) \\ &= T_{Sn} - T_0, \end{aligned} \tag{4.3.109}$$

where x_{Cn}, y_{Cn} are the coordinates of node C_n. Taking into account properties of the line integral of the first kind, we transform Eq. (4.3.109):

$$\frac{1}{2\pi\lambda h}\sum_{m=1}^{N} q_{1m}(1+k_m^2)^{1/2}\int_{x_m}^{x_{m+1}} \exp\Big(-\frac{v}{2a}(x_{C_n}-\xi)\Big)$$
$$\times \mathrm{K}_0\Big(\mu\frac{v}{2a}\{(x_{C_n}-\xi)^2+[y_{C_n}-(k_m\xi+b_m)]^2\}^{1/2}\Big)d\xi$$
$$= T_{S\,n}-T_0,\quad n=1,\ldots,N. \tag{4.3.110}$$

The derived system of Eq. (4.3.110) for q_{1n} ($n = 1, \ldots, N$) can be solved using the known methods. The temperature in any point x, y inside the solid can be found by a common integration:

$$T(x,y)-T_0=\frac{1}{2\pi\lambda h}\sum_{m=1}^{N} q_{1m}(1+k_m^2)^{1/2}\int_{x_m}^{x_{m+1}} \exp\Big(-\frac{v}{2a}(x-\xi)\Big)$$
$$\times \mathrm{K}_0\Big(\mu\frac{v}{2a}\{(x-\xi)^2+[y-(k_m\xi+b_m)]^2\}^{1/2}\Big)d\xi. \tag{4.3.111}$$

A formula (4.3.111) allows calculating all properties of the temperature field (peak temperature, temperature gradient, cooling rate, etc.).

Further on, let us consider the problem of heat conduction for a thick plate (slab) with a three-dimensional temperature field. The algorithms of solving two- and three-dimensional problems are the same.

Applying the source method (Sect. 4.2.1.1), the solution to the problem (4.3.99)–(4.3.103) will be the following:

$$T(x,y,z)-T_0=\int_{\Gamma} q_2(\xi,\eta,\xi)G(x,y,z,\xi,\eta,\zeta)d\Gamma(\xi,\eta,\zeta), \tag{4.3.112}$$

where Γ is the two-dimensional interface of the weld pool with the given solidus temperature T_S (Fig. 4.24b); $q_2(\xi,\ \eta,\ \zeta)$ is the unknown distribution of the plane power density on the inner surface of the pool Γ at the infinite domain $S + L$; $G(x, y, z, \xi, \eta, \zeta)$ is the fundamental solution to the basic differential equation (4.3.99). With no surface heat transfer (the assumption $\alpha_1 = \alpha_2 = 0$ is valid for the thick bodies) Green's function becomes the following (see Sect. 5.1.3.2):

$$G(x,y,z,\xi,\eta,\zeta)=\frac{1}{4\pi\lambda}\exp\Big(-\frac{v(x-\xi)}{2a}\Big)$$
$$\times\sum_{i=-\infty}^{\infty}\sum_{j=-1,1}\frac{1}{R_{i,j}}\exp\Big(-\frac{vR_{i,j}}{2a}\Big);$$
$$R_{i,j}=[(x-\xi)^2+(y-\eta)^2+(z-j\zeta-2ih)^2]^{1/2}. \tag{4.3.113}$$

Allowing for the surface heat transfer (when $\alpha_1 \neq 0$ and/or $\alpha_2 \neq 0$) Green's function becomes more complicated (see Sect. 5.1.3.2).

The curvilinear surface Γ can be approximated by the set of plane triangular elements (Fig. 4.24b). Within the mth element density of the fictitious source $q_{2\mathrm{m}}$ is taken as constant. Temperature T_{Sm} is prescribed in the centre of the element. The surface integration of each element is carried out numerically (Banerjee and Butterfield 1981; Brebbia et al. 1984; Golovanov 2002).

If the shape of the weldment differs from the plate with constant thickness (e.g. T-joint), the application of the BEM becomes challenging. The equation for a point source of unit power in an infinite body (4.2.82) can perform as Green's function, meanwhile all prescribed boundary conditions are met using special numerical methods (Hang and Okada 1993).

Example *Required*: temperature gradient grad$T$, the cooling rate w ($w = -\partial T/\partial t$) on the boundary of the weld pool, heat source efficiency η_h and temperature field around the pool during the argon tungsten arc welding of the plates made of aluminium alloy 1565 chMU (4 mm thick) under the conditions providing complete fusion: current $I = 110$ A, voltage $U = 12$ V, welding speed $v = 1.83$ mm s^{-1}, $T_0 = 293$ K, $\alpha_1 = \alpha_2 = 0$. Chemical composition of the alloy: 5.5–6.0% Mg, 0.7–0.95% Mn, 0.55–0.9% Zn, 0.05–0.1% Cu, 0.08–0.13% Zr, 0.05–0.1% Cr, 0.22% Fe, 0.12% Si, 0.02% V, 0.05% Ni, 0.0005% Be, 0.02–0.05% Ti. Properties of the alloy: $a = 46$ mm^2 s^{-1}, $\lambda = 0.139$ W mm^{-1} K^{-1}, $T_S = 848$ K.

To solve the two-dimensional problem, the contour of the weld pool (half-width $W = 4.04$ mm, front length $L_f = 5.57$ mm, rear length $L_r = 6.64$ mm) were divided into 200 line elements ($N = 200$). The power density distribution q_{1n} ($n = 1,\ldots, N$) and total power q ($q = 776$ W) were obtained.

The ratio of efficient power q to total power $q_{gross} = UI = 1320$ W defines heat source efficiency, $\eta_h = 776/1320 = 0.5879$ [see formula (1.2.15)].

Value q_{1n} allows calculating the temperature field by using formula (4.3.111) and all its properties including cooling rate w and temperature gradient grad$T$ (grad$T = q_{1n}/(\lambda h)$ in each element n. Figure 4.25 shows that the temperature gradient varies along the interface of the weld pool, it is a maximum at the melting front and minimum at the pool tail end. The cooling rate is changing, and it is a maximum near the weld centreline. The local microstructure and mechanical properties of the metal can be predicted according to the thermal cycle.

The boundary element method has become widespread in solving problems of mathematical physics in the past decades (Banerjee and Butterfield 1981; Brebbia and Walker 1980; Brebbia et al. 1984; Ramachandran 1994). The method is especially effective in solving problems in continuum mechanics, including welded joints (Breyev and Karkhin 1989; Karkhin 1988).

It can be noted that there are other numerical methods for solving problems of heat conduction (those of control volumes, conformal mapping, etc.). The control volume method (Patankar 1980) is very effective when solving both the problem of heat conduction and the problem of hydrodynamics to determine the temperature field and velocity field, for example, in a weld pool and a plasma jet (Frolov et al. 2010; Kumar et al. 2005; Kumar and DebRoy 2007). The method of conformal mapping

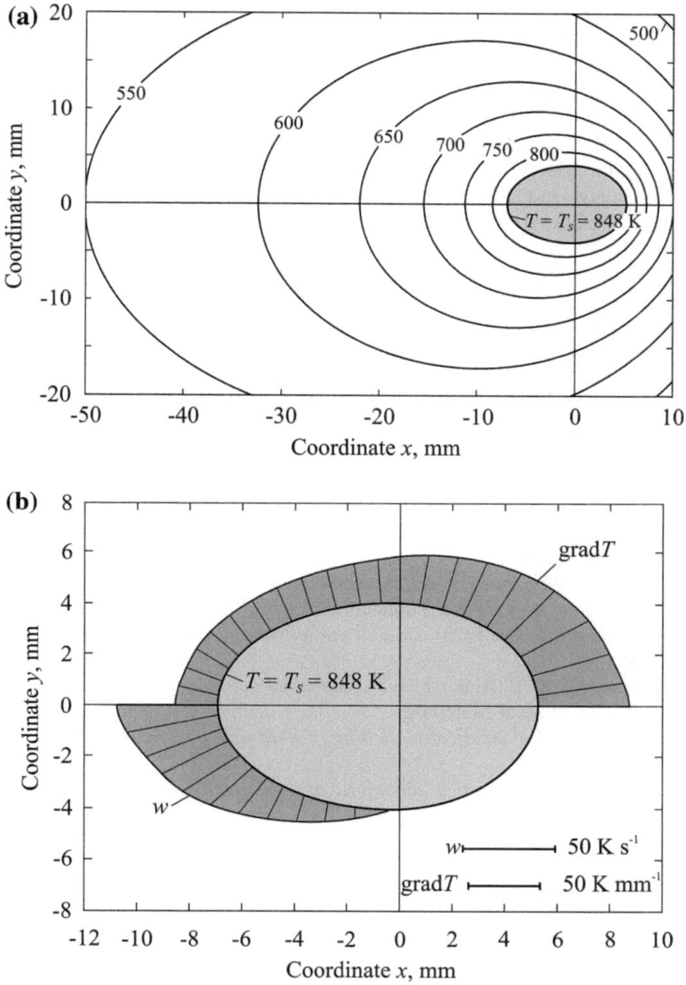

Fig. 4.25 Temperature field (**a**), temperature gradient grad$T$ and cooling rate w along the pool interface (**b**) during the argon tungsten arc welding of a 4 mm thick aluminium alloy plate

is applicable only for the solution of two-dimensional steady-state problems (Boley and Weiner 1960). It has been applied to the solution of problems of heat conduction for an angle welded joint (Jeong and Cho 1997).

The mathematical modelling of welding processes, including thermal, has been developing intensely in the past 30 years (Cerjak and Easterling 1993; Cerjak 1995, 1997, 1998, 2001, 2002; Cerjak et al. 2005, 2007; Cerjak and Enzinger 2010; Sommitsch and Enzinger 2013; Sommitsch et al. 2016). The rapid development of computer technology enables the development of new approaches to solving problems of heat conduction in welding.

References

Abramowitz, M., & Stegun, I. (1965). *Handbook of mathematical functions* (1043 pp.). New York: Dover Publications.

Banerjee, P. K., & Butterfield, R. (1981). *Boundary element methods in engineering science* (452 pp.). London: McGraw-Hill Book Co.

Bateman, H., & Erdelyi, A. (1954). *Tables of integral transforms* (Vol. 1, 454 pp. and Vol. 2, 391 pp.). New York: McGraw-Hill Book Co.

Berezovsky, B. M. (2006). *Mathematical models for arc welding: Vol. 4. Principles of thermal processes in weldments* (547 pp.). Chelyabinsk: Scientific and Technical Information Centre Publishing (in Russian).

Belyaev, N. M., & Ryadno, A. A. (1982). *Methods in theory of heat conduction* (327 pp. (Part 1), 304 pp. (Part 2)). Moscow: Vysshaya Shkola (in Russian).

Boley, B. A., & Weiner, J. H. (1960). *Theory of thermal stresses* (586 pp.). Wiley.

Brebbia, C. A., Telles, J. C. F., & Wrobel, L. C. (1984). *Boundary elements techniques. Theory and applications in engineering* (464 pp.). Berlin: Springer.

Brebbia, C. A., & Walker, S. (1980) *Boundary element techniques in engineering* (210 pp.). London: Newnes Butterworths.

Breyev, V. K., & Karkhin, V. A. (1989). Boundary element analyses of the effects of geometrical shape of welded joints on the crack trajectory and fracture mechanics parameters. *Automatic Welding,* (1), 12–18 (in Russian).

Carslaw, H. S. (1945). *Introduction to the mathematical theory of the conduction of heat in solids* (2nd ed., 264 pp.). New York: Dover Publications.

Carslaw, H. S., & Jaeger, J. C. (1973). *Conduction of heat in solids* (510 pp.). London: Oxford University Press.

Cerjak, H., & Easterling, K. E. (Eds.). (1993). *Mathematical modelling of weld phenomena* (369 pp.). London: The Institute of Materials.

Cerjak, H. (Ed.). (1995). *Mathematical modelling of weld phenomena* (Vol. 2, 287 pp.). London: The Institute of Materials.

Cerjak, H. (Ed.). (1997). *Mathematical modelling of weld phenomena* (Vol. 3, 969 pp.). London: The Institute of Materials.

Cerjak, H. (Ed.). (1998). *Mathematical modelling of weld phenomena* (Vol. 4, 683 pp.). London: IOM Communications.

Cerjak, H. (Ed.). (2001). *Mathematical modelling of weld phenomena* (Vol. 5, 1081 pp.). London: IOM Communications.

Cerjak, H. (Ed.). (2002). *Mathematical modelling of weld phenomena* (Vol. 6, 1140 pp.). London: Maney Publishing.

Cerjak, H., Bhadeshia, H. K. D. H., & Kozeschnik, E. (Eds.). (2005). *Mathematical modelling of weld phenomena* (Vol. 7, 1130 pp.). Graz: Verlag der Technischen Universitaet Graz.

Cerjak, H., Bhadeshia, H. K. D. H., & Kozeschnik, E. (Eds.). (2007). *Mathematical modelling of weld phenomena* (Vol. 8, 1169 pp.). Graz: Verlag der Technischen Universitaet Graz.

Cerjak, H., & Enzinger, N. (Eds.). (2010). *Mathematical modelling of weld phenomena* (Vol. 9, 1048 pp.). Graz: Verlag der Technischen Universitaet Graz.

Frolov, V. Y., Karkhin, V. A., & Ivanov, D. V. (2010). Modelling of plasma torch jet for heating the fine-dispersed powders. In *Proceedings of St. Petersburg State Technical University* (No. 95, pp. 255–264) (in Russian).

Gallagher, R. H. (1975). *Finite element analysis: Fundamentals* (416 pp.). Englewood Cliffs: Prentice-Hall.

Godunov, S. K., & Ryabenky, V. S. (1973). *Difference schemes* (400 pp.). Moscow: Nauka (in Russian).

Goldak, J., Asadi, M., & Karlsson, L. (2011). Numerical aspects of modeling welds. In T. J. Lienert, S. S. Babu, T. A. Siewert, & V. L. Acoff (Eds.), *ASM handbook: Vol. 06A. Welding fundamentals and processes* (pp. 789–796). Materials Park: ASM International.

Goldak, J., Mocanita, M., Aldea, V., Zhou, J., Downey, D., & Zypchen, A. (2001). Computational weld mechanics: Is real-time CWM feasible? Recent progress in CWM. In H. Cerjak (Ed.), *Mathematical modeling of weld phenomena* (Vol. 5, pp. 421–430). London: IOM Communications.

Goldak, J., Zhou, J., Tchernov, S., & Downey, D. (2007). Designer driven analysis of welded structures. In H. Cerjak, H. K. D. H. Bhadeshia, & E. Kozenschnik (Eds.), *Mathematical modelling of weld phenomena* (Vol. 8, pp. 1025–1038). Graz: Verlag der Technischen Universitaet Graz.

Goldak, J. A., & Akhlaghi, M. (2005). *Computational welding mechanics* (321 pp.). Springer.

Golovanov, N. N. (2002). *Geometrical modelling* (472 pp.). Moscow: Fizmatlit (in Russian).

Hang, M., & Okada, A. (1993). Computation of GMAW welding heat transfer with boundary element method. *Advances in Engineering Software, 16*, 1–5.

Janke, E., Emde, F., & Losch, F. (1960). *Tafeln hoeherer Funktionen* (318 pp.). Stuttgart: B. G. Teubner (in German).

Jeong, S. K., & Cho, H. S. (1997). An analytical solution to predict the transient temperature distribution in fillet arc welds. *Welding Journal,* (6), 223-s–232-s.

Kalitkin, N. N. (1978). *Numerical methods* (512 pp.). Moscow: Nauka (in Russian).

Karkhin, V. A. (1988). Application of boundary element method for the calculation of crack trajectory, stress concentration and intensity factors in welded joints. In *Application of mathematical methods in welding* (pp. 40–47). Kiev: The Paton Institute Publishing (in Russian).

Karkhin, V. A. (1990). *Principles of heat transfer in weldments* (100 pp.). Leningrad: Leningrad State Technical University Publishing (in Russian).

Karkhin, V. A. (1995). The WELDEF software for analysis of temperature, stress and strain fields in welding, cladding, cutting, heat treatment and mechanical loading. In *Proceedings of Tula Technical University "CAD/CAM and Expert Systems in Welding"* (pp. 51–59) (in Russian).

Karkhin, V. A., Khomich, P. N., Fedotov, B. V., & Rajamaki, P. (2008). Analysis of thermal cycles in resistance flash welding of steels. *Welding International, 22*(12), 903–908.

Karkhin, V. A., Khomich, P. N., & Ivanov, S. Y. (2007) Calculation of temperature distribution during flash welding of steel. In *Proceedings of 2nd International Conference on Computer Technologies in Joining of Materials-2007* (7 pp.). Tula: Tula State University (in Russian).

Karkhin, V. A., Khomich, P. N., & Ivanov, S. Y. (2016). Calculation of thermal processes around moving weld pool by boundary element method. *Welding and Diagnostics,* (4), 32–35 (in Russian).

Karkhin, V. A., Khomich, P. N., Ginzburg, S. A., & Michailov, V. G. (2017a). Study of thermal processes around moving weld pool by boundary element method. *Welding Production,* (8), 3–8 (in Russian).

Karkhin, V. A., Khomich, P. N., Panchenko, O. V., & Ivanov, S. Y. (2017b) Calculation of thermal processes around moving weld pool using boundary element method. *Voprosy Materialovedeniya,* (2), 208–216 (in Russian).

Karkhin, V. A., & Marx, O. Y. (1994). Analysis of influence of welded joint heterogeneity on hydrogen concentration in heat-affected zone. *Automatic Welding,* (2), 9–14 (in Russian).

Kartashov, E. M. (2001). *Analytical methods in theory of heat conduction in solids* (550 pp.). Moscow: Vysshaya Shkola (in Russian).

Kiselev, S. N., Shevelev, G. N., Roshchin, V. V., Zelentsov, G. N., & Khavanov, V. A. (1981). *Tube joints of dissimilar metals* (176 pp.). Moscow: Mashinostroenie (in Russian).

Kozdoba, L. A. (1976). *Solution of non-linear problems of heat conduction* (136 pp.). Kiev: Naukova Dumka (in Russian).

Krutz, G. W., & Segerling, L. J. (1978). Finite element analysis of welding structures. *Welding Journal,* (7), 211-s–216-s.

Kumar, A., & DebRoy, T. (2007). Heat transfer and fluid flow during gas-metal-arc fillet welding for various joint configurations and welding positions. *Metallurgical and Materials Transactions A, 38*(3), 506–519.

Kumar, A., Zhang, W., Kim, C.-H., & DebRoy, T. (2005). A smart bi-directional model of heat transfer and free surface flow in gas metal arc fillet welding for practicing engineers. In H. Cerjak,

H. K. D. H. Bhadeshia, & E. Kozeschnik (Eds.), *Mathematical modelling of weld phenomena* (Vol. 7, pp. 3–37). Graz: Verlag der Technischen Universitaet Graz.
Kurkin, S. A., Khovov, V. M., & Aksenov, Y. N. (2002). *Computer-aided design and preparation for production of welded structures* (464 pp.). Moscow: Moscow State Technical University Publishing (in Russian).
Lindgren, L.-E. (2001a). Finite element modelling and simulation of welding. Part 1: Increased complexity. *Journal of Thermal Stresses, 24,* 141–192.
Lindgren, L.-E. (2001b). Finite element modelling and simulation of welding. Part 2: Improved material modeling. *Journal of Thermal Stresses, 24,* 195–231.
Lindgren, L.-E. (2001c). Finite element modelling and simulation of welding. Part 3: Efficiency and integration. *Journal of Thermal Stresses, 24,* 305–334.
Lindgren, L.-E. (2007). Computational welding mechanics. *Thermomechanical and microstructural simulations* (248 pp.). Cambridge: Woodhead Publishing Ltd.
Luikov, A. (1968). *Analytical heat conduction theory* (702 pp.). Elsevier.
Makarov, E. L., Kurkin, A. S., Kiselev, A. S., & Kiselev, A. S. (2002). The "WELD.PM" software suite for modelling of thermal strain and fracture processes during the manufacture and service of welded structures. In H. Cerjak (Ed.), *Mathematical modelling of weld phenomena* (Vol. 6, pp. 1101–1111). London: Maney Publishing.
Makhnenko, V. I. (1967). Calculation of thermal processes in butt welding of dissimilar plates. *Physics and Chemistry of Materials Processing,* (6), 78–82 (in Russian).
Makhnenko, V. I. (1976). *Calculation methods for the investigation of the kinetics of welding stresses and deformations* (320 pp.). Kiev: Naukova Dumka (in Russian).
Makhnenko, V. I., & Kravtsov, T. G. (1976). *Thermal processes in mechanized deposition on circular cylinder-shaped workpieces* (159 pp.). Kiev: Naukova Dumka (in Russian).
Makhnenko, V. I., Petun, L. A., Prilutsky, V. P., & Zamkov V. M. (1969). Estimation of thermal processes in the vicinity of moving weld pool. *Automatic Welding,* (11), 1–6 (in Russian).
Negoda, E. N., & Karkhin, V. A. (1977). Calculation by finite element method of temperature field in welded structures. *Electrotechnical Industry, Series Welding, 6*(45), 5–6 (in Russian).
Nickell, R. E., & Hibbitt, H. D. (1975). Thermal and mechanical analysis of welded structures. *Nuclear Engineering and Design,* (32), 110–120.
Ottosen, N. S., & Petersson, H. (1992). *Introduction to the finite element method* (410 pp.). Harlow: Prentice Hall.
Patankar, S. (1980). *Numerical heat transfer and fluid flow* (214 pp.). CRC Press.
Pavelic, V., Tanbakuchi, R., Uyehara, O. A., & Myers, P. S. (1969). Experimental and computed temperature histories in gas tungsten-arc welding of thin plates. *Welding Journal,* (7), 295-s–305-s.
Prudnikov, A. P., Brychkov, Y. A., & Marichev, O. I. (1986). *Integrals and series. Elementary functions* (798 pp.). London: Taylor and Francis.
Ramachandran, P. A. (1994). *Boundary element method in transport phenomena* (406 pp.). Southampton: Computation Mechanics Publication.
Rao, S. S. (1999). *The finite element method in engineering* (3rd ed., 556 pp.). Boston: Butterworth Heinemann.
Roberts, O. F. T. (1923). The theoretical scattering of smoke in a turbulent atmosphere. *Proceeding of the Royal Society (London), Series A, 104,* 640–655.
Rosenthal, D. (1935). Etude theoretique du regime thermique pendant la soudure a l'arc. In *2 Congres National des Sciences* (pp. 1277–1292). Brussels (in French).
Rosenthal, D. (1941). Mathematical theory of heat distribution during welding and cutting. *Welding Journal, 20*(5), 220-s–234-s.
Rykalin, N. N. (1938a). Temperature distribution in welded joints. *Avtogennoe Delo,* (5), 6–14 (in Russian).
Rykalin, N. N. (1938b). Temperature distribution in welded joints. *Avtogennoe Delo,* (7), 19–24 (in Russian).
Rykalin, N. N. (1941). *Three-dimensional distribution in arc welding* (55 pp.). Moscow: Publication of the USSR Academy of Sciences.

Rykalin, N. N. (1947). *Thermal principles in welding. Part 1. Heat flow in arc welding* (271 pp.). Moscow: Publication of the USSR Academy of Sciences (in Russian).

Rykalin, N. N. (1951). *Calculation of heat flow in welding* (Z. Paley & C. M. Adams, Jr., Trans.) (337 pp.). Moscow.

Rykalin, N. N. (1952). *Die Waermegrundlagen des Schweissvorganges* (265 pp.). Berlin: Verlag Technik (in German).

Rykalin, N. N. (1957). *Berechnung der Waermevorgaenge beim Schweissen* (326 pp.). Berlin: VEB Verlag Technik (in German).

Samarskii, A. A. (1987). *An introduction to numerical methods* (286 pp.). Moscow: Nauka (in Russian).

Samarskii, A. A. (2001). *The theory of difference schemes* (788 pp.). CRC Press Book.

Samarskii, A. A., & Vabishchevich, P. N. (1995a). *Computational heat transfer: Vol. 1. Mathematical modelling* (406 pp.). Chichester: Wiley.

Samarskii, A. A., & Vabishchevich, P. N. (1995b). *Computational heat transfer: Vol. 2. The finite difference methodology* (421 pp.). Chichester: Wiley.

Samarskii, A. A., & Vabishchevich, P. N. (2000). *Numerical methods for solving inverse problems of mathematical physics* (438 pp.). Berlin: Walter de Gruyter GmbH.

Segerlind, L. J. (1984). *Applied finite element analysis* (2nd ed., 425 pp.). New York: Wiley.

Sommitsch, C., & Enzinger, N. (Eds.). (2013). *Mathematical modelling of weld phenomena* (Vol. 10, 755 pp.). Graz: Verlag der Technischen Universitaet Graz.

Sommitsch, C., Enzinger, N., & Mayr, P. (Eds.). (2016). *Mathematical modelling of weld phenomena* (Vol. 11, 897 pp.). Graz: Verlag der Technischen Universitaet Graz.

Strang, G., & Fix, G. J. (1973). *An analysis of the finite element method* (306 pp.). Englewood Cliffs: Prentice-Hall.

Sudnik, V. A., & Yerofeyev, V. A. (1986). *Computer calculations of welding processes* (100 pp.). Tula: Publishing House of the Technical University (in Russian).

Sudnik, V. A., & Yerofeyev, V. A. (1988). *Computer methods for research of welding processes* (94 pp.). Tula: Publishing House of the Technical University (in Russian).

Tikhonov, A. N., & Samarskii, A. A. (1990). *Equations of mathematical physics* (765 pp.). New York: Dover Publications.

Wichelhaus, G. (1975). Berechnung der Stossflachentemperatur beim Reibschweissen. *Schweissen und Schneiden*, (1), 9–10 (in German).

Wilson, H. A. (1904). On convection of heat. *Cambridge Philosophical Society, 12*, 406–423.

Wrobel, L. C., & Brebbia, C. A. (Eds.). (1992). *Boundary element methods in heat transfer* (296 pp.). London: Elsevier Applied Science.

Yazovskikh, V. M., & Belenky, V. Y. (2011). Thermal processes in surfacing of cylindrical solids with strip electrode. *Welding Production*, (12), 20–24 (in Russian).

Zienkiewicz, O. C. (1971). *The finite element method in engineering science* (521 pp.). London: McGraw-Hill.

Zienkiewicz, O. C., & Morgan, K. (1983). *Finite elements and approximation* (328 pp.). New York: Wiley-Interscience Publication.

Zienkiewicz, O. C., & Taylor, R. L. (2000). *The finite element method* (5th ed., Vol. 1, 689 pp., Vol. 2, 459 pp.). Oxford: Butterworth Heinemann.

Chapter 5
Temperature Fields in Fusion Welding

5.1 Temperature Fields of Concentrated Sources

5.1.1 Instantaneous Concentrated Sources

Instantaneous concentrated sources are extremely important in solving the problems of heat conduction, since they are the basis for the construction of welding heat sources, distributed in space and time. Besides, they can be models for highly concentrated pulsed sources, whose action time is relatively small (fractions of a second), and the power is large enough to melt the metal. Below we consider five examples: an instantaneous point source on a semi-infinite solid, an instantaneous point source on a slab, an instantaneous line source in a thin plate, an instantaneous plane source in a rod of a finite length and a composite rod.

5.1.1.1 Instantaneous Point Source on a Semi-infinite Solid

The temperature field due to an instantaneous point source on a semi-infinite solid with a heat-insulated surface is described by Eq. (4.2.11). This equation is represented in dimensionless form:

$$\theta_3(\tau) = \frac{1}{\tau^{3/2}} \exp\left(-\frac{1}{4\tau}\right), \tag{5.1.1}$$

where

$$\theta_3 = \frac{4\pi^{3/2} c\rho R^3}{Q_0}(T - T_0); \quad \tau = \frac{at}{R^2}. \tag{5.1.2}$$

V. A. Karkhin, *Thermal Processes in Welding*, Engineering Materials,
https://doi.org/10.1007/978-981-13-5965-1_5

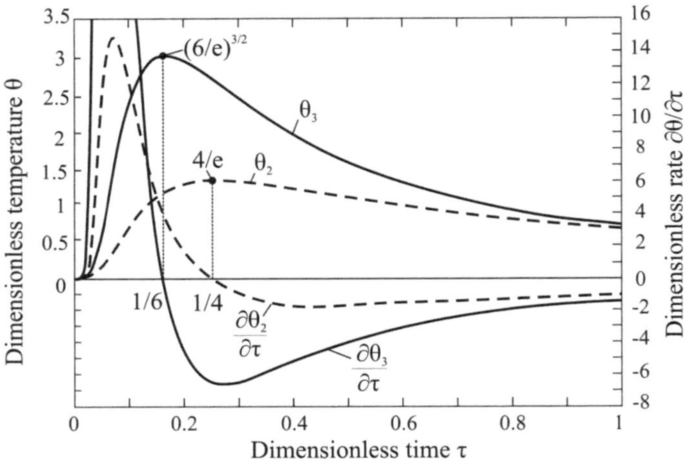

Fig. 5.1 Thermal cycles θ (τ) and dimensionless heating/cooling rates $\partial\theta/\partial\tau$ under heating by an instantaneous point source on a semi-infinite solid (θ_3) and an instantaneous line source in a plate (θ_2)

Here θ_3 is the dimensionless temperature (proportional to real temperature T), τ is the dimensionless time (the Fourier number Fo). The function $\theta_3(\tau)$ is shown in Fig. 5.1. It generalises surfaces, one of which is presented in Fig. 4.3b.

A dimensionless heating and cooling rate can be easily obtained by differentiating expression (5.1.1):

$$\frac{\partial\theta_3}{\partial\tau} = \frac{1}{\tau^{5/2}}\left(\frac{1}{4\tau} - \frac{3}{2}\right)\exp\left(-\frac{1}{4\tau}\right); \tag{5.1.3}$$

$$\frac{\partial\theta_3}{\partial\tau} = \frac{4\pi^{3/2}c\rho R^5}{Q_0 a}\cdot\frac{\partial T}{\partial t}. \tag{5.1.4}$$

The time of reaching the peak temperature and its value are determined from the condition $\partial\theta_3/\partial\tau = 0$. Hence, we obtain the root $\tau = 1/6$ and $\theta_{3\,\max} = (6/\mathrm{e})^{3/2}$ (Fig. 5.1). By substituting $\theta_{3\,\max}$ in (5.1.2), the radial distribution of the peak temperature is obtained:

$$T_{\max}(R) - T_0 = \frac{1}{4}\left(\frac{6}{\pi \mathrm{e}}\right)^{3/2}\frac{Q_0}{c\rho}\frac{1}{R^3}. \tag{5.1.5}$$

It is seen that the radial distribution is described by a cubic hyperbola.

5.1.1.2 An Instantaneous Point Source on a Slab

The temperature field due to an instantaneous point source on a heat-insulated slab is described by Eq. (4.2.28). This equation is represented in dimensionless form:

$$\theta_3(\rho_2, \zeta, \tau) = \frac{1}{\tau}\exp\left(-\frac{\rho_2^2}{4\tau}\right)\left[1 + 2\sum_{i=1}^{\infty}\cos(\pi i \zeta)\exp(-i^2\pi^2\tau)\right], \tag{5.1.6}$$

where

$$\theta_3 = 4\pi c\rho\frac{h^3}{Q_0}(T - T_0); \tag{5.1.7}$$

$$\rho_2 = \frac{\sqrt{x^2 + y^2}}{h}; \quad \zeta = \frac{z}{h}; \quad \tau = \frac{at}{h^2}. \tag{5.1.8}$$

Here θ_3 is the dimensionless temperature, ρ_2 is the dimensionless plane radius vector, ζ is the dimensionless depth of an observation point, τ is the dimensionless time (Fourier number Fo).

The thermal cycles of some points in upper ($\zeta = 0$) and lower ($\zeta = 1$) surfaces of the slab are shown in Fig. 5.2. It can be seen that as the distance to the source increases (with increasing ρ_2), the temperature drops. The peak temperature in the upper surface is higher than that in the lower surface, and it occurs the later, the farther the observation point is from the source.

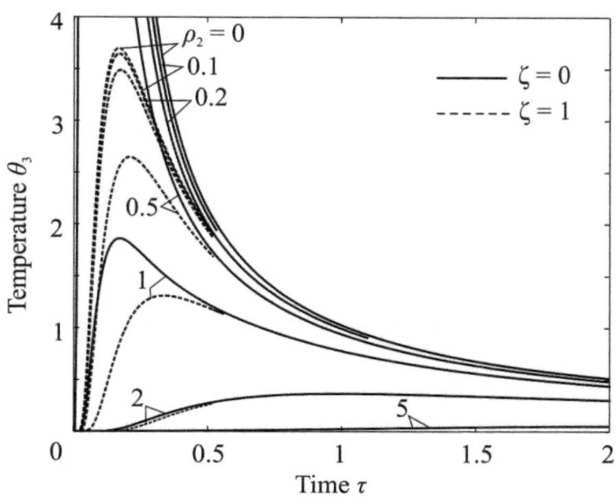

Fig. 5.2 Thermal cycles at the upper ($\zeta = 0$) and lower ($\zeta = 1$) surfaces of a slab after heating by an instantaneous point source

5.1.1.3 Instantaneous Line Source in a Plate

The temperature field around an instantaneous line source in a plate with a heat-insulated surface is described by Eq. (4.2.37). This equation is represented in dimensionless form:

$$\theta_2(\tau) = \frac{1}{\tau}\exp\left(-\frac{1}{4\tau}\right), \tag{5.1.9}$$

where

$$\theta_2 = \frac{4\pi c\rho r^2}{Q/h}(T - T_0);\ \tau = \frac{at}{r^2}. \tag{5.1.10}$$

The function $\theta_2(\tau)$ generalises surfaces, one of which is shown in Fig. 4.9b.

The heating and cooling rate follows from Eq. (5.1.9):

$$\frac{\partial\theta_2}{\partial\tau} = \frac{1}{\tau^2}\left(\frac{1}{4\tau} - 1\right)\exp\left(-\frac{1}{4\tau}\right); \tag{5.1.11}$$

$$\frac{\partial\theta_2}{\partial\tau} = \frac{4\pi c\rho r^4}{Q/ha}\cdot\frac{\partial T}{\partial t}. \tag{5.1.12}$$

It follows from (5.1.11) that the heating (cooling) rate $\partial\theta_2/\partial\tau = 0$ at $\tau = 1/4$ and the peak temperature $\theta_{2\,max} = \theta_2(1/4) = 4/\mathrm{e}$ (Fig. 5.1). By substituting $\theta_{2\,max}$ in (5.1.10), the radial distribution of the peak temperature is obtained:

$$T_{max} - T_0 = \frac{1}{\pi \mathrm{e}}\frac{Q/h}{c\rho}\frac{1}{r^2}. \tag{5.1.13}$$

It can be seen that the radial distribution is described by a hyperbola squared.

The temperature characteristics are shown in Fig. 5.1. It can be seen, that the cooling rate in the plate $\partial\theta_2/\partial\tau$ is much lower, than in the semi-infinite solid $\partial\theta_3/\partial\tau$.

Similarly, the cooling rate and peak temperature can be found from Eq. (4.2.39), taking into account the surface heat transfer.

5.1.1.4 Instantaneous Plane Source in a Rod of Finite Length

The temperature distribution in a rod of an infinite length due to an instantaneous plane source is described by Eq. (4.2.42). By using the method of images, this equation allows us to obtain a solution for a rod of the finite length L with heat-insulated ends (Fig. 5.3).

Let the instantaneous plane source of energy Q be located in the cross section $x = l$. We extend the length of the rod to infinity ($-\infty < x < \infty$) and introduce additional sources into it so that the planes $x = 0$ and $x = L$ will be the planes of

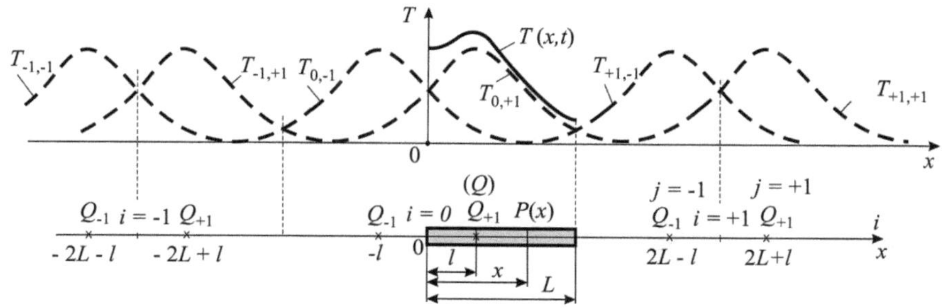

Fig. 5.3 Temperature distribution in a bounded rod

symmetry (the actual source is mirrored from the planes $x = 0$ and $x = L$) (Fig. 5.3). As a result we obtain an infinite series of source pairs Q_1 and Q_{-1} ($Q_1 = Q_{-1} = Q$) on the x-axis ($-\infty < x < \infty$). All fluxes on the left and on the right are the same on the planes of symmetry $x = 0$ and $x = L$ (the algebraic sum of the fluxes is zero), so the boundary conditions are satisfied. Subsequently, the temperature at any instant of time t at any point $P(x)$ in the bounded rod is defined by the sum of temperature increments induced by all sources in the infinite rod $T_{i,j} - T_0$ (Karkhin 1990):

$$T(x,t) - T_0 = \frac{Q/A}{c\rho(4\pi at)^{1/2}} \exp(-bt) \sum_{i=-\infty}^{\infty} \sum_{j=-1,1} \exp\left(-\frac{[x - jl - 2iL]^2}{4at}\right). \tag{5.1.14}$$

5.1.1.5 Instantaneous Plane Source in a Composite Rod

The temperature fields have certain specific features if dissimilar metals are welded. Consider the thermal processes under the action of an instantaneous plane source in a rod consisting of two dissimilar semi-bounded rods. This combination of a source and a solid can occur both in the case of butt welding of a composite plate by a rapidly moving line source (Sect. 5.1.4.4) or in the case of pressure welding of dissimilar rods.

Generally, the cross-sectional areas of rods A_1 and A_2, volumetric heat capacities $c_1\rho_1$ and $c_2\rho_2$, thermal conductivities λ_1 and λ_2, thermal diffusivities a_1 and a_2, and the coefficients of heat loss of material of the rods b_1 and b_2 may differ. Let the energy Q be generated on the contact surface $x = 0$ at the time $t = 0$ (Fig. 5.4a). The temperature distribution is found at any instant of time t, in accordance with the paper (Okamoto 1929).

The initial temperature T_0 is assumed to be constant and there is no thermal resistance on the contact surfaces $x = 0$, i.e. the temperature and flux are continuous:

$$T_1(0,t) = T_2(0,t); \tag{5.1.15}$$

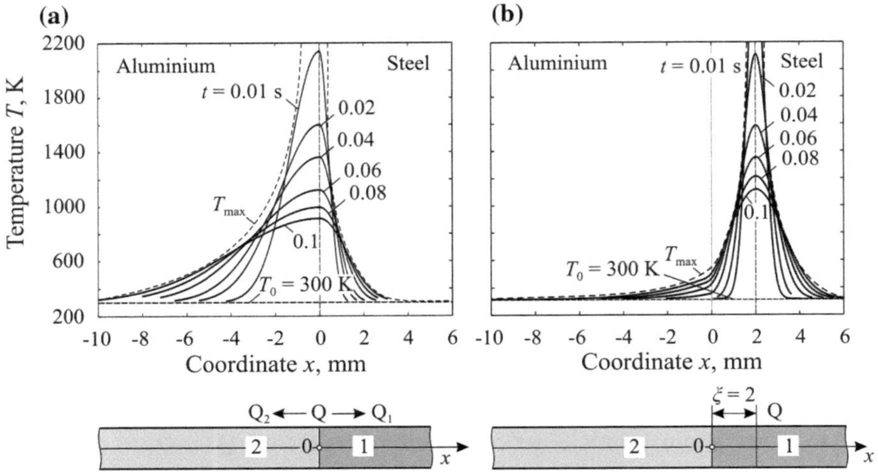

Fig. 5.4 Temperature distribution in a composite rod with an instantaneous plane source without displacement ($\xi = 0$) (**a**) and with displacement ($\xi = 2$ mm) (**b**) from the aluminium/steel interface ($x = 0$) at heat density $Q/A = 12.73$ J mm^{-2}

$$A_1\lambda_1\frac{dT_1(0,t)}{dx} = A_2\lambda_2\frac{dT_2(0,t)}{dx}. \tag{5.1.16}$$

Energy Q is divided into two parts, $Q = Q_1 + Q_2$, with each of them going into the corresponding rod (Fig. 5.4a). Then, according to Eq. (4.2.42), the temperature in each semi-bounded rod is expressed as follows:

$$T_1(x,t) - T_0 = \frac{2Q_1/A_1}{c_1\rho_1(4\pi a_1 t)^{1/2}}\exp\left(-\frac{x^2}{4a_1 t} - b_1 t\right); \tag{5.1.17}$$

$$T_2(x,t) - T_0 = \frac{2Q_2/A_2}{c_2\rho_2(4\pi a_2 t)^{1/2}}\exp\left(-\frac{x^2}{4a_2 t} - b_2 t\right). \tag{5.1.18}$$

The condition is found under which the temperature in the section $x = 0$ (5.1.15) is the same. It follows from (5.1.17) and (5.1.18) that

$$\frac{Q_1}{Q_2} = \frac{A_1(c_1\rho_1\lambda_1)^{1/2}}{A_2(c_2\rho_2\lambda_2)^{1/2}}\exp((b_1 - b_2)t). \tag{5.1.19}$$

It can be seen that the energy distribution between the rods depends on time t if $b_1 \neq b_2$. The solution to the considered problem is given in book (Kiselev et al. 1981).

If it is assumed that $b_1 = b_2 = b$, then the fraction of energy in the jth rod ($j = 1, 2$) will be constant:

$$Q_j = Q\frac{A_j\sqrt{c_j\rho_j\lambda_j}}{A_1\sqrt{c_1\rho_1\lambda_1} + A_2\sqrt{c_2\rho_2\lambda_2}}. \tag{5.1.20}$$

By substituting (5.1.20) in (5.1.17) and (5.1.18), the temperature distribution in the composite rod is obtained:

$$T_j(x,t) - T_0 = \frac{Q}{A_1\sqrt{c_1\rho_1\lambda_1} + A_2\sqrt{c_2\rho_2\lambda_2}}\frac{1}{\sqrt{\pi t}}\exp\left(\frac{x^2}{4a_jt} - bt\right). \tag{5.1.21}$$

If all the characteristics of the rods are equal, then (5.1.21) turns into (4.2.42).

The distribution of the peak temperature, the cooling rate, and other characteristics of the temperature field can be obtained from formula (5.1.21).

Figure 5.4a shows the temperature distribution for welding steel and aluminium rods of the same cross section ($A_1 = A_2 = A$). The thermophysical properties of the materials are very different ($a_1/a_2 = 8/85 = 0.094, \lambda_1/\lambda_2 = 0.04/0.23 = 0.17$). The fraction of energy that is accounted for aluminium is 64% according to Eq. (5.1.20). It can be seen from the figure that at any instant of time t the distributions of the current temperature $T(x)$ and the peak temperature $T_{\max}(x)$ are asymmetric with respect to the material interface. The distribution is gently sloping in aluminium due to its higher thermal diffusivity (the heated zone is wider) and it is steep in steel.

In order to use the method of sources to solve the problem for a composite rod with a source distributed along the axis, we must have a solution for an instantaneous concentrated source located at any distance from the material interface. Assume that the cross-section of the rod is constant ($A_1 = A_2 = A$) and there is no heat transfer ($b_1 = b_2 = 0$). Suppose that an instantaneous plane source of energy Q acts in the region $x > 0$ in plane ξ at the instant of time $t = 0$. Then the solution to the problem with boundary conditions (5.1.15) and (5.1.16) are the following expressions (Carslaw and Jaeger 1973):

$$\begin{aligned} T_1(x,t) - T_0 &= \frac{Q/A}{c_1\rho_1}\frac{1}{\sqrt{4\pi a_1 t}}\exp\left(-\frac{(x-\xi)^2}{4a_1t}\right) \\ &+ \frac{Q/A}{c_1\rho_1}\frac{\left(\lambda_1 a_2^{1/2} - \lambda_2 a_1^{1/2}\right)}{\left(\lambda_1 a_2^{1/2} - \lambda_2 a_1^{1/2}\right)\sqrt{4\pi a_1 t}}\exp\left(-\frac{(x+\xi)^2}{4a_1t}\right); \end{aligned} \tag{5.1.22}$$

$$\begin{aligned} T_2(x,t) - T_0 = \frac{2Q/A}{c_1\rho_1}&\frac{\lambda_1 a_2/a_1^{1/2}}{\left(\lambda_1 a_2^{1/2} + \lambda_2 a_1^{1/2}\right)\sqrt{4\pi a_2 t}} \\ &\times \exp\left(-\frac{\left(x - \xi\sqrt{a_2/a_1}\right)^2}{4a_2t}\right). \end{aligned} \tag{5.1.23}$$

Solution (5.1.22) in the region $x > 0$ is the sum of the temperature increments in the infinite homogeneous rod with properties λ_1, a_1, c_1, and ρ_1 due to the energy source Q acting in the cross section $x = \xi$, and the additional energy source

$Q(\lambda_1 a_2^{1/2} - \lambda_2 a_1^{1/2})/(\lambda_1 a_2^{1/2} + \lambda_2 a_1^{1/2})$ acting in the cross section $x = -\xi$ (in the "image" position). Solution (5.1.23) in the region $x < 0$ is the solution for the energy source $2Q\lambda_1 a_2/a_1^{1/2}/\left(\lambda_1 a_2^{1/2} + \lambda_2 a_1^{1/2}\right)$ acting in the cross section $x = \xi(a_2/a_1)^{1/2}$.

If the source is located in the boundary plane $x = 0$ ($\xi = 0$), then Eqs. (5.1.22) and (5.1.23) turn into (5.1.21).

Figure 5.4b shows the distributions of the current and peak temperatures in the composite rod with a displaced source. It is seen that the temperature gradients change sharply at the material interface.

Solutions are also known to similar heat conduction problems (1) for an unbounded composite rod, when the instantaneous source is located in the plane $x = \xi$ and there is a thermal resistance at the contact surface $x = 0$, (2) for a rod consisting of two bounded rods, when the source is located in the plane $x = \xi$ and there is no thermal resistance on the contact surface $x = 0$ (Carslaw and Jaeger 1973), (3) for an unbounded composite rod consisting of three dissimilar materials (Sommerfeld 1894).

5.1.2 Stationary Continuous Concentrated Sources

A distinguishing feature of fusion welding is that heat is generated for a certain period of time by the source being either moving or stationary.

Let us define the temperature field at the moment when the source is acting. A continuously acting source can be represented as a set of consecutive elementary instantaneous sources (Fig. 5.5). Subsequently, the temperature field is determined by summing the fields caused by all instantaneous sources (by integration with respect to the time). If integrals cannot be taken at a varying power $q(t)$, they are calculated numerically. Further it is assumed that the power of the source is constant, $q = \text{const}$, unless specified otherwise.

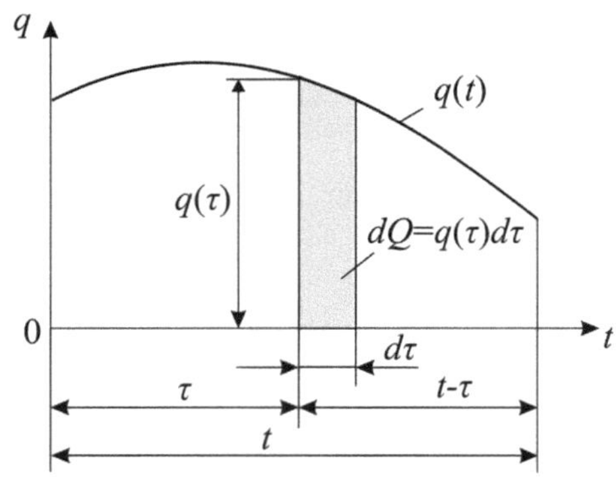

Fig. 5.5 Subdivision of the time during which a continuous source acts

5.1.2.1 Stationary Continuous Point Source in a Semi-infinite Body

Let a point source of power q_0 act at time τ over period $d\tau$ on the surface of a semi-infinite solid at the origin of coordinates ($\xi = \eta = \zeta = 0$). The temperature increment due to this elementary instantaneous source at time t is determined by formula (4.2.11):

$$dT(R,t) = \frac{2dQ_0}{c\rho[4\pi a(t-\tau)]^{3/2}} \exp\left(-\frac{R^2}{4a(t-\tau)}\right), \tag{5.1.24}$$

where $dQ_0 = q_0 d\tau$ is the heat of an elementary source, R is the distance from the point of observation to the source, $t - \tau$ is the heat diffusion time of this source (see Fig. 5.5). Then the temperature is determined by integrating all sources over time:

$$\begin{aligned} T(R,t) - T_0 &= \int_0^t \frac{2dQ_0}{c\rho[4\pi a(t-\tau)]^{3/2}} \exp\left(-\frac{R^2}{4a(t-\tau)}\right) d\tau \\ &= \frac{q_0}{2\pi\lambda R} \Phi^*\left(\frac{R}{\sqrt{4at}}\right). \end{aligned} \tag{5.1.25}$$

As time passes, the temperature increases monotonically, approaching the asymptote (Fig. 5.6, curve 1). In the limit (stationary) state ($t = \infty$)

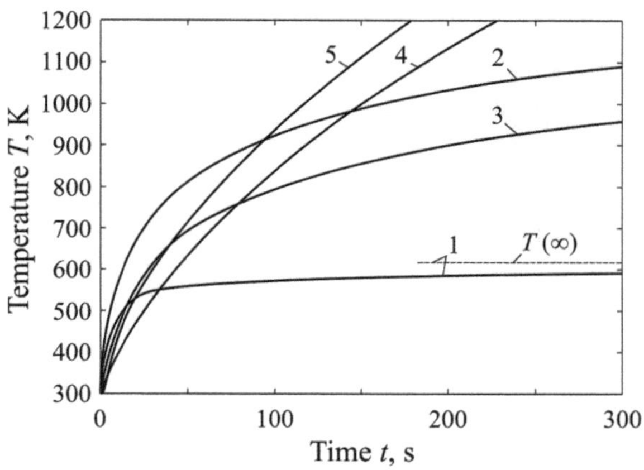

Fig. 5.6 The variation of the temperature with the time in a surface point located at 10 mm from a stationary continuous heat source ($q = 1000$ W, $\lambda = 0.04$ W mm^{-1} K^{-1}, $a = 8$ mm^2 s^{-1}, $\alpha = 0$, $T_\infty = 300$ K): (1) a point source on a semi-infinite solid; (2) a point source on a 10 mm thick infinite slab; (3) a line source in a 10 mm thick infinite plate; (4) a plane source in an infinite rod of diameter 25 mm; (5) a point source on a solid cylinder of diameter 25 mm. The source and the point under study are located in the same longitudinal plane

$$T(R,\infty) - T_0 = \frac{q_0}{2\pi\lambda R}, \tag{5.1.26}$$

i.e. the distribution is described by an equilateral hyperbola.

If the source is at point ξ, η, ζ of a semi-infinite solid, then, similarly, by using (4.2.10) the following can be obtained

$$T(x,y,z,t) - T_0 = \frac{q_0}{4\pi\lambda} \sum_{j=-1,1} \frac{1}{R_j} \Phi^*\left(\frac{R_j}{\sqrt{4at}}\right); \tag{5.1.27}$$

$$T(x,y,z,\infty) - T_0 = \frac{q_0}{4\pi\lambda} \sum_{j=-1,1} \frac{1}{R_j}, \tag{5.1.28}$$

where $R_j = \sqrt{(x-\xi)^2 + (y-\eta)^2 + (z-j\zeta)^2}$.

5.1.2.2 Stationary Continuous Point Source in an Infinite Slab

Similarly to a stationary continuous point source in a semi-infinite solid, by using (4.2.21), we obtain the temperature field due to a source acting at point ξ, η, ζ of a slab with heat-insulated surfaces:

$$T(x,y,z,t) - T_0 = \frac{q_0}{4\pi\lambda} \sum_{i=-\infty}^{\infty} \sum_{j=-1,1} \frac{1}{R_{i,j}} \Phi^*\left(\frac{R_{i,j}}{\sqrt{4at}}\right), \tag{5.1.29}$$

where $R_{i,j} = \sqrt{(x-\xi)^2 + (y-\eta)^2 + (z-j\zeta-2ih)^2}$.

If the source is on the surface of the slab at the origin of coordinates ($\xi = \eta = \zeta = 0$), then from (5.1.29) we obtain

$$T(x,y,z,t) - T_0 = \frac{q_0}{2\pi\lambda} \sum_{i=-\infty}^{\infty} \frac{1}{R_i} \Phi^*\left(\frac{R_i}{\sqrt{4at}}\right), \tag{5.1.30}$$

where $R_i = \sqrt{x^2 + y^2 + (z-2ih)^2}$.

At first, the temperature in the slab grows with time in the same way as in a semi-infinite solid, since the lower surface $z = h$ is not heated, there is practically no heat flux going through this surface and for this reason the surface virtually makes no impact (Fig. 5.6, curve 2). In contrast to the semi-infinite solid, the temperature in the slab unlimitedly increases with time, as the heat extends predominantly in the plane of the solid (plane $x0y$).

5.1.2.3 A Stationary Continuous Line Source in an Infinite Plate

The temperature field around a line source in a plate of thickness h without surface heat transfer is obtained by using (4.2.37) (Carslaw and Jaeger 1973):

$$T(r,t) - T_0 = \int_0^t \frac{q/h}{c\rho[4\pi a(t-\tau)]} \exp\left(-\frac{r^2}{4a(t-\tau)}\right) d\tau = -\frac{q/h}{4\pi\lambda} \mathrm{Ei}\left(-\frac{r^2}{4at}\right), \tag{5.1.31}$$

where r is the distance from the observation point to the origin, Ei is the exponential integral:

$$-\mathrm{Ei}(-u) = \int_u^\infty \exp(-x)/x dx, \quad x > 0. \tag{5.1.32}$$

The form of function Ei is shown in Fig. 3.14. Its representation in the form of tables or series is used in calculations. The function $\mathrm{Ei}(-u) = \ln(u) + \gamma$ (where $\gamma = 0.5772\ldots$ is Euler constant) for small values of u and therefore for large values of time t we approximately have (Carslaw and Jaeger 1973)

$$T(r,t) - T_0 = \frac{q/h}{4\pi\lambda}\left[\ln\left(\frac{4at}{r^2}\right) - \gamma\right]. \tag{5.1.33}$$

From the first term of expression (5.1.33), it is seen that the temperature in the plate grows unlimitedly with time t according to the logarithmic law (Fig. 5.6, curve 3). In the plate it is higher than the temperature in the lower surface of the slab, but lower than in its upper surface, where the point source is located (Fig. 5.6, curves 3 and 2).

5.1.2.4 Stationary Continuous Area Source in an Infinite Rod

By using formula (4.2.41), the substitution $u^2 = x^2/[4a(t-\tau)]$ and the integral (Gradshteyn and Ryzhik 2014)

$$\int_u^\infty \exp(-\mu^2 u^2)\frac{du}{u^2} = \frac{1}{u}\exp(-\mu^2 u^2) - \mu\sqrt{\pi}\Phi^*(\mu u), \tag{5.1.34}$$

we obtain the temperature field of a plane source in a rod without surface heat transfer:

$$T(x,t)-T_0=\int\limits_0^t \frac{q/A}{c\rho[4\pi a(t-\tau)]^{1/2}}\exp\left(-\frac{x^2}{4a(t-\tau)}\right)d\tau$$

$$=\frac{q/A|x|}{2\lambda}\left[\frac{1}{\sqrt{\pi}}\sqrt{\frac{4at}{x^2}}\exp\left(-\frac{x^2}{4at}\right)-\Phi^*\left(\sqrt{\frac{x^2}{4at}}\right)\right], \quad (5.1.35)$$

where $|x|$ is the distance from the observation point to the origin of coordinates. This formula can be represented in another form:

$$T(x,t)-T_0=q/A\frac{\sqrt{t}}{\sqrt{\lambda c\rho}}\mathrm{i}\Phi*\left(\sqrt{\frac{x^2}{4at}}\right), \quad (5.1.36)$$

where $\mathrm{i}\Phi^*(u)$ is the integral error function:

$$\mathrm{i}\Phi*(u)=\frac{1}{\sqrt{\pi}}\exp\left(-u^2-u\Phi*(u)\right). \quad (5.1.37)$$

The function $\mathrm{i}\Phi^*(u)$ is shown in Fig. 3.14.

With time, the temperature in the rod without surface heat transfer increases indefinitely (Fig. 5.6, curve 4).

5.1.2.5 Stationary Continuous Point Source on an Infinite Cylinder

A continuous point source of constant power is located on the surface of an infinitely long cylinder. Its temperature field can be obtained by integrating of the expression (4.2.43) over time or from Eq. (5.1.105) for a source moving along the generatrix of the cylinder at the passage to the limit (velocity $v \to 0$).

It follows from the comparison of the plane source and the point source that the temperature near the point source in the cylinder is greater than near the plane source in the rod (Fig. 5.6, curves 4 and 5).

As time passes, the temperature in the slab, plate, and cylinder without surface heat transfer increases indefinitely (Fig. 5.6).

It should be mentioned that in order to calculate the temperature (using the given formulas) at time t after the source is switched off ($t > t_w$), the simple method given in Sect. 5.1.3.12 can be used.

5.1.3 *Moving Concentrated Sources*

Moving heat sources are typical for fusion welding, so there is a great practical interest in determining temperature fields caused by moving sources.

We represent moving sources as a series of elementary instantaneous sources acting sequentially and displaced relative to each other. The solution to the heat conduction problem will be found in the moving coordinate system associated with the source, assuming, unless specified otherwise, that power q is constant and the source moves rectilinearly with a constant speed v. Varying power sources and sources with a complex trajectory will be considered in Chap. 7.

5.1.3.1 Moving Point Source in a Semi-infinite Body

At first, let us consider a simpler model for a source on the surface of a semi-infinite solid without surface heat transfer. The source is at the origin of the moving coordinate system and moves in the direction of the x-axis (Fig. 5.7a). The temperature increment will be obtained at an arbitrary point of observation P at any time t due to an elementary point source that operated at the instant of time τ at point O_1 at distance $v\tau$ from the beginning of path O_0, by using (4.2.11) and assuming that $Q_0 = q_0 d\tau$:

$$dT(x, y, z, t) = \frac{2q_0 d\tau}{c\rho[4\pi a(t-\tau)]^{3/2}} \exp\left(-\frac{R_1^2}{4a(t-\tau)}\right), \tag{5.1.38}$$

where $R_1^2 = [x + v(t-\tau)]^2 + y^2 + z^2$, $t - \tau$ is the duration of heat propagation of the elementary instantaneous source.

We sum the temperature increments caused by all the elementary sources that acted during time t on the path O_0O (Fig. 5.7a) and obtain the following (Rykalin 1947, 1952):

$$\begin{aligned} T(x, y, z, t) - T_0 &= \int_0^t \frac{2q_0}{c\rho[4\pi a(t-\tau)]^{3/2}} \\ &\quad \times \exp\left(-\frac{[x + v(t-\tau)]^2 + y^2 + z^2}{4a(t-\tau)}\right) d\tau \\ &= \frac{2q_0}{4\pi\lambda} \exp\left(-\frac{vx}{2a}\right) \frac{1}{R} \exp\left(-\frac{vR}{2a}\right) \\ &\quad \times \frac{1}{2}\left[\Phi^*\left(\frac{R}{\sqrt{4at}} - \sqrt{\frac{v^2 t}{4a}}\right) + \exp\left(\frac{vR}{a}\right)\Phi^*\left(\frac{R}{\sqrt{4at}} + \sqrt{\frac{v^2 t}{4a}}\right)\right], \end{aligned} \tag{5.1.39}$$

where $R^2 = x^2 + y^2 + z^2$ is the square of the distance from the moving observation point P to the position of the source (the origin of the moving coordinate system) (Fig. 5.7a). Here we use the substitution $u^2 = R^2/[4a(t-\tau)]$ and the known integral (Prudnikov et al. 1986):

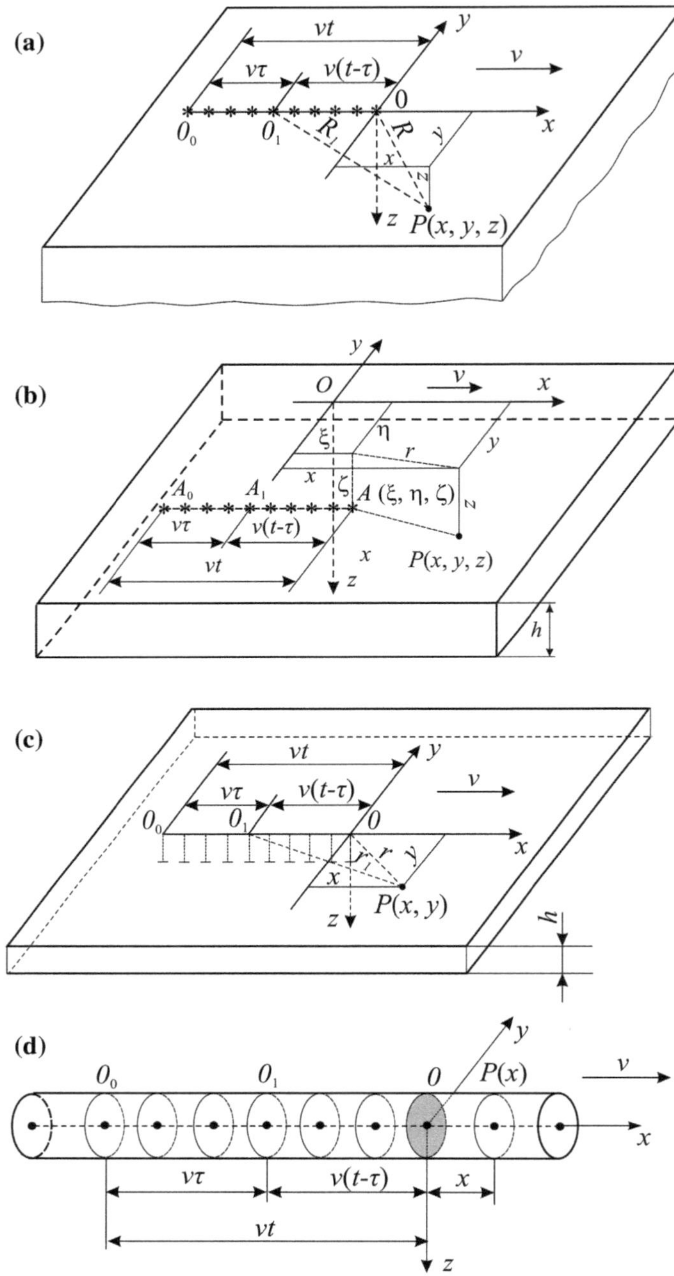

Fig. 5.7 Subdivision of continuous heat sources travelling along the x-axis with speed v into a sequence of instantaneous sources: a point source on a semi-infinite solid (**a**), a point source in an infinite slab at a depth of ζ (**b**), a line source in an infinite plate (**c**) and a plane source in an infinite rod (**d**)

$$\int_0^x \exp(-a^2x^2 - b^2/x^2)dx = \frac{\sqrt{\pi}}{4a}[\exp(2ab)\Phi(ax + b/x)$$
$$+ \exp(-2ab)\Phi(ax - b/x) - \exp(2ab) + \exp(-2ab)]. \qquad (5.1.40)$$

Function $T(t)$ (5.1.39) is a monotonically increasing one. As the source moves (its action time t passes), the temperature at all points of the solid in the moving coordinate system increases, approaching the limit value. At $t \to \infty$ the temperature field stops varying in time (but only in the moving coordinate system!), i.e. it becomes *quasi-stationary*:

$$T(x, R, \infty) - T_0 = \frac{q_0}{2\pi\lambda R}\exp\left(-\frac{v(x+R)}{2a}\right). \qquad (5.1.41)$$

This formula was obtained earlier by the method of separation of variables [Eq. (4.2.82)].

If the source is stationary ($v = 0$), then

$$T(R, \infty) - T_0 = \frac{q_0}{2\pi\lambda R}, \qquad (5.1.42)$$

which coincides with (5.1.26).

If the point source is not on the surface of a semi-infinite solid at the origin of the moving coordinate system, but is constantly offset along the coordinates by ξ, η and ζ, then we have to use formula (4.2.10) as the basis instead of (4.2.11). As a result, we obtain:

$$T(x, y, z, t) - T_0 = \frac{q_0}{4\pi\lambda}\exp\left(-\frac{v(x-\xi)}{2a}\right)\sum_{j=-1,1}\frac{1}{R_j}\exp\left(-\frac{vR_j}{2a}\right)$$
$$\times\frac{1}{2}\left[\Phi^*\left(\frac{R_j}{\sqrt{4at}} - \sqrt{\frac{v^2t}{4a}}\right) + \exp\left(\frac{vR_j}{a}\right)\Phi^*\left(\frac{R_j}{\sqrt{4at}} + \sqrt{\frac{v^2t}{4a}}\right)\right], \qquad (5.1.43)$$

where $R_j = \sqrt{(x-\xi)^2 + (y-\eta)^2 + (z-j\zeta)^2}$. Hence, it appears that the rate of temperature growth with time depends not only on the distance between the source and the observation point, but also on the depth of the source location.

In the case of a quasi-stationary state ($t = \infty$):

$$T(x, y, z, \infty) - T_0 = \frac{q_0}{4\pi\lambda}\exp\left(-\frac{v(x-\xi)}{2a}\right)\sum_{j=-1,1}\frac{1}{R_j}\exp\left(-\frac{vR_j}{2a}\right). \qquad (5.1.44)$$

If the source is stationary ($v = 0$) and the temperature field is stationary ($t = \infty$), then

$$T(x, y, z, \infty) - T_0 = \frac{q_0}{4\pi\lambda} \sum_{j=-1,1} \frac{1}{R_j}, \tag{5.1.45}$$

which, naturally, coincides with (5.1.28).

When relatively low-heat-conducting metals (for example, steel and titanium alloys) are welded at usual welding speeds, in the area close to the source the temperature state which is close to the limit one occurs shortly after the heating starts. Therefore, the study of the limit state is of the greatest interest. When short joints are welded, as well as when high-heat-conducting metals (for example, aluminium alloys) are welded, it is possible that the heat-saturation process can continue and the limit state is not reached.

Let us analyse the temperature field around a surface source in the limit state. We see from formula (5.1.41) and Fig. 5.8 that the field is symmetrical with respect to the longitudinal x-axis, that is, the isothermal surfaces are the surfaces of rotation around the x-axis. The further the point of the heated solid is from the source, the lower the peak temperature is and the later it is reached (Fig. 5.8b). The dashed line that connects the points at the instant of maximum heating separates the heating and cooling regions (Fig. 5.8d). Behind the source the components of the temperature gradient vector along the axis of source motion are much smaller than in the transverse direction, which means that the heat flux along the x-axis is much smaller than in the transverse direction, i.e. the heat predominantly diffuses in the y- and z-directions. This statement becomes even more valid, as the asymmetry of the field becomes greater with respect to the y-axis (the higher speed v is), and it will be used in developing a model for rapidly moving sources (Sect. 5.1.4).

Let us analyse the influence of welding parameters (power q and speed v) on the quasi-stationary temperature field.

At $v = \text{const}$, the temperature increment is directly proportional to power q, as can be seen from (5.1.41). This follows from the linearity of the heat conduction equation, when the material properties are constant.

At $q = \text{const}$, the temperature gradient in front of the source ($x > 0$) increases as speed v increases, but the temperature behind the source ($x < 0$) does not change (Fig. 5.9a). In the cross section, the temperature gradient increases, and the heated region narrows (Fig. 5.9b).

Let us represent the quasi-stationary temperature field of a moving point source in the surface of a semi-infinite solid (5.1.41) in dimensionless form:

$$\theta_3(\xi, \rho_3) = \frac{1}{\rho_3} \exp(-\xi - \rho_3) \tag{5.1.46}$$

or

$$\theta_3(\xi, \rho_2) = \frac{1}{\sqrt{\xi^2 + \rho_2^2}} \exp\left(-\xi - \sqrt{\xi^2 + \rho_2^2}\right), \tag{5.1.47}$$

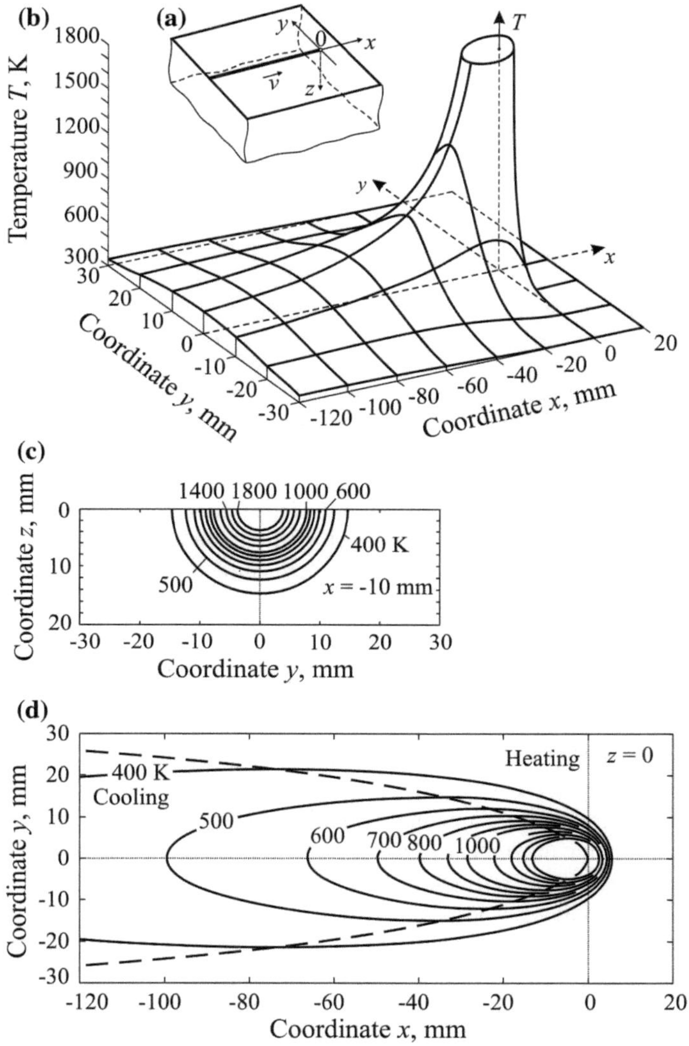

Fig. 5.8 Quasi-stationary temperature field around a moving point source on a semi-infinite solid ($q = 5000$ W, $v = 5$ mm s^{-1}, $\lambda = 0.04$ W mm^{-1} K^{-1}, $a = 8$ mm^2 s^{-1}, $T_0 = 300$ K): reference frame (**a**); temperature distribution at the surface $z = 0$ (**b**); isotherms in cross section $x = -10$ mm (**c**); isotherms at surface $z = 0$ (**d**)

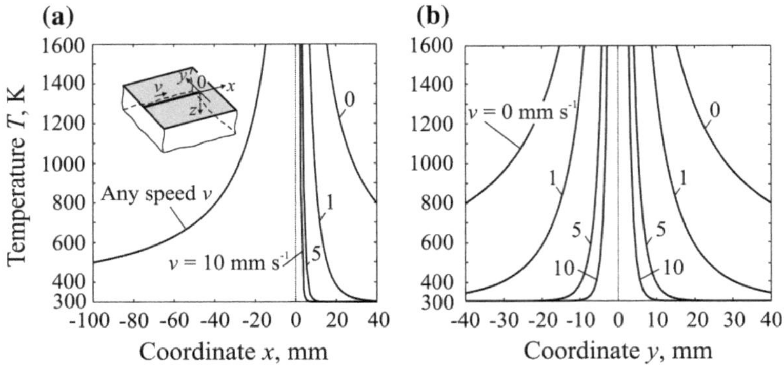

Fig. 5.9 The effect of the speed of a point source on the steady state temperature distribution over the x-axis (**a**) and the y-axis (**b**) in a semi-infinite solid

where θ_3 is the dimensionless temperature, ξ is the dimensionless longitudinal coordinate, ρ_3 is the dimensionless radius vector (the dimensionless distance in space from the source to the point under study), ρ_2 is the dimensionless plane radius vector (the dimensionless distance between the point under study and the longitudinal axis):

$$\theta_3 = \frac{4\pi a\lambda}{q_0 v}(T - T_0);\ \xi = \frac{vx}{2a};\ \rho_3 = \frac{vR}{2a};\ \rho_2 = \frac{v\sqrt{y^2 + z^2}}{2a}. \tag{5.1.48}$$

The single field $\theta_3(\xi, \rho_2)$ (Fig. 5.10a) includes all possible combinations of power q_0, speed v, and material properties (a and λ), so it includes the field shown in Fig. 5.8d. The longitudinal temperature distribution $\theta_3(\xi)$ at a different distance from the axis of the source movement ρ_2 is shown in Fig. 5.10b. As the source moves, the temperature around the fixed point of the solid increases, reaches a maximum and then decreases, tending towards the average temperature of the solid. Thus, curves $\theta(\xi)$ can be interpreted as thermal cycles of fixed points of the solid, spaced from the ξ-axis by distance ρ_2. This follows from the equation $x = -vt$, where the moment the source passes through the cross section in which the point under study is located is taken as time $t = 0$. Therefore

$$\xi = -\tau = -v^2 t/(2a) \text{ and } \theta_3(\xi, \rho_2) = \theta_3(-\tau, \rho_2). \tag{5.1.49}$$

The dimensionless heating and cooling rate $\partial\theta_3/\partial\tau$ at a fixed point can be found by differentiating function (5.1.47) with respect to ξ and by changing the variable $\xi = -\tau$:

$$\frac{\partial\theta_3}{\partial\xi} = -\left(1 + \frac{\xi}{\sqrt{\xi^2 + \rho_2^2}} + \frac{\xi}{\xi^2 + \rho_2^2}\right)\frac{\exp\left(-\xi - \sqrt{\xi^2 + \rho_2^2}\right)}{\sqrt{\xi^2 + \rho_2^2}} \tag{5.1.50}$$

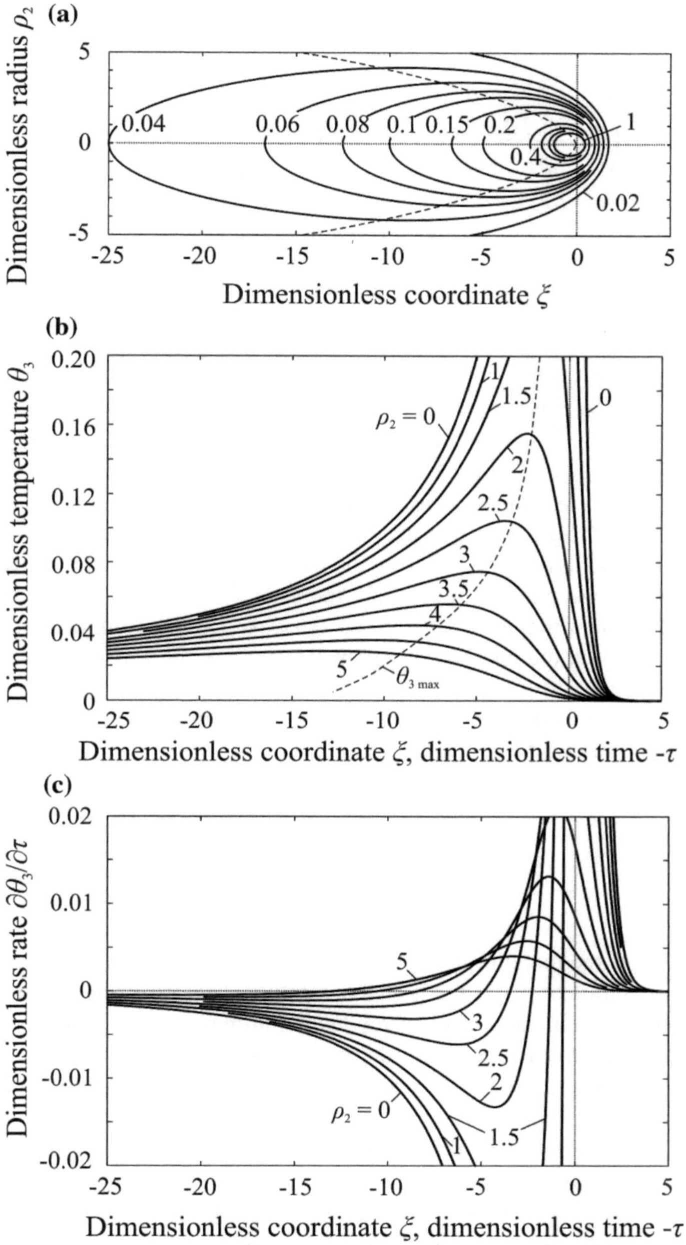

Fig. 5.10 Steady temperature field (**a**), longitudinal distribution of dimensionless temperature θ_3 (**b**) and heating/cooling rate $\partial\theta_3/\partial\tau$ (**c**), moving point source on a semi-infinite solid

and

$$\frac{\partial\theta_3}{\partial\tau} = \left(1 - \frac{\tau}{\sqrt{\tau^2 + \rho_2^2}} - \frac{\tau}{\tau^2 + \rho_2^2}\right)\frac{\exp\left(\tau - \sqrt{\tau^2 + \rho_2^2}\right)}{\sqrt{\tau^2 + \rho_2^2}}, \tag{5.1.51}$$

where

$$\frac{\partial\theta_3}{\partial\tau} = \frac{8\pi a^2\lambda}{q_0 v^3} \cdot \frac{\partial T}{\partial t}; \quad \tau = \frac{v^2 t}{2a}. \tag{5.1.52}$$

It can be seen from Fig. 5.10c that the heating rate ($\partial\theta_3/\partial\tau > 0$) is much higher than the cooling rate ($\partial\theta_3/\partial\tau < 0$).

From the metallurgical point of view, the cooling rate at a given temperature is important. Curves $\theta_3(\tau)$ in Fig. 5.10b and $\partial\theta_3/\partial\tau(\tau)$ in Fig. 5.10c allow us to find the dependence of the dimensionless cooling rate on the dimensionless temperature $\partial\theta_3/\partial\tau(\theta_3)$ (Fig. 5.11a). The cooling rate is 0, when the temperature reaches the maximum value $\theta_{3\,\max}$. It follows from the figure that the cooling rates for all points at a given temperature θ_3 are practically the same if the peak temperature $\theta_{3\,\max}$ is much higher than the set temperature, $\theta_3 \ll \theta_{3\,\max}$. This implies an important practical conclusion: the cooling rate of such points can be calculated at $\rho_2 = 0$ (at $y^2 + z^2 = 0$).

The transition from dimensionless quantities to dimensional ones is made according to (5.1.48) and (5.1.52):

$$\begin{aligned} T - T_0 &= \frac{q_0 v}{4\pi a\lambda}\theta_3; \quad \frac{\partial T}{\partial t} = \frac{q_0 v^3}{8\pi a^2\lambda} \cdot \frac{\partial\theta_3}{\partial\tau}; \\ x &= \frac{2a}{v}\xi; \quad r = \sqrt{y^2 + z^2} = \frac{2a}{v}\rho_2; \quad t = \frac{2a}{v^2}\tau. \end{aligned} \tag{5.1.53}$$

The diffusion and metallurgical processes near the weld interface depend heavily on the retention time of metal at a high temperature. Therefore, the determination of the retention time of metal above a given temperature represents a practical interest.

As can be seen from Fig. 5.12a, the dimensionless retention time τ above the dimensionless temperature θ_3 is greater, the smaller the dimensionless distance to the weld axis ρ_2 is. The cooling time from the peak temperature $\theta_{3\,\max}$ is shown in Fig. 5.12b. This time is shorter than the retention time above the set temperature. The dependence of the cooling time of metal τ on the weld axis ($\rho_2 = 0$) on the temperature θ_3 is described by a linear function in logarithmic coordinates. If our interest is with the time the metal cools from temperature θ_3' down to temperature θ_3'', this time can be determined from Fig. 5.12b. For example, the final steel microstructure after welding is predicted from the time the metal cools in the range of 800–500 °C (1073–773 K) or 850–500 °C (1123–773 K) (Seyffarth 1978; Seyffarth and Kuscher 1982; Seyffarth et al. 1992).

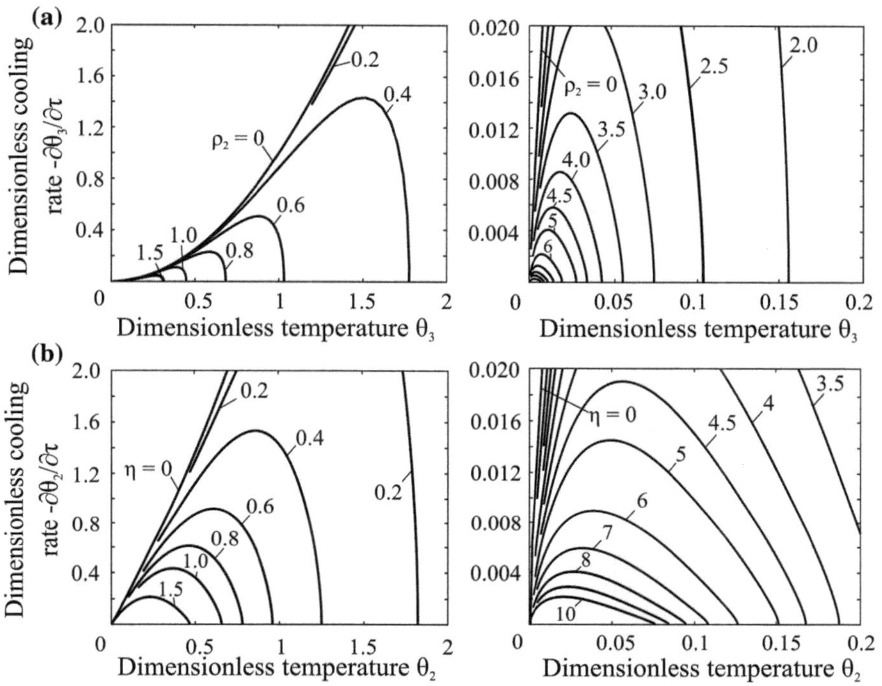

Fig. 5.11 Dimensionless cooling rate $\partial\theta/\partial\tau$ versus dimensionless temperature θ in the points located at a distance of ρ_2 and η from the weld centreline and heated by a moving point source on a semi-infinite solid (**a**) and a moving line source in an infinite plate with impermeable surface (**b**)

The dimensionless and dimensional quantities of time and temperature are connected by Eqs. (5.1.48) and (5.1.52).

While the source moves along the solid, the temperature of the points of the solid changes continuously (Fig. 5.10b). The dashed line $\theta_{3\,\max}$ passes through the points of peak temperature. Let us find the distribution of the peak temperature in the cross section $\theta_{3\,\max}(\rho_{2\,\max})$ (Fig. 5.13a).

The condition for the maximum temperature on the surface $\rho_{2\,\max} = \text{const}$ is that the derivative is zero with respect to coordinate $\xi\,(\partial\theta_3/\partial\xi = 0)$ or time $\tau(\partial\theta_3/\partial\tau = 0)$. It follows from (5.1.50) that this condition is satisfied at $\xi = \infty$, $\xi = -\infty$ and

$$1 + \frac{\xi}{\sqrt{\xi^2 + \rho_2^2}} + \frac{\xi}{\xi^2 + \rho_2^2} = 0 \quad \text{or} \quad 1 + \frac{\xi}{\rho_3} + \frac{\xi}{\rho_3^2} = 0, \tag{5.1.54}$$

from which we obtain the root

$$\xi_{\max} = -\frac{\rho_{3\,\max}^2}{1 + \rho_{3\,\max}}. \tag{5.1.55}$$

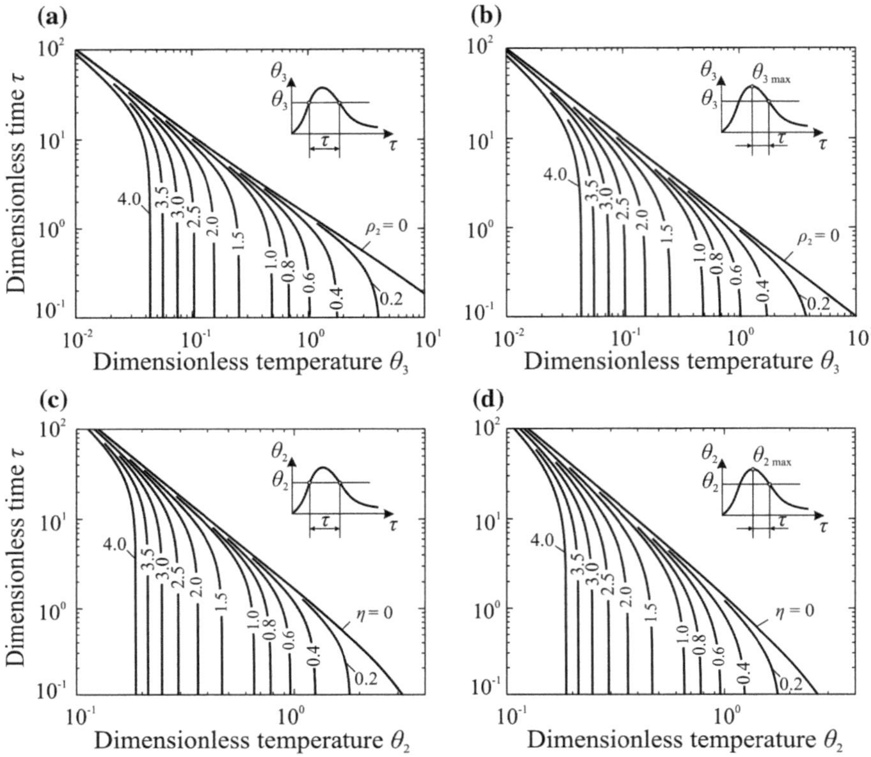

Fig. 5.12 Dimensionless retention time τ versus dimensionless temperature θ in the points located at a distance of ρ_2 and η from the weld centreline and heated by a moving point source on semi-infinite solid (**a**, **b**) and a moving line source in infinite plate (**c** and **d**)

By substituting (5.1.55) in (5.1.47) we obtain the dependence of the peak temperature on coordinate $\rho_{3\,\max}$ (Rykalin 1947, 1952):

$$\theta_{3\,\max} = \frac{1}{\rho_{3\,\max}} \exp\left(-\frac{\rho_{3\,\max}}{1+\rho_{3\,\max}}\right). \tag{5.1.56}$$

The radius of the surface of the peak temperature is (Fig. 5.13a) (Grong 1994):

$$\begin{aligned}\rho_{2\,\max} &= \eta_{\max} = \zeta_{\max} = \sqrt{\rho_{3\,\max}^2 - \xi_{\max}^2} \\ &= \frac{\rho_{3\,\max}}{1+\rho_{2\,\max}}\sqrt{1+2\rho_{3\,\max}}.\end{aligned} \tag{5.1.57}$$

Thus, the relationship between the peak temperature $\theta_{3\,\max}$ and coordinates $\xi_{\max}$ and $\rho_{2\,\max}$ is determined in the sequence: $\rho_{3\,\max}$ is given, $\theta_{3\,\max}$ is calculated by (5.1.56), $\xi_{\max}$ is calculated by (5.1.55), and $\rho_{2\,\max}$ is calculated by (5.1.57). The

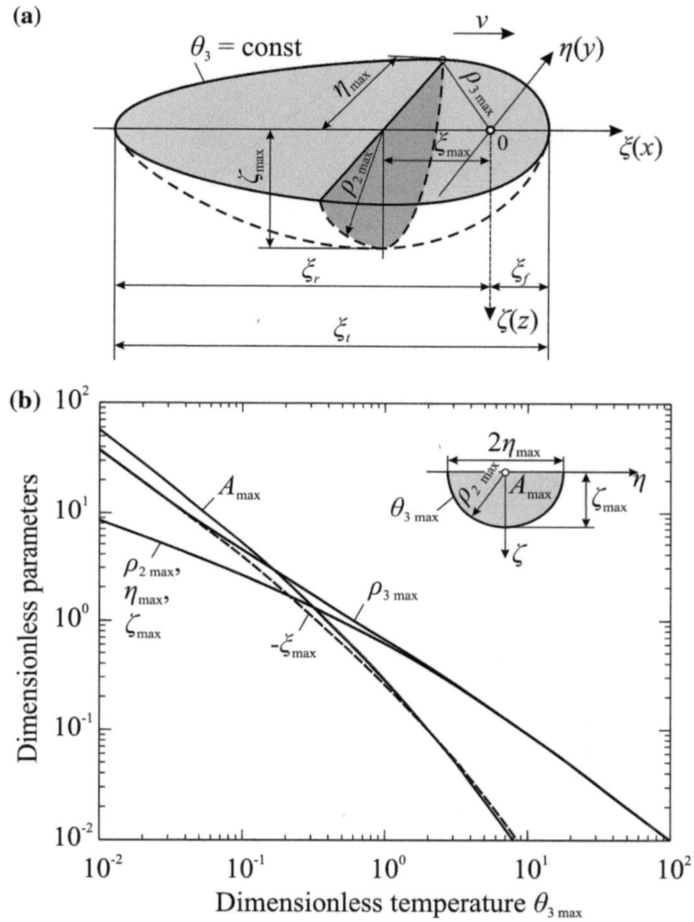

Fig. 5.13 Schematic representation of 3D solution for a moving point source on a semi-infinite solid (**a**) and dimensionless distance $\xi_{\max}$, half width $\eta_{\max}$ (depth $\zeta_{\max}$), radius vector $\rho_{3\,\max}$ and cross-sectional area $A_{\max}$ as functions of a specified dimensionless steady temperature $\theta_{3\,\max}$ (**b**) (Grong 1994; Rykalin 1947, 1952)

results of calculations are shown in Fig. 5.13b. The transition from dimensionless to dimensional quantities is made according to formulas (5.1.53).

Now let us determine the length of the zone ξ_t heated to a given temperature θ_3 (Fig. 5.13a):

$$\xi_t = \xi_f - \xi_r, \tag{5.1.58}$$

where ξ_r and ξ_f are the coordinates of the beginning and the end of the heated zone. It follows from (5.1.46) that at $\rho_3 = \pm\xi$ we obtain

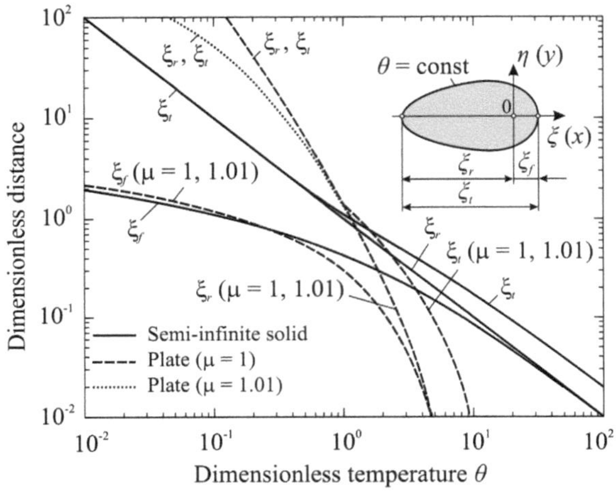

Fig. 5.14 Relationships between dimensions of a heated zone and the temperature in a semi-infinite solid and plate

$$\xi_f = -\frac{1}{2}\ln(\xi_f\theta_3), \quad \xi_r = -\theta_3^{-1}. \tag{5.1.59}$$

Curves ξ_r, ξ_f and ξ_t are shown in Fig. 5.14.

The temperature field of the buried source located at point ξ, η, ζ [formula (5.1.44)] differs only in terms that the plane $y = \eta$ is a plane of symmetry, and the plane $z = \zeta$ is not (isothermal surfaces deviate from the surfaces of rotation, extending to the side of the adiabatic surface $z = 0$). The heating and cooling rate also depends on the depth of the source location.

If we take into account the surface heat transfer, the method for finding the temperature field is analogous, but expression (4.2.18) is used instead of (4.2.11) in formula (5.1.38). The solution is obtained in quadratures. However, the integral cannot be expressed as tabulated functions (Rykalin 1947, 1952).

5.1.3.2 Moving Point Source in a Slab

First, it is assumed that there is no surface heat transfer. Then a solution for a moving point source in a slab with heat-insulated surfaces can be obtained in two ways: it is possible either to take an instantaneous point source in an infinite slab as a basis and to construct a solution for a moving source (as is done in the case of a semi-infinite solid) or to take the known solution for a moving point source in an infinite solid, and then satisfy all the boundary conditions by the method of images. We take a shorter second way. By having the solution (5.1.43) as a basis and using the method of images (Sect. 4.2.1.3, Fig. 4.5), the solution is obtained for a moving point source spaced from the origin of the moving coordinate system by ξ, η and ζ:

$$T(x, y, z, t) - T_0 = \frac{q_0}{4\pi\lambda} \exp\left(-\frac{v(x-\xi)}{2a}\right) \sum_{i=-\infty}^{\infty} \sum_{j=-1,1} \frac{1}{R_{i,j}} \exp\left(-\frac{vR_{i,j}}{2a}\right)$$
$$\times \frac{1}{2}\left[\Phi^*\left(\frac{R_{i,j}}{\sqrt{4at}} - \sqrt{\frac{v^2 t}{4a}}\right) + \exp\left(\frac{vR_{i,j}}{a}\right)\Phi^*\left(\frac{R_{i,j}}{\sqrt{4at}} + \sqrt{\frac{v^2 t}{4a}}\right)\right], \tag{5.1.60}$$

where $R_{i,j} = \sqrt{(x-\xi)^2 + (y-\eta)^2 + (z - j\zeta - 2ih)^2}$.

If the welding time is great (in theory $t = \infty$), Eq. (5.1.60) gets simplified:

$$T(x, y, z, \infty) - T_0 = \frac{q_0}{4\pi\lambda} \exp\left(-\frac{v(x-\xi)}{2a}\right) \sum_{i=-\infty}^{\infty} \sum_{j=-1,1} \frac{1}{R_{i,j}} \exp\left(-\frac{vR_{i,j}}{2a}\right). \tag{5.1.61}$$

If the point source is on the surface of a slab at the origin ($\xi = \eta = \zeta = 0$), then formula (5.1.61) becomes even simpler:

$$T(x, y, z, \infty) - T_0 = \frac{q_0}{2\pi\lambda} \exp\left(-\frac{vx}{2a}\right) \sum_{i=-\infty}^{\infty} \frac{1}{R_i} \exp\left(-\frac{vR_i}{2a}\right), \tag{5.1.62}$$

where $R_i = \sqrt{x^2 + y^2 + (z - 2ih)^2}$.

By using formula (4.2.25) and the integral (Gradshteyn and Ryzhik 2014)

$$\int_0^\infty \exp(-\beta\sqrt{\gamma^2 + u^2})(\gamma^2 + u^2)^{-1/2} \cos(au)du = \mathrm{K}_0(\gamma\sqrt{a^2 + \beta^2}), \tag{5.1.63}$$

formula (5.1.61) is represented in another form:

$$T(x, r, z, \infty) - T_0 = \frac{q_0/h}{2\pi\lambda} \exp\left(-\frac{v(x-\xi)}{2a}\right) \mathrm{K}_0\left(\frac{vr}{2a}\right) F_0(r, z), \tag{5.1.64}$$

where

$$F_0(r, z) = 1 + 2\sum_{i=1}^{\infty} \cos\frac{\pi i z}{h} \cos\frac{\pi i \zeta}{h} \mathrm{K}_0\left(\frac{vr}{2a}\sqrt{1 + \left(\frac{2\pi a i}{vh}\right)^2}\right) / \mathrm{K}_0\left(\frac{vr}{2a}\right);$$
$$r = \sqrt{(x-\xi)^2 + (y-\eta)^2}. \tag{5.1.65}$$

Here, K_0 is the modified Bessel function of the second kind of order zero [it is tabulated and represented in the form of a series (Abramowitz and Stegun 1965; Janke et al. 1960)]. Function $\mathrm{K}_0(u)$ tends to infinity with argument u tending to zero ($\mathrm{K}_0(u) \to \infty$ at $u \to 0$), and tends to zero with argument u tending to infinity

($K_0(u) \to 0$ at $u \to \infty$) (Fig. 3.14). Formulas (5.1.61) and (5.1.64) are equivalent. They only differ in the rate of convergence of the series.

Figure 5.15 shows a quasi-stationary temperature field around a moving point source on the surface of an infinite slab ($\xi = \eta = \zeta = 0$). On the lower surface ($z =$ 10 mm), the temperature is lower than on the upper one and the peak temperature is reached later. Behind the source, as we move away from it, the temperature drop across the thickness decreases, i.e. the field becomes plane. It is seen from the comparison of Figs. 5.8 and 5.15 that under the same heating conditions near the source, the temperature in the semi-infinite solid and the slab is practically the same, but at a distance from the source the temperature in the slab is higher.

Formula (5.1.62) is represented in dimensionless form:

$$\theta_3(\xi, \eta, \zeta, \delta) = \sum_{i=-\infty}^{\infty} \frac{1}{\rho_{3i}} \exp(-\xi - \rho_{3i}), \tag{5.1.66}$$

where

$$\theta_3 = \frac{4\pi a\lambda}{q_0 v}(T - T_0);\ \rho_{3i} = \sqrt{\xi^2 + \eta^2 + \zeta^2 + 4i\delta(i\delta - \zeta)};$$
$$\xi = \frac{vx}{2a};\ \eta = \frac{vy}{2a};\ \zeta = \frac{vz}{2a};\ \delta = \frac{vh}{2a}. \tag{5.1.67}$$

It is clear that in comparison with a semi-infinite solid [formulas (5.1.47)–(5.1.48)] an additional variable $R^2 = x^2 + y^2 + z^2$ appears that takes into account the thickness of the solid. If the thickness is relatively large ($h \to \infty$ and $\delta \to \infty$), only the term at $i = 0$ remains a nonzero term in the series and formula (5.1.66) turns into (5.1.46). Figure 5.16a shows the quasi-stationary field on the upper and lower surfaces of the slab.

The heating and cooling rate is obtained by differentiating Eq. (5.1.66) with respect to ξ and substituting $\xi = -\tau$ (Myers et al. 2007):

$$\frac{\partial\theta_3}{\partial\tau} = \sum_{i=-\infty}^{\infty} \left(\frac{1}{\rho_{3i}} + \frac{\xi}{\rho_{3i}^2} + \frac{\xi}{\rho_{3i}^3}\right) \exp(-\xi - \rho_{3i}) \tag{5.1.68}$$

and

$$\frac{\partial\theta_3}{\partial\tau} = \sum_{i=-\infty}^{\infty} \left[\frac{1}{\left(\tau^2 + \rho_{2i}^2\right)^{1/2}} - \frac{\tau}{\tau^2 + \rho_{2i}^2} - \frac{\tau}{\left(\tau^2 + \rho_{2i}^2\right)^{3/2}}\right] \exp\left(\tau - \left(\tau^2 + \rho_{2i}^2\right)^{1/2}\right), \tag{5.1.69}$$

where

$$\frac{\partial\theta_3}{\partial\tau} = \frac{8\pi a^2\lambda}{q_0 v^3}\frac{\partial T}{\partial t};\quad \rho_{2i}^2 = \eta^2 + \zeta^2 + 4i\delta(i\delta - \zeta). \tag{5.1.70}$$

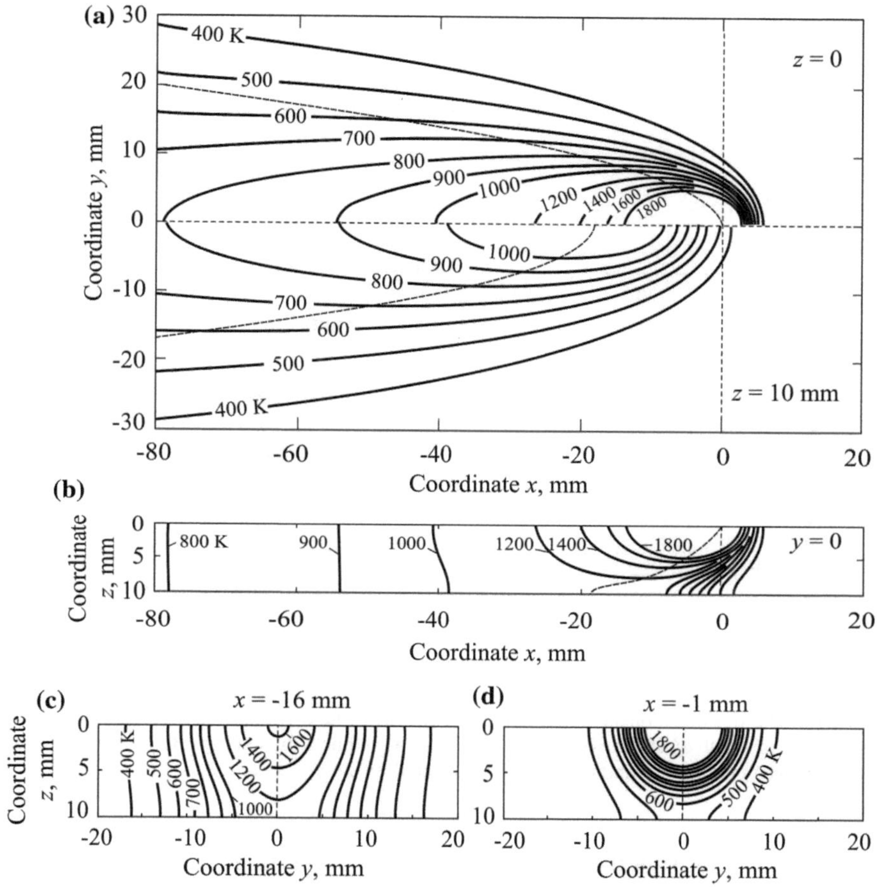

Fig. 5.15 The temperature field around the moving point source on one side of a 10 mm thick slab, quasi-stationary state in moving coordinate system x, y, z ($q = 5000$ W, $v = 5$ mm s^{-1}, $\lambda = 0.04$ W mm^{-1} K^{-1}, $a = 8$ mm^2 s^{-1}, $\alpha_1 = \alpha_2 = 0$, $T_0 = 300$ K): on topside ($z = 0$) and bottom side ($z = 10$ mm) (**a**) in longitudinal section (**b**), in cross-section $x = -16$ mm (**c**) and in cross-section $x = -1$ mm (**d**)

With the help of (5.1.67) and (5.1.70), it is easy to pass on to dimensional temperatures, heating (cooling) rates and coordinates.

Figure 5.16b shows the heating (in front of the dashed line) and cooling (behind the dashed line) rate fields. Obviously, the rate of heating (cooling) decreases as the distance from the ξ-axis increases. From Fig. 5.16a, b, the rate can be found at a given temperature θ_3 at points with a different distance η from the ξ-axis. Evidently, it depends on the thickness of slab δ.

The cooling rate at a temperature well below the maximum one is only slightly dependent on the distance from the weld axis. That is why in approximate analysis

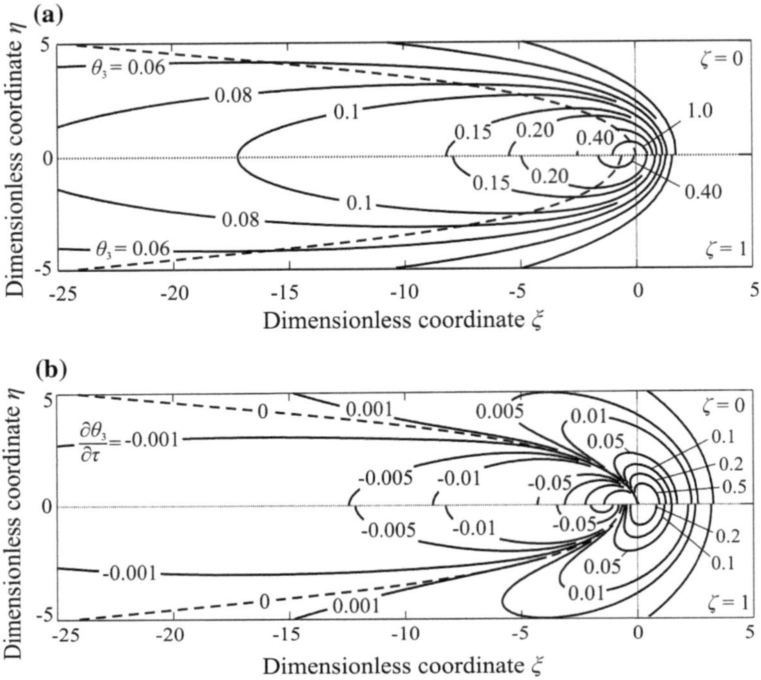

Fig. 5.16 The quasi-stationary dimensionless temperature field around the moving point source on one side of a slab of thickness $\delta = vh/(2a) = 3$ (**a**) and heating/cooling rate $\partial\theta_3/\partial\tau$ (**b**) in the upper (heated) plane $\zeta = 0$ and in the lower (unheated) plane $\zeta = 1$

of the thermal processes the axis of weld $\xi(\eta = \zeta = 0)$ is taken as a characteristic of the entire weld. It follows from Fig. 5.17 that the cooling rate at a given temperature falls with a decrease in the thickness of the slab. The curve $\delta = \infty$ corresponds to the semi-infinite solid, and Eqs. (5.1.68) and (5.1.69) turn into Eqs. (5.1.50) and (5.1.51).

The practical question arises: when can a simpler formula be used for the semi-infinite solid $T\ (h = \infty)$ to calculate the temperature in the slab $T\ (h < \infty)$ and what will be the error in this case? Let us estimate error ε for the steady state of the temperature field around a moving point source on the upper surface at the origin of the coordinate system $(\xi = \eta = \zeta = 0)$ using expressions (5.1.62) and (5.1.41):

$$\varepsilon = \frac{T(h < \infty) - T(h = \infty)}{T(h < \infty) - T_0}$$
$$= 1 - \left\{ \sum_{i=-\infty}^{\infty} \frac{\rho_3}{\sqrt{\rho_3^2 + 4i(i-\zeta)}} \exp\left(\mathrm{Pe}\left[\rho_3 - \sqrt{\rho_3^2 + 4i(i-\zeta)} \right] \right) \right\}^{-1}, \tag{5.1.71}$$

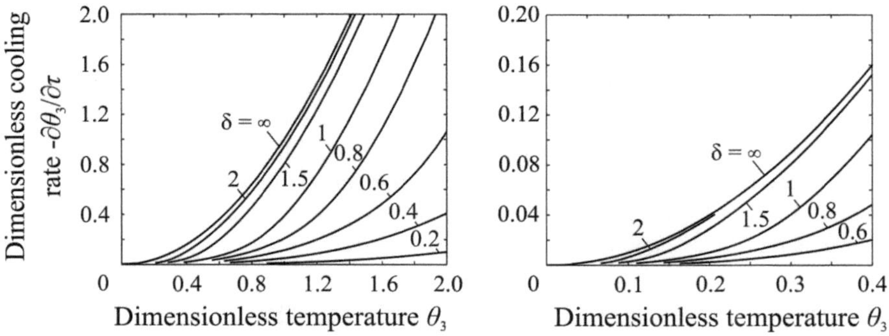

Fig. 5.17 The dependence of dimensionless cooling rate $-\partial\theta/\partial\tau$ at the weld centreline on dimensionless temperature θ under heating of a slab of thickness δ by a surface moving point source

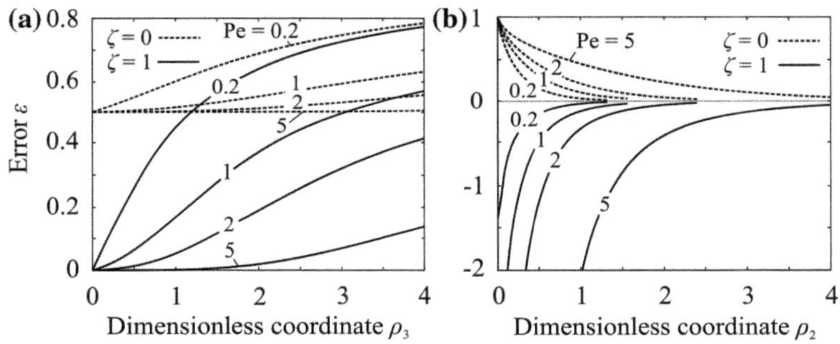

Fig. 5.18 The relative error of the calculated steady temperature in the slab by using formulas for the semi-infinite solid (**a**) and the plate (**b**)

where ρ_3 is the dimensionless distance, ζ is the dimensionless depth of the point under study, Pe is the dimensionless thickness (the Peclet number):

$$\rho_3 = \frac{\sqrt{x^2 + y^2 + z^2}}{h}; \quad \zeta = \frac{z}{h}; \quad \mathrm{Pe} = \frac{vh}{2a}. \tag{5.1.72}$$

Error ε is a function of three dimensionless parameters (Fig. 5.18a). The error is small near the source on the upper surface ($\zeta = 0$, ρ_3 is small), that is the temperature fields in the slab and the semi-infinite solid coincide. As the distance from the source increases, the error grows. The error is significant on the lower surface ($\zeta = 1$, $z = h$) (the temperature in the slab is much larger than in the semi-infinite solid). The error is the smaller, the greater the Peclet number is (the higher the speed and/or thickness are/is).

Note that with the method of images the temperature field can be found in the slab if the lower surface is isothermal ($T\ (x, y, h, t) = T_0$). When solving the problem, combinations of additional heat sources and sinks are introduced (Sect. 4.2.1.3). Such a method will be used later (Sect. 1.2.4).

Now we take into account the heat transfer from the lower and upper surfaces of the slab according to Newton's law [see formulas (4.2.29)]. The source is at point ξ, η, ζ of the moving coordinate system and moves along line A_0A parallel to the x-axis (Fig. 5.7b). The increment of temperature is obtained at an arbitrary point P at any instant of time t from an elementary point source that operated at instant τ at point A_1, by using (4.2.30) and assuming that $Q_0 = q_0 d\tau$:

$$dT(x,y,z,t) = \frac{q_0 d\tau/h}{c\rho[4\pi a(t-\tau)]}\exp\left(-\frac{[x-\xi+v(t-\tau)]^2+(y-\eta)^2}{4a(t-\tau)}\right) \times \sum_{n=1}^{\infty} A_n B_n(\zeta) B_n(z) \exp\left(-\mu_n^2 \frac{a(t-\tau)}{h^2}\right). \quad (5.1.73)$$

Here the factors A_n, the functions B_n and the roots μ_n are known [see (4.2.30)–(4.2.32)].

By integrating this equation from 0 to t, after the transformations we obtain:

$$T(x,y,z,t) = \frac{q_0/h}{4\pi\lambda}\exp\left(-\frac{v(x-\xi)}{2a}\right) \times \int_0^t \frac{1}{t-\tau}\exp\left(-\frac{(x-\xi)^2+(y-\eta)^2}{4a(t-\tau)} - \frac{v^2}{4a}(t-\tau)\right) \times \sum_{n=1}^{\infty} A_n B_n(\zeta) B_n(z) \exp\left(-\mu_n^2 \frac{a(t-\tau)}{h^2}\right) d\tau;$$

$$r = [(x-\xi)^2 + (y-\eta)^2]^{1/2}. \quad (5.1.74)$$

In the limiting case (at $t \to \infty$), by using the substitution $u = t - \tau$ and integral (Gradshteyn and Ryzhik 2014; Prudnikov et al. 1986)

$$\int_0^\infty u^{-1}\exp(-pu - q/u)du = 2\mathrm{K}_0(2\sqrt{pq}) \quad (5.1.75)$$

we obtain

$$T(x,y,z,\infty) - T_0 = \frac{q_0/h}{2\pi\lambda}\exp\left(-\frac{v(x-\xi)}{2a}\right) \times \sum_{n=1}^{\infty} A_n B_n(\zeta) B_n(z) \mathrm{K}_0\left(\frac{vr}{2a}\sqrt{1+\left(\mu_n \frac{2a}{vh}\right)^2}\right), \quad (5.1.76)$$

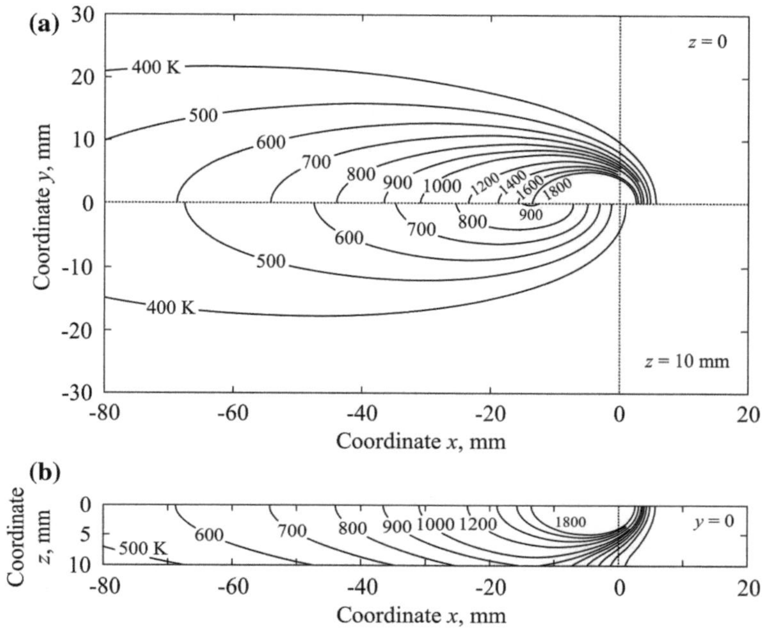

Fig. 5.19 The temperature field around the moving point source on a 10 mm thick slab, quasi-stationary state in moving coordinate system x, y, z ($q = 5000$ W, $v = 5$ mm s^{-1}, $\lambda = 0.04$ W mm^{-1} K^{-1}, $a = 8$ mm^2 s^{-1}, $\alpha_1 = 0$, $\alpha_2 = 0.004$ W mm^{-2} K^{-1}, $T_0 = 300$ K): on topside ($z = 0$) and bottom side ($z = 10$ mm) (**a**), in longitudinal section ($y = 0$) (**b**)

where K_0 is the modified Bessel function of the second kind of order zero.

Figure 5.19 shows a quasi-stationary temperature field due to a moving point source on the surface of an infinite slab ($\xi = \eta = \zeta = 0$) with intensive forced cooling of the lower surface with the help of copper water-cooled clamps ($\alpha_2 = 0.004$ W mm^{-2} K^{-1}). It follows from the comparison of Fig. 5.19 with Fig. 5.15 that under the same welding conditions, the dimensions of the weld pool (isotherms $T_m = 1800$ K) decreased very slightly, and the temperature on the lower surface of the slab decreased significantly. In the weld behind the pool, the distance between the isotherms reduced sharply (the cooling rate increased dramatically). The temperature drop across the thickness remains significant even far behind the source (Fig. 5.19b). In contrast to the heat-insulated upper surface ($\alpha_1 = 0$), the isotherms on the lower surface are not perpendicular to the surface. Their slope characterises the heat flux through the lower surface of the slab.

The efficiency of heat transfer is determined by the Biot numbers $Bi_1 = \alpha_1 h/\lambda$ and $Bi_2 = \alpha_2 h/\lambda$ [formulas (4.2.32)]. The solution obtained is applicable for the analysis of thermal processes in underwater welding (Parshin 2016).

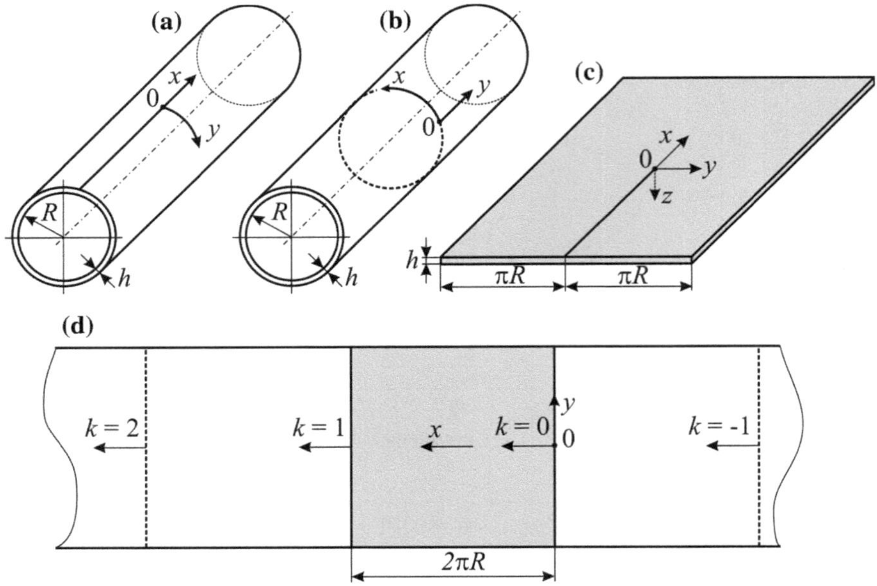

Fig. 5.20 Schematic diagrams for calculation of the temperature fields in thin-walled cylinders: a longitudinal source (**a**); a ring source (**b**); an equivalent pattern (developed cylinder) for the longitudinal source (**c**); an equivalent pattern for the ring source (**d**)

5.1.3.3 Moving Point Source on a Thin-Walled Pipe

Let the thickness of the cylindrical tube h be much less than its radius R, $h \ll R$. Then it can be assumed that the heat in the tube will be diffused according to the same laws as in the slab (Sect. 5.1.3.2).

Consider the two most common types of the source movement: the source moves along the generator of the cylinder (Fig. 5.20a) and the source moves along the ring as in girth welding (Fig. 5.20b). The surface heat transfer is negligible.

Suppose that the surface point source is at the origin of a moving coordinate system (Fig. 5.20a). If the source moves along the generator of the cylinder, then in the longitudinal section, which is diametrically opposite to the source, the algebraic sum of the transverse heat fluxes q_{2y} is zero because of symmetry, i.e. this section is adiabatic and can be used to split and unfold the cylinder into a slab with adiabatic lateral surfaces ($\partial T/\partial y = 0$ at $y = -\pi R$ and $y = \pi R$, Fig. 5.20c). Then, on the basis of Eq. (5.1.60) with $\xi = \eta = \zeta = 0$, using the method of images, the solution can be obtained for an infinitely long cylinder:

$$T(x, y, z, t) - T_0 = \frac{q_0}{2\pi\lambda} \exp\left(-\frac{vx}{2a}\right) \sum_{i=-\infty}^{\infty} \sum_{j=-\infty}^{\infty} \frac{1}{R_{i,j}} \exp\left(-\frac{vR_{i,j}}{2a}\right)$$

$$\times \frac{1}{2}\left[\Phi^*\left(\frac{R_{i,j}}{\sqrt{4at}} - \sqrt{\frac{v^2t}{4a}}\right) + \exp\left(\frac{vR_{i,j}}{a}\right)\Phi^*\left(\frac{R_{i,j}}{\sqrt{4at}} + \sqrt{\frac{v^2t}{4a}}\right)\right], \tag{5.1.77}$$

where $R_{i,j} = \sqrt{x^2 + (y - j2\pi R)^2 + (z - 2ih)^2}$.

It is easy to obtain solutions for the steady state (at $t = \infty$) from Eq. (5.1.77), as it was done in Sect. 5.1.3.2. If the source is in the pipe wall ($\zeta \neq 0$), then the corresponding solution can be obtained on the basis of Eq. (5.1.60).

Now consider an option of the point source moving along the ring. It is assumed that the speed of the source is constant, $v = \text{const}$, and power q can depend on time, $q = q(t)$. This case occurs in practice in girth welding of thin-walled pipes, when it is required to obtain a weld of constant width (Sect. 1.2.1).

When the source moves along the ring (Fig. 5.20b), the heat-impermeability condition (the sum of the longitudinal fluxes q_{2x} is zero) is satisfied in a longitudinal cross-section diametrically opposite to the source at each instant of time t. An equivalent model where the source starts moving at point O is shown in Fig. 5.20d.

Using the source method and Eq. (4.2.21) as the fundamental one, a solution is obtained to the problem for a source buried at a depth of ζ, in a stationary (fixed) coordinate system:

$$T(x, y, z, t) - T_0 = \int_0^t \frac{q(\tau)}{c\rho[4\pi a(t-\tau)]^{3/2}} \sum_{i=-\infty}^{\infty} \sum_{j=-1,1} \sum_{k=-\infty}^{\infty} \exp\left(-\frac{R^2_{i,j,k}}{4a(t-\tau)}\right) d\tau, \tag{5.1.78}$$

where $R^2_{i,j,k} = (x - v\tau - k2\pi R)^2 + y^2 + (z - j\zeta - 2ih)^2$.

If the source is of constant power, $q(\tau) = \text{const}$, the integral is expressed as the known functions (Sect. 5.1.3.1).

5.1.3.4 Moving Line Source in a Plate

The procedure for determining the temperature in a plate is the same as for a semi-infinite solid (Fig. 5.7c). By using (4.2.39) in the moving reference frame, we obtain

$$dT(x, y, t) = \frac{q/h\, d\tau}{c\rho[4\pi a(t-\tau)]} \exp\left(-\frac{[x + v(t-\tau)]^2 + y^2}{4a(t-\tau)} - b(t-\tau)\right). \tag{5.1.79}$$

We integrate this equation from 0 to t, and, after some transformations, obtain

$$T(x, r, t) - T_0 = \frac{q/h}{4\pi\lambda} \exp\left(-\frac{vx}{2a}\right)$$

$$\times \int_0^t \frac{1}{t-\tau} \exp\left(-\frac{r^2}{4a(t-\tau)} - \left(\frac{v^2}{4a} + b\right)(t-\tau)\right) d\tau. \quad (5.1.80)$$

In the limiting case (at $t \to \infty$) by substitution $u = (v^2/(4a) + b)(t-\tau)$ and integral (5.1.75) we obtain (Rykalin 1938a, b)

$$T(x, r, \infty) - T_0 = \frac{q/h}{2\pi\lambda} \exp\left(-\frac{vx}{2a}\right) \mathrm{K}_0\left(\frac{vr}{2a}\sqrt{1 + \frac{4ab}{v^2}}\right). \quad (5.1.81)$$

Formula (5.1.81) without heat transfer ($b = 0$) could be obtained differently: by integrating (5.1.44) over ζ from 0 to ∞ or (5.1.64) from 0 to h. In the latter case, $F_0(r, z) = 1$, since

$$\int_0^h \cos(\pi i \zeta / h) d\zeta = 0. \quad (5.1.82)$$

The quasi-stationary temperature field in the plate (Fig. 3.1b) is qualitatively the same as in a semi-infinite solid (Fig. 5.8b). However, the isotherms in the plate are even more elongated in the direction opposite to the direction of the source motion, and the heated regions are larger than in case of a bead on a massive solid (Figs. 3.1c and 5.8d).

It is interesting to note that in a plate with a heat-insulated surface ($b = 0$), the heat content of all cross sections behind the source ($x < 0$) is constant, i.e. it does not depend on coordinate x:

$$\int_{-\infty}^{\infty} c\rho[T(x, y, \infty) - T_0] h dy = q/v. \quad (5.1.83)$$

This follows from the property of the Bessel function (Rosenthal 1946):

$$\int_0^{\infty} \mathrm{K}_0\left(\frac{v\sqrt{x^2 + y^2}}{2a}\right) dy = \frac{\pi a}{v} \exp\left(-\frac{v|x|}{2a}\right). \quad (5.1.84)$$

By substituting (5.1.81) and (5.1.84) into the left-hand side of expression (5.1.83), its right-hand side is obtained.

Consider the effect of the welding parameters (the effective power of the welding source q and its speed v) on the temperature field in the steady state (Fig. 5.21). With a decreasing speed v and $q = \text{const}$, the heated regions expand and lengthen (Fig. 5.21a, c). With an increasing power q and $v = \text{const}$, the heated regions increase in length and width (see Fig. 5.21a, b). As for a quantitative analysis, it is convenient

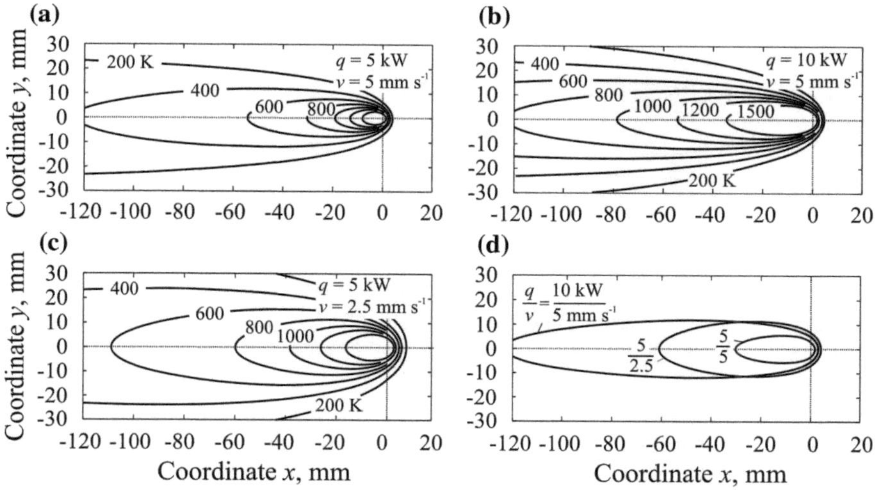

Fig. 5.21 The effect of welding conditions on the quasi-stationary temperature field $T(x, y, \infty) - T_0$ in a 10 mm thick plate (**a**, **b**, **c**) and the dimensions of isotherm $T(x, y, \infty) - T_0 = 800$ K (**d**) ($\lambda = 0.04$ W mm^{-1} K^{-1}, $a = 8$ mm^2 s^{-1}, $b = 0$)

to consider the dependence of the temperature field on the parameters of the welding conditions using an isotherm $T - T_0 = 800$ K as an example (see Fig. 5.21d). The doubled power (from 5 to 10 kW at a constant speed $v = 5$ mm s^{-1}) resulted in an increase in the length of the zone heated above 800 K by 4 times and width by 2 times, approximately. The dependence of the width and length of the isotherm on v at $q = \text{const}$ is inversely proportional. At a constant heat input per unit length ($Q_1 = q/v = 10/5 = 5/2.5$ kJ mm^{-1}), the speed has a practical effect only on the length of the heated zone (proportional dependence).

Let us consider the influence of the thermophysical properties of the metal on the quasi-stationary temperature field under the same welding conditions using the example of low-carbon steel, Cr–Ni austenitic steel, an aluminium alloy and titanium alloy (Fig. 5.22). Their properties are shown in Table 2.2.

In steels, the volume-specific heat capacity $c\rho$ is approximately the same while the thermal diffusivity a and the thermal conductivity λ are different (Table 2.2). Because of the reduced thermal conductivity, the heated region (for example, the region bounded by an isotherm of $T - T_0 = 1000$ K) in austenitic steel is longer than in low-carbon steel (Fig. 5.22a, c). Owing to the very high thermal conductivity of the aluminium alloy, the regions heated to a high temperature are much smaller than in steels and are close to a circle in shape (Fig. 5.22b).

It is convenient to use dimensionless criteria for a generalised analysis of thermal processes. The quasi-stationary temperature field (5.1.81) is represented in the form:

$$\theta_2(\xi, \rho_2) = \exp(-\xi)\mathrm{K}_0(\mu\rho_2);$$

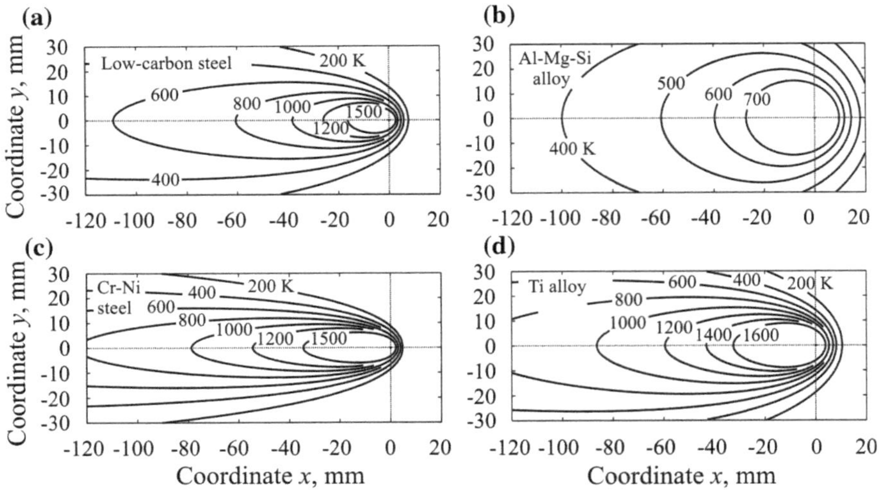

Fig. 5.22 The effect of the thermophysical properties of a metal on the temperature field $T\ (x, y, \infty) - T_0$ around a moving line source in a 10 mm thick plate: low –carbon steel ($\lambda = 0.04$ W mm^{-1} K^{-1}, $a = 8$ mm^2 s^{-1}) (**a**); Al–Mg–Si alloy ($\lambda = 0.167$ W mm^{-1} K^{-1}, $\mathbf{a} = 62$ mm^2 s^{-1}) (**b**); Cr–Ni austenitic steel ($\lambda = 0.02$ W mm^{-1} K^{-1}, $a = 4$ mm^2 s^{-1}) (**c**) and titanium alloy ($\lambda = 0.03$ W mm^{-1} K^{-1}, $a = 10$ mm^2 s^{-1}) (**d**) ($q = 5000$ W, $v = 2.5$ mm s^{-1})

$$\rho_2 = \sqrt{\xi^2 + \eta^2}, \tag{5.1.85}$$

where θ_2 is the dimensionless temperature, ξ is the dimensionless longitudinal coordinate, η is the dimensionless transversal coordinate, ρ_2 is the dimensionless plane radius vector (the dimensionless distance between the point under study and the source):

$$\theta_2 = \frac{2\pi\lambda}{q/h}(T - T_0);\ \xi = \frac{vx}{2a};\ \eta = \frac{vy}{2a};\ \mu = \sqrt{1 + \frac{4ab}{v^2}}. \tag{5.1.86}$$

The surface heat transfer is accounted by just one parameter μ. It is seen from (5.1.85) and (5.1.86) that temperature drops with an increasing coefficient of heat loss b, since function K_0 goes down with an increasing argument (see Fig. 3.14).

Figure 5.23a shows the dimensionless temperature field in a moving coordinate system, and Fig. 5.23b shows the temperature distribution $\theta_2(\xi)$ at various distances η or, which is the same, the thermal cycle $\theta(\tau)$ of fixed points located at distance η from the ξ-axis.

The dimensionless heating and cooling rate $\partial\theta_2/\partial\tau$ is determined by differentiating function (5.1.85) with respect to ξ and by changing the variable $\xi = -\tau$:

$$\frac{\partial\theta_2}{\partial\xi} = -\exp(-\xi)\left[\mathrm{K}_0(\mu\rho_2) + \mu\frac{\xi}{\rho_2}\mathrm{K}_1(\mu\rho_2)\right] \tag{5.1.87}$$

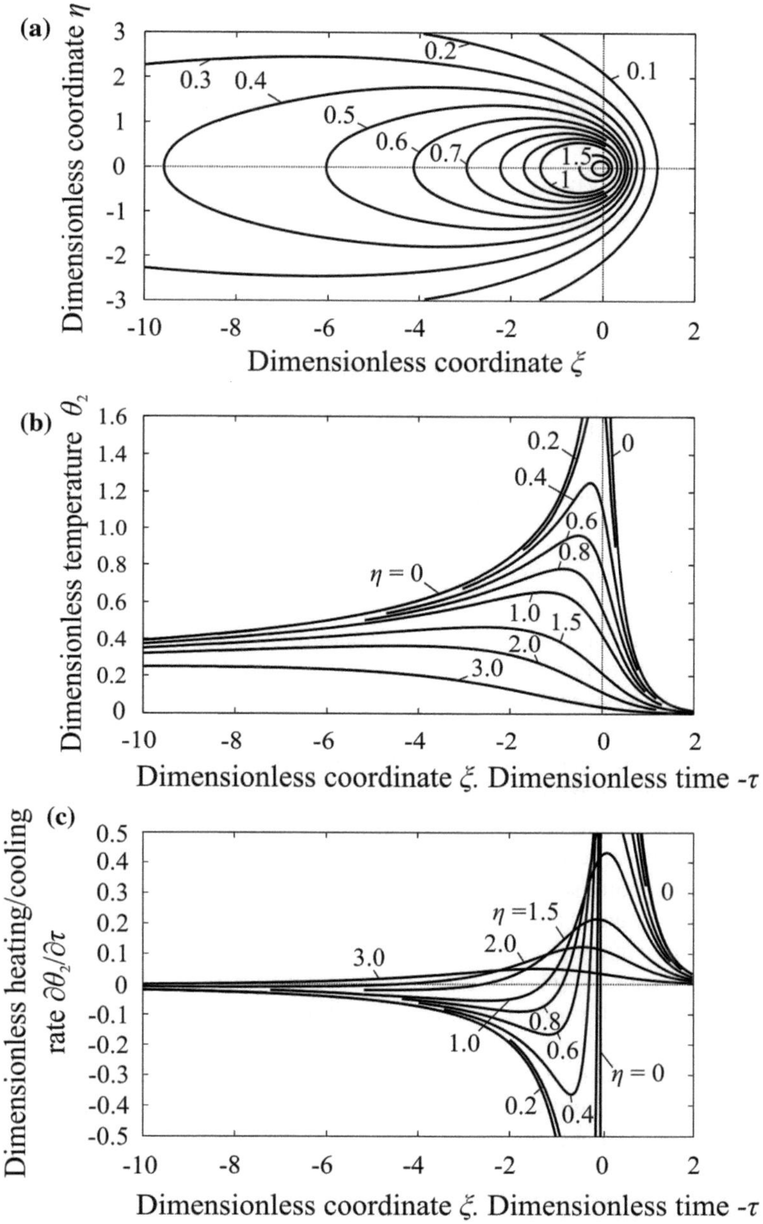

Fig. 5.23 The quasi-stationary dimensionless temperature field (**a**), the longitudinal distribution of dimensionless temperature θ_2 (**b**) and the heating/cooling rate $\partial\theta_2/\partial\tau$ (**c**) for a moving line source in an infinite plate with impermeable surface (Grong 1994; Myers et al. 2007)

and

$$\frac{\partial\theta_2}{\partial\tau} = \exp(\tau)\left[\mathrm{K}_0\left(\mu\sqrt{\tau^2+\eta^2}\right) - \mu\frac{\tau}{\sqrt{\tau^2+\eta^2}}\mathrm{K}_1\left(\mu\sqrt{\tau^2+\eta^2}\right)\right], \tag{5.1.88}$$

where $\mathrm{K}_1(u)$ is the modified Bessel function of the second kind of order 1 (Fig. 3.14), $\mathrm{K}_1(u) = -d\mathrm{K}_0(u)/du$. Here

$$\frac{\partial\theta_2}{\partial\tau} = \frac{4\pi a\lambda}{v^2 q/h}\cdot\frac{\partial T}{\partial t};\ \tau = \frac{v^2 t}{2a}. \tag{5.1.89}$$

It is seen from the comparison of Fig. 5.23 with Fig. 5.10 that the temperature fields and thermal cycles are similar.

By using Fig. 5.23b, c we build up the dependence of the cooling rate of various points of the solid on the current temperature (Fig. 5.11b). The transition from dimensionless quantities to dimensional ones is made according to formulas:

$$T - T_0 = \frac{q/h}{2\pi\lambda}\theta_2;\quad \frac{\partial T}{\partial t} = \frac{v^2 q/h}{4\pi a\lambda}\cdot\frac{\partial\theta_2}{\partial\tau}. \tag{5.1.90}$$

Formula (5.1.85) and Fig. 5.23b, corresponding to it, allow us to find the retention time above a given temperature (Fig. 5.12c). The closer the point is to the x-axis (the smaller η is), the longer this time is. The cooling time in the given temperature range can be found from the curves in Fig. 5.12d.

The algorithm for calculating the overall dimensions of the region heated above the given temperature (isotherm $\theta_2 = \theta_{2\,\max}$) in an infinite plate is analogous to that for a semi-infinite solid (Fig. 5.24).

We have considered the two extreme idealised combinations of moving concentrated sources: a point source on a semi-infinite solid and a line source in an infinite plate. The above figures allow us to obtain a range of basic characteristics of thermal processes, where many practical cases fit.

It is surface sources, which are mainly used in fusion welding. When calculating the temperature in thin plates, a model for a line source, evenly distributed across thickness, is often used. The question arises: what is the error of calculation when the surface source is replaced with a penetrating one? Let us estimate error ε provided that the temperature field is steady and there is no heat transfer. By using formulas (5.1.64) and (5.1.81), we obtain:

$$\begin{aligned}\varepsilon &= \frac{T(x,r,z,\infty) - T(x,r,\infty)}{T(x,r,z,\infty) - T_0}\\ &= 1 - \left[1 + \frac{2}{\mathrm{K}_0(\rho_2\mathrm{Pe})}\sum_{i=1}^{\infty}\cos(\pi i\zeta)\mathrm{K}_0\left(\rho_2\mathrm{Pe}\sqrt{1+\left(\frac{\pi i}{\mathrm{Pe}}\right)^2}\right)\right]^{-1},\end{aligned} \tag{5.1.91}$$

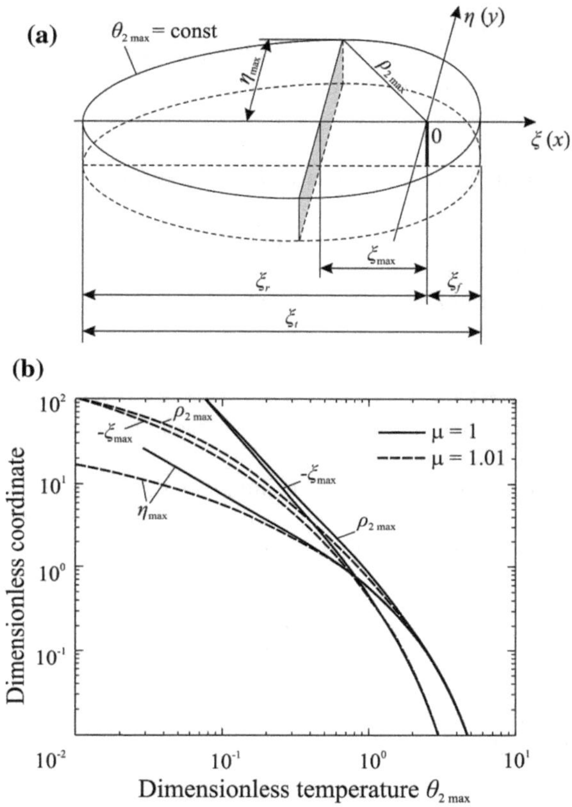

Fig. 5.24 Schematic representation of the 2D solution for a moving line source in an infinite plate (**a**) and dimensionless distance $\xi_{\max}$, half width $\eta_{\max}$ and radius vector $\rho_{2\,\max}$ as functions of a specified dimensionless steady temperature $\theta_{2\,\max}$ (**b**) (Grong 1994; Rykalin 1947, 1952)

where ρ_2 is the dimensionless distance (dimensionless plane radius vector), ζ is the dimensionless depth of the point under study, Pe is the dimensionless thickness (the Peclet number):

$$\rho_2 = \frac{\sqrt{x^2 + y^2}}{h}; \quad \zeta = \frac{z}{h}; \quad \text{Pe} = \frac{vh}{2a}. \tag{5.1.92}$$

Curves $\varepsilon\ (\rho_2)$ for several Pe values on the upper ($\zeta = 0, z = 0$) and lower ($\zeta = 1$, $z = h$) surfaces are shown in Fig. 5.18b. It is seen that near the source on the upper (heated) surface, the error is close to 1, and on the lower surface it is negative. It means that the temperature is higher on the upper surface. However, on the lower surface it is below the average temperature $T(x, r, \infty)$ determined by expression (5.1.81). With moving away from the source, the non-uniformity of the temperature distribution across the thickness decreases rapidly and does so the faster, the smaller the Peclet number Pe (lower thickness and/or speed) is. At a distance $\rho_2 = 4$ (four thicknesses) under the considered Peclet numbers, the temperature of the surfaces is practically no different from the average temperature.

5.1.3.5 Moving Line Source in a Plate of Finite Width

In Sect. 5.1.3.4, a plate with an infinite length and an infinite width was considered, and the solution to the stationary problem in a moving coordinate system was obtained. The problem remains stationary if the plate is limited by the planes parallel to the source movement.

Let the line source move in an infinitely long heat-insulated plate at distance W from the longitudinal edges. This case corresponds, for example, to the welding of two W-wide plates, the thermal cutting of a $2W$-wide plate, deposit welding over the longitudinal edge of a W-wide plate.

The quasi-stationary field in a plate of an infinite width is described by Eq. (5.1.81) at $b = 0$. To take into account the boundary conditions, the method of images based on this equation is used. As a result we obtain

$$T(x, y, \infty) - T_0 = \frac{q/h}{2\pi\lambda} \exp\left(-\frac{vx}{2a}\right) \sum_{i=-\infty}^{\infty} \mathrm{K}_0\left(\frac{v\sqrt{x^2 + (y - 2iW)^2}}{2a}\right). \quad (5.1.93)$$

Series (5.1.93) is substituted by Fourier series according to formula (4.2.25). It is known (Rosenthal 1946) that

$$\int_0^{\infty} \mathrm{K}_0\left(\frac{v\sqrt{x^2 + y^2}}{2a}\right) dy = \frac{\pi a}{v} \exp\left(-\frac{v|x|}{2a}\right); \quad (5.1.94)$$

$$\int_0^{\infty} \mathrm{K}_0\left(\frac{v\sqrt{x^2 + y'^2}}{2a}\right) \cos\frac{\pi i y'}{W} dy' = \frac{\pi a}{v\mu_i} \exp\left(-\frac{v|x|\mu_i}{2a}\right);$$

$$\mu_i = \sqrt{1 + \left(\frac{2\pi i a}{vW}\right)^2}. \quad (5.1.95)$$

Then, taking into account (4.2.25), (5.1.94) and (5.1.95), we obtain from (5.1.93)

$$T(x, y, \infty) - T_0 = \frac{q/(h2W)}{c\rho v} \exp\left(-\frac{v(x + |x|)}{2a}\right) \times \left[1 + 2\sum_{i=1}^{\infty} \frac{1}{\mu_i} \exp\left(-\frac{v|x|(\mu_i - 1)}{2a}\right) \cos\left(\frac{\pi i y}{W}\right)\right]. \quad (5.1.96)$$

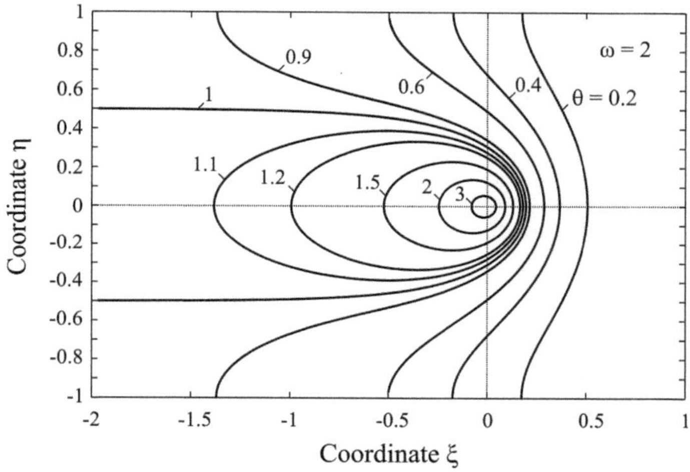

Fig. 5.25 The quasi-stationary dimensionless temperature field around a moving line source in a narrow plate (Rosenthal 1946)

The value of the first fraction in formula (5.1.96) is equal to the average temperature in the cross section of the plate behind the source. As shown in Sect. 5.1.3.10, the expression before the square brackets describes the axial temperature distribution caused by the transverse plane source in the rod without heat transfer, and the expression in the square brackets takes into account the non-uniformity of temperature distribution over the width of the plate. Formulas (5.1.93) and (5.1.96) are equivalent. They differ only in the rate of convergence: the first is preferable for a large width, and the second for a small width W.

Formula (5.1.96) can be represented in dimensionless form

$$\theta(\xi,\eta) = \exp(-\omega[\xi + |\xi|])\left\{1 + 2\sum_{i=1}^{\infty}\frac{1}{\mu_i}\exp(-\omega|\xi|[\mu_i - 1])\cos(\pi i \eta)\right\};$$

$$\theta = \frac{c\rho v}{q/(h2W)}(T - T_0);\ \xi = \frac{x}{W};\ \eta = \frac{y}{W};\ \omega = \frac{vW}{2a};$$

$$\mu_i = \sqrt{1 + (\pi i/\omega)^2}. \tag{5.1.97}$$

Here, the size of plate W is taken as a unit length.

In Fig. 5.25 the field of the dimensionless temperature θ is shown at a relative width of the plate $\omega = 2$. At some distance behind the source, the temperature field becomes linear and this distance is the lower, the smaller criterion ω is (speed v is lower, or width W is smaller). Near the source, the temperature is much higher than the average temperature $\theta = 1$.

Solutions (5.1.93) and (5.1.96) are fundamental. They can be used to solve a stationary problem for distributed sources in a plate of a finite width. The advantage

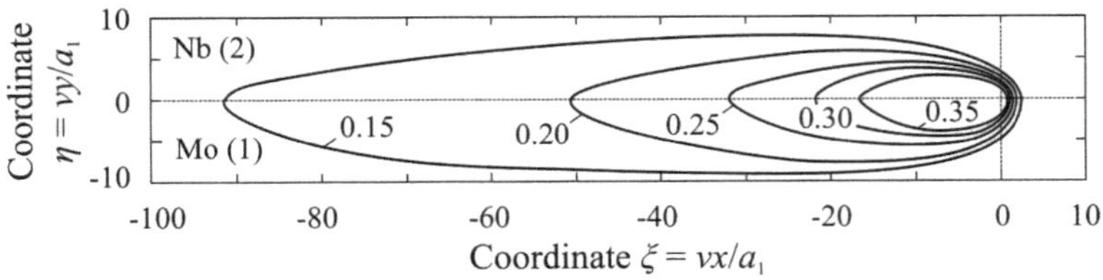

Fig. 5.26 Isotherms $\theta_j = 2\pi\lambda_1 h_1/q(T_j - T_0)$ in a composite molybdenum/niobium plate (Makhnenko 1976)

of the solution in the form of (5.1.96) is that when the method of sources is used, the integration of the product of the distribution function and the exponential function or cosine is much easier. For example, if the source is uniformly distributed over a certain interval along the x-axis, then the solution is expressed by using elementary functions in the form of an absolutely convergent series (Rosenthal 1946).

5.1.3.6 Moving Line Source in a Composite Plate

A uniformly distributed concentrated line heat source of the constant power q moves at a constant speed v along the rectilinear butt of two dissimilar plates ($y = 0$) with an ideal contact between them (Fig. 5.26). The volumetric specific heat capacities $c_1\rho_1$ and $c_2\rho_2$, thermal conductivities λ_1 and λ_2, thermal diffusivities a_1 and a_2, as well as thicknesses h_1 and h_2 within each of the two semi-bounded plates are constant. The moving temperature field at the stage of the limit state, i.e. the quasi-stationary state, is considered.

At the specified conditions, the problem can be solved using Fourier transformation method (Makhnenko 1967):

$$T_j(x, y) - T_0 = \frac{q}{\pi}\int_0^\infty u^{-1/2}\exp\left(-n_j|y|u^{1/2}\right) \times\left[K\cos\left(ux + \frac{v|y|}{2a_j}\frac{u^{1/2}}{n_j}\right) - Q\sin\left(ux + \frac{v|y|}{2a_j}\frac{u^{1/2}}{n_j}\right)\right]du, \tag{5.1.98}$$

where

$$K = \frac{\lambda_1 h_1 n_1 + \lambda_2 h_2 n_2}{(\lambda_1 h_1 n_1 + \lambda_2 h_2 n_2)^2 + v^2/4\left[\lambda_1 h_1/(n_1 a_1) + \lambda_2 h_2/(n_2 a_2)\right]^2};$$

$$Q = \frac{v}{2}\frac{\lambda_1 h_1/(n_1 a_1) + \lambda_2 h_2/(n_2 a_2)}{(\lambda_1 h_1 n_1 + \lambda_2 h_2 n_2)^2 + v^2/4\left[\lambda_1 h_1/(n_1 a_1) + \lambda_2 h_2/(n_2 a_2)\right]^2};$$

$$n_j = 2^{-1/2}\left[u + \left(u^2 + v^2/a_j^2\right)^{1/2}\right]^{1/2}.$$

Solution (5.1.98) can be also presented in dimensionless form. Figure 5.26 shows the field of dimensionless temperature in welding of a molybdenum (1) and niobium alloy (2). It is apparent that the temperature field is asymmetric with respect to the butt. In the material with a higher thermal conductivity and thermal diffusivity, isotherms expand and become more convex ($a_1 = 37$ mm^2 s^{-1} and $\lambda_1 = 0.1$ W mm^{-1} K^{-1} for molybdenum and $a_2 = 24$ mm^2 s^{-1} and $\lambda_2 = 0.06$ W mm^{-1} K^{-1} for niobium) (Makhnenko 1976). The peak temperature in cross-sections ahead of the moving source may slightly shift from the joint centreline to a material with a higher thermal conductivity. Behind the source, the peak temperature is on the axis of the source travel.

Similar results were obtained in butt gas tungsten arc welding of carbon steel and ferritic stainless steel (Ranjbarnodeh et al. 2010).

Numerical methods allow us to obtain temperature fields for different combinations of dissimilar metals, taking into consideration the temperature dependence of properties and the displacement of the source axis from the butt (Makhnenko 1976; Ryabov et al. 1984).

5.1.3.7 Moving Line Source in a Thin-Walled Pipe

As in Sect. 5.1.3.3, the thickness of the cylinder-shaped pipe h is much less than its radius R, $h \ll R$.

If the line source moves along the cylinder generatrix (Fig. 5.20a), its equivalent model will look as a plate of a limited width (see Fig. 5.20c). This case is discussed in Sect. 5.1.3.5.

If the line source moves along the ring of an infinite cylinder (see Fig. 5.20b), its equivalent model is an infinite plate (see Fig. 5.20d). The temperature field in the fixed coordinate system can be found by using (4.2.39) (Karkhin 1990):

$$T(x, y, t) - T_0 = \sum_{i=-\infty}^{\infty} \int_0^t \frac{q/h}{c\rho[4\pi a(t-\tau)]} \times \exp\left(-\frac{(x - v\tau - k2\pi R)^2 + y^2}{4a(t-\tau)} - b(t-\tau)\right) d\tau. \quad (5.1.99)$$

5.1.3.8 Moving Line Source in a Thin-Walled Cone

Like in the cases of the thin-walled cylindrical pipe (Sects. 5.1.3.3 and 5.1.3.7), the cone can be unfold into a tapered plate (Fig. 5.27). The angle at the vertex of the plate is equal to $\beta = \pi/n$, where $n = 1, 2, \ldots$ The temperature field for any path of source AB in the cone (see Fig. 5.27a) will be the same as the temperature field in the sector

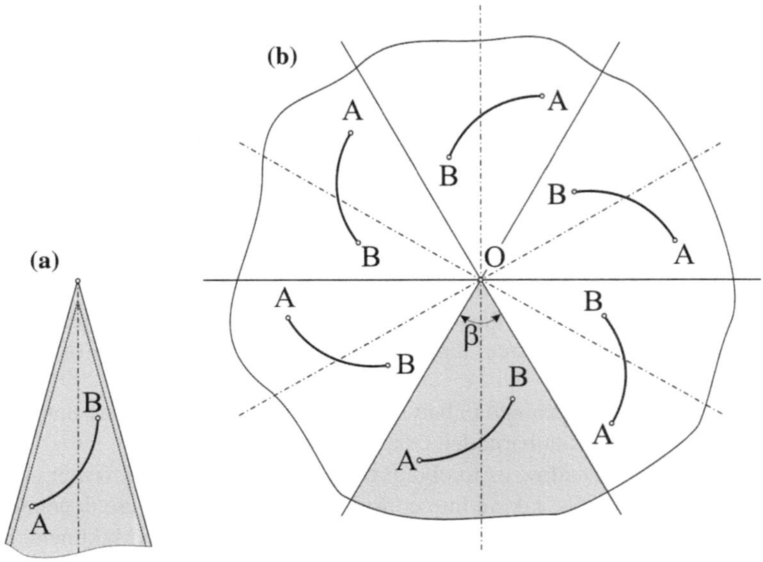

Fig. 5.27 The heat source trajectory in thin-walled cone *AB* (**a**) and an equivalent calculation model (**b**)

of the infinite plate with angle β (darkened area) due to real or fictitious sources (see Fig. 5.27b). The total number of sources equals $2n$. The temperature from each source can be found using the method of sources, based on solution (4.2.39) for the instantaneous line source in an infinite plate.

5.1.3.9 Moving Non-uniformly Distributed Line Source in a Slab

The line source with power q in a heat-insulated slab of thickness h is distributed along the z-axis according to any specified law $q_1(z)$ and move along the x-axis at a constant speed v. This source schematically describes penetrating concentrated weld heat sources (e.g., electron beam, laser beam, plasma jet).

The temperature field is found in the moving coordinate system in the limit state ($t \to \infty$). The problem will be solved by the method of sources, based upon the equation for the moving point source in a slab (5.1.64) (Karkhin 1990; Karkhin and Pilipenko 1997; Michailov and Karkhin 1997):

$$T(x, y, z, \infty) - T_0 = \frac{q/h}{2\pi\lambda} \exp\left(-\frac{vx}{2a}\right) \mathrm{K}_0\left(\frac{vr}{2a}\right) F_1(r, z), \qquad (5.1.100)$$

where

$$F_1(r,z) = 1 + 2\sum_{i=1}^{\infty} \cos\left(\frac{\pi i z}{h}\right) \mathrm{K}_0\left(\frac{vr}{2a}\sqrt{1+\left(\frac{2\pi a i}{vh}\right)^2}\right) I_i \Big/ \mathrm{K}_0\left(\frac{vr}{2a}\right);$$

$$r = \sqrt{x^2+y^2};\ I_i = \int_0^h \frac{q_1(\zeta)}{q}\cos\left(\frac{\pi i \zeta}{h}\right) d\zeta. \tag{5.1.101}$$

The right side of the expression (5.1.100) up to F_1 describes the two-dimensional temperature field due to the moving line source in the plate without surface heat transfer (5.1.81). Function $F_1(r, z)$ takes into account the deviation of the considered source from the line source of power q, equally distributed over the entire thickness of the slab (in the latter case $F_1(r, z) = 1$). The expansion of power distribution $q_1(z)$ into Fourier series in terms of cosines (5.1.101) presents no special difficulties as the Dirichlet's conditions are met in all cases of welding. For instance, for the point source in the slab $I_i = 1$ according to (1.2.7). If $q_1(z)/q = A + Bz$ in the interval z', z'', then

$$I_i = \frac{h}{\pi i}(A + Bz'')\sin\left(\frac{\pi i z''}{h}\right) + \left(\frac{h}{\pi i}\right)^2 B\cos\left(\frac{\pi i z''}{h}\right)$$
$$- \frac{h}{\pi i}(A + Bz')\sin\left(\frac{\pi i z'}{h}\right) - \left(\frac{h}{\pi i}\right)^2 B\cos\left(\frac{\pi i z'}{h}\right). \tag{5.1.102}$$

The expansions of many other functions into Fourier series can be found in professional literature (Bateman and Erdelyi 1954; Gradshteyn and Ryzhik 2014; Prudnikov et al. 1986).

It is necessary to emphasise that the piecewise linear function can be used to describe almost any distribution of power $q_1(z)$ to the specified accuracy. The source can be also distributed along x, like in multiple-electrode welding (Pilipenko et al. 1999; Pilipenko 2001). The temperature will be a sum of temperature increments produced by all separate sources. As an example, Fig. 5.28 shows the temperature field in electron beam welding of a 20 mm thick slab made of low-carbon steel under the conditions: $q = 5400$ W, $v = 8$ mm s^{-1} (Karkhin and Pilipenko 1997). The distribution of the beam effective power across the thickness was approximated by three line pieces (see Fig. 5.28a). The three-dimensional temperature field is characterised by high non-uniformity (see Fig. 5.28b–d). Since the temperature field is known, it is easy to find the change of temperature in time t at any individual points using the transformation $t = -x/v$ (see Fig. 5.28e). It follows from the figure that points B and D with the same peak temperature, thermal cycles $T(t)$, and consequently, cooling rates $\partial T/\partial t$ are significantly different. Unlike the uniformly distributed sources, curve $T(t)$ can have several maxima (sign $\partial T/\partial t$ can change several times).

The replacement of the non-uniform power distribution by a uniform one can lead to big errors in the calculation of the temperature field (Michailov and Karkhin 1997).

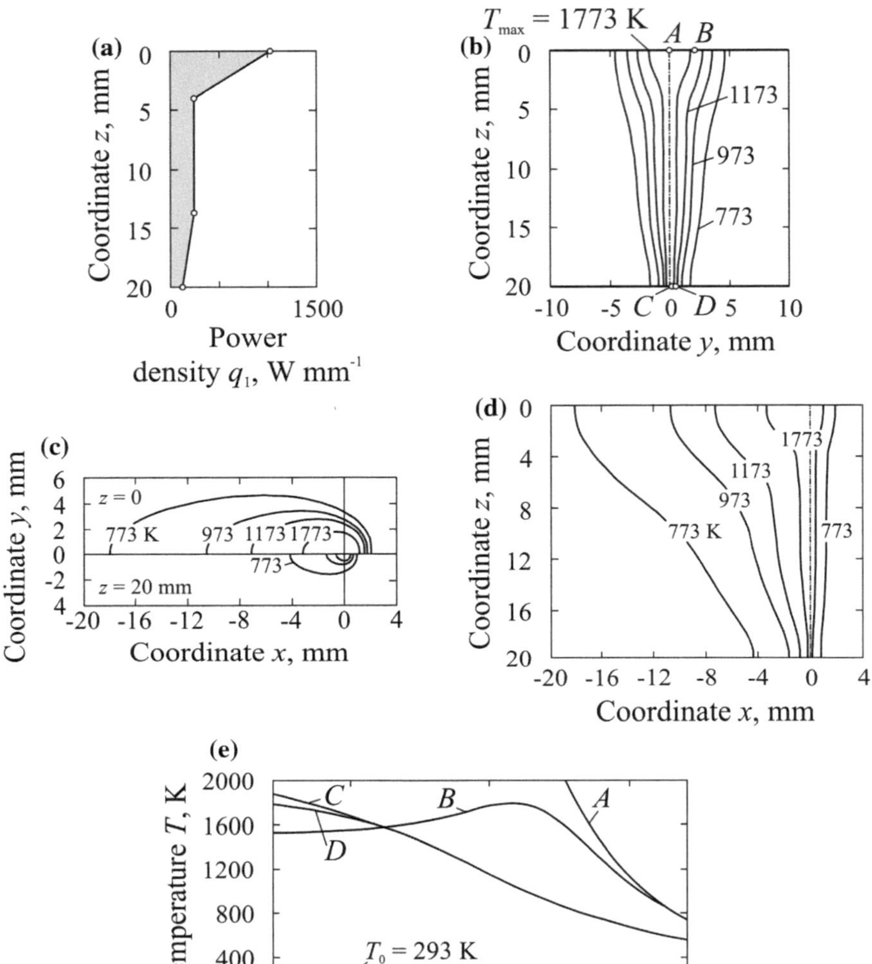

Fig. 5.28 The diagram of heat supply (**a**), peak temperature field in the cross section (**b**), quasi – stationary temperature field on the upper ($z = 0$) and lower ($z = 20$ mm) surfaces of the plate (**c**) and in the longitudinal section ($y = 0$) (**d**), thermal cycles in points A, B, C and D at the cooling stage (**e**) in electron beam welding of a 20 mm thick low–carbon steel joint (Karkhin and Pilipenko 1997)

5.1.3.10 Moving Plane Source in a Rod

It is possible to find the temperature field in a rod similarly to the cases of a semi-infinite solid and plate (Fig. 5.7d). By using (4.2.42), we obtain (Rykalin 1938a, b, 1947, 1951, 1952, 1957):

$$T(x,t) - T_0 = \int_0^t \frac{q/A}{c\rho[4\pi a(t-\tau)]^{1/2}} \exp\left(-\frac{[x+v(t-\tau)]^2}{4a(t-\tau)} - b(t-\tau)\right)d\tau$$
$$= \frac{q/A}{c\rho v\sqrt{1+\frac{4ab}{v^2}}} \exp\left(-\frac{v}{2a}\left[x + |x|\sqrt{1+\frac{4ab}{v^2}}\right]\right)$$
$$\times \frac{1}{2}\left[\Phi^*\left(\frac{\rho_1}{2\sqrt{\tau}} - \sqrt{\tau}\right) - \exp(2\rho_1)\Phi^*\left(\frac{\rho_1}{2\sqrt{\tau}} + \sqrt{\tau}\right)\right];$$
$$\rho_1 = \frac{|x|v}{2a}\sqrt{1+\frac{4ab}{v^2}};\ \tau = \frac{v^2 t}{4a}\left(1+\frac{4ab}{v^2}\right). \qquad (5.1.103)$$

The steady-state temperature ($t = \infty$)

$$T(x,\infty) - T_0 = \frac{q/A}{c\rho v\sqrt{1+\frac{4ab}{v^2}}} \exp\left(-\frac{v}{2a}\left[x + |x|\sqrt{1+\frac{4ab}{v^2}}\right]\right). \qquad (5.1.104)$$

The distribution of the steady-state temperature $T(x, \infty)$ is shown in Fig. 5.29. It follows that in a rod without surface heat transfer (with $b = 0$), the temperature in front of the source falls sharply [according to the law $\exp(-vx/a)$], while behind it is constant and equals to $q/(Ac\rho v)$. The heat transfer reduces the temperature all over the rod.

5.1.3.11 Moving Point Source in a Solid Cylinder

It is assumed that a point source with coordinates ρ, θ', ζ moves parallel to the axis of cylinder z (Fig. 3.16f). Represent this moving continuous source as a set of elementary instantaneous point sources in an infinite solid cylinder (4.2.43). By integrating with respect to time from 0 to t for temperature increments induced by elementary sources and by using integral (5.1.40), we obtain the solution in a moving reference frame (Carlson et al. 2013):

$$T(r,\theta,z,t) - T_0 = \frac{q/(\pi R)^2}{c\rho v} \exp(-\frac{v}{2a}[(z-\zeta) + |z-\zeta|])$$
$$\times {\sum_{n=0}^{\infty}}' \cos(n(\theta-\theta')) \sum_{m=1}^{\infty} \frac{A_{n,m}}{B_{n,m}} \frac{\mathrm{J_n}(\mu_{n,m}\frac{r}{R})\mathrm{J_n}(\mu_{n,m}\frac{\rho}{R})}{\mathrm{J_n}(\mu_{n,m})}$$

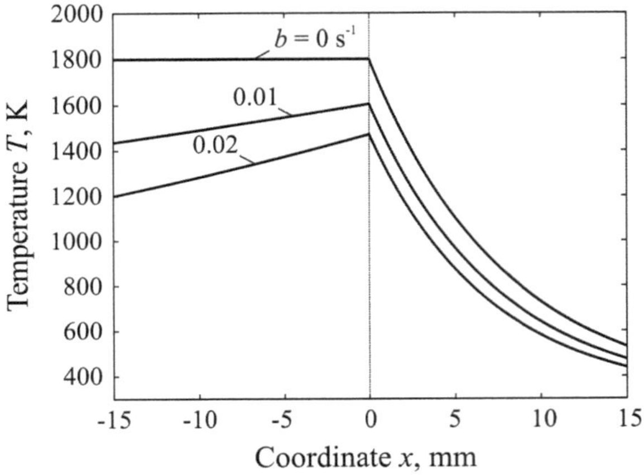

Fig. 5.29 Steady temperature distribution $T\ (x, \infty)$ due to a plane source in an infinite rod at various coefficients of heat loss b ($q/A = 7.5$ W mm^{-2}, $v = 1$ mm s^{-1}, $c\rho = 0.005$ J mm^{-3} K^{-1}, $a = 8$ mm^2 s^{-1}, $T_0 = 300$ K)

$$\times\left[\frac{1}{2}\exp\left(\frac{v|z-\zeta|}{2a}(1-B_{n,m})\right)\Phi^*\left(\frac{|z-\zeta|}{\sqrt{4at}}-B_{n,m}\sqrt{\frac{v^2t}{4a}}\right)\right.$$
$$\left.-\frac{1}{2}\exp\left(\frac{v|z-\zeta|}{2a}(1+B_{n,m})\right)\Phi^*\left(\frac{|z-\zeta|}{\sqrt{4at}}+B_{n,m}\sqrt{\frac{v^2t}{4a}}\right)\right];$$
$$B_{n,m}=\sqrt{1+\left(\frac{2\mu_{n,m}a}{Rv}\right)^2} \qquad (5.1.105)$$

If the heating is quite long, (theoretically $t = \infty$), the cylinder quasi-stationary state is described by the expression:

$$T(r,\theta,z,\infty)-T_0=\frac{q/(\pi R)^2}{c\rho v}\exp\left(-\frac{v}{2a}[(z-\zeta)+|z-\zeta|]\right)$$
$$\times\sum_{n=0}^{\infty}{}'\cos(n(\theta-\theta'))\sum_{m=1}^{\infty}\frac{A_{n,m}}{B_{n,m}}\frac{\mathrm{J_n}\left(\mu_{n,m}\frac{r}{R}\right)\mathrm{J_n}\left(\mu_{n,m}\frac{\rho}{R}\right)}{\mathrm{J_n}(\mu_{n,m})}$$
$$\times\exp\left(\frac{v|z-\zeta|}{2a}(1-B_{n,m})\right). \qquad (5.1.106)$$

The effect of the heat transfer surfaces is manifested through coefficients $A_{n,\,m}$ [see (4.2.43)].

Laser beam which heats solid copper wire during laser brazing (Fig. 5.30) can be an example of a moving point source on the solid cylinder. It follows from the figure that under the specified heating conditions, the temperature is equalised in the wire

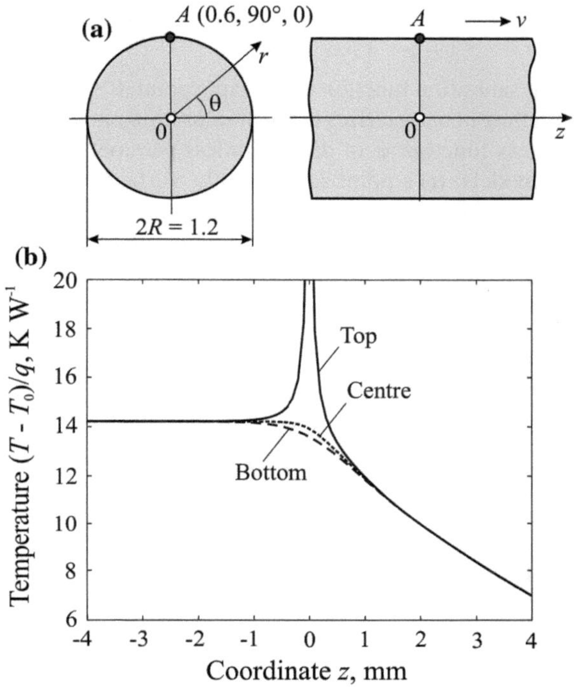

Fig. 5.30 The steady temperature distribution along the wire axis due to a moving point heat source at point *A* ($v = 16.7$ mm s^{-1}, $c\rho = 0.0037$ J mm^{-3} K^{-1}, $a = 94$ mm^2 s^{-1}, $\alpha = 0$, $T_0 = 300$ K): a moving reference frame (**a**) and the temperature distribution at top ($r = 0.6$ mm, $\theta = 90°$), the centre ($r = 0$) and the bottom ($r = 0.6$ mm, $\theta = 270°$) (**b**) (Carlson et al. 2013)

cross-section behind the beam at a distance equal to the diameter of wire (with $z =$ 1.2 mm).

5.1.3.12 Heat Saturation and Temperature Equalisation

Heat saturation period

During a long exposure to a concentrated constant power source, moving rectilinearly at a constant speed, the moving temperature field, associated with the source, tends to the limit quasi-stationary state. The duration of heat propagation from the beginning of the exposure ($t = 0$) till the establishment of the limit state is known as the *heat saturation* period. In theory, this period is infinitely long; in practice, in most cases of welding it lasts only a few seconds for a high-heated area. Therefore, the study of the limit state is of great practical interest.

The temperature increment is presented at a point in the heat saturation period in a form of product:

$$T(t) - T_0 = \psi(t)[T(\infty) - T_0], \tag{5.1.107}$$

where ψ is the heat saturation function of the same point, $0 \leq \psi(t) \leq 1$. The growth of function ψ with time t characterises the heat saturation rate for a point of the solid.

The dimensionless function ψ of dimensionless parameters is expressed for the following simple models: (i) a point source on the surface of a semi-infinite solid, (ii) a line source in the plate, and (iii) a plane source in a rod.

(i) The three-dimensional (volumetric) temperature field due to the surface point source on the semi-infinite solid (5.1.39) is described by the formula for the limit state (5.1.41) and the heat saturation function (Rykalin 1947, 1952):

$$\begin{aligned}\psi_3(R,t) &= \frac{T(R,x,t) - T_0}{T(R,x,\infty) - T_0}\\ &= \frac{1}{2}\left[\Phi^*\left(\frac{R}{\sqrt{4at}} - \sqrt{\frac{v^2 t}{4a}}\right) + \exp\left(\frac{vR}{a}\right)\Phi^*\left(\frac{R}{\sqrt{4at}} + \sqrt{\frac{v^2 t}{4a}}\right)\right]\end{aligned} \tag{5.1.108}$$

or in dimensionless form (Rykalin 1947, 1952):

$$\psi_3(\rho_3, \tau) = \frac{1}{2}\left[\Phi^*\left(\frac{\rho_3}{2\sqrt{\tau}} - \sqrt{\tau}\right) + \exp(2\rho_3)\Phi^*\left(\frac{\rho_3}{2\sqrt{\tau}} + \sqrt{\tau}\right)\right], \tag{5.1.109}$$

where

$$\rho_3 = \frac{vR}{2a}; \quad \tau = \frac{v^2 t}{4a}.$$

Here, ρ_3 is the dimensionless coordinate (the Peclet number), τ is the dimensionless time ($\tau = \mathrm{Fo}(\mathrm{Pe})^2$, Fo is the Fourier number, $\mathrm{Fo} = at/R^2$). The curves $\psi_3(\rho_3, \tau)$ are shown in Fig. 5.31. The closer the points are to the source (smaller distance ρ_3), the earlier their temperature approaches the limit. At the application point of the source ($\rho_3 = 0$), the limit state occurs instantly ($\psi_3 = 1$ with $\tau = 0$). In practice, the heat saturation effect is clearly visible, for example, in the increased depth of fusion at the beginning of the weld.

(ii) The two-dimensional (plane) temperature field due to a line source in an infinite plate (5.1.80) is described by limit state formula (5.1.81) and heat saturation function:

$$\begin{aligned}\psi_2(r,t) = \frac{T(r,x,t) - T_0}{T(r,x,\infty) - T_0} &= \frac{1}{2}\left[\mathrm{K}_0\left(\frac{vr}{2a}\sqrt{1 + \frac{4ab}{v^2}}\right)\right]^{-1}\\ &\times \int_0^t \frac{1}{t}\exp\left(-\frac{r^2}{4at} - \frac{v^2}{4a}\left(1 + \frac{4ab}{v^2}\right)t\right)dt\end{aligned} \tag{5.1.110}$$

or in dimensionless form

$$\psi_2(\rho_2, \tau) = \frac{1}{2\mathrm{K}_0(\rho_2)} \int\limits_0^{2\tau/\rho_2} \frac{1}{u} \exp\left(-\frac{\rho_2}{2}\left(u + \frac{1}{u}\right)\right) du, \tag{5.1.111}$$

where ρ_2 and τ are the dimensionless coordinate and time:

$$\rho_2 = \frac{vr}{2a}\sqrt{1 + \frac{4ab}{v^2}}; \quad \tau = \frac{v^2 t}{4a}\left(1 + \frac{4ab}{v^2}\right). \tag{5.1.112}$$

The curves $\psi_2(\rho_2, \tau)$ are shown in Fig. 5.32 (solid lines). Function $\psi_2(\rho_2)$ has the plane symmetry with respect to the line source. Heat saturation processes in the plate (see Fig. 5.32) and those of the semi-infinite solid (see Fig. 5.31) are similar.

It is convenient to use the approximation formula in calculation

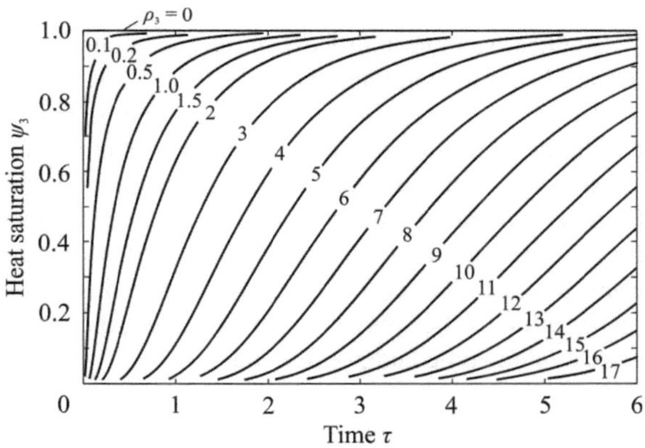

Fig. 5.31 Heat saturation function ψ_3 over dimensionless time $\tau = v^2t/(4a)$ for dimensionless distance $\rho_3 = vR/(2a)$, the moving point source on a semi-infinite solid

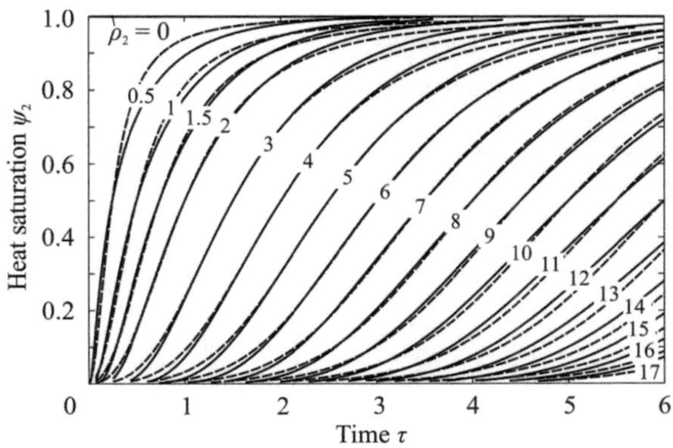

Fig. 5.32 Heat saturation function ψ_2 over dimensionless time $\tau = v^2t/(4a)(1 + 4ab/v^2)$ for dimensionless distance $\rho_2 = vr/(2a)(1{+}4ab/v^2)^{1/2}$, the moving line source in an infinite plate

$$\psi_2(\rho_2, \tau) = \frac{1}{1 + \left(\frac{\rho_2}{2\tau}\right)^{1.7+0.4\rho_2}}. \tag{5.1.113}$$

The curves, calculated on the basis of this formula, are shown in Fig. 5.32 as broken lines. This approximate function $\psi_2(\rho_2, \tau)$ coincides with the exact function (5.1.111) in limiting cases: $\psi_2(\rho_2, 0) = 0$, $\psi_2(\rho_2, \infty) = 1$, $\psi_2(\infty, \tau) = 0$.

(iii) The one-dimensional temperature field due to the plane source in the infinite rod (5.1.103) is described by the limit state formula (5.1.104) and the heat saturation function (Rykalin 1947, 1952):

$$\psi_1(\rho_1, \tau) = \frac{1}{2}\left[\Phi^*\left(\frac{\rho_1}{2\sqrt{\tau}} - \sqrt{\tau}\right) - \exp(2\rho_1)\Phi^*\left(\frac{\rho_1}{2\sqrt{\tau}} + \sqrt{\tau}\right)\right], \tag{5.1.114}$$

where ρ_1 and τ are the dimensionless coordinate and time:

$$\rho_1 = \frac{v|x|}{2a}\sqrt{1 + \frac{4ab}{v^2}}; \quad \tau = \frac{v^2 t}{4a}\sqrt{1 + \frac{4ab}{v^2}}. \tag{5.1.115}$$

The heat saturation process in the rod occurs qualitatively in the same way as in the semi-infinite solid and the plate (Fig. 5.33). From the comparison of Figs. 5.31, 5.32 and 5.33, it follows that the heat saturation in the plate occurs more slowly than that in the semi-infinite solid, but quicker than that in the rod.

Note that using the above-mentioned method, heat saturation functions can be found for the point source, buried in the semi-infinite solid, slab or cylinder, those for the line source in the plate of a limited width and other source-solid combinations, but in these cases the functions depend on additional geometrical parameters (penetration of the source, size of the solid, etc.), and, thus, such formulas turn out to be cumbersome and not illustrative.

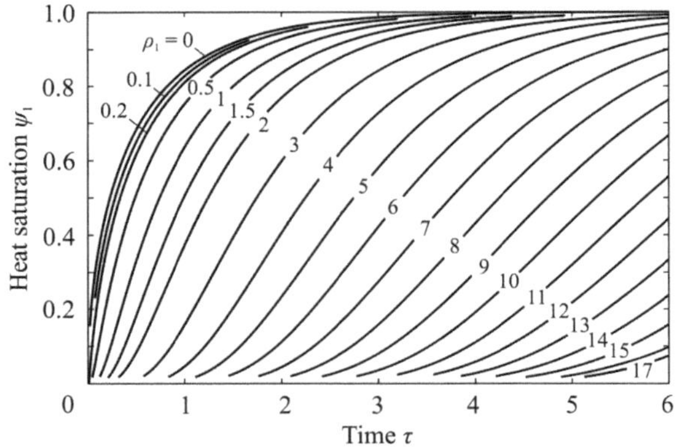

Fig. 5.33 Heat saturation function ψ_1 over dimensionless time $\tau = v^2 t/(4a)(1 + 4ab/v^2)$ for dimensionless distance $\rho_1 = v|x|/(2a)(1 + 4ab/v^2)^{1/2}$, the moving plane source in an infinite rod

Temperature equalisation period

Due to the heating by the source, the temperature rises (region I, Fig. 5.34a). When the source ceases to act ($t > t_w$), the temperature equalises (region II, Fig. 5.34a). At any time t of the equalisation period, it is possible to find the temperature by formulas of type (5.1.39), (5.1.80) or (5.1.103) depending on the calculation model, but time t in the integration limit should be replaced by the actual action time of the source t_w:

$$T(t) - T_0 = \int_0^{t_w} f(t - \tau)d\tau, \qquad (5.1.116)$$

where f is the corresponding integrand. Integral (5.1.116) can be presented in another way:

$$T(t) - T_0 = \int_0^{t} f(t - \tau)d\tau - \int_{t_w}^{t} f(t - \tau)d\tau. \qquad (5.1.117)$$

The physical interpretation of the last expression is in the presentation of the actual source (see Fig. 5.34b) as a pair: the source (see Fig. 5.34c), continuously

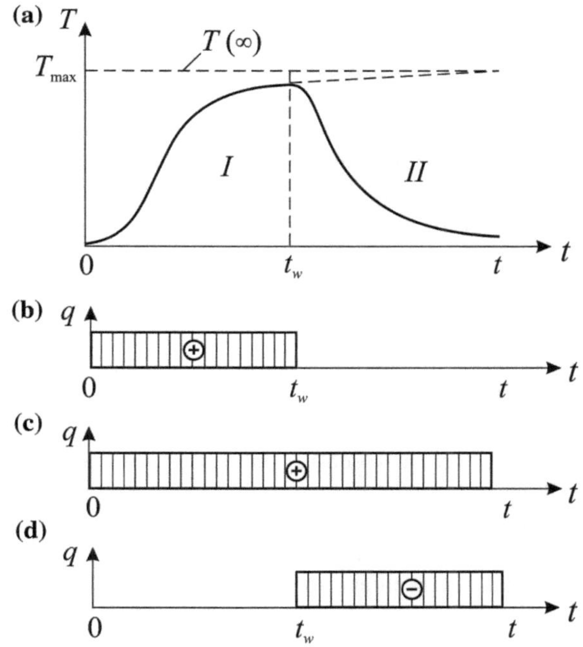

Fig. 5.34 A model for analysis of the heat diffusion process in the equalising period, after the action of a source of constant power has ceased: the temperature variation over time in a moving reference frame (**a**), an actual source (**b**), the introduction of an imaginary source (**c**) and an imaginary sink (**d**)

acting within the interval (0, t), and a sink (see Fig. 5.34d), continuously acting within the interval (t_w, t).

If the moving or stationary source is pulsed, then

$$T(t) - T_0 = \sum_{i=1}^{n} \int_{t_{s\,i}}^{\min(t_{f\,i},\, t)} f(t-\tau)d\tau, \qquad (5.1.118)$$

where n is the number of pulses per time t; $t_{s\,i}$ and t_{fi} are the start and end times of the ith pulse respectively.

If the mentioned integrals are not expressed as the known functions, it is advisable to use integrals in form (5.1.116) and determine them by using numerical methods; if they are, then integrals in the form (5.1.117) should be used. In the latter case, the temperature can be expressed by using the temperature of the limit state and the corresponding heat saturation functions:

$$\begin{aligned} T(t) - T_0 &= [T(\infty) - T_0]\,\psi(t) - [T(\infty) - T_0]\,\psi(t - t_w) \\ &= [T(\infty) - T_0]\,[\psi(t) - \psi(t - t_w)]. \end{aligned} \qquad (5.1.119)$$

The above-mentioned approach can be applied to the stationary sources. In this case, the corresponding formulas are used: (5.1.25), (5.1.27), (5.1.29), (5.1.31), (5.1.35), etc.

Note that formulas (5.1.41), (5.1.81) and (5.1.104), which describe quasi-stationary temperature fields around point, line and plane sources in a semi-infinite solid, plate and rod, are the most cited in the literature dedicated to thermal welding processes. Following N. N. Rykalin (Rykalin 1938a, b, 1941, 1947, 1951, 1952, 1957) they were derived from general transient solutions (5.1.39), (5.1.80) and (5.1.103) and in the Russian-language literature they are known as ‘Rykalin’s solutions’). Probably, for the first time, they were obtained by H. A. Wilson over 100 years ago in his solution of the stationary equation

$$\nabla^2 T + \frac{v}{a}\frac{\partial T}{\partial x} + \frac{q_3}{\lambda} = 0 \qquad (5.1.120)$$

by using the method of separation of variables and the method of sources (Wilson 1904). Later, O. F. T. Roberts solved the stationary equation for the cases of point and line sources, moving with respect to the isotropic and orthotropic medium (Roberts 1923). Later, D. Rosenthal published (without references to other authors’ works published earlier) solutions to the stationary equation for the point source on a semi-infinite solid, the point source in an infinite slab, the line source in an infinite plate, and the plane source in an infinite rod. In the English-language literature, these solutions are called ‘Rosenthal’s solutions’.

5.1.4 Rapidly Moving Concentrated Heat Sources

Earlier, in Sect. 5.1.3.4, as exemplified by the line source in a plate, it was shown that with an increasing source speed v and a proportional growth in its power q (heat input per unit length $q/v = \text{const}$), the size of the area heated above the certain temperature, increases: the area length increases in proportion to the speed, while its width varies insignificantly (Fig. 5.21). Behind the source, the distance between isotherms in the longitudinal direction (on the x-axis) is much higher than in the transverse direction (on the y-axis), that is $|\partial T/\partial x| \ll |\partial T/\partial y|$. It means that longitudinal fluxes along the x-axis are much smaller than the fluxes along the y-axis according to Fourier's law, $q_{2x} \ll q_{2y}$. The higher the speed v is, the more valid this inequality is.

The models for rapidly moving sources are based on the additional assumption that there are no longitudinal fluxes ($q_{2x} = 0$), which, in turn, makes it possible to reduce the dimensionality of the heat conduction problem by one and, consequently, greatly simplify the calculation models. This advantage of rapidly moving sources has made them popular for calculating temperature fields in welding. The main source-solid combinations reviewed in previous Sect. 5.1.3 are discussed below.

5.1.4.1 Rapidly Moving Point Source in a Semi-infinite Body

First, consider the point source on a semi-infinite solid. Separate a thin transverse layer (with the thickness of $h = 1$) out of the solid, with the introduced heat input per unit length $Q_1 = q/v$ (J m^{-1}) at the instant the heat source crosses the layer (Fig. 5.35a). According to the accepted assumption, the heat propagates only in the radial direction r, which corresponds to the model for the instantaneous line heat source Q_1 in a semi-infinite plate $h = 1$ thick or, which is the same, that for source $2Q_1$ in an infinite plate without surface heat transfer (Fig. 5.35c). By substituting $2q/v$ for Q/h in formula (4.2.37), we obtain

$$T(r,t) - T_0 = \frac{q/v}{2\pi\lambda t}\exp\left(-\frac{r^2}{4at}\right), \tag{5.1.121}$$

where r is the distance from the point under study P to the axis of the source movement x, $r^2 = y^2 + z^2$; t is the time counted from the instant when the source crosses the section, in which point P is located.

It is possible to present the solid behind the source as a set of thin transverse layers with coordinate x, which differ only in the reference time t (see Fig. 5.35a). By substituting $t = -x/v$ into (5.1.121), an equation is obtained to describe the volumetric temperature field in the moving coordinate system:

$$T(x,r) - T_0 = -\frac{q}{2\pi\lambda x}\exp\left(\frac{r^2 v}{4ax}\right). \tag{5.1.122}$$

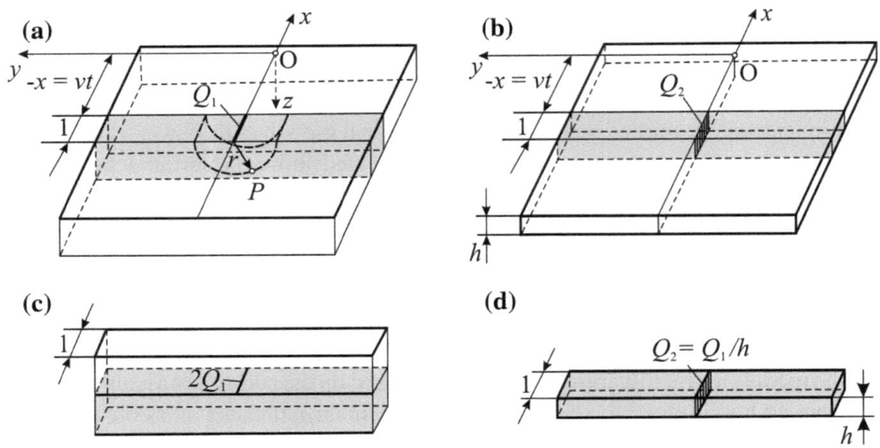

Fig. 5.35 A rapidly moving point source on a semi-infinite solid (**a**) and a line source in an infinite plate (**b**), separated transverse layers of unit thickness with line (**c**) and plane (**d**) sources

Here $x < 0$ (the formula with $x > 0$ does not make sense as, according to the assumption, the solid does not heat up ahead of the source).

Note that formula (5.1.122) can be derived from (5.1.41) using the passage to the limit at $v \to \infty$ providing that $Q_1 = q/v = \text{const}$ (the heat input per unit length is constant), when the assumption concerning the absence of longitudinal fluxes is strictly satisfied (the weld is made simultaneously along its entire length). For this, let us substitute $x = -vt$ into Eq. (5.1.41), where t is the time counted from the instant when the source crosses the cross section where the point under study is located (Fig. 5.35a) (Frolov 1988; Negoda 2008):

$$T(y, z, t) - T_0 = \frac{q/v}{2\pi\lambda t\sqrt{1 + (y^2 + z^2)/(vt)^2}} \times \exp\left(-\frac{v^2 t}{2a}\left[-1 + \sqrt{1 + \frac{y^2 + z^2}{(vt)^2}}\right]\right). \qquad (5.1.123)$$

At $v \to \infty$, the fraction under the radical sign is much less than 1, so

$$\sqrt{1 + \frac{y^2 + z^2}{(vt)^2}} \approx 1 + \frac{1}{2}\frac{y^2 + z^2}{(vt)^2}. \qquad (5.1.124)$$

Taking into account the approximation (5.1.124), we get from (5.1.123)

$$T(y, z, t) - T_0 = \frac{q/v}{2\pi\lambda t}\exp\left(-\frac{y^2 + z^2}{4at}\right), \qquad (5.1.125)$$

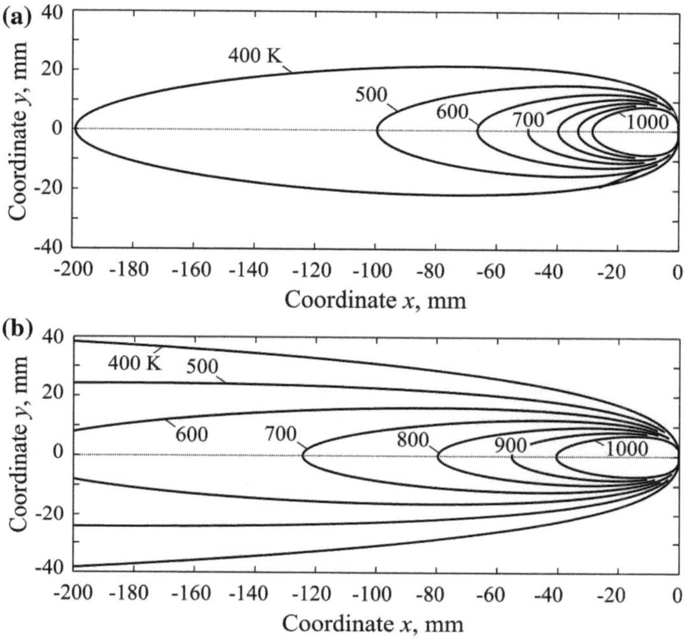

Fig. 5.36 Temperature fields around a rapidly moving heat source ($q = 5000$ W, $v = 5$ mm s^{-1}, $\lambda = 0.04$ W mm^{-1} K^{-1}, $a = 8$ mm^2 s^{-1}, $\alpha = 0$, $T_0 = 300$ K): a point source on a semi-infinite solid (at $z = 0$) (**a**); a line source in a 10 mm thick infinite plate (**b**)

which coincides with Eq. (5.1.121).

Figure 5.36a shows the field around the rapidly moving point source on the surface of semi-infinite solid $T(x, y, 0)$ in the moving coordinate system, associated with the source. Isotherms are stretched along the longitudinal x-axis and converge to the point of the source position (to the origin of coordinates).

For the longitudinal axis ($r = 0$, $x < 0$), expression (5.1.122) is simplified:

$$T(x, 0) - T_0 = -\frac{q}{2\pi\lambda x}, \tag{5.1.126}$$

which coincides with the general formula (5.1.41). From (5.1.126) it is possible to find the length of the liquid pool L_m, by substituting the melting temperature T_m:

$$L_m = \frac{q}{2\pi\lambda(T_m - T_0)}. \tag{5.1.127}$$

It can be seen that the pool length is proportional to the source power and does not depend on the welding speed.

Let us find the radial distribution of the peak temperature $T_{\max}(r)$, width W_m, and depth H_m for the area heated above the predetermined temperature.

Differentiate function $T(r, t)$ (5.1.121) with respect to time:

$$\frac{\partial T}{\partial t} = \frac{q/v}{2\pi\lambda t^2}\exp\left(-\frac{r^2}{4at}\right)\left(\frac{r^2}{4at} - 1\right). \tag{5.1.128}$$

Set this derivative equal to zero, which is the condition for the maximum of the function, and find the root $t_{\max}$: $t_{\max} = r^2/(4a)$. By substituting $t_{\max}$ in (5.1.121), we obtain

$$T_{\max}(r) - T_0 = T(r, t_{\max}) - T_0 = \frac{2}{\pi \mathrm{e}}\frac{q/v}{c\rho}\frac{1}{r^2}. \tag{5.1.129}$$

It follows that the function $T_{\max}(r)$ is a hyperboloid of rotation (Fig. 5.37a). It allows us to determine the width of the area heated above the predetermined temperature. For example, having $T_{\max}$ equal to the melting temperature T_m, $T_{\max}(r_m) = T_m$, it is possible to find the width $W_m = 2r_m$ and the depth $H_m = r_m$ of the liquid pool:

$$W_m = 2H_m = 2\sqrt{\frac{2}{\pi \mathrm{e}}\frac{q/v}{c\rho(T_m - T_0)}}. \tag{5.1.130}$$

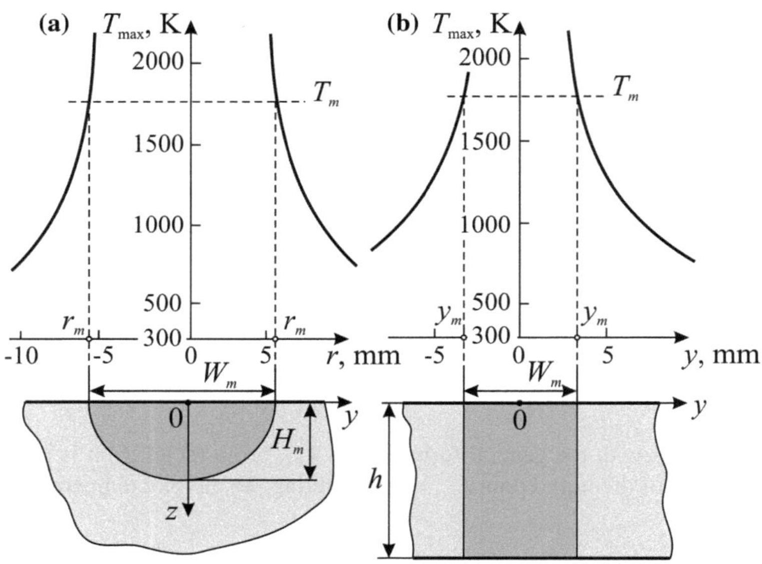

Fig. 5.37 The peak temperature distribution ($q = 5000$ W, $v = 5$ mm s^{-1}, $a = 8$ mm^2 s^{-1}, $c\rho = 0.005$ J mm^{-3} K^{-1}, $T_m = 1800$ K, $\alpha = 0$, $T_0 = 300$ K): a rapidly moving point source on a semi-infinite solid (**a**) and a rapidly moving line source in a 10 mm thick infinite plate (**b**)

It is seen that the pool width and depth depend on the heat input per unit length *q*/*v* according to the square root law. The liquid pool length was found earlier [see Eq. (5.1.127)].

In dimensionless form, it is possible to obtain temperature θ_2, thermal cycles θ_2 (τ) and peak temperature $\theta_{2\,\max}$ in a manner similar to formulas (5.1.9)–(5.1.12), taking into account the condition $Q/h = 2\,q/v$:

$$\theta_2(\tau) = \frac{1}{\tau}\exp\left(-\frac{1}{4\tau}\right); \tag{5.1.131}$$

$$\frac{\partial\theta_2}{\partial\tau} = \frac{1}{\tau^2}\left(\frac{1}{4\tau} - 1\right)\exp\left(-\frac{1}{4\tau}\right); \tag{5.1.132}$$

$$\theta_{2\,\max} = 4/\mathrm{e}, \tag{5.1.133}$$

where

$$\theta_2 = \frac{2\pi c\rho r^2}{q/v}(T - T_0); \quad \frac{\partial\theta_2}{\partial\tau} = \frac{2\pi c\rho r^4}{q/va}\cdot\frac{\partial T}{\partial t}; \quad \tau = \frac{at}{r^2}. \tag{5.1.134}$$

The form of curves θ_2 and $\partial\theta_2/\partial\tau$ is shown in Fig. 5.1. They allow us to define the time the metal dwells above the given temperature, the time the metal retentions within the given temperature interval, and the cooling rate at the given temperature for any point, distant from the longitudinal axis at a given distance.

The analysed formulas (5.1.121) and (5.1.128) show that the cooling rate $\partial T/\partial t$ at the given temperature T has only a weak dependence on the coordinates of point x, y, z, if its peak temperature $T_{\max}$ (x, y, z) is much higher than temperature T (see Fig. 5.11). Therefore, we shall determine the cooling rate only for the axis of the source (weld), by using the models for rapidly moving sources, as in this case it is possible to find the explicit dependence $\partial T/\partial t$ $(0, 0, 0, T)$.

With $r = 0$ from (5.1.128), we have

$$\frac{\partial T(0,t)}{\partial t} = -\frac{q/v}{2\pi\lambda t^2}. \tag{5.1.135}$$

The time to attain temperature T can be found from (5.1.121) at $r = 0$:

$$t = \frac{q/v}{2\pi\lambda(T - T_0)}. \tag{5.1.136}$$

By substituting (5.1.136) in (5.1.135), we have the cooling rate for the weld:

$$\frac{\partial T(0,T)}{\partial t} = -2\pi\lambda\frac{(T - T_0)^2}{q/v}. \tag{5.1.137}$$

It follows from (5.1.137) that the cooling rate depends on the welding conditions (q/v), initial temperature (T_0) and material property (λ). With an increase in the heat input per unit length q/v and initial temperature T_0, the cooling rate drops sharply. For example, the preheat temperature T_0 increased from 300 to 550 K reduces the cooling rate at the temperature of 800 K (the temperature of minimum stability of austenite in quenched steels) by 4 times.

It is possible to find the cooling time of the weld axis Δt from temperature T_1 to temperature T_2 from (5.1.136):

$$\Delta t = \frac{q/v}{2\pi\lambda}\left(\frac{1}{T_2 - T_0} - \frac{1}{T_1 - T_0}\right). \tag{5.1.138}$$

With an increase in the heat input per unit length q/v and initial temperature T_0, the cooling time gets longer.

If the point source is located in a semi-infinite solid at a depth of ζ, its temperature field can be found using the method of images:

$$T(y, z, t) - T_0 = \frac{q/v}{4\pi\lambda t}\sum_{j=-1,1}\exp\left(-\frac{r_j^2}{4at}\right); \tag{5.1.139}$$

$$T(x, y, z) - T_0 = -\frac{q}{4\pi\lambda x}\sum_{j=-1,1}\exp\left(\frac{r_j^2 v}{4ax}\right), \tag{5.1.140}$$

where $r_j^2 = y^2 + (z - j\zeta)^2$. All the temperature field characteristics additionally depend on depth ζ, so all formulas for calculating the peak temperature, cooling rate, and so on, get more intricate, while their graphical representation is devoid of the simplicity inherent in case of a surface source. Isothermal surfaces are not surfaces of rotation; they are stretched towards the surface of the solid. In a special case, with $\zeta = 0$ (surface source) we get formula (5.1.122).

This naturally brings up the question of the calculation error for the temperature $T(q_{2x} = 0)$, estimated from the approximate model of the rapidly moving source (Sect. 5.1.4.1), compared to the temperature estimated from the general equation (Sect. 5.1.3.1):

$$\varepsilon = \frac{T(q_{2x} \neq 0) - T(q_{2x} = 0)}{T(q_{2x} \neq 0) - T_0}. \tag{5.1.141}$$

Consider the simplest case of the surface source. By substituting (5.1.122) and (5.1.41) in (5.1.141), we obtain

$$\varepsilon = 1 + \frac{\sqrt{\xi^2 + \rho_2^2}}{\xi}\exp\left(\xi + \frac{\rho_2^2}{2\xi} + \sqrt{\xi^2 + \rho_2^2}\right), \tag{5.1.142}$$

where ξ and ρ_2 are the dimensionless coordinates:

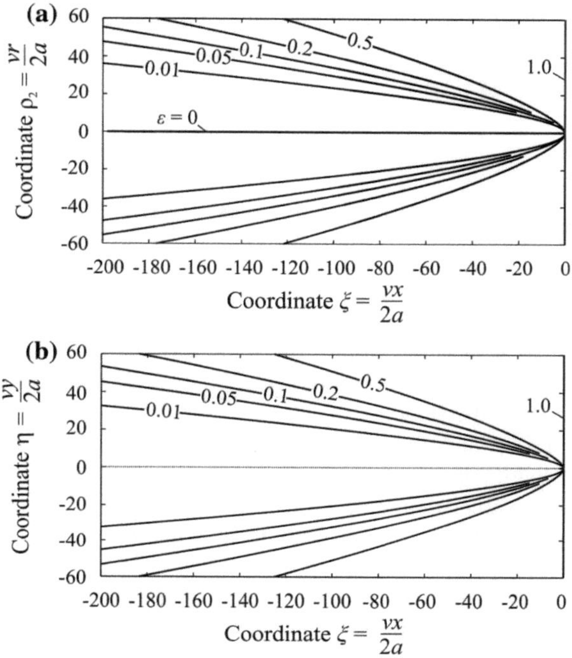

Fig. 5.38 The error of computational models for a rapidly moving point source on a semi-infinite solid (**a**) and a rapidly moving line source in a plate (**b**)

$$\xi = \frac{vx}{2a}; \quad \rho_2 = \frac{vr}{2a}. \tag{5.1.143}$$

The calculation error field for the temperature in case we utilise the model for the rapidly moving point source on a semi-infinite solid behind the source is shown in Fig. 5.38a. It is apparent that the error is always positive, as when longitudinal fluxes q_{2x} are accepted, the metal begins to heat up earlier. Error ε in the proximity of the weld is small, but on its axis $\varepsilon = 0$. On the y-axis $\varepsilon = 1$ as according to the model for the rapidly moving source, the temperature on it does not have time to increase. Thus, for the approximate calculation of the temperature in the area close to the source travel path, Eqs. (5.1.121) and (5.1.122) can be used. See the error analysis for models for rapidly moving sources in Refs. (Berezovsky 2008; Makhnenko and Yegorova 1975; Rykalin 1947, 1952).

5.1.4.2 Rapidly Moving Point Source in a Slab

Let us assume that the surface of the slab of thickness h is heat insulated. Then, based on Eqs. (5.1.121) and (5.1.122), and using the method of images, it is possible to obtain the temperature field due to the rapidly moving point source with coordinates η, ζ:

$$T(y, z, t) - T_0 = \frac{q/v}{4\pi\lambda t} \sum_{i=-\infty}^{\infty} \sum_{j=-1,1} \exp\left(-\frac{r_{i,j}^2}{4at}\right); \tag{5.1.144}$$

$$T(x, y, z) - T_0 = -\frac{q}{4\pi\lambda x} \sum_{i=1}^{\infty} \sum_{j=-1,1} \exp\left(\frac{r_{i,j}^2 v}{4ax}\right), \tag{5.1.145}$$

where $r_{i,j}^2 = (y-\eta)^2 + (z-j\zeta-2ih)^2$. The larger thickness h is, the faster the series converge. At $h \to \infty$ (a semi-infinite solid), the formulas transform into (5.1.139) and (5.1.140). The structure of formulas (5.1.121), (5.1.139), (5.1.144) and (5.1.122), (5.1.140), (5.1.145) is the same.

Formulas (5.1.144) and (5.1.145) allow us to calculate all the characteristics of the temperature field: peak temperature, cooling rate, etc. These characteristics are within the range of values that the corresponding characteristics have for analytical models for the point source in a semi-infinite solid and the line source in a plate.

5.1.4.3 Rapidly Moving Line Source in a Plate

Let us separate a thin transverse layer out of the infinite plate (Fig. 5.35b). Assuming, like in the case of the rapidly moving point source, that heat propagates only in the transverse direction (along the y-axis), we obtain the model for the instantaneous plane source in the rod of section $A = 1 \times h$ with the heat transfer through upper and lower surfaces. Replacing Q/A in (4.2.42) by $q/(vh)$, we obtain

$$T(y, t) - T_0 = \frac{q/(vh)}{c\rho(4\pi at)^{1/2}} \exp\left(-\frac{y^2}{4at} - bt\right), \tag{5.1.146}$$

where the coefficient of heat loss for a thin plate $b = 2\alpha/(c\rho h)$.

The temperature field $T(x, y)$ is obtained from (5.1.146) by substituting $t = -x/v$:

$$T(x, y) - T_0 = \frac{q/h}{c\rho(-4\pi avx)^{1/2}} \exp\left(\frac{y^2 v}{4ax} + \frac{b}{v}x\right). \tag{5.1.147}$$

The temperature field, calculated from (5.1.147) is shown in Fig. 5.36b. It follows that isotherms in the plate are more elongated than those in the semi-infinite solid as the plate cools down more slowly.

It is possible to obtain the length of the liquid weld pool (length of isotherm T_m) from (5.1.147) when $b = 0$:

$$L_m = \frac{1}{4\pi a}\left[\frac{q/h}{c\rho(T_m - T_0)}\right]^2 \frac{1}{v}. \tag{5.1.148}$$

It is seen that the pool length is proportional to the power squared and inversely proportional to the speed. It follows from here that the influence of the welding conditions in case of a plate is stronger than in case of a massive solid [cf. with (5.1.127)].

Let the transverse peak temperature distribution $T_{\max}(y)$ be found together with width W_m of the area heated above the predetermined temperature, using (5.1.146) similarly to the case of the rapidly moving point source on a semi-infinite solid:

$$\frac{\partial T}{\partial t} = -\frac{q/(vh)}{c\rho(4\pi a)^{1/2}t^{5/2}}\exp\left(-\frac{y^2}{4at} - bt\right)\left(bt^2 + \frac{1}{2}t - \frac{y^2}{4a}\right); \tag{5.1.149}$$

$$t_{\max} = \frac{1}{4b}\left(\sqrt{1 + \frac{4by^2}{a}} - 1\right)\text{for } b > 0; \tag{5.1.150}$$

$$t_{\max} = \frac{y^2}{2a}\text{for } b = 0; \tag{5.1.151}$$

$$T_{\max}(y) - T_0 = T(y, t_{\max}) - T_0 = \frac{q/(vh)}{c\rho\left[\frac{\pi a}{b}\left(\sqrt{1 + \frac{4by^2}{a}} - 1\right)\right]^{1/2}}$$
$$\times \exp\left(-\frac{y^2}{\frac{a}{b}\left(\sqrt{1 + \frac{4by^2}{a}} - 1\right)} - \frac{1}{4}\left(\sqrt{1 + \frac{4by^2}{a}} - 1\right)\right)\text{for } b > 0; \tag{5.1.152}$$

$$T_{\max}(y) - T_0 = T(y, t_{\max}) - T_0 = \frac{1}{\sqrt{2\pi \mathrm{e}}}\frac{q/(vh)}{c\rho}\frac{1}{|y|}\text{for } b = 0. \tag{5.1.153}$$

The function $T_{\max}(y)$ for $b = 0$ is an equilateral hyperbola. The width of the liquid pool (Fig. 5.37b)

$$W_m = 2y_m = \frac{2}{\sqrt{2\pi \mathrm{e}}}\frac{q/(vh)}{c\rho(T_m - T_0)}. \tag{5.1.154}$$

The length of the liquid pool was found earlier, Eq. (5.1.148).

Conclusions on the effect of the welding parameters q and v on the dimensions of the weld pool L_m and W_m, which follow from formulas (5.1.148) and (5.1.154), fully correspond to the findings from the isotherm analysis, obtained from a more precise formula (5.1.81) and Fig. 5.21d in Sect. 5.1.3.4.

Similarly to the model for the point source on the semi-infinite solid, it is possible to find the cooling rate of the weld centre in the plate without surface heat transfer ($b = 0$). At $y = 0$ from (5.1.146), we obtain

$$\frac{\partial T(0, t)}{\partial t} = -\frac{q/(vh)}{4c\rho(\pi a)^{1/2}t^{3/2}}; \tag{5.1.155}$$

$$t = \frac{1}{4\pi a}\left(\frac{q/(vh)}{c\rho(T - T_0)}\right)^2; \tag{5.1.156}$$

$$\frac{\partial T(0, T)}{\partial t} = -2\pi\lambda c\rho\frac{(T - T_0)^3}{[q/(vh)]^2}. \tag{5.1.157}$$

It follows form the comparison of (5.1.157) with (5.1.137) that with an increase in heat input per unit length q/v and initial temperature T_0, there is a decrease in the cooling rate, especially sharp in case of the plate, as exemplified by the exponents of power. For example, the increase in the preheat temperature T_0 from 300 to 550 K decreases the cooling rate at a temperature of 800 K by 8 times.

It is easy to find the cooling time for the weld axis Δt from temperature T_1 to temperature T_2 from (5.1.156):

$$\Delta t = \frac{1}{4\pi c\rho\lambda}\left(\frac{q}{vh}\right)^2\left[\frac{1}{(T_2 - T_0)^2} - \frac{1}{(T_1 - T_0)^2}\right]. \tag{5.1.158}$$

This time is the longer, the higher is the heat input per unit of the plate thickness $q/(vh)$ and the initial temperature T_0, and this effect is more pronounced than in case of the point source on the infinite solid.

Let the calculation error for the temperature $T(q_{2x} = 0)$ be estimated, obtained from the approximate model for the rapidly moving source (5.1.147), compared to the temperature $T(q_{2x} \neq 0)$ calculated from the general Eq. (5.1.81):

$$\begin{aligned}\varepsilon &= \frac{T(q_{2x} \neq 0) - T(q_{2x} = 0)}{T(q_{2x} \neq 0) - T_0}\\ &= 1 - \sqrt{\frac{\pi}{2}} \cdot \frac{1}{\sqrt{-\xi}}\exp\left(\xi + \frac{\eta^2}{2\xi}\right)\Big/ \mathrm{K}_0\left(\sqrt{\xi^2 + \eta^2}\right).\end{aligned} \tag{5.1.159}$$

where ξ and η are the dimensionless coordinates:

$$\xi = \frac{vx}{2a}; \quad \eta = \frac{vy}{2a}. \tag{5.1.160}$$

As it follows from Fig. 5.38b, the nature of the temperature calculation error, when the model for the rapidly moving line source is used, is the same as in case of the point source.

Thus, for an approximate calculation of the temperature in the area close to the source travel path, it is possible to use Eqs. (5.1.146) and (5.1.147).

It should be emphasised that formulas for rapidly moving sources are much simpler than general equations that take into account longitudinal heat fluxes. Therefore, they are widely used in the analysis of the effect of welding parameters on thermal processes and properties of joint weld metal (see Chaps. 8, 9, and 13).

5.1.4.4 Rapidly Moving Line Source in a Composite Plate

Thermal processes in butt welding of plates with different thermal and physical properties have their specific features. Let the rapidly moving line source be located on the boundary between two semi-bounded plates. Like in the case of similar plates, the plane transverse layer is separated (Fig. 5.35b). Then, heat diffusion in the transverse direction (along the y-axis) will correspond to the model for the instantaneous plane source in a composite rod with the power density $q/(vh)$, J m^{-2} (see Sect. 5.1.1.5). The temperature distribution is described by the expression similar to (5.1.21):

$$T_j(y,t) - T_0 = \frac{q/v}{(h_1\sqrt{c_1\rho_1\lambda_1} + h_2\sqrt{c_2\rho_2\lambda_2})\sqrt{\pi t}} \times \exp\left(-\frac{y^2}{4a_j t} - bt\right) \quad (j = 1, 2). \tag{5.1.161}$$

Here h_1 and h_2 stand for plate thickness; b is the coefficient of heat loss (average for plates).

From formula (5.1.161), the peak temperature distribution can be obtained, taking into account the plate dissimilarity. If there is no surface heat transfer ($b = 0$), then

$$T_{\max j}(y) - T_0 = \sqrt{\frac{2}{\pi e}} \frac{q}{v} \frac{\sqrt{a_j}}{(h_1\sqrt{c_1\rho_1\lambda_1} + h_2\sqrt{c_2\rho_2\lambda_2})} \frac{1}{|y|}. \tag{5.1.162}$$

It follows from the equation that the peak temperature falls according to the hyperbolic law and its falling rate depends both on the difference of properties and thickness, and a local thermal diffusivity of the material. The material with a higher thermal diffusivity has a lower peak temperature gradient.

Note that it is possible to reduce rapidly moving sources in the cylinder to simplified computational models for instantaneous sources. For example, it is possible to reduce a point source, moving along the cylinder generatrix, to an instantaneous line source on the disk with the heat input per unit length Q_1 (Fig. 4.11b). One can reduce a point source, moving along the cylinder ring, to an instantaneous ring source with heat input per unit length Q_1 (Fig. 4.11c); a source, moving along a helical path, to an instantaneous plane source with power density Q_2 (Fig. 4.11d). This approach reduces the problem dimensionality and simplifies its solution.

5.2 Temperature Fields of Distributed Sources

In the calculation of thermal processes we have so far assumed that a heat source is a point source, a uniformly distributed line source, or a uniformly distributed plane source. In fact, the welding sources of heat are distributed according to a more complex law. For example, a welding arc, a gas flame, a plasma jet, an electron beam, etc. are satisfactorily described by normally distributed circular (disk) sources (Chap. 1).

When we choose a mathematical model for the source, we should use the local influence principle, which in the context of welding heating can be formulated as follows: the law of heat source distribution significantly affects the temperature field only at a distance of the same order of magnitude as the dimensions of the source area. The temperature field in the area sufficiently distant from the source does not change if we replace an arbitrarily distributed heat source with a concentrated source of the same magnitude and location. For example, the temperature field at a distance from the welding arc is determined by the shape of the workpiece, i.e. it is three-dimensional in a workpiece of considerable thickness, two-dimensional in a plate, one-dimensional in a rod, three-, two- and one-dimensional in a cylinder.

Next, we will consider some cases of distributed sources in a semi-infinite body, a slab, a plate and a cylinder, which have practical importance.

5.2.1 *Instantaneous Distributed Sources*

Using instantaneous point, line and plane sources in a semi-infinite body, a slab, a plate, a rod and a cylinder (see Sect. 4.2.1), various computational models (continuous stationary and moving, rapidly moving and other sources in the corresponding bodies) were developed. Similarly, using the method of sources, similar models can be developed based on instantaneous uniformly and normally distributed line, ring, disk and other sources that represent a wide range of welding heat sources (see Chap. 1).

5.2.1.1 Instantaneous Uniformly Distributed Plane Rectangular Source on the Surface of a Semi-infinite Body and a Slab

Let us consider the temperature field in a semi-infinite body due to an instantaneous surface source whose energy Q is released at the time $t = 0$ and uniformly distributed over a rectangular part of the size $2W \times 2l$ (Fig. 5.39). The uniformly distributed area-specific energy density is $Q_2 = Q/(2W \times 2l)$, J m^{-2}. Let us divide the source into elementary point sources, with their energy being $dQ = Q_2 dx'dyz'$, J. According to (4.2.9), the temperature increment due to the instantaneous elementary point source at the point $P(x, y, z)$ is defined by the expression:

$$dT(x, y, z, t) = \frac{2Q/(2W2l)dx'dy'}{c\rho(4\pi at)^{3/2}} \exp\left(-\frac{(x-x')^2 + (y-y')^2 + z^2}{4at}\right). \tag{5.2.1}$$

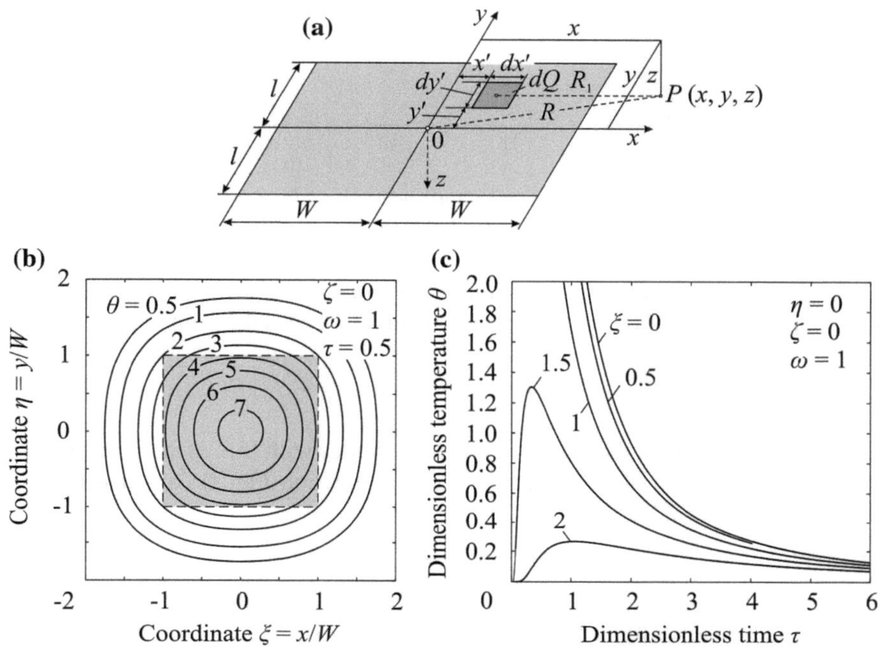

Fig. 5.39 Computational model for an instantaneous plane uniformly distributed rectangular source with its centre at point O (**a**), dimensionless temperature field θ at the surface of a semi-infinite body at dimensionless time $\tau = 0.5$ (**b**) and thermal cycles (**c**)

By integrating the expression (5.2.1) over interval $[-W, W]$ with respect to x' and over interval $[-l, l]$ with respect to y', using the substitution $u = (x - x')(4at)^{-1/2}$ and integral (Prudnikov et al. 1986)

$$\int_0^x \exp(-a^2x^2)dx = \frac{\sqrt{\pi}}{2a}\Phi(ax) \tag{5.2.2}$$

we obtain (Berezovsky 2006)

$$T(x, y, z, t) - T_0 = \frac{Q/(2W2l)}{2c\rho(4\pi at)^{1/2}}\exp\left(-\frac{z^2}{4at}\right) \times \left[\Phi\left(\frac{x+W}{\sqrt{4at}}\right) - \Phi\left(\frac{x-W}{\sqrt{4at}}\right)\right]\left[\Phi\left(\frac{y+l}{\sqrt{4at}}\right) - \Phi\left(\frac{y-l}{\sqrt{4at}}\right)\right]. \tag{5.2.3}$$

This equation is representable in a dimensionless form:

$$\theta(\xi,\eta,\zeta,\tau)=\frac{1}{\tau}\exp\left(-\frac{\zeta^2}{\tau}\right)\left[\Phi\left(\frac{\xi+1}{\sqrt{\tau}}\right)-\Phi\left(\frac{\xi-1}{\sqrt{\tau}}\right)\right]\left[\Phi\left(\frac{\eta+\omega}{\sqrt{\tau}}\right)-\Phi\left(\frac{\eta-\omega}{\sqrt{\tau}}\right)\right];\qquad(5.2.4)$$

$$\theta=\frac{8\pi^{1/2}c\rho W^2 l}{Q}(T-T_0);\ \tau=\frac{4at}{W};\quad \xi=\frac{x}{W};\quad \eta=\frac{y}{W};\quad \zeta=\frac{z}{W};\quad \omega=\frac{l}{W}.$$

Figure 5.39 shows temperature field and thermal cycles on the surface of a semi-infinite body. In the centre of the square heating zone and far from it, the isotherms have shapes close to concentric circles.

If the source is located on a slab of thickness h ($h < \infty$) with heat-insulated planes, then we apply the method of images and formula (5.2.3) and obtain

$$T(x,y,z,t)|_{h<\infty}-T_0=\left[T(x,y,z,t)|_{h=\infty}-T_0\right]F(z,t),\qquad(5.2.5)$$

where the expression in the square brackets coincides with (5.2.3), and the function $F(z, t)$ coincides with (4.2.23). As $h \to \infty$ the function $F(z, t) \to 1$ and formula (5.2.5) goes into (5.2.3).

5.2.1.2 Instantaneous Normally Distributed Circular Source in a Semi-infinite Body, a Slab and a Plate

Consider successively the schemes of instantaneous distributed sources in an infinite body, on a semi-infinite body, in a slab and in a plate.

Let the normal distribution circular (disk) source of energy Q and with the concentration factor C be applied at the time $t = 0$ to the plane $z = 0$ of the infinite body (Fig. 5.40). For a source we introduce a coordinate system ξ, η whose centre coincides with the centre of the source. The radial energy density distribution is described by an equation similar to (1.2.3):

$$Q_2(\rho)=Q_{2\,\max}\exp\left(-C\rho^2\right)=C/\pi\,Q\exp\left(-C\rho^2\right),\qquad(5.2.6)$$

where ρ is the distance in the plane from the centre, $\rho^2 = \xi^2 + \eta^2$.

Let us split the normal distribution source into elementary point sources $dQ = Q_2(\rho)d\xi d\eta$. According to (4.2.9), the temperature increment at arbitrary point $P(x, y, z)$ due to an elementary source, located at arbitrary point A (Fig. 5.40) will be:

$$dT(x,y,z,t)=\frac{Q_2(\rho)d\xi d\eta}{c\rho(4\pi at)^{3/2}}\exp\left(-\frac{R_1^2}{4at}\right),\qquad(5.2.7)$$

where $R_1^2 = (x-\xi)^2 + (y-\eta)^2 + z^2$.

By substituting (5.2.6) in (5.2.7), we obtain

$$dT(x,y,z,t)=\frac{CQ}{\pi c\rho(4\pi at)^{3/2}}\exp\left(-\frac{x^2+y^2+z^2}{4at}\right)$$

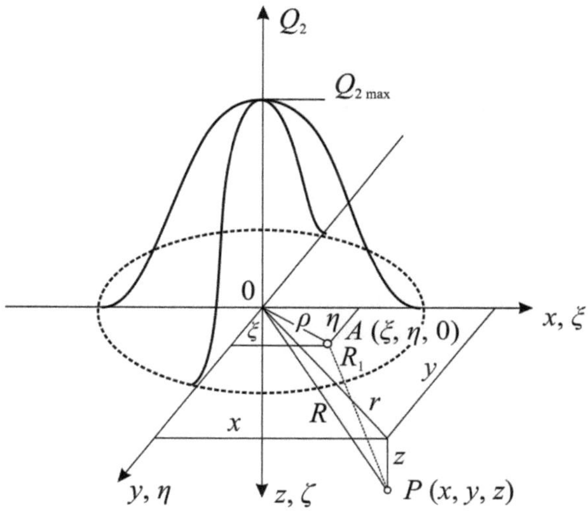

Fig. 5.40 Model for heating an infinite solid by a normal circular heat source

$$\times \exp\left(-\frac{4atC+1}{4at}\xi^2 + \frac{2x}{4at}\xi\right)\exp\left(-\frac{4atC+1}{4at}\eta^2 + \frac{2y}{4at}\eta\right)d\xi d\eta. \tag{5.2.8}$$

By integrating this expression with respect to ξ and η from $-\infty$ to $+\infty$, using the integral (Prudnikov et al. 1986)

$$\int_{-\infty}^{\infty} \exp(-pu^2 - qu)du = \sqrt{\frac{\pi}{p}}\exp\left(\frac{q^2}{4p}\right) \tag{5.2.9}$$

we obtain (Rykalin 1951, 1957)

$$T(r,z,t) - T_0 = \frac{Q}{c\rho}\left[\frac{1}{4\pi a(t+t_0)}\exp\left(-\frac{r^2}{4a(t+t_0)}\right)\right] \times \left[\frac{1}{(4\pi at)^{1/2}}\exp\left(-\frac{z^2}{4at}\right)\right], \tag{5.2.10}$$

where $r^2 = x^2 + y^2$; $t_0 = 1/(4aC)$. In this equation, the first term in square brackets with accuracy up to a constant factor describes the process of heat propagation of a fictitious instantaneous line source coinciding with the z-axis in an infinite plate (4.2.37), which acted on t_0 earlier than the real source. The second term in the square brackets expresses the process of heat propagation of a fictitious instantaneous source in an infinite rod (4.2.41), which acted at the time $t = 0$, as a real normal distribution disk source. Thus, the process of heat propagation of an instantaneous normal distribution source in an infinite body is expressed as the product of expressions for the plane and linear processes of heat propagation.

A generalisation of a disk source is an elliptic source whose energy distribution is described by the expression

$$
\begin{aligned}
Q_2(\xi,\eta) &= Q_{2\max}\exp\left(-C_x\xi^2 - C_y\eta^2\right) \\
&= \sqrt{C_xC_y}\Big/\pi\; Q\exp\left(-C_x\xi^2 - C_y\eta^2\right).
\end{aligned} \tag{5.2.11}
$$

As before, by splitting the distributed source into elementary point sources and summing the temperature increments produced by all elementary sources, we obtain

$$
\begin{aligned}
T(x,y,z,t) - T_0 &= \frac{Q}{c\rho}\frac{1}{[4\pi a(t+t_{0x})(t+t_{0y})t]^{3/2}} \\
&\quad\times\exp\left(-\frac{x^2}{4a(t+t_{0x})} - \frac{y^2}{4a(t+t_{0y})} - \frac{z^2}{4at}\right); \\
t_{0x} &= 1/(4aC_x); \quad t_{0y} = 1/(4aC_y).
\end{aligned} \tag{5.2.12}
$$

In a special case, if $C_x = C_y = C$, we obtain the solution for a normal distribution disk source (5.2.10).

In the following, solutions (5.2.10) and (5.2.12) will be used as fundamental for the corresponding sources in other bodies.

Consider a semi-infinite body. Let a normal distribution disk source be applied at the time $t = 0$ to the plane $z = \xi$. As in the case of a concentrated source (see Sect. 4.2.1.2), temperature increments can be found by the method of images, provided that the surface is heat-impermeable:

$$
\begin{aligned}
T(r,z,t) - T_0 &= \frac{Q}{c\rho}\frac{1}{4\pi a(t+t_0)(4\pi at)^{1/2}} \\
&\quad\times\exp\left(-\frac{r^2}{4a(t+t_0)}\right)\sum_{j=-1,1}\exp\left(-\frac{(z-j\zeta)^2}{4at}\right).
\end{aligned} \tag{5.2.13}
$$

If the source is on the surface ($\zeta = 0$), then

$$
\begin{aligned}
T(r,z,t) - T_0 &= \frac{2Q}{c\rho}\frac{1}{4\pi a(t+t_0)(4\pi at)^{1/2}} \\
&\quad\times\exp\left(-\frac{r^2}{4a(t+t_0)} - \frac{z^2}{4at}\right).
\end{aligned} \tag{5.2.14}
$$

Obviously, if the source is extremely concentrated ($C = \infty$, $t_0 = 0$), then formulas (5.2.13) and (5.2.14) turn into formulas (4.2.10) and (4.2.11) for the instantaneous point source.

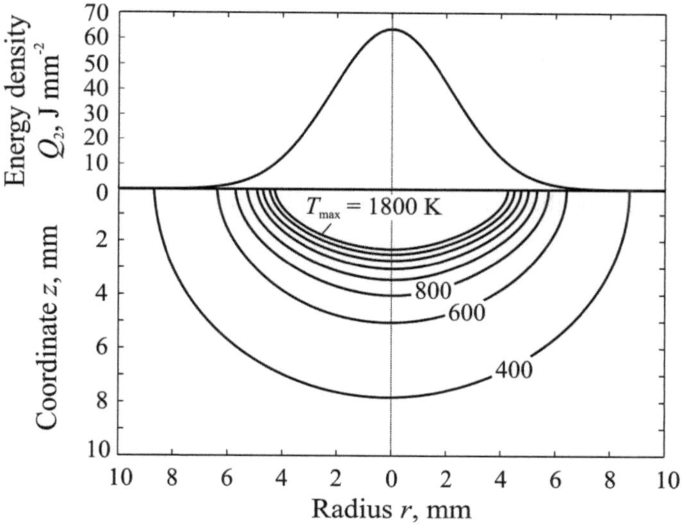

Fig. 5.41 Peak temperature field due to an instantaneous normally distributed circular source on a semi-infinite solid ($Q = 2000$ J, $C = 0.1$ mm^{-2}, $T_0 = 300$ K, $a = 8$ mm^2 s^{-1}, $c\rho = 0.005$ J mm^{-3} K^{-1})

As an example, Fig. 5.41 shows the peak temperature field. It can be seen that the isotherms are not semicircular, as in the case of a point source, but are elongated in the radial direction (in the direction of lower distribution). On the surface in the centre of the source, the peak temperature is finite, in contrast to the point source, where the peak temperature is infinite. Thus, the calculation model for the distributed source is more realistic than the model for the point source.

The temperature field generated by a surface instantaneous normal distribution elliptical source is described by expression (5.2.12), if we substitute $2Q$ instead of Q.

Let us consider a slab with heat-insulated surfaces. The temperature problem can be solved using the method of images, as in the case of a point source [formula (4.2.21)]. Taking Eq. (5.2.10) as a basis, we obtain a solution for a source located at the depth ζ:

$$T(r, z, t) - T_0 = \frac{Q}{c\rho} \frac{1}{4\pi a(t + t_0)(4\pi at)^{1/2}} \times \exp\left(-\frac{r^2}{4a(t + t_0)}\right) \sum_{i=-\infty}^{\infty} \sum_{j=-1,1} \exp\left(-\frac{(z - j\zeta - 2ih)^2}{4at}\right). \tag{5.2.15}$$

This solution can be expanded in Fourier series by means of the transformation (4.2.26):

$$T(r,z,t) - T_0 = \frac{Q/h}{4\pi\lambda(t+t_0)}\exp\left(-\frac{r^2}{4a(t+t_0)}\right) \times \left[1 + 2\sum_{i=1}^{\infty}\cos\left(\frac{\pi i z}{h}\right)\cos\left(\frac{\pi i \zeta}{h}\right)\exp\left(-\pi^2 i^2 \frac{at}{h^2}\right)\right]. \quad (5.2.16)$$

If the source is on the surface of a slab ($\zeta = 0$), formulas (5.2.15) and (5.2.16) are slightly simplified:

$$T(r,z,t) - T_0 = \frac{2Q}{c\rho}\frac{1}{4\pi a(t+t_0)(4at)^{1/2}} \times \exp\left(-\frac{r^2}{4a(t+t_0)}\right)\sum_{i=-\infty}^{\infty}\exp\left(-\frac{(z-2ih)^2}{4at}\right) \quad (5.2.17)$$

and

$$T(r,z,t) - T_0 = \frac{Q/h}{4\pi\lambda(t+t_0)}\exp\left(-\frac{r^2}{4a(t+t_0)}\right) \times \left[1 + 2\sum_{i=1}^{\infty}\cos\left(\frac{\pi i z}{h}\right)\exp\left(-\pi^2 i^2 \frac{at}{h^2}\right)\right]. \quad (5.2.18)$$

The last two formulas are equivalent, they differ only in the rate of convergence: the first series converges faster for small Fourier numbers $\mathrm{Fo} = at/h^2$ and the second one for large numbers.

If Figs. 5.42 and 5.43 are compared, we see that under the same heating conditions the temperature in the slab is greater than in a semi-infinite solid, especially near the heat-impermeable bottom surface.

Similarly, we can obtain a solution for a normal distribution elliptic source, taking as a basis the formula (5.2.12).

Now let us consider an instantaneous volumetric source normally distributed in an infinite plate. The radial distribution of the source is described by Eq. (5.2.6). We split the source into elementary sources with an area of $\rho d\varphi d\rho$, where the amount of heat released at the moment $t = 0$ was $dQ(\rho) = C/\pi\, Q\, \exp(-C\rho^2)\rho d\varphi d\rho$ (Fig. 5.43).

According to (4.2.39), the temperature increment due to this source at an arbitrary point of the plate $P(r)$ will be:

$$dT(r,t) = \frac{dQ(\rho)/h}{4\pi\lambda t}\exp\left(-\frac{r_1^2}{4at} - bt\right), \quad (5.2.19)$$

where $r_1^2 = \rho^2 + r^2 - 2r\rho\cos(\varphi)$ (Fig. 5.43). Summing all elementary sources by using an integral (Gradshteyn and Ryzhik 2014)

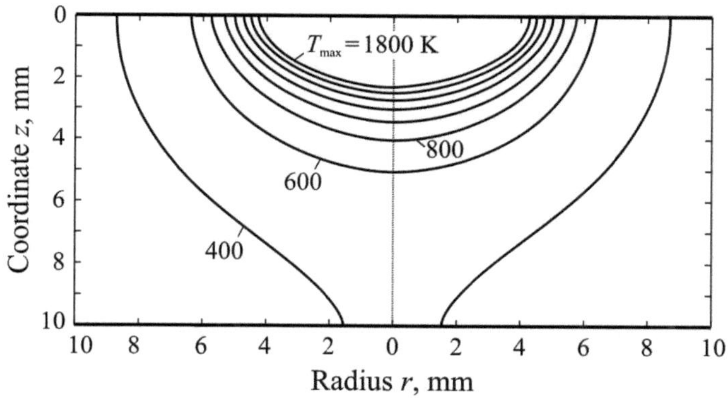

Fig. 5.42 Peak temperature field around an instantaneous normally distributed circular source on a 10 mm thick slab ($Q = 2000$ J, $C = 0.1$ mm^{-2}, $T_0 = 300$ K, $a = 8$ mm^2 s^{-1}, $c\rho = 0.005$ J mm^{-3} K^{-1})

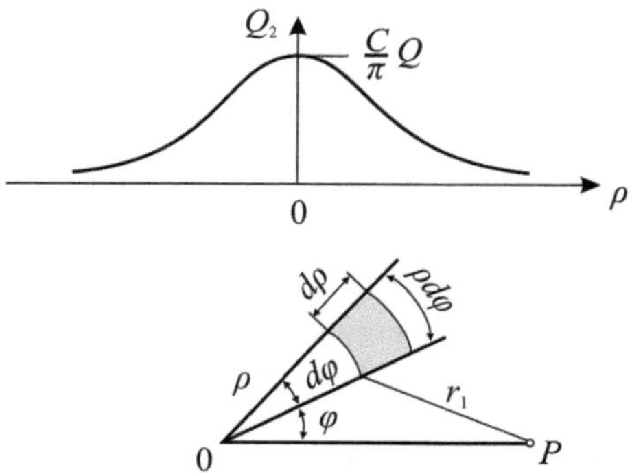

Fig. 5.43 Model for a normally distributed circular source

$$\int_0^\infty u\ \exp(-\alpha u^2)\mathrm{I}_0(\beta u)du = 1/(2\alpha)\exp(\beta^2/(4\alpha)); \tag{5.2.20}$$

$$\mathrm{I}_0(\beta u) = 1/(2\pi)\int_0^{2\pi} \exp(-\beta u\cos(\varphi))d\varphi \tag{5.2.21}$$

we obtain the temperature in an infinite plate:

$$T(r,t) - T_0 = \int_0^\infty \int_0^{2\pi} \frac{CQ/h}{4\pi^2\lambda t}\exp\left(-C\rho^2 - \frac{\rho^2 + r^2 - 2r\rho\ \cos(\varphi)}{4at} - bt\right)d\varphi\rho d\rho$$

$$= \frac{Q/h}{4\pi\lambda(t+t_0)} \exp\left(-\frac{r^2}{4a(t+t_0)} - bt\right), \tag{5.2.22}$$

where $t_0 = 1/(4aC)$. In the limit, when $C \to \infty$ ($t_0 \to 0$) the normal distribution disk source is defined by δ-function (1.2.6), and formula (5.2.22) turns into (4.2.39). Obviously, when we take the distribution of the considered source into account, we formally add t_0 to the time of the action of the line concentrated source.

If the source is elliptic (5.2.11), the temperature field can be found by the same algorithm:

$$T(x, y, t) - T_0 = \frac{Q/h}{4\pi\lambda[(t+t_{0x})(t+t_{0y})]^{1/2}} \times \exp\left(-\frac{x^2}{4a(t+t_{0x})} - \frac{y^2}{4a(t+t_{0y})} - bt\right), \tag{5.2.23}$$

where $t_{0x} = 1/(4aC_x)$; $t_{0y} = 1/(4aC_y)$.

5.2.1.3 Instantaneous Uniformly Distributed Ring Source on a Semi-infinite Body and in a Slab and Cylindrical Surface Source in a Plate

Axisymmetric heat sources are widely used in welding and related technologies: uniformly distributed disk sources, normally distributed radius-limited sources, etc. They can be represented as a set of radially concentrated ring sources whose energy density depends on the radius, but does not depend on the azimuth angle (the energy density is constant over the ring). So, the solution of the heat conduction problem for uniformly distributed ring sources is of practical interest.

Let the heat Q, J be uniformly distributed along a circle of the radius r_0 with the centre at the origin (Fig. 5.44a) on the thermally insulated surface of a semi-infinite solid at the initial time $t = 0$. This initial distribution can be considered as a set of instantaneous point sources of heat $dQ = Q/(2\pi r_0) r_0 d\phi$, J. The increment of temperature due to an elementary point source on a circle (r_0, φ) is defined by Eq. (4.2.11)

$$dT(R_1, t) = \frac{2Q/(2\pi r_0) r_0 d\varphi}{c\rho(4\pi at)^{3/2}} \exp\left(-\frac{R_1^2(\varphi)}{4at}\right), \tag{5.2.24}$$

where R_1 is the distance between observation point P and the elementary source (Fig. 5.44a).

We sum the temperature increments caused by elementary sources:

$$T(R_1, t) - T_0 = \frac{2Q}{c\rho(4\pi at)^{3/2}} \frac{1}{2\pi} \int_0^{2\pi} \exp\left(-\frac{R_1^2(\varphi)}{4at}\right) d\varphi. \tag{5.2.25}$$

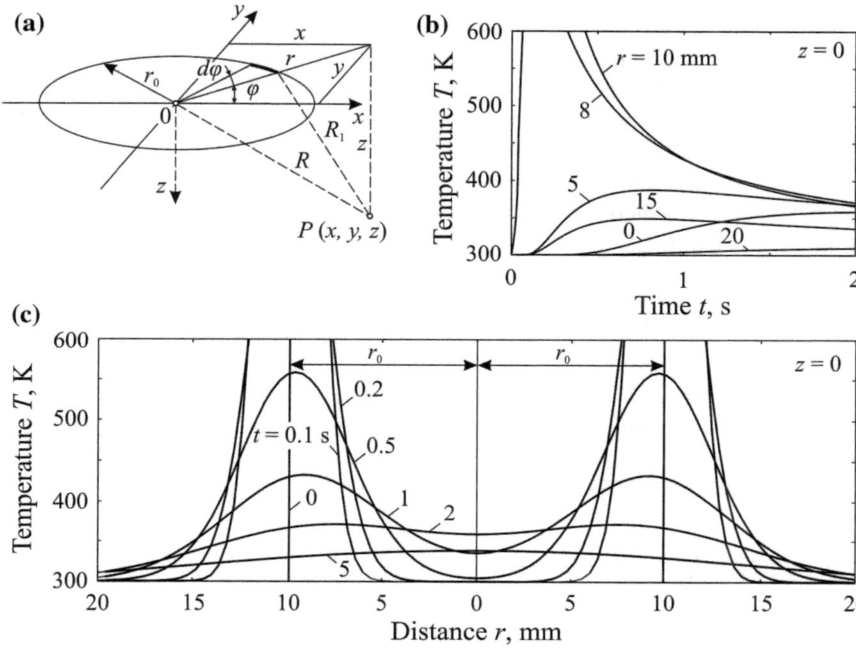

Fig. 5.44 Model for source energy (**a**), thermal cycles of surface points (**b**) and temperature distribution on a semi-infinite solid due to instantaneous uniformly distributed ring source (**c**) ($Q = 2000$ J, $c\rho = 0.005$ J mm^{-3} K^{-1}, $a = 8$ mm^2 s^{-1}, $r_0 = 10$ mm, $T_0 = 300$ K)

If we observe that

$$R_1^2 = r_0^2 + R^2 - 2rr_0 \cos\varphi, \tag{5.2.26}$$

the integral can be expressed in terms of the modified Bessel function of the first kind of order zero:

$$\mathrm{I}_0(u) = \frac{1}{2\pi} \int_0^{2\pi} \exp(u \cos\varphi) d\varphi. \tag{5.2.27}$$

By substituting (5.2.26) and (5.2.27) in (5.2.25), we obtain (Carslaw and Jaeger 1973; Rykalin 1947, 1952)

$$T(r, R, t) - T_0 = \frac{2Q}{c\rho(4\pi at)^{3/2}} \exp\left(-\frac{r_0^2 + R^2}{4at}\right) \mathrm{I}_0\left(\frac{rr_0}{2at}\right). \tag{5.2.28}$$

As the radius of the circle decreases ($r_0 \to 0$), this equation turns into the equation for the instantaneous point source on the semi-infinite solid (4.2.11), since $I_0(0) = 1$ (see Fig. 3.14). On the z-axis, the temperature drops according to the normal law.

Figure 5.44b, c show the thermal cycles of points and the radial temperature distribution in the surface of the body. Clearly, thermal cycles are more complex than in the case of a point source (see Fig. 4.3).

If the source is located in an infinite body, then the right-hand side of Eq. (5.2.28) should be divided by 2:

$$T(r, R, t) - T_0 = \frac{Q}{c\rho(4\pi at)^{3/2}} \exp\left(-\frac{r_0^2 + R^2}{4at}\right) I_0\left(\frac{rr_0}{2at}\right). \tag{5.2.29}$$

Let the instantaneous uniformly distributed ring source be located at depth ζ in an infinite slab of thickness h with heat-insulated surfaces. Using the formula for an infinite body (5.2.29), it is easy to obtain a solution for a slab by means of the method of images (see Sect. 4.2.1.3):

$$\begin{aligned} T(r, z, t) - T_0 = {} & \frac{Q}{c\rho(4\pi at)^{3/2}} \exp\left(-\frac{r_0^2 + r^2}{4at}\right) \\ & \times I_0\left(\frac{rr_0}{2at}\right) \sum_{i=-\infty}^{\infty} \sum_{j=-1,1} \exp\left(-\frac{(z - j\zeta - 2ih)^2}{4at}\right). \end{aligned} \tag{5.2.30}$$

If $h \to \infty$ (the solid is semi-infinite), in the series with respect to i there will be only one term for $i = 0$.

Suppose now that in an infinite heat-insulated plate of thickness h at the initial time $t = 0$ the heat Q is uniformly distributed over the surface of a circular cylinder of the radius r_0 with the centre at the origin (Fig. 5.45a). As before, we split the distributed source into elementary line sources and sum the temperature increments generated by elementary sources. As a result, we obtain (Carslaw and Jaeger 1973; Kartashov 2001; Rykalin 1947, 1952)

$$T(r, t) - T_0 = \frac{Q/h}{c\rho(4\pi at)} \exp\left(-\frac{r_0^2 + r^2}{4at}\right) I_0\left(\frac{rr_0}{2at}\right). \tag{5.2.31}$$

In the centre of the circle ($r = 0$) Eq. (5.2.31) has the form (4.2.37), since $I_0(0) = 1$. This means that at the origin the temperature produced by the instantaneous source Q which is uniformly distributed along the cylindrical surface of radius r_0, is equal to temperature (4.2.37) on this surface produced by the instantaneous line source located on the axis of the cylinder.

At the beginning of the temperature equalisation process, i.e. when time t is small, we use the asymptotic decomposition for a large argument ($u > 10$) of function $I_0(u) = \exp(u)(2\pi u)^{-1/2}$ and from (5.2.31) we obtain a simplified expression (Rykalin 1947, 1952):

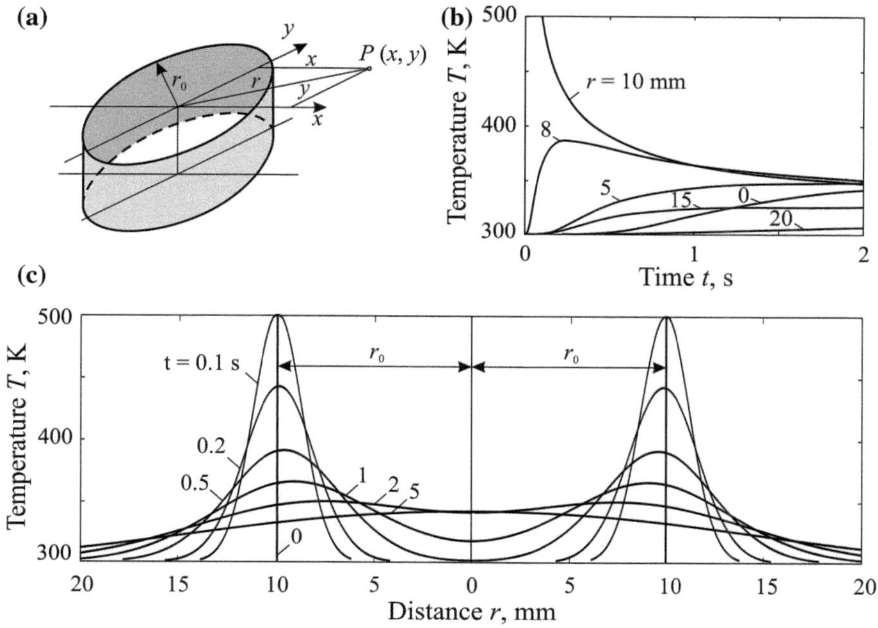

Fig. 5.45 Model for instantaneous uniformly distributed cylindrical surface heat source (**a**), thermal cycles (**b**) and temperature distribution in a 10 mm thick infinite plate (**c**) ($Q = 2000$ J, $c\rho = 0.005$ J mm^{-3} K^{-1}, $a = 8$ mm^2 s^{-1}, $r_0 = 10$ mm, $T_0 = 300$ K)

$$T(r,t) - T_0 = \frac{Q/h}{4\pi^{3/2} c\rho\sqrt{r r_0 a t}} \exp\left(-\frac{(r_0 - r)^2}{4at}\right). \tag{5.2.32}$$

It follows that at the beginning of the equalisation the heat from an instantaneous source distributed over a cylindrical surface propagates according to the law of a plane source in a rod (the structure of formula (5.2.32) is similar to (4.2.41)). From Fig. 5.45c it follows that for $t \leq 0.2$ s the curvature of the plane field does not have a noticeable effect on the temperature distribution.

After a long time, the fluxes of heat are directed from the axis of the cylinder, as in the case of a line source located at the origin (curve $t = 5$ s in Fig. 5.45c). The thermal cycles of the plate points are similar to those of a semi-infinite solid (see Figs. 5.44b and 5.45b).

5.2.1.4 Instantaneous Uniformly Distributed Disk Source on a Semi-infinite Body and in a Plate

The heat Q, J at the initial time $t = 0$ is uniformly distributed on a semi-infinite solid with a heat-insulated surface over the area of the circle of radius r_0 with the

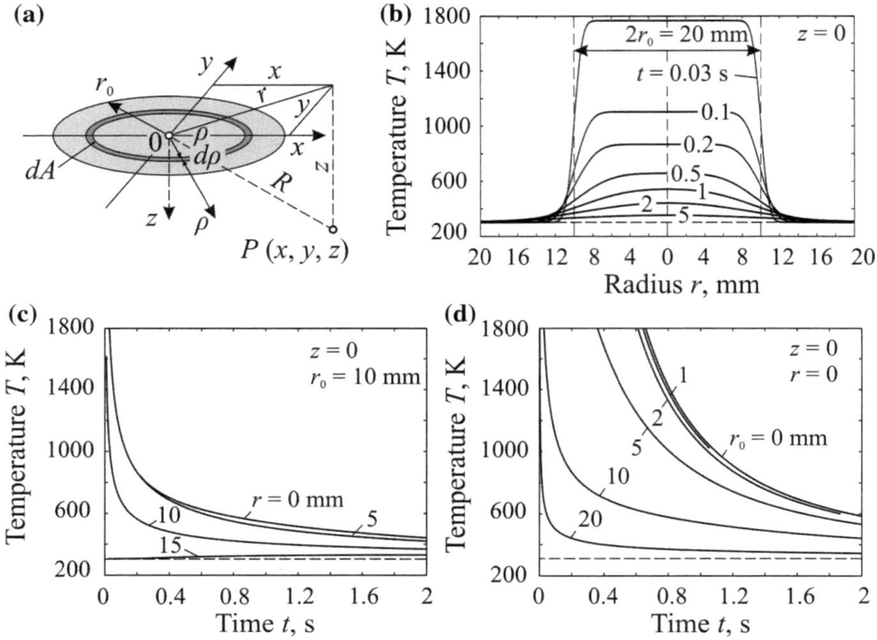

Fig. 5.46 Model for instantaneous uniformly distributed disk source (**a**), temperature distribution (**b**) and thermal cycles of surface points (**c**, **d**) due to the surface source on a semi-infinite solid ($Q = 2000$ J, $c\rho = 0.005$ J mm^{-3} K^{-1}, $a = 8$ mm^2 s^{-1}, $T_0 = 300$ K)

centre at the origin (Fig. 5.46a). Let us represent the source as a set of elementary ring heat sources $dQ = Q_2 dA$, where $Q_2 = Q/\left(\pi r_0^2\right)$ is the energy density, J m^{-2}, $dA = 2\pi\rho d\rho$ is the area of the ring of radius ρ (Fig. 5.46a). Using Eq. (5.2.28) and summing the temperature increments generated by all elementary sources, we obtain (Carslaw and Jaeger 1973)

$$T(r, z, t) - T_0 = \frac{4Q}{r_0^2 c\rho (4\pi a t)^{3/2}} \exp\left(-\frac{r^2 + z^2}{4at}\right) \times \int_0^{r_0} \exp\left(-\frac{\rho^2}{4at}\right) \mathrm{I}_0\left(\frac{r\rho}{2at}\right) \rho d\rho. \tag{5.2.33}$$

This expression can be represented in another form (Carslaw and Jaeger 1973):

$$T(r, z, t) - T_0 = \frac{2Q}{\pi r_0 c\rho \sqrt{4\pi a t}} \exp\left(-\frac{z^2}{4at}\right)$$

$$\times \int_0^\infty \exp(-atu^2) \mathrm{J}_0(ur) \mathrm{J}_1(ur_0) du, \tag{5.2.34}$$

where J_0 and J_1 are Bessel functions of a real argument of the first kind of orders 0 and 1 (see Fig. 3.14).

The temperature at the origin ($r = 0$) is (Carslaw and Jaeger 1973)

$$T(0, z, t) - T_0 = \frac{2Q/\left(\pi r_0^2\right)}{c\rho\sqrt{4\pi at}} \exp\left(-\frac{z^2}{4at}\right)\left[1 - \exp\left(-\frac{r_0^2}{4at}\right)\right]. \tag{5.2.35}$$

If the body is infinite, then expressions (5.2.33)–(5.2.35) should be divided by 2.

In an infinite heat-insulated plate of thickness h at the initial time $t = 0$, the distribution of the volumetric heat density is defined by the function $Q_3(\rho)$, J m^{-3} in cylindrical coordinates. It follows from the relation (5.2.31) that the temperature at point r at time t is written in the form (Carslaw and Jaeger 1973)

$$T(r, t) - T_0 = \frac{1}{c\rho} \frac{1}{2at} \exp\left(-\frac{r^2}{4at}\right)$$
$$\times \int_0^\infty Q_3(\rho) \exp\left(-\frac{\rho^2}{4at}\right) \mathrm{I}_0\left(\frac{r\rho}{2at}\right) \rho d\rho. \tag{5.2.36}$$

If instantaneous energy source Q is uniformly distributed over the volume of a circular cylinder of radius r_0 with centre at the origin, then (5.2.36) implies (Carslaw and Jaeger 1973; Radaj 2003)

$$T(r, t) - T_0 = \frac{Q/\left(\pi r_0^2 h\right)}{c\rho} \frac{1}{2at} \exp\left(-\frac{r^2}{4at}\right)$$
$$\times \int_0^{r_0} \exp\left(-\frac{\rho^2}{4at}\right) \mathrm{I}_0\left(\frac{r\rho}{2at}\right) \rho d\rho. \tag{5.2.37}$$

The first fraction in this expression represents the initial distribution of temperature or, just the same, the distribution of the peak temperature in a cylindrical source $T_{\max} - T_0$.

The integral should be determined by a numerical method, with the exception of the case $r = 0$, when the temperature is equal to (Carslaw and Jaeger 1973)

$$T(0, t) - T_0 = \frac{Q/\left(\pi r_0^2 h\right)}{c\rho}\left[1 - \exp\left(-\frac{r_0^2}{4at}\right)\right]. \tag{5.2.38}$$

If $r_0 \to 0$, then, by expanding the exponential function in formula (5.2.38) into a power series, and using the passage to the limit, we obtain

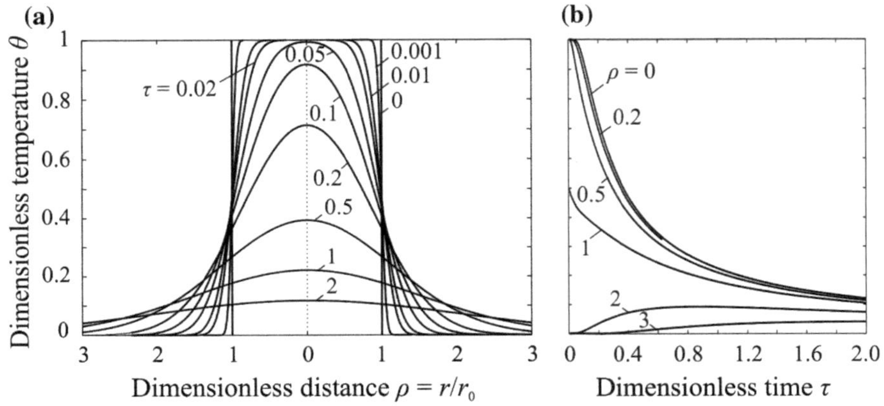

Fig. 5.47 Radial distribution of dimensionless temperature $\theta = (T - T_0)/(T_{\max} - T_0)$ at different instants of dimensionless time $\partial\theta_3/\partial\tau$ (**a**) and thermal cycles (**b**) due to an instantaneous uniformly distributed disk source of radius r_0 in the plate (Carslaw and Jaeger 1973)

$$T(0,t) - T_0 = \frac{Q/h}{c\rho(4\pi a t)}, \tag{5.2.39}$$

which for $r = 0$ coincides with formula (4.2.37) for the instantaneous line source in the plate.

Equation (5.2.37) can be represented in a dimensionless form

$$\theta(\rho,\tau) = \frac{T(r,t) - T_0}{T_{\max} - T_0} = \exp\left(-\frac{\rho^2}{4\tau}\right) \int\limits_0^{1/\sqrt{2\tau}} \exp\left(-\frac{u^2}{2}\right) \mathrm{I}_0\left(\frac{\rho u}{\sqrt{2\tau}}\right) u du$$

$$\rho = \frac{r}{r_0}; \quad \tau = \frac{at}{r_0^2}. \tag{5.2.40}$$

Figure 5.47 shows temperature distributions and thermal cycles. In the beginning the temperature is equal to the peak temperature within the source and the initial temperature outside the source.

5.2.1.5 Instantaneous Uniformly Distributed Spherical Surface Source

In an infinite solid at the initial time $t = 0$, heat Q is uniformly distributed over the surface of a sphere of radius R_0 with the centre at the origin (Fig. 5.48a). The energy density is equal to $Q_2 = Q/\left(4\pi R_0^2\right)$, J m^{-2}.

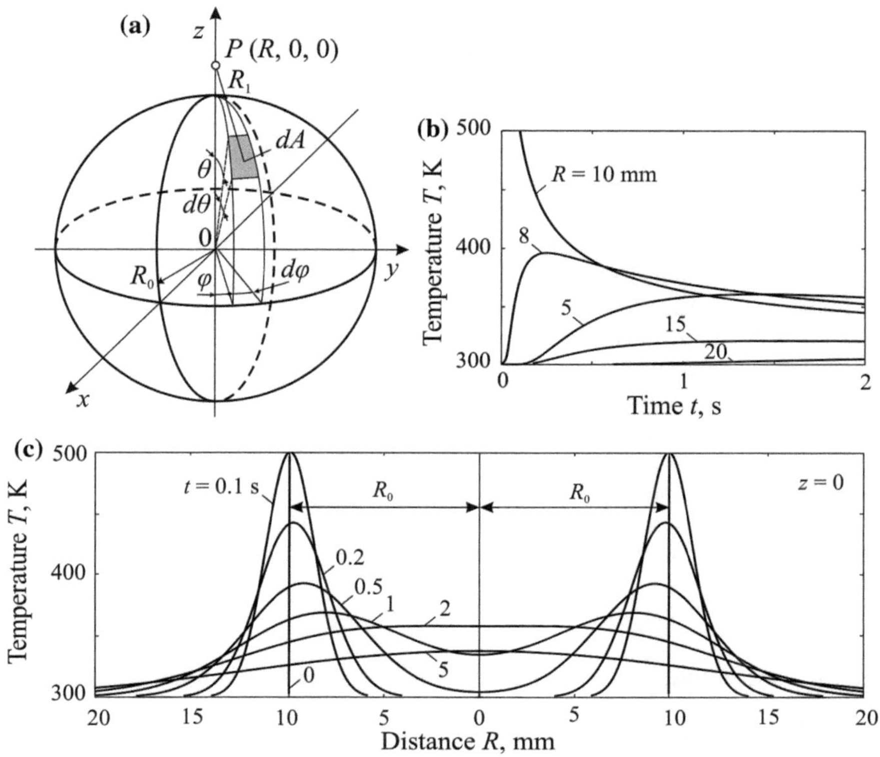

Fig. 5.48 Model for spherical surface source (**a**), thermal cycles (**b**) and temperature distribution in infinite solid around instantaneous uniformly distributed spherical surface source (**c**) ($Q = 4000$ J, $c\rho = 0.005$ J mm^{-3} K^{-1}, $a = 8$ mm^2 s^{-1}, $R_0 = 10$ mm, $T_0 = 300$ K)

Let us use spherical coordinates R, θ, φ (Fig. 5.48). Consider the elementary area of the sphere $dA = R_0^2 \sin\theta d\theta d\phi$ and the point source $dQ = Q_2 dA$ that acts on it. According to Eq. (4.2.4), the increment of temperature caused by this source at point $P(R, 0, 0)$ is equal to

$$dT(R,t) = \frac{Q \sin\theta d\theta d\phi}{4\pi c\rho (4\pi at)^{3/2}} \exp\left(-\frac{R_1^2}{4at}\right). \tag{5.2.41}$$

Then the temperature at point P due to the entire surface source is determined by integration over angles θ and φ (Carslaw and Jaeger 1973; Kartashov 2001; Rykalin 1947, 1952):

$$T(R,t) - T_0 = \frac{Q}{4\pi c\rho (4\pi at)^{3/2}} \int_0^{2\pi}\int_0^{\pi} \exp\left(-\frac{R_0^2 + R^2 - 2RR_0\cos\theta}{4at}\right) \sin\theta d\theta d\varphi$$

$$= \frac{Q}{4\pi c\rho R R_0 \sqrt{4\pi at}} \left[\exp\left(-\frac{(R - R_0)^2}{4at} \right) - \exp\left(-\frac{(R + R_0)^2}{4at} \right) \right]. \tag{5.2.42}$$

When $R_0 \to 0$, Eq. (5.2.42) turns into Eq. (4.2.4) for the instantaneous point source. The temperature at the origin ($R = 0$) of the instantaneous source Q which is uniformly distributed over the surface of a sphere of radius R_0 is equal to the temperature (4.2.4) on the surface of sphere R_0 generated by instantaneous point source Q at the origin (Rykalin 1947, 1952).

Equation (5.2.42) allows us to obtain a solution by the method of sources for any radial distribution of a volumetric source in an infinite body or for a semi-infinite solid with a source whose centre is located on its surface (in the latter case, the expression should be multiplied by 2).

The temperature field (5.2.42) is three-dimensional radial. The temperature at a point depends on the radius of sphere R_0, the distance from the point to the centre of source R, and the time after the action of source t.

If the spherical surface source is completely located in a slab, then we can easily obtain the solution to the problem on the basis of Eq. (5.2.42) by using the method of images.

5.2.1.6 Instantaneous Uniformly Distributed Spherical Volume Source

In an infinite body at the initial time $t = 0$, the distribution of the volumetric heat density is defined by function $Q_3(\rho)$, J m^{-3}, in spherical coordinates. It follows from relation (5.2.42) that the temperature at point R at time t is written in the form

$$T(R,t) - T_0 = \frac{1}{c\rho R\sqrt{4\pi at}} \int_0^\infty Q_3(\rho)\rho \times \left[\exp\left(-\frac{(R - \rho)^2}{4at} \right) - \exp\left(-\frac{(R + \rho)^2}{4at} \right) \right] d\rho. \tag{5.2.43}$$

If energy source Q is uniformly distributed over the volume of the sphere of radius R_0 with the centre at the origin, then the energy density is $Q_3 = 3Q/\left(4\pi R_0^3\right)$. Then from (5.2.43) by using integral (5.2.2) and the integral (Prudnikov et al. 1986)

$$\int x \exp(-ax^2)dx = -\frac{1}{2a} \exp(-ax^2) \tag{5.2.44}$$

we obtain

$$T(R,t) - T_0 = \frac{3Q}{4\pi R_0^3 c\rho} \left\{ \frac{1}{2} \left[\Phi\left(\frac{R + R_0}{\sqrt{4at}} \right) - \Phi\left(\frac{R - R_0}{\sqrt{4at}} \right) \right] \right.$$

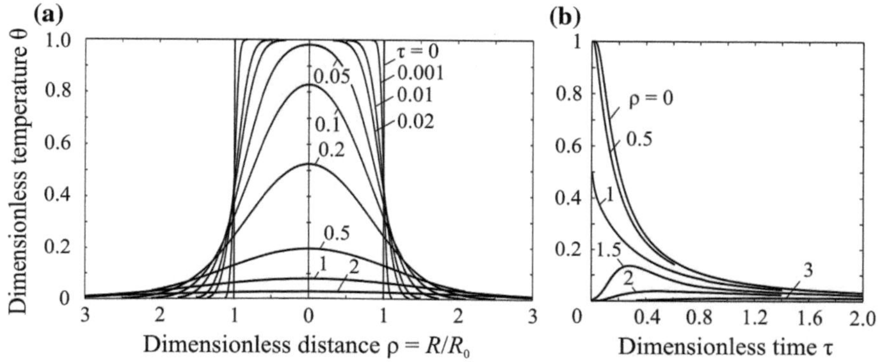

Fig. 5.49 Radial distribution of dimensionless temperature $\theta = (T - T_0)/(T_{\max} - T_0)$ at different instants of dimensionless time $\tau = at/R_0^2$ (**a**) and thermal cycles (**b**) due to the instantaneous uniformly distributed spherical volume source of radius R_0 in an infinite solid

$$+\frac{\sqrt{at}}{\sqrt{\pi}R}\left[\exp\left(-\frac{(R+R_0)^2}{4at}\right)-\exp\left(-\frac{(R-R_0)^2}{4at}\right)\right]\Bigg\}. \quad (5.2.45)$$

This equation can be represented in a dimensionless form

$$\theta(\rho,\tau)=\frac{T-T_0}{T_{\max}-T_0}=\frac{1}{2}\left[\Phi\left(\frac{\rho+1}{2\sqrt{\tau}}\right)-\Phi\left(\frac{\rho-1}{2\sqrt{\tau}}\right)\right]$$
$$+\frac{\sqrt{\tau}}{\sqrt{\pi}\rho}\left[\exp\left(-\frac{(\rho+1)^2}{4\tau}\right)-\exp\left(-\frac{(\rho-1)^2}{4\tau}\right)\right]$$
$$\rho=\frac{R}{R_0};\quad \tau=\frac{at}{R_0^2}. \quad (5.2.46)$$

Here $T_{\max}$ is the peak (initial) temperature in the sphere.

If the centre of the spherical source is on the surface of a heat-insulated semi-infinite solid, the right-hand side of Eq. (5.2.45) should be multiplied by 2. If the source is completely in a semi-infinite solid or slab, the solution to the problem can be easily obtained by the method of images in the form of a series (see Sects. 4.2.1.2 and 4.2.1.3).

Figure 5.49 shows the temperature distribution over the radius and time. The process of the temperature equalisation is more intensive in comparison to the cylindrical source in the plate (see Fig. 5.47).

5.2.1.7 Instantaneous Bounded Volume Sources

With a rigorous approach, spatially unbounded distributed sources cannot be used in computational models if the body is bounded and the method of images is used to

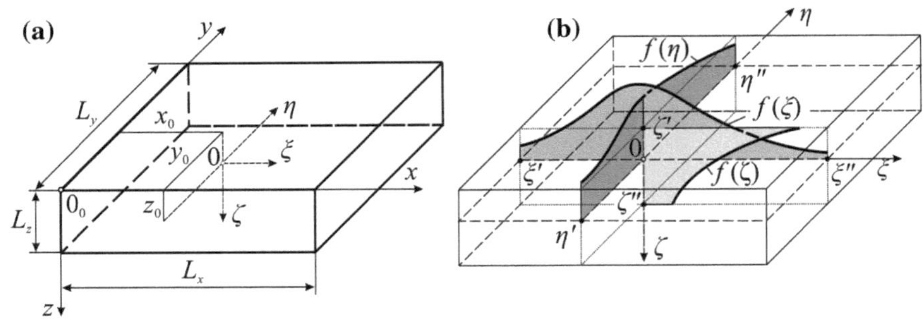

Fig. 5.50 Global (*x*, *y*, *z*) and local (*ξ*, *η*, *ζ*) reference frames (**a**) and bounded unit volumetric energy distribution (**b**) (Karkhin et al. 2010, 2011a, b; Pittner et al. 2015)

take into account its boundedness. For example, if the body is limited in thickness, the source cannot be unlimited in thickness. This follows from the fact that fictitious sources introduced into the extended unbounded body to meet the condition of heat impermeability of the boundary of the real body must be completely outside the real body (they cannot act in a real body). For concentrated sources, this requirement is fulfilled by definition, for distributed sources it is satisfied only if the directions of the unboundedness of the source and body coincide.

On the other hand, the welding source of heat can be significantly bounded by its nature. For example, in laser welding with deep penetration, the beam, as a heat source, is volumetric, but limited both in the radial direction and in the depth of penetration into the welded body. Therefore, taking into account the limitation of the source in the calculation of thermal processes is of practical interest.

In Sect. 3.7 bounded sources of unit power distributed along the *s*-axis ($s = x$, y, z) according to the normal, exponential and parabolic laws $f(s)$ were considered (see Fig. 3.7.7). As it will be shown below, combinations of such distributions allow us to describe a wide class of welding volumetric heat sources and to find analytical solutions to heat conduction problems for bodies of simple geometric shapes.

Let the body occupy the region in the form of a cuboid $[0, L_x] \times [0, L_y] \times [0, L_z]$ in the global coordinate system x, y, z associated with the body (Fig. 5.50a).

Let us introduce a local coordinate system for source ξ, η, ζ, whose origin is offset relative to the global system by x_0, y_0, z_0 (see Fig. 5.50a). The action area of the instantaneous source in the body is limited to a cuboid of the size $[\xi', \xi''] \times [\eta', \eta''] \times [\zeta', \zeta'']$ (see Fig. 5.50b) and its volumetric energy density Q_3 (J m^{-3}) is represented as a product

$$Q_3(\xi, \eta, \zeta) = Q f(\xi) f(\eta) f(\zeta), \tag{5.2.47}$$

where Q is the source energy (J), f is the functions of the distribution of unit energy along the corresponding axis (m^{-1}). It follows from the energy balance condition that:

$$\int_{-\infty}^{\infty}\int_{-\infty}^{\infty}\int_{-\infty}^{\infty} Q_3(\xi,\eta,\zeta)d\xi d\eta d\zeta = Q\int_{\xi'}^{\xi''} f(\xi)d\xi \int_{\eta'}^{\eta''} f(\eta)d\eta \int_{\zeta'}^{\zeta''} f(\zeta)d\zeta = Q, \tag{5.2.48}$$

since the area under curve f (i.e. the energy) is 1.

The temperature field due to the instantaneous volumetric source can be represented in the form (Carslaw and Jaeger 1973)

$$T(x,y,z,t) - T_0 = \frac{Q}{c\rho}\theta(x,t)\theta(y,t)\theta(z,t), \tag{5.2.49}$$

where $\theta(s,t)$ is the relative temperature produced by corresponding source $f(s)$, $s = x, y, z$; its dimension is m^{-1}. Function θ can be understood as the temperature in a rod of length L_s due to the instantaneous source of unit energy distributed along the s-axis and released at the time $t = 0$.

Let us find function $\theta(s,\ t)$ for the unbounded ($-\infty < s < \infty$) and bounded ($0 < s < L_s$, $L_s < \infty$) domains. We shall assume that the source of unit energy is distributed according to the law of δ-function (the source is concentrated) or to a normal, exponential or parabolic law (sources are distributed). We take up the conditions of heat impermeability (heat transfer coefficients $\alpha_1 = \alpha(s = 0) = 0$ and $\alpha_2 = \alpha(s = L_s) = 0$) or heat exchange ($\alpha_1$ and/or α_2 are not equal to 0) on both domain boundaries ($s = 0$ and $s = L_s$).

Let the instantaneous plane source be *concentrated*, its relative energy is 1 ($Q/(Ac\rho) = 1$), the body is unbounded along the s-axis. The source is at point $s = s_0$, its action boundaries coincide, $\sigma' = \sigma'' = 0$ (Fig. 5.51a). The heat is released at the time $t = 0$. Then the distribution of the relative temperature along the s-axis is determined by formula (4.2.41):

$$\theta(s,t) = \frac{1}{\sqrt{4\pi at}}\exp\left(-\frac{[s - s_0]^2}{4at}\right). \tag{5.2.50}$$

If the body is bounded ($0 \le s \le L_s$) and its boundaries are heat-impermeable (Fig. 5.51b), the temperature field can be found by the method of images using formula (5.2.50). For this purpose, we extend the bounded body to infinity and introduce pairs of fictitious sources into an unbounded body at a pitch of $2L_s$ so that the surfaces of the real body are symmetry planes in an unbounded body (see Fig. 5.51b). The total flux of heat on the planes of symmetry from the entire set of sources (real and fictitious) is zero, i.e. the condition of heat impermeability of the boundaries of the real body is satisfied. Therefore, the temperature field in a real body is described by a series

$$\theta(s,t) = \frac{1}{\sqrt{4\pi at}}\sum_{i=-\infty}^{\infty}\sum_{j=-1,1}\exp\left(-\frac{\left[s - (js_0 + 2iL_s)\right]^2}{4at}\right). \tag{5.2.51}$$

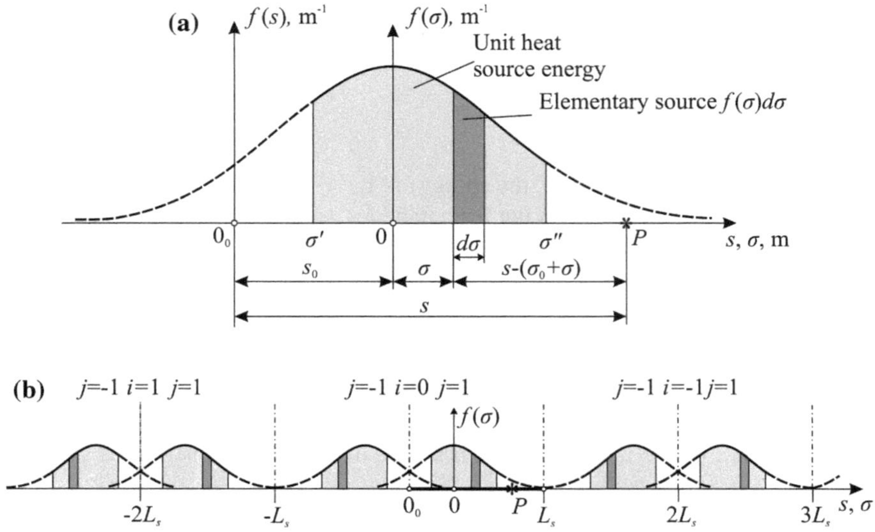

Fig. 5.51 Model for calculating the temperature at point P due to a distributed unit energy in unbounded (**a**) and bounded (**b**) bodies

If the body is unbounded ($L_s = \infty$), then for $i \neq 0$ all terms of the series disappear and the expression (5.2.51) is simplified.

Equation (5.2.51) can be represented in the form of a trigonometric series by means of transformation (4.2.26) (Carslaw and Jaeger 1973; Karkhin et al. 2011a):

$$\theta(s,t) = \frac{1}{L_s}\left[1 + 2\sum_{i=1}^{\infty}\cos\left(\frac{i\pi s}{L_s}\right)\cos\left(\frac{i\pi s_0}{L_s}\right)\exp\left(-i^2\pi^2\frac{at}{L_s^2}\right)\right]. \tag{5.2.52}$$

Now let us take into account heat transfer. Suppose that at boundary $s = 0$ the coefficient of heat transfer is equal to α_1, and at boundary $s = L_s$ is equal to α_2. The temperature due to the concentrated source of unit energy, released at the point $s = s_0$ at the time $t = 0$, is described by the following series (Carslaw and Jaeger 1973):

$$\theta(s,t) = \frac{1}{L_s}\sum_{n=1}^{\infty} A_n B_n(s_0) B_n(s)\exp\left(-\mu_n^2\frac{at}{L_s^2}\right), \tag{5.2.53}$$

where

$$A_n = \frac{2}{\left(\mu_n^2 + \mathrm{Bi}_1^2\right)\left(1 + \frac{\mathrm{Bi}_2}{\mu_n^2 + \mathrm{Bi}_2^2}\right) + \mathrm{Bi}_1};$$

$$B_n(u) = \mu_n\cos\left(\mu_n\frac{u}{L_s}\right) + \mathrm{Bi}_1\sin\left(\mu_n\frac{u}{L_s}\right);$$

$$\mathrm{Bi}_1 = \frac{\alpha_1 L_s}{\lambda}; \quad \mathrm{Bi}_2 = \frac{\alpha_2 L_s}{\lambda}.$$

Here μ_n are positive roots of Eq. (4.2.31); some roots are shown in Fig. 4.8. The structure of formula (5.2.53) is similar to (4.2.30).

It should be noted that we cannot obtain the solution (5.2.52) by setting $\mathrm{Bi}_1 = \mathrm{Bi}_2 = 0$ in formula (5.2.53), as in the latter case there are no roots of Eq. (4.2.31) (see Fig. 4.8).

Formulas (5.2.50)–(5.2.53) are the main ones for solving the problems by the method of sources for bounded and unbounded bodies with spatially bounded and unbounded instantaneous volumetric sources. As an example, let us consider the solution sequence of the problem in the case of a distributed source with boundaries σ' and σ'' in an unbounded body (see Fig. 5.51a).

Let us split the distributed source into elementary concentrated sources with energy $f(\sigma)d\sigma$ (Fig. 5.51a). According to (5.2.50), the increment of temperature caused by an elementary source is equal to

$$d\theta(s,t) = \frac{f(\sigma)d\sigma}{\sqrt{4\pi a t}} \exp\left(-\frac{[s-(s_0+\sigma)]^2}{4at}\right). \tag{5.2.54}$$

By summing the increments (5.2.54) produced by elementary sources over the entire interval $\sigma' - \sigma''$, we obtain the temperature in an unbounded body produced by the entire distributed source:

$$d\theta(s,t) = \frac{1}{\sqrt{4\pi a t}} \int\limits_{\sigma'}^{\sigma''} f(\sigma) \exp\left(-\frac{[s-(s_0+\sigma)]^2}{4at}\right) d\sigma. \tag{5.2.55}$$

Similarly, we can find the temperature in a bounded body resulted from a source of unit energy distributed in interval $\sigma' - \sigma''$ according to any given law $f(\sigma)$. On the basis of Eqs. (5.2.51) and (5.2.52) we obtain two types of solutions for a bounded body with heat-insulated boundaries:

$$\theta(s,t) = \frac{1}{\sqrt{4\pi a t}} \int\limits_{\sigma'}^{\sigma''} f(\sigma) \sum_{i=-\infty}^{\infty} \sum_{j=-1,1} \exp\left(-\frac{\{s-[j(s_0+\sigma)+2iL_s]\}^2}{4at}\right) d\sigma \tag{5.2.56}$$

and

$$\theta(s,t) = \frac{1}{L_s}\left[1 + 2\sum_{i=1}^{\infty} \cos\left(\frac{i\pi s}{L_s}\right) \exp\left(-i^2\pi^2 \frac{at}{L_s^2}\right)\right.$$

$$\times \int_{\sigma''}^{\sigma'} f(\sigma) \cos\left(\frac{i\pi(s_0 + \sigma)}{L_s}\right) d\sigma \Bigg] \tag{5.2.57}$$

In the same way, on the basis of Eq. (5.2.53), we obtain a solution to the problem for a bounded body with heat transfer at the boundary:

$$\theta(s,t) = \frac{1}{L_s} \sum_{n=1}^{\infty} A_n B_n(s) \exp\left(-\mu_n^2 \frac{at}{L_s^2}\right)$$
$$\times \left[\mu_n \int_{\sigma'}^{\sigma''} f(\sigma) \cos\left(\mu_n \frac{s_0 + \sigma}{L_s}\right) d\sigma + \mathrm{Bi}_1 \int_{\sigma'}^{\sigma''} f(\sigma) \sin\left(\mu_n \frac{s_0 + \sigma}{L_s}\right) d\sigma\right]. \tag{5.2.58}$$

Therefore, the calculation by formula (5.2.49) of the temperature field produced by a volume source in a body that is bounded or unbounded along the x-, y- and z-axes has reduced to determining the integrals in Eqs. (5.2.55)–(5.2.58). The ability to express the integrals as the known functions is determined by the form of energy distribution curve $f(\sigma)$. We note that the series of elementary functions f when integrals (5.2.55) and (5.2.56) are expressed as the known functions is rather short. Integrals (5.2.57) and (5.2.58) are taken by the methods of functional analysis in many variants of function f: parabolic, hyperbolic, exponential, trigonometric, etc. (Gradshteyn and Ryzhik 2014; Prudnikov et al. 1986). If the integrals are not expressed as the known functions, they can be determined by numerical methods (Simpson, Newton, Gauss, etc.).

Let us give the solutions $\theta(s,t)$ for the cases of a bounded source distributed along the s-axis according to a normal, exponential or parabolic law (Fig. 3.13).

The normal energy distribution is described by Eq. (3.7.3). Let the body be bounded and its boundaries are heat impermeable. To determine the temperature inside the body, we substitute Eq. (3.7.3) in (5.2.56) and use the integral (Prudnikov et al. 1986)

$$\int \exp(-[ax^2 + bx + c])\, dx = \frac{1}{2}\sqrt{\frac{\pi}{a}} \exp\left(\frac{b^2}{4a} - c\right) \Phi\left(x\sqrt{a} + \frac{b}{2\sqrt{a}}\right), \tag{5.2.59}$$

as a result, we obtain (Karkhin et al. 2010, 2011a):

$$\theta(s,t) = \frac{1}{\sqrt{\pi(4at + \sigma_e^2)}\left[\Phi(\sigma''/\sigma_e) - \Phi(\sigma'/\sigma_e)\right]}$$
$$\times \sum_{i=-\infty}^{\infty} \sum_{j=-1,1} \exp\left(-\frac{(s - js_0 - 2iL_s)^2}{4at + \sigma_e^2}\right)\left[F(\sigma'') - F(\sigma')\right], \tag{5.2.60}$$

where

$$F(u) = \Phi\left(\frac{u}{\sigma_e}\sqrt{1 + \frac{\sigma_e^2}{4at}} - \frac{j(s - js_0 - 2iL_s)\sigma_e}{\sqrt{4at\left(4at + \sigma_e^2\right)}}\right).$$

Here σ_e is the distribution parameter, $\sigma_e = \xi_e, \eta_e, \zeta_e$ (this is the coordinate at which the function falls in e times, i.e. $f(\sigma_e) = f(0)/e$).

If the body is semi-bounded, only the term of series at $i = 0$ should be left in Eq. (5.2.60).

After substituting (3.7.3) in (5.2.57) by utilising the integral (Prudnikov et al. 1986)

$$\int \exp(-x^2)\begin{Bmatrix} \sin(ax) \\ \cos(ax) \end{Bmatrix} dx = \frac{\sqrt{\pi}}{2}\exp\left(-\frac{a^2}{4}\right)\begin{Bmatrix} \operatorname{Im}\Phi(x - \mathrm{i}a/2) \\ \operatorname{Re}\Phi(x - \mathrm{i}a/2) \end{Bmatrix} \quad (5.2.61)$$

we obtain the following solution for a bounded body (Karkhin et al. 2010, 2011a):

$$\theta(s, t) = \frac{1}{L_s}\left\{1 + \frac{2}{\sqrt{\pi}\left[\Phi(\sigma''/\sigma_e) - \Phi(\sigma'/\sigma_e)\right]} \times \sum_{k=1}^{\infty}\cos\left(\frac{k\pi s}{L_s}\right)\exp\left(-\frac{k^2\pi^2}{L_s^2}\left(at + \frac{\sigma_e^2}{4}\right)\right)\left[F(\sigma'') - F(\sigma')\right]\right\}, \quad (5.2.62)$$

where

$$F(u) = \cos\left(\frac{k\pi s_0}{L_s}\right)\operatorname{Re}\left(\Phi\left(\frac{u}{\sigma_e} - \mathrm{i}\frac{k\pi\sigma_e}{2L_s}\right)\right) - \sin\left(\frac{k\pi s_0}{L_s}\right)\operatorname{Im}\left(\Phi\left(\frac{u}{\sigma_e} - \mathrm{i}\frac{k\pi\sigma_e}{2L_s}\right)\right);$$

$$\Phi(w) = \frac{2}{\sqrt{\pi}}\int_0^w \exp(-z^2)\,dz; \quad z = x + \mathrm{i}y; \quad \mathrm{i} = \sqrt{-1}.$$

The exponential energy distribution is described by Eq. (3.7.5). Let the source be in a body with heat-insulated boundaries. By substituting Eq. (3.7.5) in (5.2.56) and using (5.2.59), we obtain (Karkhin et al. 2010, 2011a):

$$\theta(s, t) = \frac{1}{2\sigma_e\left[\exp(-\sigma'/\sigma_e) - \exp(-\sigma''/\sigma_e)\right]}\exp\left(\frac{at}{\sigma_e^2}\right) \times \sum_{i=-\infty}^{\infty}\sum_{j=-1,1}\exp\left(-\frac{j(s - js_0 - 2iL_s)}{\sigma_e}\right)\left[F(\sigma'') - F(\sigma')\right], \quad (5.2.63)$$

where

$$F(u) = \Phi\left(\frac{u - j(s - js_0 - 2iL_s)}{\sqrt{4at}} + \frac{\sqrt{at}}{\sigma_e}\right).$$

If the body is semi-bounded, then the terms of the series for $i \neq 0$ in Eq. (5.2.63) should be dropped.

If we substitute (3.7.5) in (5.2.57) and use the integral (Prudnikov et al. 1986)

$$\int \exp(ax)\left\{\begin{matrix}\sin(bx)\\ \cos(bx)\end{matrix}\right\}dx = \frac{\exp(ax)}{a^2+b^2}\left[a\left\{\begin{matrix}\sin(bx)\\ \cos(bx)\end{matrix}\right\} \pm b\left\{\begin{matrix}\cos(bx)\\ \sin(bx)\end{matrix}\right\}\right], \quad (5.2.64)$$

then we obtain

$$\theta(s,t) = \frac{1}{L_s}\left\{1 + \frac{2}{\exp(-\sigma'/\sigma_e) - \exp(\sigma''/\sigma_e)}\right.$$
$$\left.\times \sum_{i=1}^{\infty} \frac{1}{1 + (i\pi\sigma_e/L_s)^2}\cos\left(\frac{i\pi s}{L_s}\right)\exp\left(-\frac{i^2\pi^2 at}{L_s^2}\right)\left[F(\sigma'') - F(\sigma')\right]\right\}, \quad (5.2.65)$$

where

$$F(u) = \exp\left(-\frac{u}{\sigma_e}\right)\left[\frac{i\pi\sigma_e}{L_s}\sin\left(\frac{i\pi(s_0+u)}{L_s}\right) - \cos\left(\frac{i\pi(s_0+u)}{L_s}\right)\right].$$

Similarly, the temperature can be found in a bounded body with surface heat transfer.

The parabolic energy distribution is described by Eq. (3.7.6). At first, let us find the temperature field in a bounded body with heat-insulated boundaries. For this purpose, we substitute Eq. (3.7.6) into (5.2.56) and use the integrals (Prudnikov et al. 1986)

$$\int f(x)\exp(-[ax^2+bx+c]) = \frac{1}{\sqrt{a}}\exp\left(\frac{b^2}{4a} - c\right)\int f\left(\frac{2\sqrt{a}t - b}{2a}\right)\exp(-t^2)dt$$
$$\left[t = \sqrt{a}\left(x + \frac{b}{2a}\right), a > 0\right];$$
$$\int x\exp(-ax^2)\,dx = -\frac{1}{2a}\exp(-ax^2);$$
$$\int x^2\exp(-a^2x^2)dx = -\frac{1}{2a^2}x\exp(-a^2x^2) + \frac{\sqrt{\pi}}{4a^3}\Phi(ax), \quad (5.2.66)$$

as a result, we obtain (Karkhin et al. 2010, 2011a):

$$\theta(s,t) = \frac{1}{a_0(\sigma'' - \sigma') + a_1/2(\sigma''^2 - \sigma'^2) + a_2/3(\sigma''^3 - \sigma'^3)} \times \sum_{i=-\infty}^{\infty} \sum_{j=-1,1} [F(\sigma') - F(\sigma')], \quad (5.2.67)$$

where

$$F(u) = \sqrt{\frac{at}{\pi}}\{-a_1 - a_2[u - j(s - js_0 - 2iL_s)] - a_2 2j(s - js_0 - 2iL_s)\} \times \exp\left(-\frac{[u - j(s - js_0 - 2iL_s)]^2}{4at}\right) + \frac{1}{2}[a_0 + a_1 j(s - js_0 - 2iL_s) + a_2 2at + a_2(s - js_0 - 2iL_s)^2]\Phi\left(\frac{u - j(s - js_0 - 2iL_s)}{\sqrt{4at}}\right).$$

If the body is semi-bounded, the terms of the series for $i \neq 0$ in Eq. (5.2.67) should be dropped.

If we substitute (3.7.6) in (5.2.57) and use the integrals (Gradshteyn and Ryzhik 2014; Prudnikov et al. 1986)

$$\int x\begin{Bmatrix}\sin(bx)\\ \cos(bx)\end{Bmatrix}dx = \frac{1}{b^2}\left[\begin{Bmatrix}\sin(bx)\\ \cos(bx)\end{Bmatrix} \mp bx\begin{Bmatrix}\cos(bx)\\ \sin(bx)\end{Bmatrix}\right]; \quad (5.2.68)$$

$$\int x^2\begin{Bmatrix}\sin(bx)\\ \cos(bx)\end{Bmatrix}dx = \frac{2}{b^2}x\begin{Bmatrix}\sin(bx)\\ \cos(bx)\end{Bmatrix} \mp \frac{1}{b}\left(x^2 - \frac{2}{b^2}\right)\begin{Bmatrix}\cos(bx)\\ \sin(bx)\end{Bmatrix}, \quad (5.2.69)$$

then we obtain the solution in another form (Karkhin et al. 2011a):

$$\theta(s,t) = \frac{1}{L_s}\left\{1 + \frac{2}{a_0(\sigma'' - \sigma') + a_1/2(\sigma''^2 - \sigma'^2) + a_2/3(\sigma''^3 - \sigma'^3)} \times \sum_{i=1}^{\infty} \frac{L_s}{i\pi}\cos\left(\frac{i\pi s}{L_s}\right)\exp\left(-\frac{i^2\pi^2 at}{L_s^2}\right)[F(\sigma'') - F(\sigma')]\right\}, \quad (5.2.70)$$

where

$$F(u) = a_0\sin\left(\frac{i\pi(s_0 + u)}{L_s}\right) + a_1\left[\frac{L_s}{i\pi}\cos\left(\frac{i\pi(s_0 + u)}{L_s}\right) + u\sin\left(\frac{i\pi(s_0 + u)}{L_s}\right)\right] + a_2\left\{\frac{2L_s}{i\pi}u\cos\left(\frac{i\pi(s_0 + u)}{L_s}\right) + \left[u^2 - 2\left(\frac{L_s}{i\pi}\right)^2\right]\sin\left(\frac{i\pi(s_0 + u)}{L_s}\right)\right\}. \quad (5.2.71)$$

Solutions (5.2.67) and (5.2.70) are equivalent. They differ only in the rate of convergence of the series.

Now we take into account the heat exchange of a bounded body with the medium for the case of a parabolic distribution of unit energy. On the basis of Eq. (5.2.58), by using integrals (5.2.68) and (5.2.69), we obtain

$$\theta(s,t) = \frac{1}{L_s}\sum_{n=1}^{\infty} A_n B_n(s)\exp\left(-\mu_n^2\frac{at}{L_s^2}\right) \times \frac{1}{a_0(\sigma''-\sigma') + \frac{1}{2}a_1\left(\sigma''^2 - \sigma'^2\right) + \frac{1}{3}a_2\left(\sigma''^3 - \sigma'^3\right)}\frac{L_s}{\mu_n}\left[F(\sigma'') - F(\sigma')\right], \tag{5.2.72}$$

where

$$\begin{aligned} F(u) = {} & a_0\left[\mu_n \sin\left(\frac{\mu_n u}{L_s}\right) - \mathrm{Bi}_1\cos\left(\frac{\mu_n u}{L_s}\right)\right] \\ & + a_1\left\{\mu_n\left[u\sin\left(\frac{\mu_n u}{L_s}\right) + \frac{L_s}{\mu_n}\cos\left(\frac{\mu_n u}{L_s}\right)\right]\right. \\ & \left. + \mathrm{Bi}_1\left[-u\cos\left(\frac{\mu_n u}{L_s}\right) + \frac{L_s}{\mu_n}\sin\left(\frac{\mu_n u}{L_s}\right)\right]\right\} \\ & + a_2\left\{\mu_n\left[u^2\sin\left(\frac{\mu_n u}{L_s}\right) + \frac{2L_s}{\mu_n}u\cos\left(\frac{\mu_n u}{L_s}\right)\right]\right. \\ & \left. + \mathrm{Bi}_1\left[-u^2\cos\left(\frac{\mu_n u}{L_s}\right) + \frac{2L_s}{\mu_n}u\sin\left(\frac{\mu_n u}{L_s}\right)\right]\right\} \end{aligned} \tag{5.2.73}$$

Note that almost any energy distribution $f(s)$ can be represented in the form of a piecewise parabolic function and, thus, we can find a temperature field due to such a source by using formulas (5.2.67), (5.2.70) and (5.2.72).

We point out that, in spite of cumbersome form of the above formulas (5.2.51)–(5.2.73), the series converge rapidly if for small Fourier numbers $\mathrm{Fo} = at/L_s^2$ we use the formulas obtained on the basis of (5.2.56), and for large numbers we use the formulas obtained on the basis of (5.2.57).

Figure 5.52 shows the distribution of dimensionless temperature $L_s\theta$ at various instants of dimensionless time $\mathrm{Fo} = at/L_s^2$ with and without heat transfer.

So, the solutions to the heat conduction problem for instantaneous line sources with different energy distributions are obtained. The method of sources makes it possible to obtain the corresponding solutions for numerous combinations of line sources according to (5.2.49). If the source is plane (two-dimensional), then the solution can be used directly to calculate the temperature due to rapidly moving distributed sources (see Sect. 5.2.4). If the source is moving and volumetric (three-dimensional) and its trajectory is given, then it is represented in the form of a series

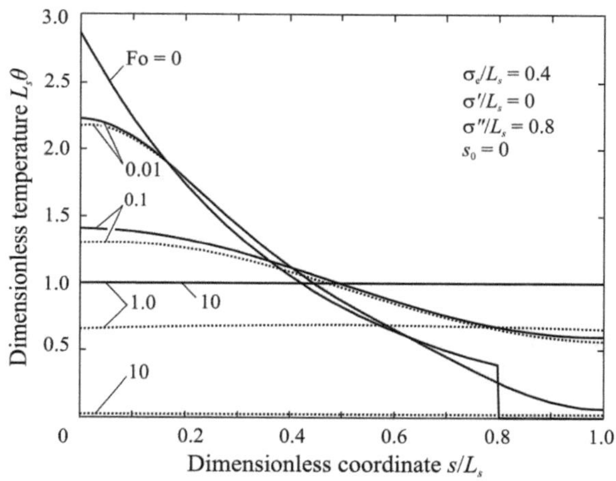

Fig. 5.52 Dependence of dimensionless temperature $L_s\theta$ on dimensionless coordinate s/L_s at different instances of dimensionless time Fo $= at/L_s^2$ with exponential energy distribution when heat loss is considered ($\mathrm{Bi}_1 = \mathrm{Bi}_2 = 0.2$; dotted lines) and heat loss is neglected ($\mathrm{Bi}_1 = \mathrm{Bi}_2 = 0$; solid lines)

of instantaneous volumetric sources and its temperature field is found by integrating Eq. (5.2.49) over time.

The above source models are a generalisation of the known special cases (Fachinotti and Cardona 2008; Goldak et al. 1984; Hou and Komanduri 2000; Karkhin et al. 2007; Kasuya and Yurioka 1993; Nguyen et al. 1999, 2004; Nguyen 2004).

5.2.2 Stationary Continuous Distributed Sources

5.2.2.1 Stationary Continuous Normally Distributed Circular Source in a Semi-infinite Body

As in the case of stationary concentrated sources (Sect. 5.1.2), we represent a continuous distributed source of power $q(t)$ as a set of consecutive instantaneous sources. Then the temperature field is determined by the summation (integration over time) of temperature increments caused by instantaneous sources. By taking Eq. (5.2.13) as a basis for the instantaneous unbounded normally distributed circular (disk) source, we obtain

$$T(r,z,t) - T_0 = \frac{1}{8c\rho(\pi a)^{3/2}} \int\limits_0^t \frac{q(\tau)}{(t + t_0 - \tau)(t-\tau)^{1/2}}$$

$$\times \exp\left(-\frac{r^2}{4a(t+t_0-\tau)}\right)\sum_{j=-1,1}\exp\left(-\frac{(z-j\zeta)^2}{4a(t-\tau)}\right)d\tau. \tag{5.2.74}$$

We point out that there can be any time law of the source power $q(t)$.

In the same way the solutions for circular and elliptic sources in a semi-infinite solid and in a slab can be obtained on the basis of Eqs. (5.2.12)–(5.2.16).

5.2.2.2 Stationary Continuous Uniformly Distributed Disk Source in an Infinite Body

Let the constant power of source q be uniformly distributed over the area of a circle of the radius r_0 lying in the plane $z = 0$ in an infinite solid. The temperature at a point with cylindrical coordinates r, z at time t is written as follows (Carslaw and Jaeger 1973):

$$T(r,z,t) - T_0 = \frac{q}{2\pi r_0\lambda}\int_0^\infty \mathrm{J}_0(ur)\mathrm{J}_1(ur_0)\left[-\exp(-uz)\Phi^*\left(\frac{z}{\sqrt{4at}} - u\sqrt{at}\right)\right.$$
$$\left. - \exp(uz)\Phi^*\left(\frac{z}{\sqrt{4at}} + u\sqrt{at}\right)\right]\frac{du}{u}. \tag{5.2.75}$$

The temperature at the points on the z-axis is

$$T(0,z,t) - T_0 = \frac{2q\sqrt{at}}{\pi r_0^2\lambda}\left[\mathrm{i}\Phi^*\left(\frac{z}{\sqrt{4at}}\right) - \mathrm{i}\Phi^*\left(\frac{\sqrt{z^2+r_0^2}}{\sqrt{4at}}\right)\right]. \tag{5.2.76}$$

Equation (5.2.76) can be represented in dimensionless form:

$$\theta = \frac{\pi r_0\lambda}{2q}(T - T_0) = \sqrt{\tau}\left[\mathrm{i}\Phi^*\left(\frac{\zeta}{2\sqrt{\tau}}\right) - \mathrm{i}\Phi^*\left(\frac{\sqrt{1+\zeta^2}}{2\sqrt{\tau}}\right)\right];$$
$$\tau = at/r_0^2;\quad \zeta = z/r_0. \tag{5.2.77}$$

Thermal cycles of the points on the z-axis are shown in Fig. 5.53a.

It is not difficult to obtain a solution for a disk source in a semi-infinite solid and a slab on the basis of Eq. (5.2.75), by using the method of images. It can be used to calculate temperature during surface treatment by an electron beam, arc welding of cylindrical rods to a massive body, spot welding of two plates, etc.

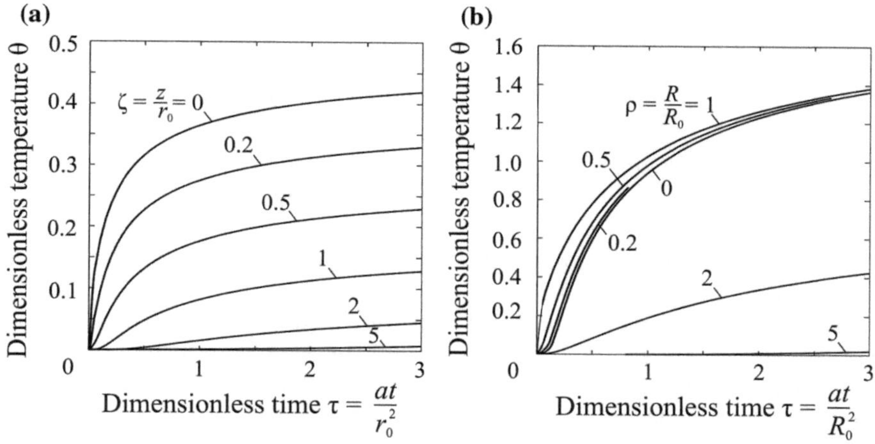

Fig. 5.53 Thermal cycles of points in an infinite solid: at the z-axis passing through the centre of a continuous uniformly distributed disk source of radius r_0 (**a**) and at distance $\rho = R/R_0$ away from the centre of a continuous uniformly distributed spherical surface source (**b**)

5.2.2.3 Stationary Continuous Uniformly Distributed Spherical Surface Source in an Infinite Body

Suppose that the source of power $q(t)$ is uniformly distributed on a spherical surface of radius R_0, starting from the time $t = 0$. Then the temperature at point R at time t is determined by integrating expression (5.2.42):

$$T(R,t) - T_0 = \frac{1}{8\pi c\rho R R_0 (\pi a)^{1/2}} \int_0^t \frac{q(\tau)}{(t-\tau)^{1/2}} \times \left[\exp\left(-\frac{(R-R_0)^2}{4a(t-\tau)}\right) - \exp\left(-\frac{(R+R_0)^2}{4a(t-\tau)}\right)\right] d\tau. \quad (5.2.78)$$

If $q = \text{const}$, the solution takes the form (Carslaw and Jaeger 1973)

$$T(R,t) - T_0 = \frac{q}{8\pi\lambda R R_0}\left\{\sqrt{4at/\pi} \times \left[\exp\left(-\frac{(R-R_0)^2}{4at}\right) - \exp\left(-\frac{(R+R_0)^2}{4at}\right)\right] - |R-R_0|\Phi^*\left(\frac{|R-R_0|}{\sqrt{4at}}\right) + (R+R_0)\Phi^*\left(\frac{R+R_0}{\sqrt{4at}}\right)\right\}. \quad (5.2.79)$$

This equation can be represented in a dimensionless form

$$\theta = \frac{\sqrt{\tau}}{\rho}\left\{\frac{2}{\sqrt{\pi}}\left[\exp\left(-\frac{(\rho-1)^2}{4\tau}\right) - \exp\left(-\frac{(\rho+1)^2}{4\tau}\right)\right]\right.$$
$$\left. - \frac{|\rho-1|}{\sqrt{\tau}}\Phi^*\left(\frac{|\rho-1|}{2\sqrt{\tau}}\right) + \frac{(\rho+1)}{\sqrt{\tau}}\Phi^*\left(\frac{\rho+1}{2\sqrt{\tau}}\right)\right\};$$
$$\theta = \frac{2\pi\lambda R_0}{q}(T - T_0); \quad \tau = \frac{at}{R_0^2}; \quad \rho = \frac{R}{R_0}. \tag{5.2.80}$$

Thermal cycles of points located at different distances from the centre of a spherical source are shown in Fig. 5.53b.

If the source is completely located in a heat-insulated semi-infinite solid or a slab, with the help of expression (5.2.79), it is easy to obtain the solution to the problems for spherical surface and spherical volumetric sources of heat by the method of images and the method of sources.

5.2.2.4 Stationary Continuous Uniformly Distributed Strip Source on a Semi-infinite Body

For time t, uniformly distributed heat flux of density q_2, W m^{-2}, is applied to the surface of a semi-infinite solid through a half-plane $x < 0$, $-\infty < y < \infty$, $z = 0$.

By integrating the instantaneous line sources positioned along the y-axis, over the half-plane with respect to x (along the x-axis), we obtain instantaneous plane sources on the half-plane; then, by integrating them over the time, we obtain the temperature field on the surface ($z = 0$) (Carslaw and Jaeger 1973):

$$T(x,t) - T_0 = \int_0^t \int_{-\infty}^0 \frac{q_2}{2\pi\lambda(t-\tau)} \exp\left(-\frac{(x-\xi)^2}{4a(t-\tau)}\right) d\xi\, d\tau$$
$$= \frac{q_2 a^{1/2}}{2\pi^{1/2}\lambda}\int_0^t \frac{1}{\sqrt{t-\tau}}\Phi^*\left(\frac{x}{\sqrt{4a(t-\tau)}}\right)d\tau$$
$$= \frac{2q_2}{\sqrt{\pi}\lambda}\sqrt{at}\left[\frac{1}{2}\Phi^*\left(\frac{x}{\sqrt{4at}}\right) + \frac{x}{4\sqrt{\pi at}}\mathrm{Ei}\left(-\frac{x^2}{4at}\right)\right]. \tag{5.2.81}$$

This equation allows us to obtain a solution to the problem for an infinite strip of limited width $2W$: $-W < x < W$, $-\infty < y < \infty$, $z = 0$. By combining sources and sinks, we get an increase in temperature at point x of the body surface at time t (Carslaw and Jaeger 1973):

$$T(x,t) - T_0 = \frac{q_2 W\sqrt{\tau}}{\sqrt{\pi}\lambda}$$
$$\times\left[\Phi\left(\frac{1+\xi}{2\sqrt{\tau}}\right) + \Phi\left(\frac{1-\xi}{2\sqrt{\tau}}\right) - \frac{1+\xi}{2\sqrt{\pi\tau}}\mathrm{Ei}\left(-\frac{(1+\xi)^2}{4\tau}\right)\right.$$

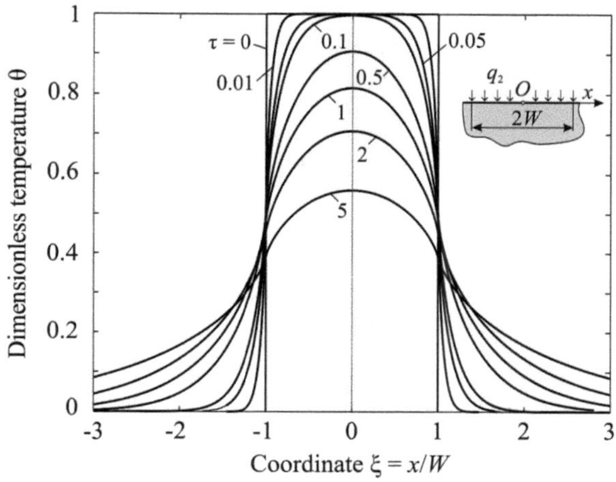

Fig. 5.54 Dimensionless temperature distribution across a continuously heated strip of width $2W$ in the surface of a semi-infinite solid (Carslaw and Jaeger 1973)

$$-\frac{1-\xi}{2\sqrt{\pi\tau}}\text{Ei}\left(-\frac{(1-\xi)^2}{4\tau}\right)\Bigg]; \quad \xi = x/W; \quad \tau = at/W^2. \tag{5.2.82}$$

It is seen that the temperature depends on the relative distance between the observation point and the centre of strip ξ and relative time τ.

We relate the temperature (5.2.82), preconditioned by heating a $2W$-wide strip, to the temperature of heating the entire surface by the source of similar power density q_2 (to doubled temperature, using formula (5.1.35) for $x = 0$). The relative temperature is described by the expression (Carslaw and Jaeger 1973)

$$\theta(\xi,\tau)=\frac{1}{2}\Bigg[\Phi\left(\frac{1+\xi}{2\sqrt{\tau}}\right)+\Phi\left(\frac{1-\xi}{2\sqrt{\tau}}\right) - \frac{1+\xi}{2\sqrt{\pi\tau}}\text{Ei}\left(-\frac{(1+\xi)^2}{4\tau}\right)-\frac{1-\xi}{2\sqrt{\pi\tau}}\text{Ei}\left(-\frac{(1-\xi)^2}{4\tau}\right)\Bigg]. \tag{5.2.83}$$

It follows from Fig. 5.54 that initially the strip boundedness has a weak effect on the temperature of the strip, but this influence increases over time.

Formulas (5.2.81) and (5.2.82) allow us to obtain a solution for a slab by the method of images.

The above-mentioned model can be used to calculate temperature field when edges of the joint are heated before welding and the structure is straightened after welding.

If the heating strip of a semi-infinite solid is limited both in width ($-W < x < W$) and in length ($-l < y < l$), the temperature field is obtained by integrating continuous point sources (5.1.25) over the area where the distributed source acts:

$$T(x, y, z, t) - T_0 = \frac{q_2}{2\pi\lambda} \int\limits_{-W}^{W} \int\limits_{-l}^{l} \frac{1}{\left[(x - x')^2 + (y - y')^2 + z^2\right]^{1/2}} \times \Phi^* \left(\left[\frac{(x - x')^2 + (y - y')^2 + z^2}{4at} \right]^{1/2} \right) dx'dy'. \quad (5.2.84)$$

In the steady state ($t \to \infty$), temperature field on the surface of a semi-infinite solid ($z = 0$) is described via the inverse hyperbolic sine (Loewen and Shaw 1954; Trigger and Chao 1951):

$$\begin{aligned} T(x, y, 0, \infty) - T_0 = \frac{q_2}{2\pi\lambda} \Bigg\{ & |x + W| \left[\mathrm{Arsh}\left(\frac{y + l}{x + W} \right) - \mathrm{Arsh}\left(\frac{y - l}{x + W} \right) \right] \\ & + |x - W| \left[\mathrm{Arsh}\left(\frac{y - l}{x - W} \right) - \mathrm{Arsh}\left(\frac{y + l}{x - W} \right) \right] \\ & + |y + l| \left[\mathrm{Arsh}\left(\frac{x + W}{y + l} \right) - \mathrm{Arsh}\left(\frac{x - W}{y + l} \right) \right] \\ & + |y - l| \left[\mathrm{Arsh}\left(\frac{x - W}{y - l} \right) - \mathrm{Arsh}\left(\frac{x + W}{y - l} \right) \right] \Bigg\}, \end{aligned} \quad (5.2.85)$$

where $\mathrm{Arsh}(u) = \ln\left(u + \sqrt{u^2 + 1}\right)$.

Equation (5.2.85) can be represented in a dimensionless form:

$$\begin{aligned} \theta = & |\xi + 1| \left[\mathrm{Arsh}\left(\frac{\eta + 1}{\omega(\xi + 1)} \right) - \mathrm{Arsh}\left(\frac{\eta - 1}{\omega(\xi + 1)} \right) \right] \\ & + |\xi - 1| \left[\mathrm{Arsh}\left(\frac{\eta - 1}{\omega(\xi - 1)} \right) - \mathrm{Arsh}\left(\frac{\eta + 1}{\omega(\xi - 1)} \right) \right] \\ & + \frac{|\eta + 1|}{\omega} \left[\mathrm{Arsh}\left(\frac{\omega(\xi + 1)}{\eta + 1} \right) - \mathrm{Arsh}\left(\frac{\omega(\xi - 1)}{\eta + 1} \right) \right] \\ & + \frac{|\eta - 1|}{\omega} \left[\mathrm{Arsh}\left(\frac{\omega(\xi - 1)}{\eta - 1} \right) - \mathrm{Arsh}\left(\frac{\omega(\xi + 1)}{\eta - 1} \right) \right]; \\ \theta = & \frac{2\pi\lambda}{q_2 W}(T - T_0); \quad \xi = \frac{x}{W}; \quad \eta = \frac{y}{W}; \quad \omega = \frac{W}{l}. \end{aligned} \quad (5.2.86)$$

Solutions to this type of problems are often used when temperature is studied in cutting tools and sliding bars (Carslaw and Jaeger 1973; Loewen and Shaw 1954; Trigger and Chao 1951).

Figure 5.55 shows the field of a dimensionless temperature due to a continuous stationary uniformly-distributed square source on a semi-infinite solid.

The maximum heated point is located at the origin (Carslaw and Jaeger 1973)

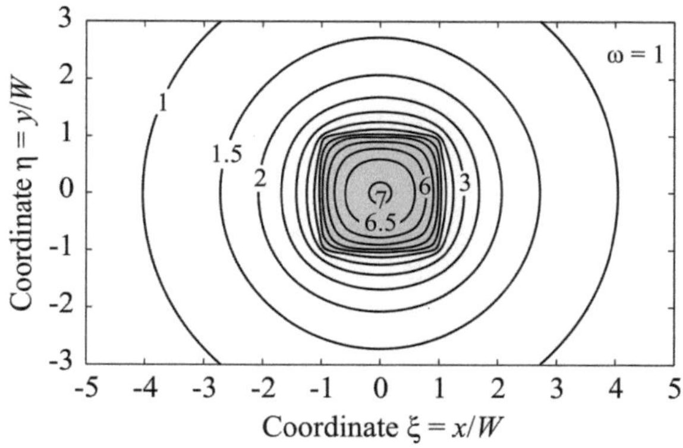

Fig. 5.55 Dimensionless steady temperature field due to a continuous stationary uniformly distributed square source on a semi-infinite solid

$$T_{\max} - T_0 = \frac{2q_2}{\pi\lambda}\left[l\,\mathrm{Arsh}\left(\frac{W}{l}\right) + W\mathrm{Arsh}\left(\frac{l}{W}\right)\right]. \tag{5.2.87}$$

The average integral temperature increase in a rectangular heating region is (Loewen and Shaw 1954)

$$\begin{aligned} T_{av} - T_0 &= \frac{1}{4Wl}\int\limits_{-W}^{W}\int\limits_{-l}^{l}[T(x,y,0,\infty) - T_0]dxdy \\ &= \frac{2q_2}{\pi l W\lambda}\left\{Wl^2\mathrm{Arsh}\left(\frac{W}{l}\right) + lW^2\mathrm{Arsh}\left(\frac{l}{W}\right)\right. \\ &\quad \left. + \frac{1}{3}\left[l^3 + W^3 - \left(W^2 + l^2\right)^{3/2}\right]\right\}. \end{aligned} \tag{5.2.88}$$

Equation (5.2.88) can be represented in a dimensionless form

$$\begin{aligned} \theta_{av} &= 4\left[\mathrm{Arsh}\left(\frac{1}{\omega}\right) + \frac{1}{\omega}\mathrm{Arsh}(\omega) + \frac{1}{3\omega^2}\right. \\ &\quad \left. + \frac{\omega}{3} - \frac{1}{3}\left(\omega + \frac{1}{\omega}\right)\left(1 + \frac{1}{\omega^2}\right)^{1/2}\right]; \\ \theta_{av} &= \frac{2\pi\lambda}{q_2 W}(T_{av} - T_0); \quad \omega = \frac{W}{l}. \end{aligned} \tag{5.2.89}$$

It can be seen that the dimensionless temperature depends only on the ratio of the source's rectangular sides. This dependence is shown in Fig. 5.56.

It should be noted that in the case a slab is continuously heated by a strip source, the limit state of the temperature field is not reached (temperature increases with no limit).

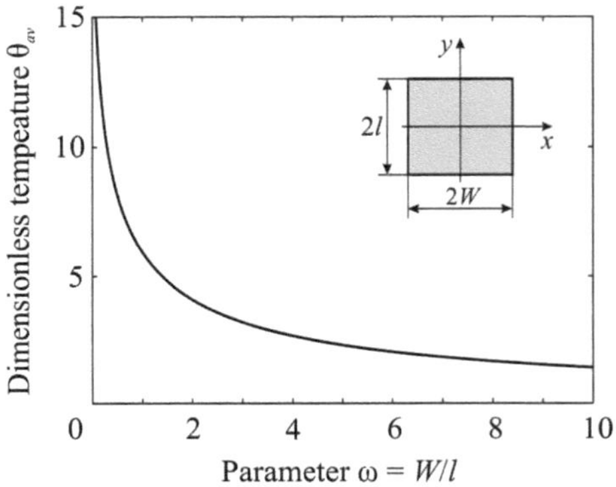

Fig. 5.56 Average dimensionless steady temperature in the heated rectangle on a semi-infinite solid as a function of the ratio of rectangular sides

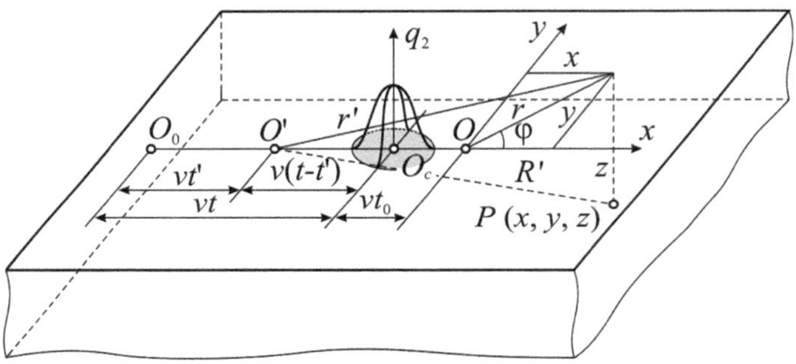

Fig. 5.57 Model for heating a semi-infinite solid by a superficial moving normally distributed circular heat source (Rykalin 1951, 1957)

5.2.3 Moving Distributed Sources

5.2.3.1 Moving Normally Distributed Circular Source on a Semi-infinite Body

We take an instantaneous normally distributed source on a semi-infinite solid as the basis (see Sect. 5.2.1.2) and use the method of sources.

Let the source move from point O_0 along the x-axis on the surface of the body with constant speed v and be at time t at point O_C (Fig. 5.57). We shift the origin of the moving coordinates from the source centre by vt_0 value, where constant t_0 characterises the distribution of the source, $t_0 = 1/(4aC)$. We split the continuous source into elementary instantaneous sources. Heat $dQ = qdt'$ introduced by the source with the centre at point O' at time t' and propagating during $t - t'$ according to Eq. (5.2.10) will increase the temperature at point $P(x, y, z)$ by time t by value

$$dT\left(r', z, t - t'\right) = \frac{2q\,dt'}{c\rho}\left[\frac{1}{4\pi a(t + t_0 - t')}\exp\left(-\frac{r'^2}{4a(t + t_0 - t')}\right)\right] \times \left[\frac{1}{\sqrt{4\pi a(t - t')}}\exp\left(-\frac{z^2}{4a(t - t')}\right)\right], \tag{5.2.90}$$

where $r'^2 = [x + vt_0 + v(t - t')]^2 + y^2$ (Fig. 5.57).

By summing all the increments of temperature at time t and using the substitution $u = t - t'$, we obtain Rykalin 1951, 1957)

$$T(x, y, z, t) - T_0 = \frac{2q}{c\rho(4\pi a)^{3/2}}\exp\left(-\frac{vx}{2a}\right) \times \int_0^t \frac{1}{u^{1/2}(u + t_0)}\exp\left(-\frac{x^2 + y^2}{4a(u + t_0)} - \frac{z^2}{4au} - \frac{v^2}{4a}(u + t_0)\right)du. \tag{5.2.91}$$

This equation can be represented in a dimensionless form

$$\theta(\xi, \eta, \zeta, \tau) = \frac{2}{\pi}\exp(-2\omega\xi)\int_0^{\sqrt{\tau}} \frac{1}{1 + u^2} \times \exp\left(-\frac{\xi^2 + \eta^2}{1 + u^2} - \frac{\zeta^2}{u^2} - \omega^2\left(1 + u^2\right)\right)du;$$

$$\theta = \frac{2\lambda\sqrt{4\pi a t_0}}{q}(T - T_0);$$

$$\xi = \frac{x}{\sqrt{4at_0}};\quad \eta = \frac{y}{\sqrt{4at_0}};\quad \zeta = \frac{z}{\sqrt{4at_0}};\quad \omega^2 = \frac{v^2 t_0}{4a};\quad \tau = \frac{t}{t_0}. \tag{5.2.92}$$

A graphical presentation of the integral for $\zeta = 0$ (for the surface of the body) is given in the books (Rykalin 1951, 1957).

Integrals (5.2.91) and (5.2.92) can be expressed in terms of the known functions only in some special cases. Examples of such cases are given below.

If the source is stationary ($v = 0$), then by using the known integral (Prudnikov et al. 1986)

$$\int \frac{dx}{x^{1/2}(ax + b)} = \frac{2}{(ab)^{1/2}}\arctan\left(\frac{ax}{b}\right)^{1/2} \tag{5.2.93}$$

the temperature at the centre of the surface source ($x = y = z = 0$) determined from (5.2.91):

$$T(0, 0, 0, t) - T_0 = \frac{q}{\pi \lambda \sqrt{4\pi a t_0}} \arctan\left(\sqrt{\frac{t}{t_0}}\right). \tag{5.2.94}$$

The limiting temperature of the central point is the maximum that can be induced by a surface normally distributed disk source on the surface of a semi-infinite solid. It is proportional to the power of source q and the square root of concentration factor C and inversely proportional to thermal conductivity λ:

$$T(0, 0, 0, \infty) - T_0 = \frac{q}{2\lambda \sqrt{4\pi a t_0}} = \frac{q}{2\lambda} \sqrt{\frac{C}{\pi}}. \tag{5.2.95}$$

At the beginning of the heating, when $t \ll t_0$ and $\arctan(t/t_0)^{1/2} \approx (t/t_0)^{1/2}$, the temperature increases according to the square root law, and then it increases by asymptotically approaching the temperature of the limit state (Rykalin 1951, 1957).

If the source is stationary ($v = 0$) and acts for a sufficiently long time ($t \to \infty$), the integral in (5.2.91) for $z = 0$ can be expressed in terms of the known functions by substitution $u' = 1/(u + t_0)$ and integral (Prudnikov et al. 1986)

$$\int_0^a x^{-1/2}(a - x)^{-1/2} \exp(-px)dx = \pi \exp\left(\frac{ap}{2}\right) \mathrm{I}_0\left(\frac{ap}{2}\right) \tag{5.2.96}$$

Then the surface temperature is

$$T(x, y, 0, \infty) - T_0 = \frac{q}{2\lambda \sqrt{4\pi a t_0}} \exp\left(-\frac{x^2 + y^2}{8 a t_0}\right) \mathrm{I}_0\left(\frac{x^2 + y^2}{8 a t_0}\right). \tag{5.2.97}$$

or in a dimensionless form

$$\frac{T(x, y, 0, \infty) - T_0}{T(0, 0, 0, \infty) - T_0} = \exp\left(-\frac{\xi^2 + \eta^2}{2}\right) \mathrm{I}_0\left(\frac{\xi^2 + \eta^2}{2}\right). \tag{5.2.98}$$

We note here that an approximate analytical approach to calculating non-stationary temperature is possible. The entire domain of integration is divided into several non-intersecting intervals, the integrand within each interval is approximated by simple analytic expressions that can be easily integrated, and then the obtained solutions are joined together (Uglov et al. 1981). At present, it is easy to define integrals using the known numerical methods, although the results of the calculation do not have generalisation, which is specific to analytical methods.

Figure 5.58 shows the steady-state field calculated from (5.2.92) for $\omega = 1$, which corresponds to steel arc welding. The origin of coordinates is coincident with the centre of the distributed source, i.e. the centre of the source is at point $\xi + \omega = 0$, $\eta = 0$, $\zeta = 0$.

The most heated area is on the longitudinal x-axis along which the source moves (Fig. 5.59). As can be seen from the figure, the higher dimensionless velocity ω is,

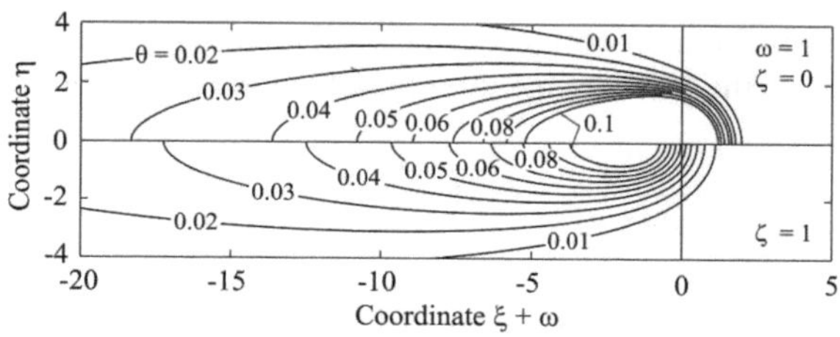

Fig. 5.58 Quasi-stationary dimensionless temperature field in the upper (heated) plane ($\zeta = 0$) and at the depth of $\zeta = 1$ of a semi-infinite solid due to a superficial normally distributed circular heat source

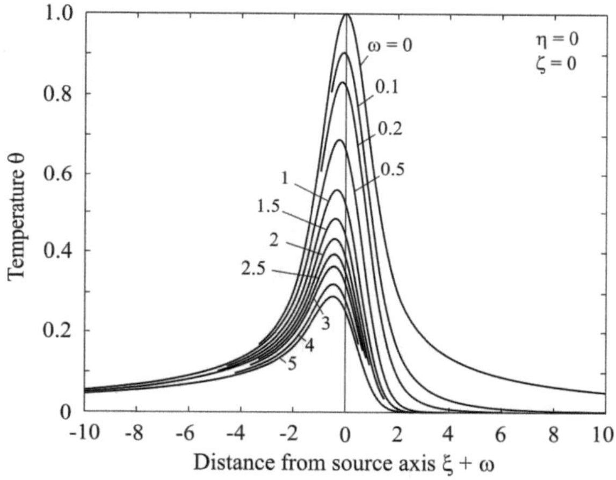

Fig. 5.59 Steady-state dimensionless temperature distribution along the centreline (x-axis) due to a moving normally distributed circular heat source on a semi-infinite solid (Rykalin 1951, 1957)

the steeper the temperature increases and the gentler it decreases, the farther from the centre of the source the maximally heated zone is and the lower its temperature is.

These equations allow us to obtain a solution to the problem by the method of images for a moving distributed source buried in a semi-infinite solid.

If the source is a normally distributed elliptic one, formula (5.2.12) should be taken as the basis instead of (5.2.10). The solution to the problem becomes more complicated here, but its structure will be retained.

5.2.3.2 Moving Normally Distributed Circular Source on a Slab

Let the source move along the x-axis over the surface of a slab with constant velocity v (see Fig. 5.57). To calculate the temperature field, we use the method of sources, as in the case of an infinite body (see Sect. 5.2.3.1). We take formulas (5.2.17) and

(5.2.18) as the basis for an instantaneous normally distributed disk source on a slab. Then the solution to the problem is representable in the following forms:

$$T(x,y,z,t) - T_0 = \frac{2q}{c\rho(4\pi a)^{3/2}} \exp\left(-\frac{vx}{2a}\right) \times \int_0^t \frac{1}{u^{1/2}(u+t_0)} \exp\left(-\frac{x^2+y^2}{4a(u+t_0)} - \frac{z^2}{4au} - \frac{v^2}{4a}(u+t_0)\right) \times \sum_{i=-\infty}^{\infty} \exp\left(-\frac{ih(ih-z)}{au}\right) du. \quad (5.2.99)$$

and

$$T(x,y,z,t) - T_0 = \frac{q/h}{4\pi\lambda} \int_0^t \frac{1}{u+t_0} \exp\left(-\frac{x^2+y^2}{4a(u+t_0)}\right) \times \left[1 + 2\sum_{i=1}^{\infty} \cos\left(\frac{\pi i z}{h}\right) \exp\left(-\pi^2 i^2 \frac{au}{h^2}\right)\right] du. \quad (5.2.100)$$

These two formulas are equivalent. Formula (5.2.99) can be represented in a dimensionless form

$$\theta(\xi,\eta,\zeta,\tau) = \frac{2}{\pi} \exp(-2\omega\xi) \int_0^{\sqrt{\tau}} \frac{1}{1+u^2} \times \exp\left(-\frac{\xi^2+\eta^2}{1+u^2} - \frac{\zeta^2}{u^2} - \omega^2\left(1+u^2\right)\right) \times \sum_{i=-\infty}^{\infty} \exp\left(-\frac{4i\delta(i\delta - z)}{u^2}\right) du; \quad (5.2.101)$$

$$\theta = \frac{2\lambda\sqrt{4\pi a t_0}}{q}(T - T_0);$$

$$\xi = \frac{x}{\sqrt{4at_0}}; \quad \eta = \frac{y}{\sqrt{4at_0}}; \quad \zeta = \frac{z}{\sqrt{4at_0}}; \quad \omega^2 = \frac{v^2 t_0}{4a}; \quad \tau = \frac{t}{t_0}; \quad \delta = \frac{h}{\sqrt{4at_0}}.$$

This formula has the structure of formula (5.2.92) and contains an additional series. As thickness increases ($\delta \to \infty$), expression (5.2.101) tends to (5.2.92).

Figure 5.60 shows the temperature fields in the upper and lower planes of a slab. The centre of the distributed source is taken as the origin of the coordinates. From the comparison of a semi-infinite solid and a slab heated under the same conditions (Figs. 5.58 and 5.60), it is seen that the temperature in a slab is much higher, especially at a depth equal to the thickness of a slab.

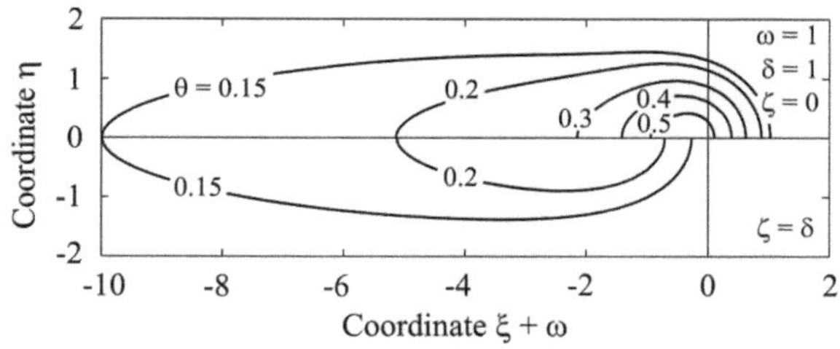

Fig. 5.60 Quasi-stationary dimensionless temperature field in the upper (heated) plane ($\zeta = 0$) and in the lower (unheated) plane ($\zeta = \delta$) of a slab around a moving superficial normally distributed circular heat source

Similarly, a solution for a moving normally distributed elliptic source can be developed.

5.2.3.3 Moving Normally Distributed Cylindrical Source in a Bounded Plate

Earlier in Sect. 5.2.1.2, a solution was obtained for an instantaneous volumetric source normally distributed in an infinite plate. It allows us to find a solution for a continuous moving normally distributed source in a bounded rectangular plate. In this case the method of sources is used to allow for continuity of action and movement of the source, and the method of images is used to consider heat-impermeable boundaries of the plate.

Let the centre of the source be in the longitudinal plane of the symmetry (on the x-axis), and move from point O_1 to point O_2 at time t (Fig. 5.61). At each moment, the power of source q is known. Let us find the temperature field at time t. We use solution (5.2.22) for an instantaneous normally distributed cylindrical source in an infinite plate as the basis. We reflect the source from the planes $x = 0$, $x = L$ and $y = -W/2$, $y = +W/2$, resulting in a system of sources, each of which is characterised by number i, j, k ($-\infty < i < \infty$; $-\infty < j < \infty$; $k = -1, 1$), and the real source has number 0,0,1. Then in a fixed coordinate system the temperature field in a bounded plate is described by the triple sum of the corresponding solutions for an unbounded plate (Karkhin 1990):

$$T(x, y, t) - T_0 = \sum_{i=-\infty}^{\infty} \sum_{j=-\infty}^{\infty} \sum_{k=-1,1} \int_0^t \frac{q(\tau)/h}{4\pi\lambda(t + t_0 - \tau)} \times \exp\left(-\frac{[x - 2iL - k(l + v\tau)]^2 + (y - jW)^2}{4a(t + t_0 - \tau)} - b(t - \tau) \right) d\tau, \tag{5.2.102}$$

where $l = OO_1$ (Fig. 5.61). If observation time t is longer than the action time of source t_w (the source is switched off), the upper limit of integration t should be replaced with t_w. If the source is offset from the axis of the plate, expression (5.2.102)

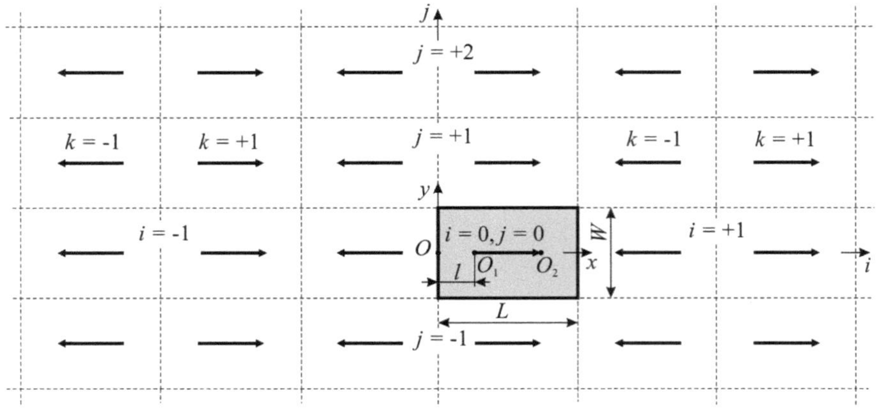

Fig. 5.61 Model of introduction of additional sources to account for finite body length and width (Karkhin 1990)

becomes more complicated (the reflection in the y-axis changes), but the structure of the formula is preserved. If the plate is unbounded in length ($L = \infty$) and / or width ($W = \infty$), only the term for $i = 0$ and / or $j = 0$ remains non-zero. If velocity v is variable and is the function of time, path $\int_0^\tau v(\tau')d\tau'$ must be substituted for term $v\tau$ in (5.2.102). Programmed heat input will be discussed in Chap. 7.

It should be kept in mind that the method of images assumes that fictitious sources do not act within a real body (a real source is concentrated or distributed, but it is not located at the boundary of the body).

Note that the solution can be obtained for a moving normally distributed disk source on a slab in the same way. For this purpose, the solution should be based on the corresponding instantaneous source (5.2.17) or (5.2.18).

5.2.3.4 Moving Uniformly Distributed Ring Source on a Semi-infinite Body

In welding with a hollow cathode electrode in vacuum (Yampolsky 1972) and when the magnetic field is applied to the welding arc, the arc acquires a special conical shape and its central part has no plasma (Gagen and Taran 1970). With high frequency circular oscillations of the electron beam (Bashkatov 1972), the surface flow inside the circle may be absent. Such sources can be considered as rings with radius r_0 and a uniformly distributed power per unit length

$$q_1 = \frac{q}{2\pi r_0}. \tag{5.2.103}$$

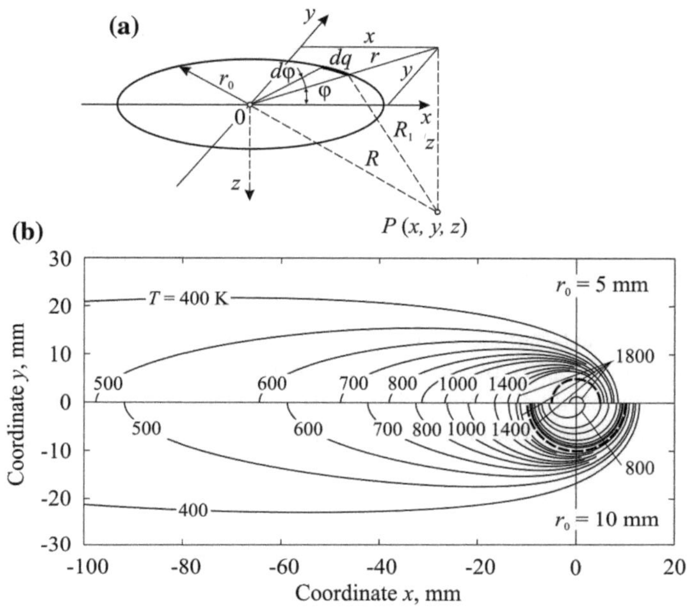

Fig. 5.62 Model of the ring source (**a**) and quasi-stationary temperature field generated by a moving uniformly distributed ring-shaped source with radius r_0 on a semi-infinite solid (**b**) ($q = 5000$ W, $v = 5$ mm s^{-1}, $\lambda = 0.04$ W mm^{-1} K^{-1}, $a = 8$ mm^2 s^{-1}, $T_0 = 300$ K)

The temperature field due to the moving ring source can be obtained by the method of sources in two ways: (1) represent the moving source as a set of consecutive instantaneous ring sources (see Sect. 5.2.1.3) and find out temperature by integrating temperature increments over time, or (2) represent the ring source as a set of elementary moving point sources (see Sect. 5.1.3.1) and find out temperature by integrating the temperature increments along the perimeter of the ring. Let us choose the second way and find out the limiting (quasi-stationary) temperature field in the moving coordinate system.

The temperature increment caused by an elementary point source of power $dq = q_1 r_0 d\phi$ on the surface of a semi-infinite solid (Fig. 5.62a) is defined by expression (5.1.42):

$$dT(x, y, z, \infty) = \frac{q_1 r_0 d\varphi}{2\pi\lambda R_1} \exp\left(-\frac{v(x - r_0\cos\varphi + R_1)}{2a}\right);$$

$$R_1 = \sqrt{(x - r_0\cos\varphi)^2 + (y - r_0\sin\varphi)^2 + z^2}. \qquad (5.2.104)$$

Integrating this equation over the entire length of the ring, we obtain

$$T(x, y, z, \infty) - T_0 = \frac{q}{4\pi^2\lambda} \int_0^{2\pi} \frac{\exp\left(-v[x - r_0 \cos\varphi + R_1(\varphi)]/(2a)\right)}{R_1(\varphi)} d\varphi. \tag{5.2.105}$$

When the radius of the ring source decreases ($r_0 \to 0$), formula (5.2.105) turns into formula (5.1.42).

Figure 5.62b shows the fields on the surface of the body at source radii $r_0 = 5$ and 10 mm. From the comparison of Fig. 5.62b with Fig. 5.8d it follows that the special feature of temperature fields in the case of a ring source is the lowered temperature around the ring. The larger the radius is, the lower the temperature is inside the ring.

Based on Eq. (5.2.105), by the method of images it is easy to obtain the solution for a moving uniformly distributed ring source in a slab.

Temperature on the z-axis ($x = y = 0$) is expressed in terms of the known functions:

$$T(0, 0, z, \infty) - T_0 = \frac{q}{2\pi\lambda} \mathrm{I}_0\left(\frac{vr_0}{2a}\right) \frac{\exp\left(-\sqrt{r_0 + r^2}/(2a)\right)}{\sqrt{r_0^2 + z^2}}, \tag{5.2.106}$$

since (Janke et al. 1960)

$$\int_0^{2\pi} \exp\left(\frac{vr_0}{2a} \cos\varphi\right) = 2\pi\ \mathrm{I}_0\left(\frac{vr_0}{2a}\right). \tag{5.2.107}$$

In a dimensionless form

$$\theta = \frac{1}{2} \mathrm{I}_0(\rho_0) \frac{\exp\left(-\rho_0\sqrt{1 + \bar{z}^2}\right)}{\sqrt{1 + \bar{z}^2}}, \tag{5.2.108}$$

where

$$\theta = \frac{\pi\lambda r_0}{q}(T - T_0); \quad \rho_0 = \frac{vr_0}{2a} \quad \bar{z} = \frac{z}{r_0}. \tag{5.2.109}$$

In the centre of the ring (at $\bar{z} = 0$), the temperature is

$$\theta = \frac{1}{2} \mathrm{I}_0(\rho_0) \exp(-\rho_0). \tag{5.2.110}$$

With a relatively small radius ($\rho_0 \to 0$), the temperature $\theta \to 1/2$ and

$$T(0, 0, 0, \infty) - T_0 = \frac{q}{2\pi\lambda r_0}, \tag{5.2.111}$$

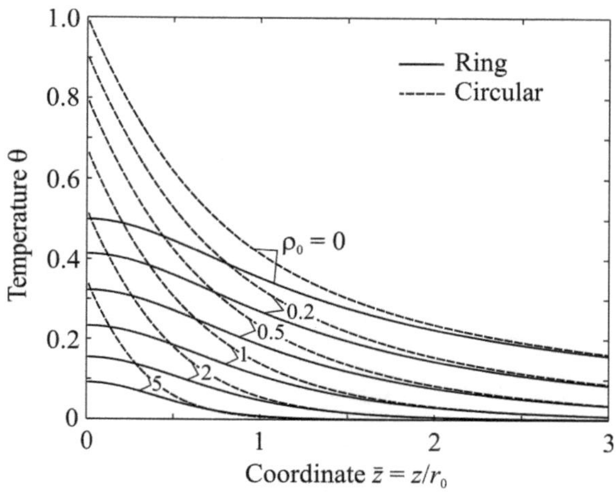

Fig. 5.63 Distribution of dimensionless temperature $\theta = \pi \lambda r_0 (T - T_0)/q$ along the z-axis as a function of dimensionless radius $\rho_0 = v r_0/(2a)$ of the uniformly distributed ring and circular sources moving on a semi-infinite body

that is, the temperature is equal to the limiting (steady) temperature generated by the stationary point source (5.1.42) at a distance from the centre equal to r_0. It does not depend on the travel speed.

The temperature distributions along the z-axis for different radii of the sources are shown in Fig. 5.63. It can be seen that the smaller the radius of the source is (the higher the power density is), the higher the temperature is. It decreases rapidly as we move away from the surface.

Note that using Eq. (5.2.105) and the method of images, it is easy to obtain the solution to the problem for a moving uniformly distributed ring source in a slab. It is possible to construct axisymmetric sources with any radial power distribution using ring sources.

5.2.3.5 Moving Uniformly Distributed Cylindrical Surface Source in a Plate

Let the source be uniformly distributed over the surface of a circular cylinder of radius r_0 and move for a sufficiently long time ($t \to \infty$) and linearly along the x-axis in an unbounded plate with boundary conditions of the third kind on the surface. Let us use the method of sources, taking a moving line source in the plate as an elementary one (5.1.81).

We split the ring source of power$q$ into elementary sources of power $dq = q_1 r_0 d\phi$, where $q_1 = q \big/ (2\pi r_0)$ (Fig. 5.64a). Then the temperature at any point P is determined by summing the temperature increments caused by all elementary sources:

$$T(x, y, \infty) - T_0 = \frac{q_1 r_0}{2\pi \lambda h} \int_0^{2\pi} \exp\left(-\frac{v(x - x_A)}{2a}\right) \mathrm{K}_0\left(\frac{v r_1}{2a} \mu\right) d\varphi;$$

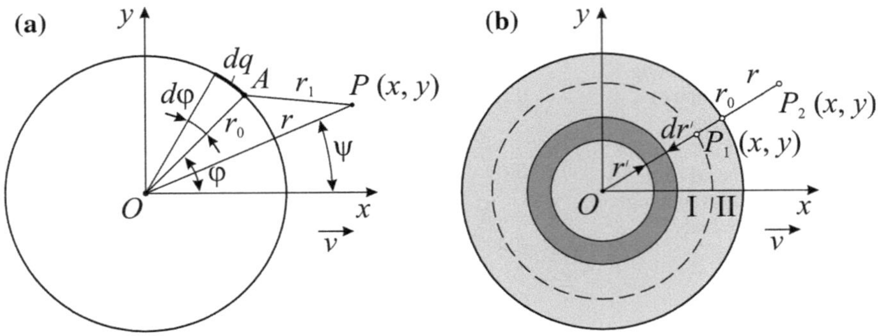

Fig. 5.64 Models for a cylindrical surface source (**a**) and a disk source (**b**) in a plate

$$\mu = \sqrt{1 + 4ab/v^2}, \tag{5.2.112}$$

where $r_1 = \sqrt{r^2 + r_0^2 - 2rr_0\cos(\psi - \varphi)}$ and $x - x_A = r\cos\psi - r_0\cos\varphi$ are distance between the elementary line source located on the ring at point A and the observation point P and its projection onto the x-axis (Fig. 5.64a).

The integrand in (5.2.112) is singular for $r_1 = 0$, since $\mathrm{K}_0(0) = \infty$. Therefore, numerical integration is difficult. We shall search the solution to the problem using the Bessel functions (Korenev 1971, 1980; Watson 1995).

Using Neumann's addition theorem (Bateman and Erdelyi 1953; Korenev 1971), we represent the Bessel function in the form:

$$\mathrm{K}_0\left(\frac{vr_1}{2a}\mu\right) = \sum_{n=0}^{\infty}{}' \mathrm{I}_n\left(\frac{vr}{2a}\mu\right)\mathrm{K}_n\left(\frac{vr_0}{2a}\mu\right)\cos(n(\psi - \varphi)),\ r \le r_0; \tag{5.2.113}$$

$$\mathrm{K}_0\left(\frac{vr_1}{2a}\mu\right) = \sum_{n=0}^{\infty}{}' \mathrm{I}_n\left(\frac{vr_0}{2a}\mu\right)\mathrm{K}_n\left(\frac{vr}{2a}\mu\right)\cos(n(\psi - \varphi)),\ r \ge r_0;$$

$$\psi = \arctan(y/x), \tag{5.2.114}$$

where I_n is the modified Bessel function of the first kind of order n, K_n is the modified Bessel function of the second kind of order n (the Macdonald function) (Fig. 3.14):

$$\mathrm{I}_n(u) = \left(\frac{u}{2}\right)^n \sum_{m=0}^{\infty} \frac{1}{m!\,\Gamma(n + m + 1)}\left(\frac{u}{2}\right)^{2m} \tag{5.2.115}$$

$$\mathrm{K}_n(u) = (-1)^{n+1}\mathrm{I}_n(u)\ln\left(\frac{\gamma u}{2}\right) + \frac{(-1)^n}{2}\sum_{k=0}^{\infty}\frac{(u/2)^{n+2k}}{k!\,(n + k)!}$$

$$\times\left(\sum_{m=1}^{k}\frac{1}{m}+\sum_{m=1}^{n+k}\frac{1}{m}+\frac{1}{2}\sum_{k=0}^{n-1}\frac{(-1)^k(n-k-1)!}{k!}\left(\frac{u}{2}\right)^{2k-n}\right),\ (n=0,1,2,\ldots) \tag{5.2.116}$$

where γ is Euler's constant ($\gamma = 0.5772\ldots$). There are tables and approximate formulas (Abramowitz and Stegun 1965; Janke et al. 1960). Symbol $\sum\limits_{n=0}^{\infty}{}'$ in formulas (5.2.113) and (5.2.114) means the sum in which the term with the index $n > 0$ is multiplied by 2.

We shall factorise the exponential contained in integral (5.2.112) and represent it in the form (Abramowitz and Stegun 1965):

$$\begin{aligned}\exp\left(-\frac{v(x-x_A)}{2a}\right) &= \exp\left(-\frac{vx}{2a}\right)\exp\left(\frac{v}{2a}r_0\cos\varphi\right)\\ &= \exp\left(-\frac{vx}{2a}\right)\sum_{k=0}^{\infty}{}'\mathrm{I_k}\left(\frac{vr_0}{2a}\right)\cos(k\varphi).\end{aligned} \tag{5.2.117}$$

Inserting (5.2.113), (5.2.114) and (5.2.117) into (5.2.112) we obtain:

$$\begin{aligned}T(x,y,\infty)-T_0 &= \frac{q_1r_0}{2\pi\lambda h}\exp\left(-\frac{vx}{2a}\right)\sum_{n=0}^{\infty}{}'\mathrm{I_n}\left(\frac{vr}{2a}\mu\right)\mathrm{K_n}\left(\frac{vr_0}{2a}\mu\right)\\ &\times\sum_{k=0}^{\infty}{}'\mathrm{I_k}\left(\frac{vr_0}{2a}\right)\int_0^{2\pi}\cos(n(\psi-\varphi))\cos(k\varphi)\,d\varphi,\ r\le r_0;\end{aligned} \tag{5.2.118}$$

$$\begin{aligned}T(x,y,\infty)-T_0 &= \frac{q_1r_0}{2\pi\lambda h}\exp\left(-\frac{vx}{2a}\right)\sum_{n=0}^{\infty}{}'\mathrm{I_n}\left(\frac{vr_0}{2a}\mu\right)\mathrm{K_n}\left(\frac{vr}{2a}\mu\right)\\ &\times\sum_{k=0}^{\infty}{}'\mathrm{I_k}\left(\frac{vr_0}{2a}\right)\int_0^{2\pi}\cos(n(\psi-\varphi))\cos(k\varphi)d\varphi,\ r\ge r_0.\end{aligned} \tag{5.2.119}$$

The definite integral in (5.2.118) and (5.2.119) is known (Korenev 1980):

$$\int_0^{2\pi}\cos(n(\psi-\varphi))\cos(k\varphi)\,d\varphi=\begin{cases}\pi\ \cos(n\psi) & \text{for}\,k=n;\\ 0 & \text{for}\,k\ne n.\end{cases} \tag{5.2.120}$$

Allowing for this integral, we finally obtain the solution to the problem (Karkhin and Zharkov 2015):

$$T(x,y,\infty)-T_0=\frac{q}{2\pi\lambda h}\exp\Big(-\frac{vx}{2a}\Big)\times\sum_{n=0}^{\infty}{}'\mathrm{I_n}\Big(\frac{vr}{2a}\mu\Big)\mathrm{K_n}\Big(\frac{vr_0}{2a}\mu\Big)\mathrm{I_n}\Big(\frac{vr_0}{2a}\Big)\cos\big(n\arctan(y/x)\big),\ r\le r_0;\tag{5.2.121}$$

$$T(x,y,\infty)-T_0=\frac{q}{2\pi\lambda h}\exp\Big(-\frac{vx}{2a}\Big)\times\sum_{n=0}^{\infty}{}'\mathrm{I_n}\Big(\frac{vr_0}{2a}\mu\Big)\mathrm{K_n}\Big(\frac{vr}{2a}\mu\Big)\mathrm{I_n}\Big(\frac{vr_0}{2a}\Big)\cos\big(n\arctan(y/x)\big),\ r\ge r_0.\tag{5.2.122}$$

It is seen that the temperature field on the action surface of the source ($r = r_0$) is continuous.

Equations (5.2.121) and (5.2.122) can be represented in a dimensionless form:

$$\theta(\xi,\eta)=\exp(-\xi)\sum_{n=0}^{\infty}{}'\mathrm{I_n}(\mu\rho)\mathrm{K_n}(\mu\rho_0)\mathrm{I_n}(\rho_0)\times\cos\big(n\arctan(\eta/\xi)\big),\quad\rho\le\rho_0;\tag{5.2.123}$$

$$\theta(\xi,\eta)=\exp(-\xi)\sum_{n=0}^{\infty}{}'\mathrm{I_n}(\mu\rho_0)\mathrm{K_n}(\mu\rho)\mathrm{I_n}(\rho_0)\times\cos\big(n\arctan(\eta/\xi)\big),\quad\rho\ge\rho_0;\tag{5.2.124}$$

$$\theta=\frac{2\pi\lambda h}{q}(T-T_0);\xi=\frac{vx}{2a};\eta=\frac{vy}{2a};\rho=\sqrt{\xi^2+\eta^2};\rho_0=\frac{vr_0}{2a}.$$

The quasi-stationary fields of the dimensionless temperature generated by the cylindrical surface ($\rho_0 = 1$) and line ($\rho_0 = 0$) sources are shown in Fig. 5.65. It is evident that the temperature gradients near the cylindrical surface source are much lower than near the line source. The maximum heated zone is at the end of the source on the x-axis. As the distance from the sources increases, the relative temperature difference, calculated for both models, decreases.

Based on the method of sources, Eqs. (5.2.121) and (5.2.122) allow us to obtain the solutions to the problems for moving axisymmetric volumetric sources with any radial distribution.

5.2.3.6 Moving Uniformly Distributed Ring Source in a Slab

The steady-state temperature field around a moving uniformly distributed ring source of radius r_0 located in a heat-insulated slab at the depth $z = \zeta$ is easy to obtain by the method of images, using the solution for a semi-infinite solid (5.2.105). In this case

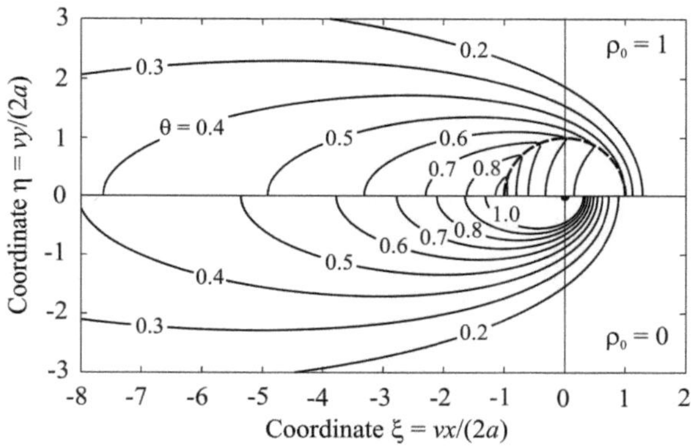

Fig. 5.65 The quasi-stationary fields of dimensionless temperature θ generated by a moving uniformly distributed cylindrical surface source (radius $\rho_0 = 1$) and a line source ($\rho_0 = 0$) in an infinite plate with surface heat transfer ($\mu = 1.01$)

it is necessary to calculate the integral with a singular integrand. A more effective way is to apply the equation for a buried moving point source in a slab (5.1.64). The integration of elementary sources over the perimeter of the action of the ring source gives (Karkhin et al. 2016):

$$T(x, y, z, \infty) - T_0 = \left[T_{ring\,0}(x, y, \infty) - T_0\right] + 2\sum_{i=1}^{\infty} \cos\left(\frac{\pi i z}{h}\right)\cos\left(\frac{\pi i \zeta}{h}\right) \times \left[T_{ring\,i}(x, y, \infty) - T_0\right], \quad (5.2.125)$$

where

$$T_{ring\,0}(x, y, \infty) - T_0 = \frac{q}{4\pi^2 \lambda h}\int_0^{2\pi} \exp\left(-\frac{v(x - r_0\cos\varphi)}{2a}\right) \mathrm{K}_0\left(\frac{vr_1}{2a}\right) d\varphi;$$

$$T_{ring\,i}(x, y, \infty) - T_0 = \frac{q}{4\pi^2 \lambda h}\int_0^{2\pi} \exp\left(-\frac{v(x - r_0\cos\varphi)}{2a}\right) \mathrm{K}_0\left(\frac{vr_1}{2a}\mu_i\right) d\varphi;$$

$$\mu_i = \sqrt{1 + \left(\frac{2\pi a i}{vh}\right)^2}.$$

Here $r_1 = [r^2 - r_0^2 - 2rr_0\cos(\psi - \varphi)]^{1/2}$ is projection of the distance between an elementary point source and the observation point on plane xOy (Fig. 5.64a). The physical content of formula (5.2.125) is obvious: $T_{ring\,0}$ (x, y, ∞) function describes a two-dimensional temperature field due to a moving uniformly distributed ring (cylindrical surface) source in a plate without heat transfer, and $T_{ring\,i}$ (x, y, ∞) function describes a two-dimensional temperature field due to a moving uniformly distributed ring (cylindrical surface) source in a plate with a coefficient of heat transfer influence $\mu_i > 1$ (Eq. (5.2.112)). Both functions are expressed as the known functions.

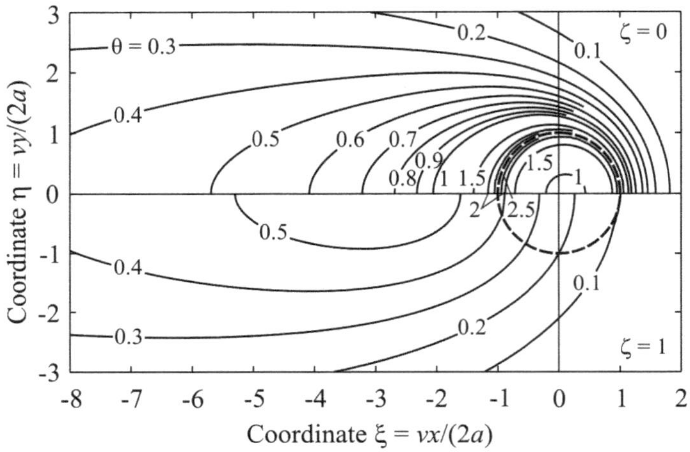

Fig. 5.66 Quasi-stationary fields of dimensionless temperature $\theta = 2\pi\lambda h(T - T_0)/q$ induced by a moving uniformly distributed ring-shaped source with radius $\rho_0 = vr_0/(2a) = 1$ on the slab with a thickness of $\delta = vh/(2a) = 2$ on the upper surface ($\zeta = z/h = 0$) and the lower surface ($\zeta = 1$) (Karkhin et al. 2016)

The corresponding formulas are given in Sect. 5.2.3.5. The non-uniformity of the temperature distribution across the thickness (along the z-axis) of the slab is allowed for by the trigonometric series; and it obviously depends on the location depth of source ζ.

Figure 5.66 shows the quasi-stationary field of dimensionless temperature $\theta = 2\pi\lambda h(T - T_0)/q$ generated by a moving uniformly distributed ring source of dimensionless radius $\rho_0 = vr_0/(2a) = 1$ on an infinite slab, whose thickness is equal to the diameter of the source. It can be seen that the area of maximum heating is on the upper surface and includes a ring source. Inside the ring the temperature is lower. On the lower surface, the heated area is significantly displaced from the z-axis of the source and its temperature is relatively low. As the distance from the source increases, the temperature drop across the thickness decreases (the isotherms on the upper and lower surfaces converge).

Note that solution (5.2.125), compared to the solution constructed on Eq. (5.2.105) for a ring source on a semi-infinite solid using the method of images, has several advantages: it is not necessary to integrate numerically the singular function over the area of action of the ring source. As for the source distributed also in depth, additional integration over z does not have any fundamental difficulties (see Sect. 5.2.1.7).

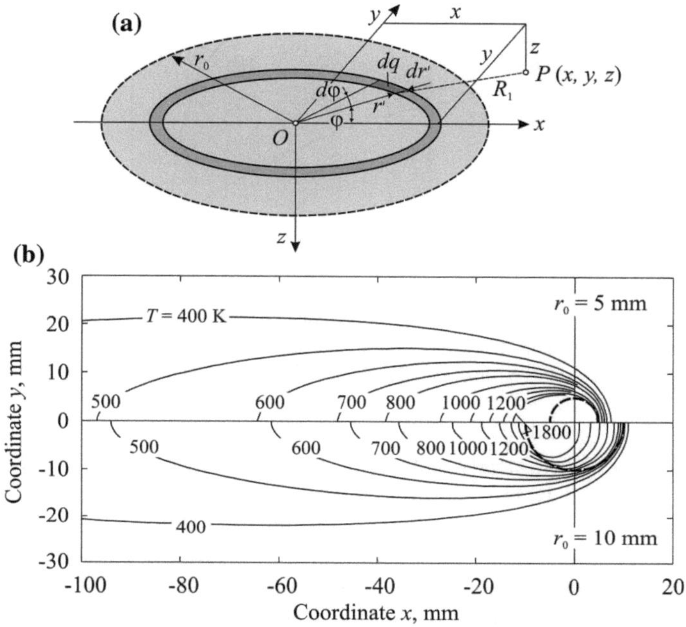

Fig. 5.67 Model for a disk source (**a**) and quasi-stationary fields due to a moving uniformly distributed disk source with radius r_0 on a semi-infinite body (**b**) ($q = 5000$ W, $v = 5$ mm s^{-1}, $\lambda = 0.04$ W mm^{-1} K^{-1}, $a = 8$ mm^2 s^{-1}, $T_0 = 300$ K)

5.2.3.7 Moving Uniformly Distributed Disk Source on a Semi-infinite Body

The solution to the problem for a uniformly distributed disk source on a semi-infinite solid can be obtained using the method of sources and taking Eq. (5.1.39) as the basis. If the heating time is long ($t \to \infty$), Eq. (5.1.41) should be taken as the basis. The power of an elementary point source is $dq = q/(\pi r_0^2) r' d\varphi dr'$ (Fig. 5.67a). Integration is performed over the area of action of the source, i.e. over the ring, as in the case of the ring source (see Sect. 5.2.3.4), and over radius r' (from 0 to the radius of the source r_0). The quasi-stationary solution to the considered problem for a moving source located on a semi-infinite solid has the form:

$$
\begin{aligned}
T(x, y, z, \infty) - T_0 = \frac{q/(\pi r_0^2)}{2\pi\lambda} \int_0^{r_0} \int_0^{2\pi} \exp\left(-\frac{v(x - r'\cos\varphi)}{2a}\right) \\
* \frac{1}{R_1} \exp\left(-\frac{vR_1}{2a}\right) r' d\varphi dr'; \\
R_1 = \sqrt{(x - r'\cos\varphi)^2 + (y - r'\sin\varphi)^2 + z^2}.
\end{aligned}
\quad (5.2.126)
$$

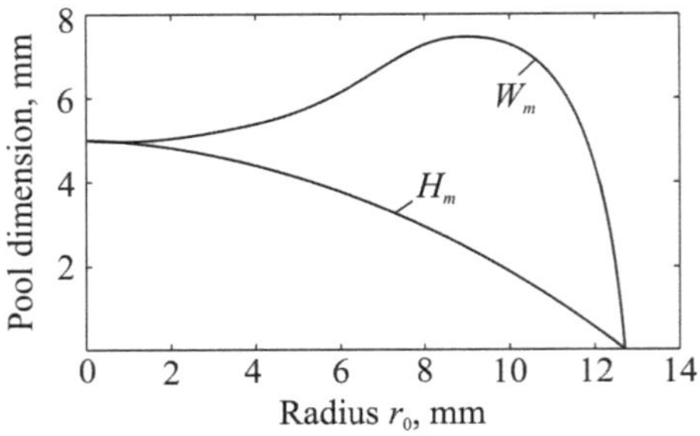

Fig. 5.68 Dependence of half width W_m and depth H_m of a weld pool on the radius of a moving uniformly distributed disk source on a semi-infinite body ($q = 5000$ W, $v = 5$ mm s^{-1}, $\lambda = 0.04$ W mm^{-1} K^{-1}, $a = 8$ mm^2 s^{-1}, $T_m = 1800$ K, $T_0 = 300$ K)

Figure 5.67b displays steady-state fields due to moving uniformly distributed disk sources of 5 mm and 10 mm radii on the surface of a semi-infinite solid. From the comparison of Figs. 5.67b and 5.62b it results that the fields differ significantly near the ring and disk sources, but at a distance from the sources fields practically coincide.

The maximum heated zone is displaced to the rear boundary of the source (Fig. 5.67b). The width and depth of the weld pool (the zone heated above the melting point $T_m = 1800$ K) depends on the radius of the source (Fig. 5.68). If the source is extremely concentrated ($r_0 = 0$), the weld cross-section is a semicircle ($H_m = W_m$). As the radius of the source increases, the depth of pool H_m decreases, width W_m first increases, and later decreases. With a relatively large radius ($r_0 > 13$ mm), the metal does not melt.

The integral (5.2.126) can be expressed as the known functions only in special cases. For example, the temperature on the z-axis (for $x = y = 0$) caused by a moving disk source on a semi-infinite solid (5.2.107) is

$$T(0, 0, z, \infty) - T_0 = \frac{q/(\pi r_0^2)}{\lambda} \int_0^{r_0} \mathrm{I}_0\left(\frac{vr'}{2a}\right) \frac{\exp(-v/(2a)\sqrt{r'^2 + z^2})}{\sqrt{r'^2 + z^2}} r' dr' \tag{5.2.127}$$

or in a dimensionless form

$$\theta = \int_0^1 \mathrm{I}_0(\rho_0 u) \frac{\exp(-\rho_0 \sqrt{u^2 + \bar{z}^2})}{\sqrt{u^2 + \bar{z}^2}} u du; \tag{5.2.128}$$

$$\theta = \frac{\pi \lambda r_0}{q}(T - T_0); \quad \rho_0 = \frac{v r_0}{2a}; \quad \bar{z} = \frac{z}{r_0}. \tag{5.2.129}$$

Figures 5.63 and 5.69 illustrate temperature distributions along the z-axis for different radii of a disk source moving along the surface of a semi-infinite solid.

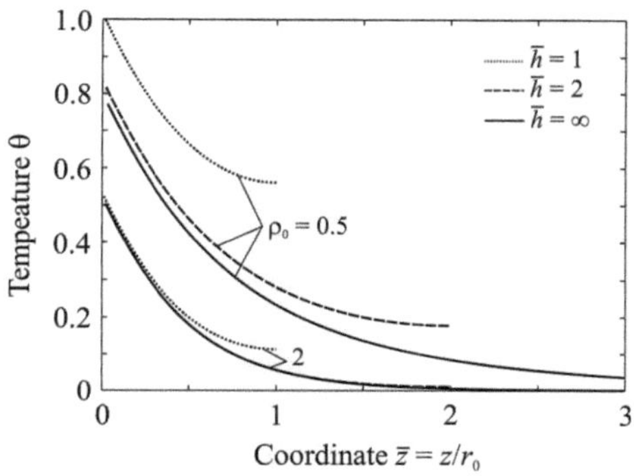

Fig. 5.69 Dimensionless temperature $\theta = 2\pi\lambda r_0(T - T_0)/q$ in the z-axis as a function of relative radius $\rho_0 = vr_0/(2a)$ of a uniformly distributed disk source moving on a slab with thickness $\bar{h}(\bar{h} = h/r_0 = 1 \text{ and } 2)$ and a semi-infinite solid ($\bar{h} = \infty$)

From Fig. 5.63 it follows, that the temperature due to the disk source is much higher near the surface of the body, but it falls faster than the temperature from the ring source. At depth $z/r_0 > 1.5$, their values have little difference.

By using integral (Prudnikov et al. 2002)

$$\int_0^x x^\nu \exp(-x)\mathrm{I}_\nu(x)dx = \frac{x^{\nu+1}}{2\nu+1}\exp(-x)[\mathrm{I}_\nu(x) + \mathrm{I}_{\nu+1}(x)] \tag{5.2.130}$$

from (5.2.127), we obtain the value of the temperature at the source centre

$$T(0, 0, 0, \infty) - T_0 = \frac{q}{\pi r_0 \lambda}\exp\left(-\frac{vr_0}{2a}\right)\left[\mathrm{I}_0\left(\frac{vr_0}{2a}\right) + \mathrm{I}_1\left(\frac{vr_0}{2a}\right)\right]. \tag{5.2.131}$$

If dimensionless radius $vr_0/(2a)$ is small, then $\exp(-vr_0/(2a)) \rightarrow 1$, $\mathrm{I}_0(vr_0/(2a)) \rightarrow 1$, $\mathrm{I}_1(vr_0/(2a)) \rightarrow 0$ and

$$T(0, 0, 0, \infty) - T_0 = \frac{q}{\pi r_0 \lambda}. \tag{5.2.132}$$

It follows that the steady temperature at the centre of a uniformly distributed disk source moving along the surface of a semi-infinite solid is independent of the speed of motion and is equal to the steady temperature induced by a stationary point source (5.1.42) at the distance from the centre equal to half radius r_0. The more distributed the source is (the larger radius r_0 is), the lower the temperature is. The temperature in the centre of the disk source is double higher than the temperature in the centre of the ring source (5.2.111). Note that this temperature is not the maximum, the maximum heated zone is located at the surface behind the source centre (Fig. 5.67b).

5.2.3.8 Moving Uniformly Distributed Cylindrical Volume Source in a Plate

Let a uniformly distributed cylindrical volume source of radius r_0 move linearly along the x-axis in an infinite plate with boundary conditions of the third kind on its surface. If the time of source action is long $(t \to \infty)$, the quasi-stationary temperature field in the moving coordinate system can be found using the method of sources, taking a moving source uniformly distributed over the surface of a circular cylinder of radius r_0 as an elementary source [Eqs. (5.2.121) and (5.2.122)].

At first, let us find the temperature at outer point P_2, which is outside the area of source action, $r > r_0$ (Fig. 5.64b). We split the entire volume source into elementary cylindrical surface sources of power $dq = q_2 2\pi r' dr'$, where $q_2 = q/\left(\pi r_0^2\right)$. Then the temperature at point P_2 is determined by summing the temperature increments due to all the elementary sources, according to Eq. (5.2.122):

$$
\begin{aligned}
T(x, y, \infty) - T_0 &= \int_0^{r_0} \frac{q_2 2\pi r'}{2\pi\lambda h} \exp\left(-\frac{vx}{2a}\right) \\
&\quad \times \sum_{n=0}^{\infty}{}' \mathrm{I_n}\left(\frac{vr'}{2a}\mu\right) \mathrm{K_n}\left(\frac{vr}{2a}\mu\right) \mathrm{I_n}\left(\frac{vr'}{2a}\right) \cos(n\psi)\, dr' \\
&= \frac{q_2}{\lambda h} \exp\left(-\frac{vx}{2a}\right) \sum_{n=0}^{\infty}{}' \mathrm{K_n}\left(\frac{vr}{2a}\mu\right) \cos(n\psi) \\
&\quad \times \int_0^{r_0} r' \mathrm{I_n}\left(\frac{vr'}{2a}\right) \mathrm{I_n}\left(\frac{vr'}{2a}\mu\right) dr'.
\end{aligned}
\tag{5.2.133}
$$

Hence, the solution reduces to calculating Lommel's integral (Korenev 1971; Watson 1995):

$$
\begin{aligned}
&\int_0^{u} r' \mathrm{I_n}\left(\frac{vr'}{2a}\right) \mathrm{I_n}\left(\frac{vr'}{2a}\mu\right) dr' \\
&\quad = \frac{2au}{v(1-\mu^2)} \left[\mathrm{I_{n+1}}\left(\frac{vu}{2a}\right) \mathrm{I_n}\left(\frac{vu}{2a}\mu\right) - \mu\, \mathrm{I_n}\left(\frac{vu}{2a}\right) \mathrm{I_{n+1}}\left(\frac{vu}{2a}\mu\right)\right].
\end{aligned}
\tag{5.2.134}
$$

By inserting (5.2.134) in (5.2.133), we obtain (Karkhin and Zharkov 2015):

$$
\begin{aligned}
T(x, y, \infty) - T_0 &= \frac{q}{\pi\lambda h} \frac{2a}{vr_0} \frac{1}{1-\mu^2} \exp\left(-\frac{vx}{2a}\right) \sum_{n=0}^{\infty}{}' \mathrm{K_n}\left(\frac{vr}{2a}\mu\right) \\
&\quad \times \left[\mathrm{I_{n+1}}\left(\frac{vr_0}{2a}\right) \mathrm{I_n}\left(\frac{vr_0}{2a}\mu\right) - \mu\, \mathrm{I_n}\left(\frac{vr_0}{2a}\right) \mathrm{I_{n+1}}\left(\frac{vr_0}{2a}\mu\right)\right] \\
&\quad \times \cos\left(n \arctan\left(y/x\right)\right), \quad r \ge r_0.
\end{aligned}
\tag{5.2.135}
$$

Equation (5.2.135) takes on an indeterminate form 0/0 for $\mu = 1$.

Passing on to dimensionless parameters (5.2.124), we result from (5.2.135):

$$\theta(\xi, \eta) = \frac{2}{\rho_0(1-\mu^2)} \exp(-\xi) \sum_{n=0}^{\infty}{}' \mathrm{K_n}(\mu\rho) \times [\mathrm{I_{n+1}}(\rho_0)\,\mathrm{I_n}(\mu\rho_0) - \mu\,\mathrm{I_n}(\rho_0)\,\mathrm{I_{n+1}}(\mu\rho_0)] \times \cos\big(n \arctan(\eta/\xi)\big), \quad \rho \geq \rho_0. \tag{5.2.136}$$

Similarly, we find the temperature at internal point $P_1(x,y)$, located in the area of the source action, $r < r_0$ (Fig. 5.64b). We divide the whole area of the source action into two parts: region I ($r' < r$) with respect to which point P_1 is external, and region II ($r' > r$) with respect to which point P_1 is internal. Then the temperature at point P_1 will be the sum of the temperature increments produced by both sources, that is:

$$T - T_0 = (T_{\mathrm{I}} - T_0) + (T_{\mathrm{II}} - T_0). \tag{5.2.137}$$

The first term in (5.2.137) is determined by the method of sources on the basis of Eq. (5.2.122) using integral (5.2.134). The second term is determined on the basis of Eq. (5.2.121) with the help of Lommel's integral for miscellaneous Bessel functions (Gradshteyn and Ryzhik 2014):

$$\int_r^{r_0} r'\,\mathrm{I_n}\left(\frac{vr'}{2a}\right)\mathrm{K_n}\left(\frac{vr'}{2a}\mu\right)dr' = \frac{2ar_0}{v(1-\mu^2)}\left[\mathrm{I_{n+1}}\left(\frac{vr_0}{2a}\right)\mathrm{K_n}\left(\frac{vr_0}{2a}\mu\right) + \mu\mathrm{I_n}\left(\frac{vr_0}{2a}\right)\mathrm{K_{n+1}}\left(\frac{vr_0}{2a}\mu\right)\right] - \frac{2ar}{v(1-\mu^2)}\left[\mathrm{I_{n+1}}\left(\frac{vr}{2a}\right)\mathrm{K_n}\left(\frac{vr}{2a}\mu\right) + \mu\mathrm{I_n}\left(\frac{vr}{2a}\right)\mathrm{K_{n+1}}\left(\frac{vr}{2a}\mu\right)\right]. \tag{5.2.138}$$

By substituting the corresponding equations in (5.2.137), we obtain the solution to the problem for the area of the source action (Karkhin and Zharkov 2015):

$$T(x,y,\infty) - T_0 = \frac{q}{\pi\lambda h}\frac{2a}{vr_0^2}\exp\left(-\frac{vx}{2a}\right)\sum_{n=0}^{\infty}{}'\cos(n\arctan(y/x))\left\{\frac{r}{1-\mu^2} \times \left[\mathrm{I_{n+1}}\left(\frac{vr}{2a}\right)\mathrm{I_n}\left(\frac{vr}{2a}\mu\right) - \mu\,\mathrm{I_n}\left(\frac{vr}{2a}\right)\mathrm{I_{n+1}}\left(\frac{vr}{2a}\mu\right)\right] \mathrm{K_n}\left(\frac{vr}{2a}\mu\right) - \frac{r_0}{1-\mu^2} \times \left[\mathrm{I_{n+1}}\left(\frac{vr_0}{2a}\right)\mathrm{K_n}\left(\frac{vr_0}{2a}\mu\right) + \mu\,\mathrm{I_n}\left(\frac{vr_0}{2a}\right)\mathrm{K_{n+1}}\left(\frac{vr_0}{2a}\mu\right)\right]\mathrm{I_n}\left(\frac{vr}{2a}\mu\right) + \frac{r}{1-\mu^2} \times \left[\mathrm{I_{n+1}}\left(\frac{vr}{2a}\right)\mathrm{K_n}\left(\frac{vr}{2a}\mu\right) + \mu\,\mathrm{I_n}\left(\frac{vr}{2a}\right)\mathrm{K_{n+1}}\left(\frac{vr}{2a}\mu\right)\right]\mathrm{I_n}\left(\frac{vr}{2a}\mu\right)\right\}, \quad r \leq r_0. \tag{5.2.139}$$

Formula (5.2.139) can be represented in a dimensionless form:

$$\theta(\xi,\eta) = \frac{2}{\rho_0^2}\exp(-\xi){\sum_{n=0}^{\infty}}'\cos(n\arctan(\eta/\xi))$$
$$\times\left\{\frac{\rho}{1-\mu^2}[\mathrm{I}_{n+1}(\rho)\mathrm{I}_n(\mu\rho)-\mu\,\mathrm{I}_n(\rho)\mathrm{I}_{n+1}(\mu\rho)]\mathrm{K}_n(\mu\rho)\right.$$
$$-\frac{\rho_0}{1-\mu^2}[\mathrm{I}_{n+1}(\rho_0)\mathrm{K}_n(\mu\rho_0)+\mu\,\mathrm{I}_n(\rho_0)\mathrm{K}_{n+1}(\mu\rho_0)]\mathrm{I}_n(\mu\rho)$$
$$\left.+\frac{\rho}{1-\mu^2}[\mathrm{I}_{n+1}(\rho)\mathrm{K}_n(\mu\rho)+\mu\,\mathrm{I}_n(\rho)\mathrm{K}_{n+1}(\mu\rho)]\mathrm{I}_n(\mu\rho)\right\},\quad \rho\le\rho_0. \tag{5.2.140}$$

The quasi-stationary fields of the dimensionless temperature due to the cylindrical volume ($\rho_0 = 1$) and line ($\rho_0 = 0$) sources are shown in Fig. 5.70a. It is seen that the temperature gradients near the cylindrical volume source (disk-shaped source) are much lower than near the line source. The maximum heated zone in the cross sections of the plate is on the x-axis. As the distance from the sources increases, the relative temperature difference calculated by both models decreases.

If there is no heat transfer ($\mu = 1$), Eqs. (5.2.135), (5.2.139) and (5.2.140) are invalid due to division by zero. In this case the sequence of solving the problem is preserved, but integrals (5.2.134) and (5.2.138) have a different form (Watson 1995):

$$\int_0^u r'\mathrm{I}_n\left(\frac{vr'}{2a}\right)\mathrm{I}_n\left(\frac{vr'}{2a}\right)dr' = \frac{u^2}{2}\left[\mathrm{I}_n^2\left(\frac{vu}{2a}\right)-\mathrm{I}_{n-1}\left(\frac{vu}{2a}\right)\mathrm{I}_{n+1}\left(\frac{vu}{2a}\right)\right]; \tag{5.2.141}$$

$$\int_r^{r_0} r'\mathrm{I}_n\left(\frac{vr'}{2a}\right)\mathrm{K}_n\left(\frac{vr'}{2a}\right)dr' = \frac{r_0^2}{4}\left[2\mathrm{I}_n\left(\frac{vr_0}{2a}\right)\mathrm{K}_n\left(\frac{vr_0}{2a}\right)+\mathrm{I}_{n-1}\left(\frac{vr_0}{2a}\right)\mathrm{K}_{n+1}\left(\frac{vr_0}{2a}\right)\right.$$
$$\left.+\mathrm{I}_{n+1}\left(\frac{vr_0}{2a}\right)\mathrm{K}_{n-1}\left(\frac{vr_0}{2a}\right)\right]$$
$$-\frac{r^2}{4}\left[2\mathrm{I}_n\left(\frac{vr}{2a}\right)\mathrm{K}_n\left(\frac{vr}{2a}\right)+\mathrm{I}_{n-1}\left(\frac{vr}{2a}\right)\mathrm{K}_{n+1}\left(\frac{vr}{2a}\right)\right.$$
$$\left.+\mathrm{I}_{n+1}\left(\frac{vr}{2a}\right)\mathrm{K}_{n-1}\left(\frac{vr}{2a}\right)\right]. \tag{5.2.142}$$

Then the temperature field in the plate without heat transfer ($\mu = 1$) outside the source area is described by the following formula (Karkhin and Zharkov 2015):

$$T(x,y,\infty)-T_0 = \frac{q}{2\pi\lambda h}\exp\left(-\frac{vx}{2a}\right){\sum_{n=0}^{\infty}}'\mathrm{K}_n\left(\frac{vr}{2a}\right)$$
$$\times\left[\mathrm{I}_n^2\left(\frac{vr_0}{2a}\right)-\mathrm{I}_{n-1}\left(\frac{vr_0}{2a}\right)\mathrm{I}_{n+1}\left(\frac{vr_0}{2a}\right)\right]$$
$$\times\cos(n\arctan(y/x)),\quad r\ge r_0. \tag{5.2.143}$$

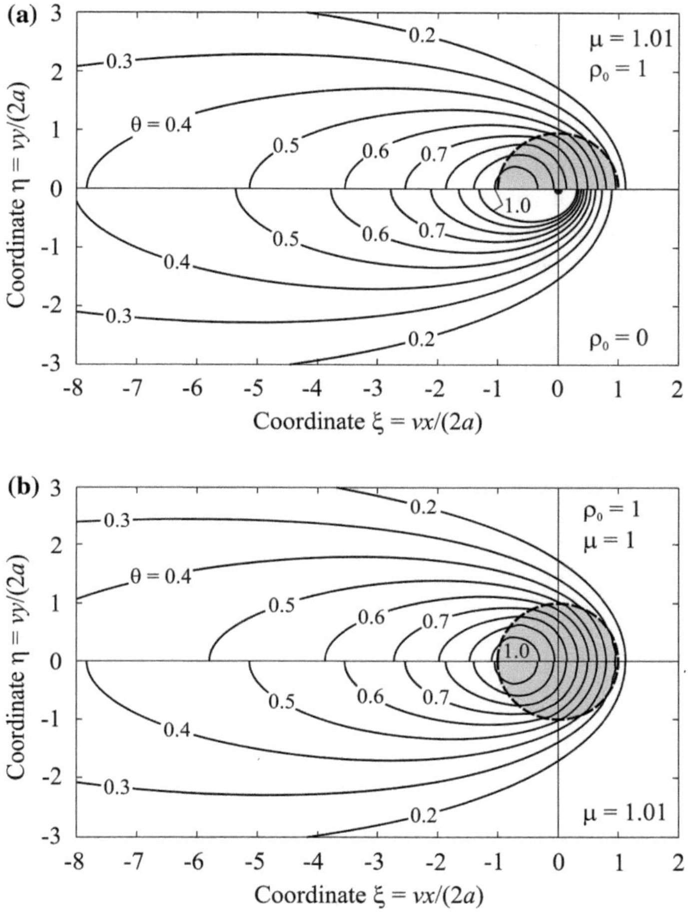

Fig. 5.70 Quasi-stationary fields of dimensionless temperature $\theta = 2\pi\lambda h(T - T_0)/q$ due to a moving uniformly distributed cylindrical volume source in an infinite plate (radius $\rho_0 = vr_0/(2a) = 1$) and a line source ($\rho_0 = 0$) in an infinite plate with surface heat transfer ($\mu = 1.01$) (**a**) and a source with radius $\rho_0 = 1$ without surface heat transfer ($\mu = 1$) and with surface heat transfer ($\mu = 1.01$) (**b**)

Similarly, the temperature in the area of the source action is defined (Karkhin and Zharkov 2015):

$$T(x, y) - T_0 = \frac{q}{2\pi\lambda h}\frac{1}{r_0^2}\exp\left(-\frac{vx}{2a}\right)\sum_{n=0}^{\infty}{}' \cos(n \arctan(y/x))$$
$$\times \left\{r^2\left[\mathrm{I}_n^2\left(\frac{vr}{2a}\right) - \mathrm{I}_{n-1}\left(\frac{vr}{2a}\right)\mathrm{I}_{n+1}\left(\frac{vr}{2a}\right)\right]\mathrm{K}_n\left(\frac{vr}{2a}\right) + \frac{r_0^2}{2}\mathrm{I}_n\left(\frac{vr}{2a}\right)\right.$$
$$\times \left[2\mathrm{I}_n\left(\frac{vr_0}{2a}\right)\mathrm{K}_n\left(\frac{vr_0}{2a}\right) + \mathrm{I}_{n-1}\left(\frac{vr_0}{2a}\right)\mathrm{K}_{n+1}\left(\frac{vr_0}{2a}\right)\right.$$

$$+ \mathrm{I}_{n+1}\left(\frac{vr_0}{2a}\right)\mathrm{K}_{n-1}\left(\frac{vr_0}{2a}\right)\Big] - \frac{r^2}{2}\mathrm{I}_n\left(\frac{vr}{2a}\right)$$
$$\times\Big[2\mathrm{I}_n\left(\frac{vr}{2a}\right)\mathrm{K}_n\left(\frac{vr}{2a}\right) + \mathrm{I}_{n-1}\left(\frac{vr}{2a}\right)\mathrm{K}_{n+1}\left(\frac{vr}{2a}\right) + \mathrm{I}_{n+1}\left(\frac{vr}{2a}\right)\mathrm{K}_{n-1}\left(\frac{vr}{2a}\right)\Big]\Big\}, \quad r \le r_0. \tag{5.2.144}$$

The quasi-stationary fields of a dimensionless temperature due to a uniformly distributed cylindrical volume source of radius $\rho_0 = 1$ for different coefficients of heat transfer influence μ are shown in Fig. 5.70b. Heat transfer leads to a decrease in temperature everywhere else.

5.2.3.9 Moving Uniformly Distributed Disk Source in a Slab

The steady-state temperature field due to a moving uniformly distributed circular (disk) source of radius r_0 located in a heat-insulated slab at the depth $z = \zeta$, can be obtained in two ways: (1) using Eq. (5.2.126) and the method of images or (2) by the method of sources, using the fundamental solution for a buried moving point source in a slab (5.1.64). Let us choose the second way, as in the case of a ring source in a slab (see Sect. 5.2.3.6). The integration of elementary sources over the perimeter of the area of the disk source action results in (Karkhin et al. 2016):

$$T(x, y, z, \infty) - T_0 = [T_{cir\,0}(x, y, \infty) - T_0] + 2\sum_{i=1}^{\infty}\cos\left(\frac{\pi i z}{h}\right)\cos\left(\frac{\pi i \zeta}{h}\right)$$
$$\times [T_{cir\,i}(x, y, \infty) - T_0], \tag{5.2.145}$$

where

$$T_{cir\,0}(x, y, \infty) - T_0 = \frac{q/(\pi r_0^2)}{2\pi\lambda h}\int_0^{r_0}\int_0^{2\pi}\exp\left(-\frac{v(x - r'\cos\varphi)}{2a}\right)\mathrm{K}_0\left(\frac{vr_1}{2a}\right)d\varphi r' dr';$$

$$T_{cir\,i}(x, y, \infty) - T_0 = \frac{q/(\pi r_0^2)}{2\pi\lambda h}\int_0^{r_0}\int_0^{2\pi}\exp\left(-\frac{v(x - r'\cos\varphi)}{2a}\right)\mathrm{K}_0\left(\frac{vr_1}{2a}\mu_i\right)d\varphi r' dr'$$

$$\mu_i = \sqrt{1 + \left(\frac{2\pi a i}{vh}\right)^2}.$$

Here $r_1 = [r^2 - r'^2 - 2rr'\cos(\psi - \varphi)]^{1/2}$ is projection of distance R_1 between the elementary point source and the observation point on the plane xOy (Fig. 5.67a). Functions $T_{cir\,0}$ (x, y, ∞) and $T_{cir\,\mathrm{i}}$ (x, y, ∞) describe two-dimensional temperature fields due to moving uniformly distributed disk (cylindrical volume) sources in a plate and they are expressed as the known functions (see Sect. 5.2.3.8).

Figure 5.71 demonstrates the temperature field in different sections of a slab whose thickness is equal to the diameter of the source. The maximum heated zone

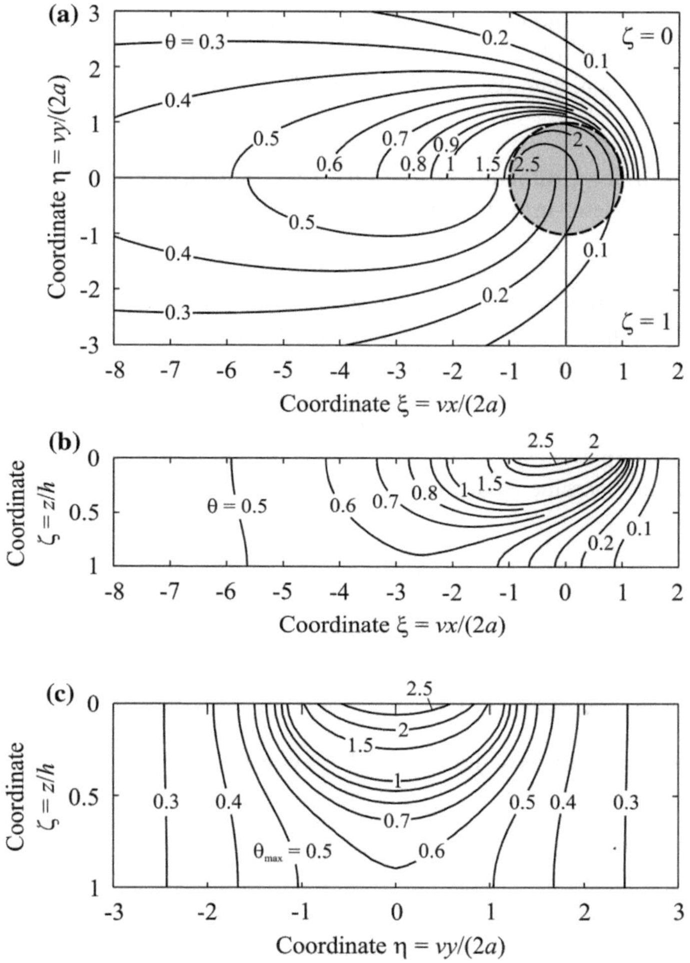

Fig. 5.71 Quasi-stationary fields of dimensionless temperature $\theta = 2\pi\lambda h(T - T_0)/q$ due to a moving surface uniformly distributed disk-shaped source with radius $\rho_0 = vr_0/(2a) = 1$ on the slab with thickness $\delta = vh/(2a) = 2$: on the upper surface ($\zeta = z/h = 0$) and lower surface ($\zeta = 1$) (**a**); in the longitudinal section ($\eta = 0$) (**b**); dimensionless peak temperature θ_{max} in the cross-section (**c**) (Karkhin et al. 2016)

is located inside the area of source action and is displaced to its tail part (Fig. 5.71a). As in the case of the ring source (Fig. 5.66), the bottom surface is heated much lower. At distance $\xi \leq -6$ there is almost no temperature difference across the thickness (Fig. 5.71b). Near the upper surface, the isotherms of the peak temperature are elongated in the transverse direction η (Fig. 5.71c). For $|\eta| > 2$ the temperature field is two-dimensional (there is no difference in thickness).

The influence of the slab thickness can be judged by temperature distribution along the source axis of a slab (Fig. 5.71). At a fixed radius and power of the source, the temperature is higher, the thickness is smaller.

5.2.3.10 Moving Uniformly Distributed Line Source of Finite Width on a Semi-infinite Body

When using strip fusion cladding and welding by a source with high frequency transverse oscillations (when the transverse velocity of the source is much higher than its longitudinal velocity), we can assume that the source is distributed uniformly along the interval of $2W$ in length (Fig. 5.72a).

Let the origin of the moving coordinate system coincide with the middle of the source (Fig. 5.72a). We represent a moving line source as a set of elementary moving point sources of power $dq = q_1 d\eta$. Then the temperature is determined by summing the temperature increments due to all point sources according to formulas (5.1.39) and (5.1.108):

$$T(x, y, z, t) - T_0 = \int_{-W}^{W} \frac{q_1}{2\pi\lambda R_1} \exp\left(-\frac{vR_1}{2a}\right)\psi_3(R_1, t)d\eta;$$

$$R_1 = \sqrt{x^2 + (y-\eta)^2 + z^2}. \quad (5.2.146)$$

If $t = \infty$ (i.e. $\psi_3 = 1$) and $q_1 = q/(2W) = \text{const}$, then

$$T(x, y, z, \infty) - T_0 = \frac{q/(2W)}{2\pi\lambda}\exp\left(-\frac{vx}{2a}\right) \times \int_{y-W}^{y+W} \frac{1}{\left(x^2+u^2+z^2\right)^{1/2}} \exp\left(-\frac{v\left(x^2+u^2+z^2\right)^{1/2}}{2a}\right)du. \quad (5.2.147)$$

Integral is not expressed in tabulated functions. If $x = z = 0$, then (Rykalin 1947, 1952)

$$T(0, y, 0, \infty) - T_0 = \frac{q/(2W)}{2\pi\lambda}\left[-\text{Ei}\left(\frac{v}{2a}(-|y| + W)\right) + \text{Ei}\left(\frac{v}{2a}(-|y| - W)\right)\right]. \quad (5.2.148)$$

Here the temperature is defined for $|y| > W$, since the real function Ei(-x) is obtained only for the positive values of argument x.

Let us represent solutions (5.2.147) and (5.2.148) in a dimensionless form:

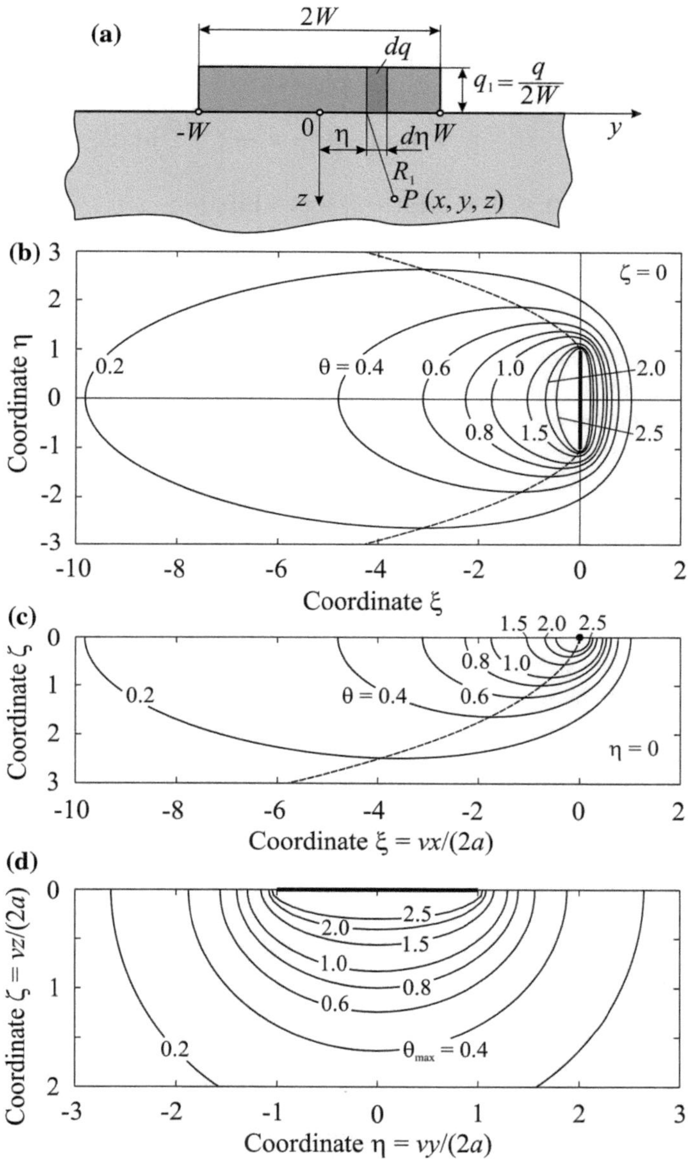

Fig. 5.72 Model for a uniformly distributed line source with width $2W$ on a semi-infinite solid (**a**), fields of dimensionless temperature on the surface (**b**) and in the longitudinal section (**c**) and dimensionless peak temperature $\theta_{\max}$ in the cross section (**d**)

$$\theta(\xi, \eta, \zeta) = \exp(-\xi) \int\limits_{\eta-\omega}^{\eta+\omega} \frac{1}{\left(\xi^2 + u^2 + \zeta^2\right)^{1/2}} \times \exp\left(-\left(\xi^2 + u^2 + \zeta^2\right)^{1/2}\right) du; \tag{5.2.149}$$

$$\theta(0, \eta, 0) = -\mathrm{Ei}(-|\eta| + \omega) + \mathrm{Ei}(-|\eta| - \omega); \tag{5.2.150}$$

$$\theta = \frac{2\pi\lambda}{q/(2W)}(T - T_0); \quad \xi = \frac{vx}{2a}; \quad \eta = \frac{vy}{2a}; \quad \zeta = \frac{vz}{2a}; \quad \omega = \frac{vW}{2a}. \tag{5.2.151}$$

Figure 5.72 displays the fields of the dimensionless temperature in the limiting state due to a moving line source of the finite width $\omega = 1$. The dashed lines indicate the demarcation between the heating and cooling zones. The isotherms near the source have the shape of an oval with a major axis directed to the width of the source (Fig. 5.72b). The lower the temperature is, the more isotherms are elongated in the direction opposite to the direction of the motion of the source. Behind, away from the source, the isotherms are similar to the isotherms around a concentrated point source (Fig. 5.10a).

In the longitudinal section, the isotherms are similar in form to the isotherms for a point source (Figs. 5.72c and 5.10a). The isotherms of the peak temperature near the source have the shape of semi-ovals elongated along the source (Fig. 5.72d). As the distance from the source increases, isotherms take the form close to semicircles.

On the basis of the obtained solutions, it is easy to derive formulas for calculating the temperature on a slab using the method of images.

If the source is non-rectilinear (the radius of curvature is finite and may vary in width), then the calculation of the temperature becomes more complicated. These cases are discussed in Sect. 12.3.

5.2.3.11 Moving Uniformly Distributed Strip Source on a Semi-infinite Body

At first, consider a moving source in the form of an infinite strip $-W < x < W$, $-\infty < y < \infty$, located in the plane $z = 0$. The area-specific source power density is $q_2 = q_1/(2W)$, where q_1 is the power per unit of the y-axis length. This scheme is applicable, for example, to wide-layer strip surfacing and is widely used to analyse the surface temperature during grinding (Loewen and Shaw 1954; Sipaylov 1978; Trigger and Chao 1951).

Earlier in Sect. 5.2.2.4 the heat conduction problem for a stationary strip source was solved. Similarly, we can obtain a solution for a moving strip source with power density q_2 for $t \to \infty$ by integrating (5.1.81) with power per unit length q_1:

$$T(x, z, \infty) - T_0 = \frac{q_1}{2\pi\lambda W}\int\limits_{-W}^{W}\exp\left(-\frac{v(x-x')}{2a}\right)\mathrm{K}_0\left(\frac{v\sqrt{(x-x')^2+z^2}}{2a}\right)dx'. \tag{5.2.152}$$

Hence, introducing the dimensionless quantities, we obtain (Carslaw and Jaeger 1973)

$$\theta(\xi, \varsigma, \infty) = \frac{1}{\omega}\int\limits_{\xi-\omega}^{\xi+\omega}\exp(-u)\mathrm{K}_0\left(\zeta^2+u^2\right)^{1/2}du;$$

$$\theta = \frac{2\pi\lambda}{q_1}(T-T_0);\quad \xi = \frac{vx}{2a};\quad \zeta = \frac{vz}{2a};\quad \omega = \frac{vW}{2a}. \tag{5.2.153}$$

This expression is easy to find by a numerical method. For the surface, $\zeta = 0$, the result can be written down using the integral (Prudnikov et al. 2002)

$$\int\limits_0^a \exp(\pm u)\mathrm{K}_0(u)du = a\ \exp(\pm a)[\mathrm{K}_0(a) \pm \mathrm{K}_1(a)] \mp 1. \tag{5.2.154}$$

Then the steady-state temperature on the longitudinal axis is

$$\begin{aligned}\theta(\xi, 0, \infty) = &\frac{1}{\omega}\{(\xi+\omega)\exp(-(\xi+\omega))[\mathrm{K}_0(\xi+\omega) - \mathrm{K}_1(\xi+\omega)]\\ &-(\xi-\omega)\exp(-(\xi-\omega))[\mathrm{K}_0(\xi-\omega) - \mathrm{K}_1(\xi-\omega)]\}.\end{aligned} \tag{5.2.155}$$

The steady-state temperature distribution at the surface at some values of the width of the strip source is shown in Fig. 5.73. It is seen that the narrower the strip is, the higher the peak temperature is.

For large values of W, the temperature is maximal for $x = -W$ (at the end of heating the fixed surface point) and is approximately equal to (Carslaw and Jaeger 1973):

$$T_{\max} - T_0 = \frac{2q_2}{\lambda}\sqrt{\frac{2aW}{\pi v}}. \tag{5.2.156}$$

This expression is in line with the solution to the problem of uniform heating of the entire surface of a semi-infinite solid during time $t = 2W/v$ by a source of power density q_2, W m^{-2}, or, in the same way, heating by a continuous stationary plane source in a semi-infinite rod [see Eq. (5.1.35)].

Now we consider a moving finite strip source $-W < x < W$, $-l < y < l$, $z = 0$. Using the method of sources, we can obtain a solution to the heat conduction problem in two ways: (1) take an instantaneous finite strip source as the basis and integrate

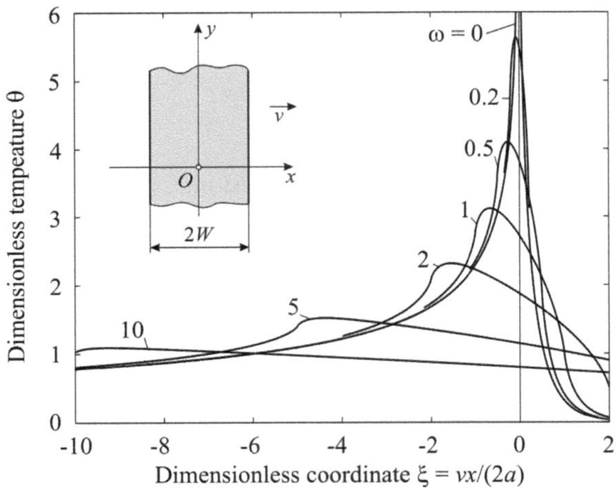

Fig. 5.73 Steady distribution of dimensionless temperature $\theta = 2\pi\lambda(T - T_0)/q_1$ on a semi-infinite solid under heating by a moving surface uniformly distributed strip-like source of dimensionless width $\omega = vW(2a)$

expression (5.2.2) over time between 0 and t in the moving coordinate system; (2) take a moving point source as an elementary source and integrate expression (5.1.39) or (5.1.41) over interval $[-W, W]$ with respect to x' and over interval $[-l, l]$ with respect to y'.

The first way gives us the following solution (Berezovsky 2006):

$$T(x, y, z, t) - T_0 = \frac{q/(2W2l)}{2c\rho(4\pi a)^{1/2}} \int_0^t \frac{1}{\sqrt{\tau}} \exp\left(-\frac{z^2}{4a\tau}\right)\left[\Phi\left(\frac{x + W + v\tau}{\sqrt{4a\tau}}\right)\right.$$
$$\left.-\Phi\left(\frac{x - W + v\tau}{\sqrt{4a\tau}}\right)\right]\left[\Phi\left(\frac{y + l}{\sqrt{4a\tau}}\right) - \Phi\left(\frac{y - l}{\sqrt{4a\tau}}\right)\right] d\tau. \tag{5.2.157}$$

The dimensionless form of Eq. (5.2.157) has the form (Carslaw and Jaeger 1973):

$$\theta(\xi, \eta, \zeta, \tau) = \int_0^\tau \exp\left(-\frac{\zeta^2}{2u}\right)\left[\Phi\left(\frac{\xi + \omega + u}{\sqrt{2u}}\right) - \Phi\left(\frac{\xi - \omega + u}{\sqrt{2u}}\right)\right]$$
$$\times \left[\Phi\left(\frac{\eta + \beta}{\sqrt{2u}}\right) - \Phi\left(\frac{\eta - \beta}{\sqrt{2u}}\right)\right] \frac{du}{\sqrt{u}};$$
$$\theta = \frac{2\sqrt{2\pi}c\rho v}{q/(2W2l)}(T - T_0);$$
$$\tau = \frac{v^2 t}{2a}; \quad \xi = \frac{vx}{2a}; \quad \eta = \frac{vy}{2a}; \quad \zeta = \frac{vz}{2a}; \quad \omega = \frac{vW}{2a}; \quad \beta = \frac{vl}{2a}. \tag{5.2.158}$$

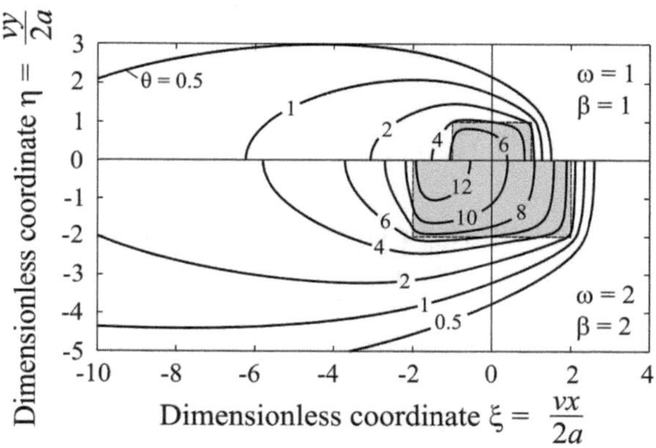

Fig. 5.74 Quasi-stationary fields of dimensionless temperature θ on a semi-infinite solid around the surface uniformly distributed rectangular source of size $\omega \times \beta$

For $t \to \infty$ and $\tau \to \infty$ equations (5.2.157) and (5.2.158) correspond to the quasi-stationary (limit) state.

The second way leads to the following solution under the quasi-stationary condition ($t \to \infty$):

$$T(x, y, z, \infty) - T_0 = \frac{q/(2W2l)}{2\pi\lambda} \int_{-W}^{W} \int_{-l}^{l} \frac{1}{\sqrt{(x - x')^2 + (y - y')^2 + z^2}} \times \exp\left(-\frac{v\left[x - x' + \sqrt{(x - x')^2 + (y - y')^2 + z^2}\right]}{2a} \right) dx'dy' \tag{5.2.159}$$

or in a dimensionless form

$$\theta(\xi, \eta, \zeta, \infty) = \frac{2\sqrt{2}}{\sqrt{\pi}} \int_{\xi-\omega}^{\xi+\omega} \int_{\eta-\beta}^{\eta+\beta} \left(\xi'^2 + \eta'^2 + \zeta^2\right)^{-1/2} \times \exp\left(-\xi' - \left[\xi'^2 + \eta'^2 + \zeta^2\right]^{1/2}\right) d\xi' d\eta'. \tag{5.2.160}$$

When we reduce only one source size ($\omega \to 0$ or $\beta \to 0$), we obtain a moving line source oriented across or along the longitudinal axis ξ. When all dimensions ($\omega \to 0$ and $\beta \to 0$) are reduced, a moving point source on the surface of a semi-infinite solid is obtained.

A quasi-stationary field at the surface of a semi-infinite solid ($\zeta = 0$) for various sizes ω and β of the source is shown in Fig. 5.74. It can be seen that the maximum heated point is behind the centre of the source and the larger the area of the distributed source is, the more this point is displaced from the centre.

It is not difficult to obtain a solution to the corresponding heat conduction problem for a moving uniformly distributed source on the surface of a slab by Eqs. (5.2.152), (5.2.157) and (5.2.159) and the method of images. In this case a series of type (4.2.23) appears in the integrand.

5.2.3.12 Moving Cylindrical Source with an Isothermal Surface in a Plate

A source distributed over the surface of a moving circular cylinder is a model of a plasma jet, an electron beam, and a laser beam when the keyhole technique is used for welding thin plates.

Let us assume that the surface of a circular cylinder of radius r_0 moves in an infinite heat-insulated plate at a constant velocity v along the x-axis and is maintained at constant temperature $T_{\max}$. The equation for heat conduction in a moving coordinate system has the form (3.3.30) and the boundary conditions are as follows:

$$\begin{aligned} &T = T_{\max} \quad \text{for} \quad x^2 + y^2 = r_0^2; \\ &T(x, y, t) = T_0 \ \text{for} \quad x = \pm\infty \ \text{and} \ y = \pm\infty. \end{aligned} \qquad (5.2.161)$$

The general solution to this problem is known (Carslaw and Jaeger 1973). If $\partial T / \partial t = 0$ (for $t = \infty$ the temperature field is quasi-stationary), the solution to the problem in the polar coordinate system ρ, φ is the following series in a dimensionless form

$$\theta(\rho, \varphi, \infty) = \exp(-\rho \mathrm{Pe}\cos\varphi) \sum_{n=0}^{\infty} \varepsilon_n \frac{\mathrm{I_n(Pe)}}{\mathrm{K_n(Pe)}} \mathrm{K_n}(\rho \mathrm{Pe}) \cos(n\varphi);$$

$$\theta = \frac{T - T_0}{T_{\max} - T_0}; \quad \mathrm{Pe} = \frac{v r_0}{2a}; \quad \rho = \frac{r}{r_0}; \quad \cos\varphi = \frac{x}{r}. \qquad (5.2.162)$$

Here θ is the dimensionless temperature, Pe is the Peclet number, ρ is a dimensionless radius vector; angle $\varphi = 0$ corresponds to the direction of the source motion (Fig. 5.75); $\varepsilon_n = 1$ for $n = 0$ and $\varepsilon_n = 2$ for $n > 0$, $\mathrm{I_n}$ and $\mathrm{K_n}$ are modified Bessel functions of the first and second kind of order n, (5.2.115) and (5.2.116).

In a Cartesian coordinate system $\xi = x/r_0$ and $\eta = y/r_0$ formula (5.2.162) has the form

$$\theta(\xi, \eta, \infty) = \exp(-\xi \mathrm{Pe}) \sum_{n=0}^{\infty} \varepsilon_n \frac{\mathrm{I_n(Pe)}}{\mathrm{K_n(Pe)}} \mathrm{K_n}\left(\mathrm{Pe}\sqrt{\xi^2 + \eta^2}\right)$$
$$\times \cos\left(n \arccos\left(\frac{\xi}{\sqrt{\xi^2 + \eta^2}}\right)\right). \qquad (5.2.163)$$

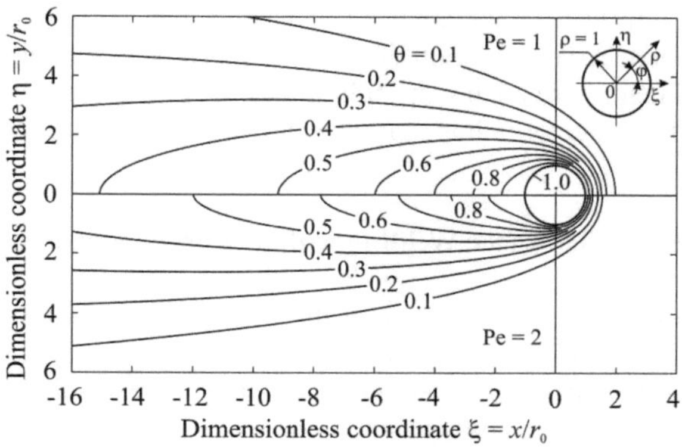

Fig. 5.75 Quasi-stationary fields of dimensionless temperature $\theta = (T - T_0)/(T_{\max} - T_0)$ around a moving cylindrical surface isotherm for Peclet numbers Pe = 1 (top) and Pe = 2 (bottom)

It can be seen that the temperature field is symmetric with respect to the longitudinal section $\eta = y/r_0 = 0$.

The quasi-stationary temperature fields for various Peclet numbers Pe are shown in Fig. 5.75.

On the cylinder surface $\rho = 1$, the temperature is constant $\theta = 1$, but the heat flux to the body is variable in angle φ. Let us find the power of the source per unit thickness q_1, W m^{-1}:

$$q_1 = \int_0^{2\pi} \left(-\lambda \left. \frac{\partial T}{\partial \rho} \right|_{\rho=1} \right) d\varphi. \tag{5.2.164}$$

By differentiating (5.2.162) and substituting the derivative in (5.2.164), we obtain (Lankalapalli et al. 1996):

$$q_1 = \lambda (T_{\max} - T_0) g(\text{Pe}), \tag{5.2.165}$$

where

$$g(\text{Pe}) = -\int_0^{2\pi} \text{Pe} \exp(-\text{Pe} \cos\varphi) \sum_{n=0}^{\infty} \varepsilon_n \text{I}_n(\text{Pe}) \times \cos(n\varphi) \left[\frac{n}{\text{Pe}} - \frac{\text{K}_{n+1}(\text{Pe})}{\text{K}_n(\text{Pe})} - \cos\varphi \right] d\varphi. \tag{5.2.166}$$

It is seen that the power per unit length q_1 depends on temperature $T_{\max}$ and Peclet number Pe (radius r_0). For Pe = 0–5, function (5.2.166) can be approximated by the polynomial (Lankalapalli et al. 1996)

$$g(\text{Pe}) = 2.1995 + 6.2962\text{Pe} - 0.4994\text{Pe}^2 + 0.0461\text{Pe}^3. \tag{5.2.167}$$

In fact, temperature T_{max} is not known, it may be close to the temperature of intensive evaporation of the material. It can be found, for example, from an experiment, normalising the solution to the problem according to the width of the weld. The value of radius r_0 is chosen from the physical conditions.

5.2.3.13 Laser Welding with a Deep-Penetration Beam

Incident laser radiation on the surface of a solid body is partially absorbed. Heat source efficiency of laser welding (the fraction of the laser radiation power absorbed by the metal) η_h depends on the nature of the metal, the state of the surface, the wavelength of the laser radiation, the power density distribution, the radiation direction, etc. The efficiency η_h is sensitive to laser welding conditions and measured over a wide range $\eta_h = 0.2$–0.7 (Grigoryants 1994).

Thermal processes during laser heating have the same physical nature as other traditional methods of metal heating. Therefore, the propagation of heat in the metal during laser processing is considered according to the classical theory of heat conduction (the Fourier's equation of heat conduction is valid).

If the laser beam is focused, it can be represented in the form of a concentrated source: point or line, instantaneous or continuous, stationary or moving. The methods for calculating the temperature fields due to concentrated sources were discussed in Sect. 5.1.

Let us consider laser welding with a deep-penetration beam in a solid. The laser beam can be focused with a power density of more than 10^4 W mm^{-2}. When heated by such a source, metal evaporates, forming a gas-plasma cavity (keyhole) surrounded by liquid metal (Fig. 5.76). The vapour pressure in the cavity is balanced by surface tension and hydrostatic pressure. As a result, a weld with a large ratio of the depth of penetration to the width of weld *H/W* is formed.

Many mathematical models of thermal processes for welding with deep penetration are proposed (Dowden et al. 1991, 1995, 1998; Dowden 2001; Kapadia and Dowden 1994; Lopota et al. 2006; Steen et al. 1988; Swift-Hook and Gick 1973; Turichin et al. 2008, 2009).

We assume the following. The gas-plasma cavity has a conical shape. The temperature of the cavity surface is constant. The cone of the cavity is determined by radius $r_0(0)$ on the surface of the body ($z = 0$) and penetration depth H (Fig. 5.76 a). The temperature on the surface of the cavity T_{max} is constant, the temperature field is steady (quasi-stationary in the moving coordinate system related to the beam).

The three-dimensional heat conduction problem with the above-mentioned assumptions can be solved by numerical methods. With additional assumptions, the problem can be solved in a closed form (Lankalapalli et al. 1996). Let us choose from the body a layer 1 of thickness H (Fig. 5.76). Since $r_0(0) \ll H$, the flux along the z-axis is much smaller than the radial flux, $q_{2\,z} \ll q_{2\,r}$. We neglect the flow $q_{2\,z}$ in layer 1, that is, we reduce the three-dimensional problem to a two-dimensional one. Let us find the distribution of temperature and introduced power in a layer 1. Then, with the power found, we solve the three-dimensional problem for layer 2.

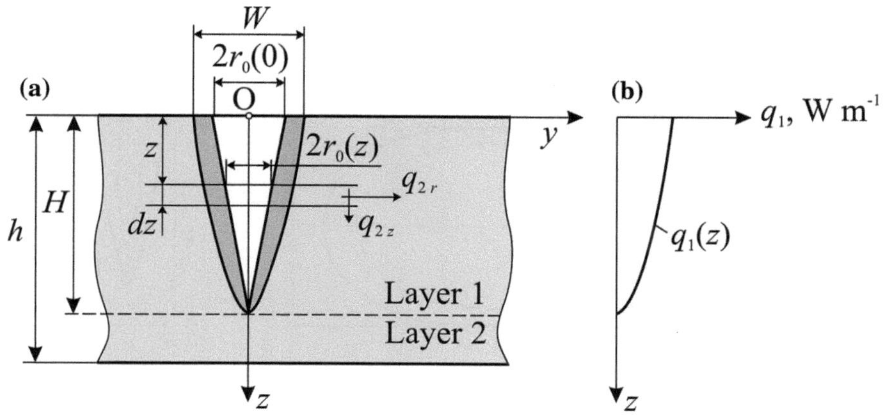

Fig. 5.76 A cross section of the keyhole and weld profile (**a**) and the power distribution across thickness (**b**)

We represent layer 1 as a set of thin layers of thickness *dz* with a given constant temperature $T_{\max}$ on the ring $r = r_0(z)$. Each thin layer is thermally insulated ($\partial T/\partial z = 0$), so the temperature field and the power per unit length (power of the source per unit thickness) $q_1(z)$ in a thin layer at depth z can be determined by formulas (5.2.162) and (5.2.165).

The source power q is determined from the known power distribution $q_1(z)$ in layer 1 (5.2.165):

$$q = \int_0^H q_1(z)dz = \int_0^H \lambda(T_{\max} - T_0)g(\mathrm{Pe})dz. \tag{5.2.168}$$

This formula is valid for any shape of keyhole. If the keyhole has a conical shape, then

$$\begin{aligned} &r_0(z) = r_0(0)\big(1 - z/H\big); \\ &\mathrm{Pe}(z) = \frac{vr_0(z)}{2a} = \frac{vr_0(0)}{2a}\left(1 - \frac{z}{H}\right) = \mathrm{Pe}_0\left(1 - \frac{z}{H}\right); \\ &\mathrm{dPe} = -\frac{\mathrm{Pe}_0}{H}\mathrm{d}z; \qquad dz = -\frac{H}{\mathrm{Pe}_0}\mathrm{dPe}. \end{aligned} \tag{5.2.169}$$

By substituting *dz* in (5.2.168), we obtain the source power as a function of Pe_0 number (Lankalapalli et al. 1996):

$$q = \frac{H\lambda(T_{\max} - T_0)}{\mathrm{Pe}_0}\int_0^{\mathrm{Pe}_0} g(\mathrm{Pe})\mathrm{dPe}_0. \tag{5.2.170}$$

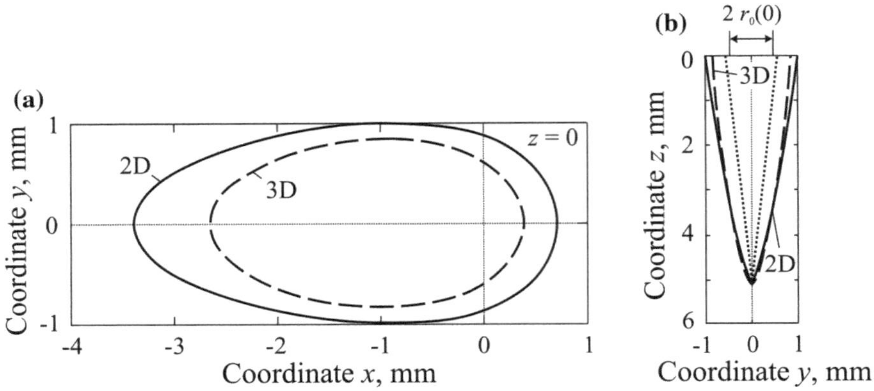

Fig. 5.77 Melting isotherm T_m on the surface (**a**) and weld cross-section (**b**) according to a two-dimensional (2D) solution (5.2.162) and a three-dimensional (3D) solution (5.2.171) ($T_{\max}$ = 3003 K, T_m = 1803 K, T_0 = 293 K, λ = 0.06 W mm^{-1} K^{-1}, a = 5 mm^2 s^{-1}, q = 4500 W, v = 20 mm s^{-1}, W = 2 mm) (Lankalapalli et al. 1996)

Hence, the power of source (the absorbed power of the laser beam) q, penetration depth H, Pe_0 number [the keyhole radius $r_0(0)$] and temperature of the metal on the surface of keyhole $T_{\max}$ are related by Eq. (5.2.170). These parameters determine the temperature field (see Sect. 5.2.3.12). Thus, by measuring, for example, the width of the weld on the surface of the solid at some distance from the keyhole during welding, it is possible to estimate the penetration depth and, as a result, to control the welding conditions.

To this point we have assumed that there are no heat fluxes across thickness, $q_{2z} = 0$. Their influence can be estimated by assumption that the source is line one with the known power distribution $q_1(z)$, and by using the solution for a non-uniformly distributed line source in a slab (Sect. 5.1.3.9), or a solution for a point source in a semi-infinite solid (5.1.44) (Lankalapalli et al. 1996):

$$T(x, y, z, \infty) - T_0 = \frac{1}{4\pi\lambda} \exp\left(-\frac{vx}{2a}\right) \int_0^H q_1(\zeta) \sum_{j=-1,1} \frac{1}{R_j}\left(-\frac{vR_j}{2a}\right) d\zeta;$$

$$R_j = \left[x^2 + y^2 + (z - j\zeta)^2\right]^{1/2}. \tag{5.2.171}$$

The neglect of thermal conductivity along the z-axis leads to an increase in the width of the weld pool (Fig. 5.77a), but almost does not affect the depth of penetration (Fig. 5.77b). With an increase in the depth to the width ratio of the weld, the difference in the shapes of the weld pool decreases.

Results of solving the heat conduction problem by numerical methods are given in numerous papers (Gebhardt et al. 2010). A numerical transient analysis is conducted to understand the overall transient characteristics of the laser-arc hybrid welding processes (Cho et al. 2010).

Like laser welding, electron-beam welding with a deep-penetration beam is used in producing structures of large thickness. When solving the heat conduction problem, various heat sources are utilized: a line source, exponentially distributed across the thickness (Piekarska 2002), a line source with a piecewise-linear power distribution across thickness (Karkhin and Pilipenko 1997); combination of surface and line buried sources (Bashenko and Petrov 1977; Michailov and Karkhin 1997; Yazovskikh et al. 2012).

A three-dimensional conical heat source model was proposed as the basis for the analysis of temperature fields in the keyhole plasma arc welding (Wu et al. 2006).

5.2.4 *Rapidly Moving Distributed Sources*

5.2.4.1 Rapidly Moving Uniformly Distributed Line Source of Finite Width on a Semi-infinite Body

The temperature field in a semi-infinite solid due to a rapidly moving line source of finite width is the limiting form, to which the quasi-stationary temperature field tends towards, under an unlimited and proportional increase in the power and velocity of the source (see Sect. 5.2.3.10). This model assumes the absence of longitudinal heat fluxes.

The considered source corresponds to a set of elementary instantaneous line sources of energy per unit length $dQ_1 = Q_2 d\eta = q/(v2W)d\eta$ (J m^{-1}), directed along the longitudinal x-axis and distributed along the y-axis across the width of $2W$ (Fig. 5.78a).

The temperature is determined by summing the temperature increments induced by all elementary line sources by formula (4.2.36) (Karkhin 1990):

$$\begin{aligned} T(y,z,t) - T_0 &= \int\limits_{y-W}^{y+W} \frac{2q/(v2W)}{c\rho(4\pi at)} \exp\left(-\frac{(y-\eta)^2+z^2}{4at}\right) d\eta \\ &= \frac{q/(v2W)}{c\rho\sqrt{4\pi at}} \exp\left(-\frac{z^2}{4at}\right)\left[\Phi\left(\frac{y+W}{\sqrt{4at}}\right) - \Phi\left(\frac{y-W}{\sqrt{4at}}\right)\right]. \end{aligned} \tag{5.2.172}$$

This formula allows us to construct thermal cycles at any point in the cross-section of the solid. To construct a temperature field, time t should be replaced by $-x/v$.

The temperature field can be represented in a dimensionless form:

$$\theta(\xi,\eta,\zeta) = \sqrt{\frac{\omega}{-\xi}} \exp\left(-\frac{\zeta^2\omega}{-\xi}\right)$$

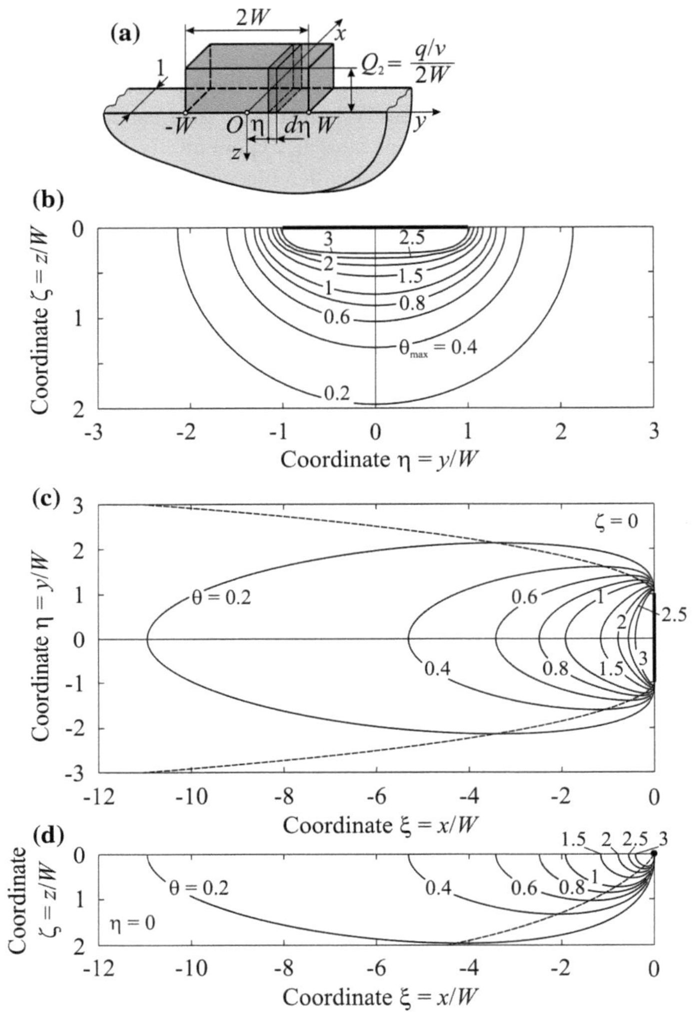

Fig. 5.78 Model for a rapidly moving uniformly distributed line source of $2W$ in width on a semi-infinite solid (**a**), dimensionless peak temperature field in cross-section (**b**), temperature fields on solid $\theta(\xi, \eta, 0)$ (**c**) and in longitudinal section $\theta(\xi, 0, \zeta)$ (**d**), dimensionless source travel speed is $\omega = vW/(4a) = 1$

$$\times\left[\Phi\left(\sqrt{\frac{\omega}{-\xi}}(\eta+1)\right)-\Phi\left(\sqrt{\frac{\omega}{-\xi}}(\eta-1)\right)\right];$$

$$\theta=\frac{2\sqrt{\pi}c\rho W^2}{q/v}(T-T_0);$$

$$\xi=\frac{x}{W};\quad \eta=\frac{y}{W};\quad \zeta=\frac{z}{W};\quad \omega=\frac{vW}{4a};\quad \frac{\omega}{-\xi}=\frac{vW}{4a(-\xi)}. \qquad (5.2.173)$$

The fields of dimensionless temperature at a dimensionless velocity (width) of line source $\omega = 1$ are shown in Fig. 5.78. The field around the source differs in shape from the field which is calculated taking into account longitudinal heat fluxes (Fig. 5.72).

Let us analyse the temperature distribution behind the source on the longitudinal x-axis. From (5.2.173) for $\eta = \zeta = 0$ we obtain:

$$\theta(\xi,0,0)=2\sqrt{\frac{\omega}{-\xi}}\Phi\left(\sqrt{\frac{\omega}{-\xi}}\right). \qquad (5.2.174)$$

For small values of ω/ξ (small width of source W or large distance from source ξ), by approximating function Φ, we obtain the model of the point source:

$$\theta(\xi,0,0)=\frac{4}{\sqrt{\pi}}\frac{\omega}{-\xi}. \qquad (5.2.175)$$

With a distance from the source, the temperature falls according to the hyperbolic law, as in the formula for a point source (5.1.122).

As the width of the source increases, the temperature tends to the limit

$$\theta(\xi,0,0)=2\sqrt{\frac{\omega}{-\xi}}. \qquad (5.2.176)$$

The structure of this formula is similar to the formula for a line source (5.1.147). At intermediate values of the source width, the temperature on the x-axis is determined by Eq. (5.2.174). It is seen from the equation that it depends only on criterion $\omega/(-\xi)$. Figure 5.79 shows the effect of this criterion.

Equations (5.2.172) and (5.2.173) allow us to obtain solutions to the problem for a source of finite width on a slab by the method of images (Kopetman 1965).

The solution to the heat conduction problem for a curvilinear source of finite width is given in Sect. 5.12.3.

5.2.4.2 Rapidly Moving Uniformly Distributed Plane Source of Finite Width in a Plate

Let a uniformly distributed plane source of finite width $2l$ move in an infinite plate along the x-axis. Such a heat input model corresponds to an instantaneous uniformly distributed source of volumetric energy density Q_3

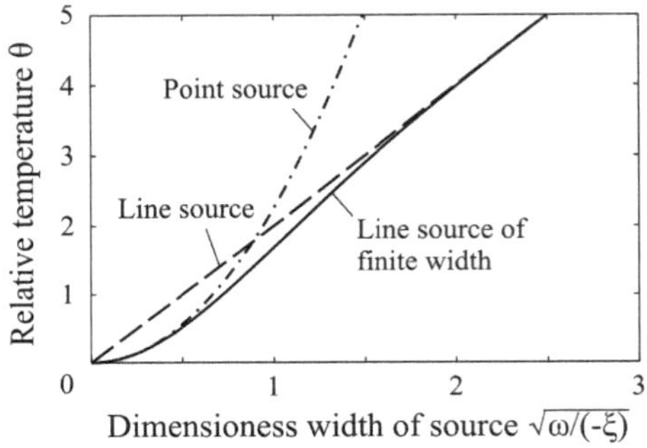

Fig. 5.79 Relative temperature distribution along the longitudinal axis behind a rapidly moving source of finite width on a semi-infinite solid (Rykalin 1947, 1952)

$$Q_3(y) = \begin{cases} \frac{q/v}{2lh} & \text{for } |y| \le l; \\ 0 & \text{for } |y| > l. \end{cases} \tag{5.2.177}$$

As in Sect. 5.2.4.1, the temperature field can be found by the method of sources, taking the fundamental solution (4.2.42) as the basis:

$$\begin{aligned} T(y,t) - T_0 &= \frac{1}{\sqrt{4\pi at}} \int_{-\infty}^{\infty} \frac{Q_3(\eta)d\eta}{c\rho} \exp\left(-\frac{(y-\eta)^2}{4at} - bt\right) d\eta \\ &= \frac{q/v}{4c\rho lh} \exp(-bt) \left[\Phi\left(\frac{y+l}{\sqrt{4at}}\right) - \Phi\left(\frac{y-l}{\sqrt{4at}}\right)\right]. \end{aligned} \tag{5.2.178}$$

It is seen from Fig. 5.80 that at any time the temperature of the material is maximal for $y = 0$.

5.2.4.3 Rapidly Moving Normally Distributed Circular Source on a Semi-infinite Body

The heating of a semi-infinite solid by a normally distributed circular (disk) source moving at a finite speed was discussed in Sect. 5.2.3.1. Let speed v of the source be relatively large and, consequently, the time of heat input on the surface transverse line AB be small (Fig. 5.81a). Then we can make two assumptions: (1) heat generation on the AB line is instantaneous (this moment is taken as the origin of time, $t = 0$) and (2) it propagates only in the direction perpendicular to the x-axis, i.e. in the plane $ABCD$ (there is no heat fluxes along the x-axis). This physical model corresponds to the heating model of a semi-infinite plate by a boundary instantaneous line source. The source power per unit length $q_1(y)$ is determined by the area-specific power density $q_2(x, y)$:

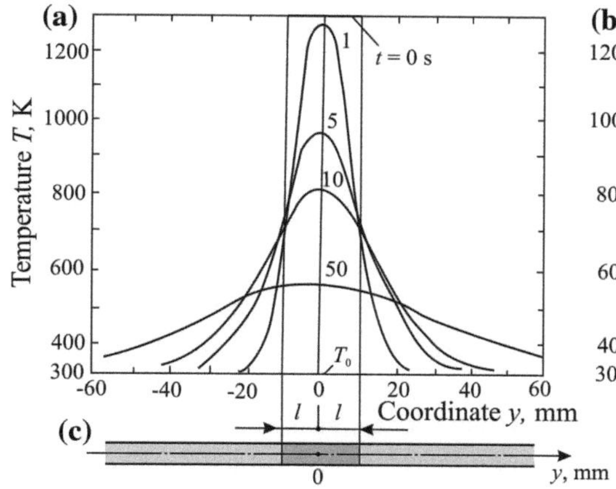

Fig. 5.80 Temperature distribution (**a**) and thermal cycles (**b**) in an infinite plate (**c**) due to a rapidly moving heat source uniformly distributed across the width $2l = 20$ mm ($q/v/(2lhc\rho) = 1000$ K, $b = 0$, $a = 10$ mm^2 s^{-1}, $T_0 = 300$ K) (Karkhin 1990)

$$
\begin{aligned}
q_1(y) &= \int_{-\infty}^{\infty} q_2(x, y)dx \\
&= \int_{-\infty}^{\infty} \frac{C}{\pi} q \exp\left(-C\left(x^2 + y^2\right)\right)dx = \sqrt{\frac{C}{\pi}} q \exp\left(-Cy^2\right).
\end{aligned}
\tag{5.2.179}
$$

It is seen that the source has become a normally distributed line source. Similarly, a normally distributed elliptic source can be oblated in the longitudinal direction (the source can be projected onto the plane yOz).

We use the method of sources. To do this, we split surface source q_1 into elementary line sources oriented along the x-axis, find the temperature increment produced by all elementary sources, sum them, and obtain as a result (Rykalin 1951, 1957)

$$
\begin{aligned}
&T(y, z, t) - T_0 = \frac{2q/v}{c\rho} \frac{\exp\left(-\frac{y^2}{4a(t+t_0)}\right)}{\sqrt{4\pi a(t + t_0)}} \frac{\exp\left(-\frac{z^2}{4at}\right)}{\sqrt{4\pi at}}; \\
&t_0 = 1/(4aC).
\end{aligned}
\tag{5.2.180}
$$

The process of heat propagation in the $ABCD$ plane at any instant of time t splits into two linear processes: in the direction of the y-axis with the duration of $t + t_0$ and in the direction of the z-axis with the duration of t.

The peak temperature behind the source on the x-axis is maximum and equal to

$$
T(0, 0, t) - T_0 = \frac{q/v}{2\pi\lambda\sqrt{(t + t_0)t}}. \tag{5.2.181}
$$

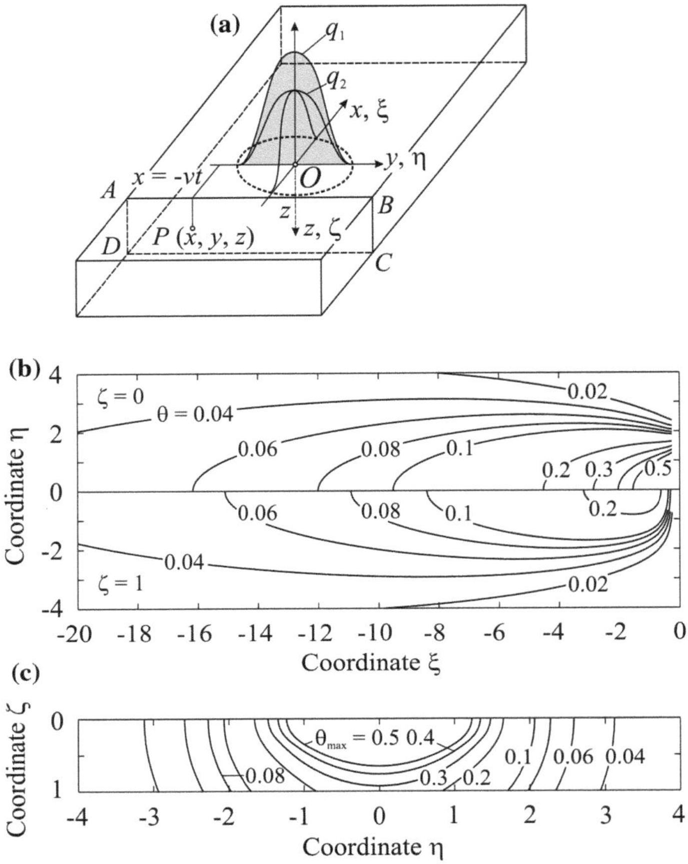

Fig. 5.81 Heating the surface of a semi-infinite solid by a rapidly moving normally distributed circular source (**a**); dimensionless temperature fields at the surface ($\zeta = 0$) and at the depth ($\zeta = 1$) (**b**) and peak temperature field in cross-section (**c**); the speed is $\omega = 1$

At the beginning of the process ($t \ll t_0$), the temperature decreases according to the $1/\sqrt{t}$ law as in the linear heat equalisation process of the instantaneous plane source [Eq. (4.2.41)]. At the end of the process ($t \gg t_0$), the temperature falls according to the $1/t$ law, as in a plane process of equalising the heat of an instantaneous line source [Eq. (4.2.37)]. In the middle part of the process, the hyperbola index varies from 1/2 to 1.

Equation (5.2.180) can be represented in a dimensionless form

$$\theta(\eta, \zeta, \tau) = \frac{2\pi\lambda t_0}{q/v}(T - T_0) = \frac{1}{\sqrt{\tau(1+\tau)}} \exp\left(-\frac{\eta^2}{1+\tau} - \frac{\zeta^2}{\tau}\right);$$

$$\eta = \frac{y}{\sqrt{4at_0}}; \quad \zeta = \frac{z}{\sqrt{4at_0}}; \quad \tau = \frac{t}{t_0}. \tag{5.2.182}$$

The three-dimensional temperature field in a moving coordinate system, whose origin coincides with the centre of the source, is easily obtained by replacing t by $-x/v$ in (5.2.180). In a dimensionless form, the three-dimensional temperature field is described by the expression:

$$\theta(\xi, \eta, \zeta) = \frac{\omega}{\sqrt{\xi(\xi - \omega)}} \exp\left(\frac{\omega}{\xi - \omega}\eta^2 + \frac{\omega}{\xi}\zeta^2\right);$$
$$\xi = \frac{x}{\sqrt{4at_0}}; \quad \omega = v\sqrt{\frac{t_0}{4a}}. \tag{5.2.183}$$

For the given dimensionless speed ω in cross section ξ = const, any isotherm θ = const has the form of a half-ellipse

$$\frac{\omega}{\xi - \omega}\eta^2 + \frac{\omega}{\xi}\zeta^2 = \ln\left(\frac{\theta\sqrt{\xi(\xi - \omega)}}{\omega}\right). \tag{5.2.184}$$

The major axis of the half-ellipse coincides with the η-axis (the y-axis), and the minor axis coincides with the ζ-axis (the z-axis). When the distance between cross-section and the source (coordinate ξ) increases, the eccentricity decreases and the half-ellipse tends to the semicircle. The field of the peak temperature in the cross section has the same property (Fig. 5.81c). The temperature field at the surface at the speed $\omega = 1$ is shown in Fig. 5.81b.

5.2.4.4 Rapidly Moving Normally Distributed Circular Source on a Slab

In Sect. 5.2.4.3 we discussed that a normally distributed circular (disk) source at a high speed is reduced to a normally distributed line source. We use the approach stated in Sect. 5.2.4.3 and take into account the effect of finite thickness of a solid.

If the surfaces of a slab are heat-impermeable, by using Eq. (5.2.180) for a semi-infinite body and applying the method of images, we obtain the solution for a slab of thickness h:

$$T(y, z, t) - T_0 = \frac{q/v}{2\pi\lambda\sqrt{t(t + t_0)}} \exp\left(-\frac{y^2}{4a(t + t_0)} - \frac{z^2}{4at}\right)$$
$$\times \sum_{i=-\infty}^{\infty} \exp\left(-\frac{ih(ih - z)}{at}\right). \tag{5.2.185}$$

This expression can be presented in a dimensionless form:

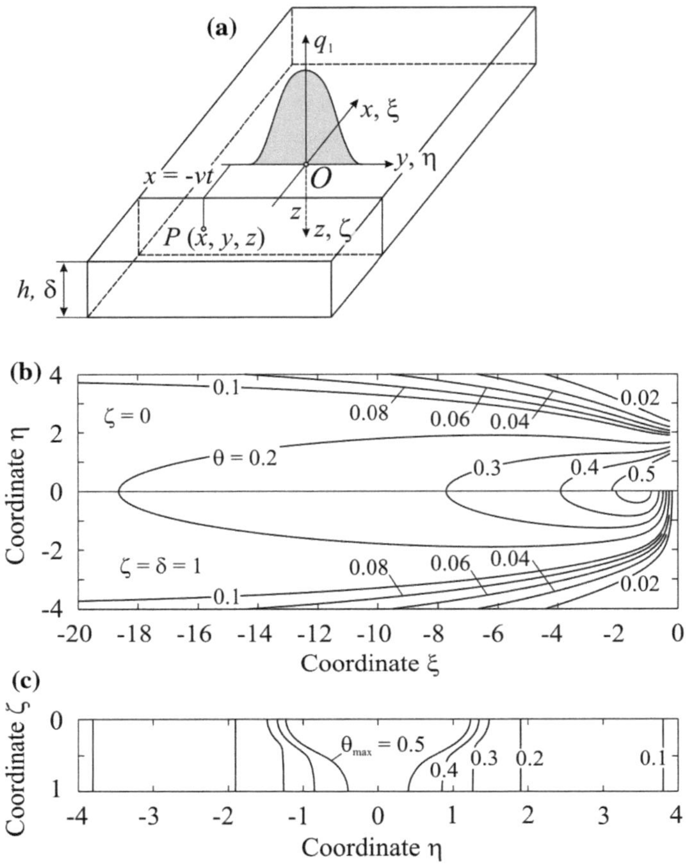

Fig. 5.82 Model for heating the surface of a slab by a rapidly moving normally-distributed line source (**a**), dimensionless temperature at the surfaces $\zeta = 0$ and $\zeta = \delta$ (**b**) and peak temperature field $\theta_{\max}$ in the cross-section (**c**) of the slab; dimensionless speed $\omega = 1$

$$\theta(\eta, \zeta, \tau) = \frac{2\pi\lambda t_0}{q/v}(T - T_0) = \frac{1}{\sqrt{\tau(1+\tau)}} \exp\left(-\frac{\eta^2}{1+\tau} - \frac{\zeta^2}{\tau}\right)$$
$$\times \sum_{i=-\infty}^{\infty} \exp\left(\frac{4i\delta(\zeta - i\delta)}{\tau}\right)$$
$$\eta = \frac{y}{\sqrt{4at_0}}; \quad \zeta = \frac{z}{\sqrt{4at_0}}; \quad \delta = \frac{h}{\sqrt{4at_0}}; \quad \tau = \frac{t}{t_0}. \tag{5.2.186}$$

Replacing time t with $-x/v$ (Fig. 5.82a), we obtain a three-dimensional temperature field:

$$\theta(\xi,\eta,\zeta) = \frac{\omega}{\sqrt{\xi(\xi-\omega)}} \exp\left(\frac{\omega}{\xi-\omega}\eta^2 + \frac{\omega}{\xi}\zeta^2\right)$$
$$\times \sum_{i=-\infty}^{\infty} \exp\left(4i\delta\left[i\frac{\delta}{\xi} - 1\right]\omega\right);$$
$$\xi = \frac{x}{\sqrt{4at_0}}; \quad \omega^2 = \frac{v^2 t_0}{4a}. \qquad (5.2.187)$$

Formulas (5.2.185)–(5.2.187) for a slab differ from formulas (5.2.180), (5.2.182) and (5.2.183) for a semi-infinite solid only by the presence of additional series, which allow for thickness of the slab. Obviously, with an unlimited increase in the thickness of the slab, these formulas are pairwise identical.

Figure 5.82 shows the temperature field at the upper and lower surfaces and the distribution of the peak temperature in the cross section of the slab. The comparison of Figs. 5.82 and 5.81 draws the conclusion that in a slab isotherms are more elongated along the longitudinal axis (the cooling rate is lower), and the peak temperature is higher.

It is significant that the presented methods for solving heat conduction problems can be used for other combinations of rapidly moving sources and bodies (uniformly distributed disk sources on a semi-infinite solid, a slab and a plate, a normally distributed source in a plate, various sources on solid and hollow cylinders, etc.). In these cases, the problem is reduced to the selection of relevant instantaneous sources.

References

Abramowitz, M., & Stegun, I. (1965). *Handbook of mathematical functions* (1043 pp.). New York: Dover Publications.

Bashkatov, A. V. (1972). Features of welding OT4 titanium alloy with weaving electron beam. *Automatic Welding, 11*, 68–69 (in Russian).

Bashenko, V. V., & Petrov, G. L. (1977). Formation of fusion zone in electron beam welding. *Automatic Welding, 9*, 23–27 (in Russian).

Bateman, H., & Erdelyi, A. (1953). *Higher transcendental functions* (Vol. 2, 396 pp.). New York: McGraw-Hill Book Co.

Bateman, H., & Erdelyi, A. (1954). *Tables of integral transforms* (Vol. 1, 454 pp. and Vol. 2, 391 pp.). New York: McGraw-Hill Book Co.

Berezovsky, B. M. (2006). *Mathematical models for arc welding: Vol. 4. Principles of thermal processes in weldments* (547 pp.). Chelyabinsk: Scientific and Technical Information Centre Publishing (in Russian).

Berezovsky, B. M. (2008). *Mathematical models for arc welding: Vol. 5. Temperature fields and thermal cycles in welded joints* (656 pp.). Chelyabinsk: Scientific and Technical Information Centre Publishing (in Russian).

Carlson, B. E., Wang, H.-P., Karkhin, V. A., Ivanov, S. Y., Turichin, G. A., & Valdaitseva Y. A. (2013). Mathematical model of filler metal heating and melting for plasma arc brazing. In C. Sommitsch & N. Enzinger (Eds.), *Mathematical modelling of weld phenomena* (Vol. 10, pp. 267–286). Graz: Varlag der Technischen Universitaet Graz.

Carslaw, H. S., & Jaeger, J. C. (1973). *Conduction of heat in solids* (510 pp.). London: Oxford University Press.

Cho, W. I., Na, S. Y., Cho, M. H., & Lee, J. S. (2010). A transient investigation of laser–arc hybrid welding by numerical simulation. In H. Cerjak & N. Enzinger (Eds.), *Mathematical modelling of weld phenomena* (Vol. 9, pp. 57–63). Graz: Verlag der Technischen Universitaet Graz.

Dowden, J., Kapadia, P., & Sibold, D. (1991). Mathematical modelling of laser welding. *International Journal for the Joining of Materials, 3*(3), 73–78.

Dowden, J. M. (2001). *The mathematics of thermal modelling: An introduction to the theory of laser material processing* (292 pp.). Chapman and Hall/CRC.

Dowden, J. M., Ducharme, R., & Kapadia, P. (1998). Time-dependent line and point sources: a simple model for time-dependent welding processes. *Laser in Engineering, 7*, 215–228.

Dowden, J. M., Kapadia, P., & Ducharme, R. (1995). An analytical model of deep penetration welding of metals with a continuous CO_2 laser. *International Journal for the Joining of Materials, 7*(2/3), 54–61.

Fachinotti, V. D., & Cardona, A. (2008). Semi-analytical solution of the thermal field induced by a moving double-ellipsoidal welding heat source in a semi-infinite body. *Mecanica Computacional*, 1519–1530.

Frolov, V. V. (Ed.). (1988). *Theory of welding processes* (559 pp.). Moscow: Vysshaya Shkola (in Russian).

Gagen, Y. G., & Taran, V. D. (1970). *Magnetically impelled arc butt welding* (159 pp.). Moscow: Mashinostroenie (in Russian).

Gebhardt, M., Quiros, V., Gumenyuk, A., & Rethmeier, M. (2010). Restraint effects on stresses and strains in single-run high power laser beam welding of thick plates. In H. Cerjak, & N. Enzinger (Eds.), *Mathematical modelling of weld phenomena* (Vol. 9, pp. 1011–1033). Graz: Verlag der Technischen Universitaet Graz.

Goldak, J., Chakravarti, A., & Bibby, M. (1984). A new finite element model for welding heat sources. *Metallurgical Transactions B, 15B*(6), 299–305.

Gradshteyn, I. S., & Ryzhik, I. M. (2014). *Table of integrals, series, and products* (8th ed., 1184 pp.). Amsterdam: Elsevier.

Grigoryants, A. G. (1994). *Basics of laser material processing* (313 pp.). Taylor and Francis Inc.

Grong, O. (1994). *Metallurgical modelling of welding* (581 pp.). London: The Institute of Materials.

Hou, Z. B., & Komanduri, R. (2000). General solutions for stationary/moving plane heat source problems in manufacturing and tribology. *International Journal of Heat and Mass Transfer, 43*, 1679–1698.

Janke, E., Emde, F., & Losch, F. (1960). Tafeln hoeherer Funktionen (318 pp.). Stuttgart: B. G. Teubner (in German).

Kapadia, P., & Dowden, J. (1994). Review of mathematical models of deep penetration laser welding. *Lasers in Engineering, 3*, 187–279.

Karkhin, V. A. (1990). *Principles of heat transfer in weldments* (100 pp.). Leningrad: Leningrad State Technical University Publishing (in Russian).

Karkhin, V. A., Khomich, P. N., & Michailov, V. G. (2007). Models for volume heat sources and functional-analytical technique for calculating the temperature fields in butt welding. In H. Cerjak, H. K. D. H. Bhadeshia, & E. Kozeschnik (Eds.). *Mathematical modelling of weld phenomena* (Vol. 8, pp. 819–834). Graz: Verlag der Technischen Universitaet Graz.

Karkhin, V. A., & Pilipenko, A. Y. (1997). Modelling thermal cycles in the weld metal and the heat affected zone in beam methods of welding thick plates. *Welding International, 11*(5), 401–403.

Karkhin, V. A., Pittner, A., Schwenk, C., & Rethmeier, M. (2010). Heat source models in simulation of heat flow in fusion welding. In *Proceedings of 5th International Conference "Mathematical Modelling and Information Technologies in Welding and Related Processes"*, 25–28 May 2010, Crimea (pp. 56–60).

Karkhin, V. A., Pittner, A., Schwenk, C., & Rethmeier, M. (2011a). Simulation of inverse heat conduction problems in fusion welding with extended analytical heat source models. *Frontiers of Materials Science, 5*(2), 119–125.

Karkhin, V. A., Pittner, A., Schwenk, C., & Rethmeier M. (2011b). Simulation of the temperature field in laser beam welding by inverse technique. In E. Halmoy (Ed.). *Proceedings of the 13th Conference on Laser Materials Processing in the Nordic Countries*, 27–29 June 2011 (pp. 223–234). Trondheim, Norway.

Karkhin, V. A., & Zharkov S. V. (2015). Calculation of thermal processes in heating plates with moving axisymmetric heat sources. *Welding International, 29*(7), 538–542.

Karkhin, V. A., Zharkov, S. V., & Mikhaylov V. G. (2016). Calculation of thermal processes in welding thick plates with moving axisymmetric heat sources. *Welding International, 30*(9), 708–711.

Kartashov, E. M. (2001). *Analytical methods in theory of heat conduction in solids* (550 pp.). Moscow: Vysshaya Shkola (in Russian).

Kasuya, T., & Yurioka, N. (1993). Prediction of welding thermal history by a comprehensive solution. *Welding Journal, 3*, 107-s–115-s.

Kiselev, S. N., Shevelev G. N., Roshchin V. V., Zelentsov G. N., & Khavanov, V. A. (1981). *Tube joints of dissimilar metals* (176 pp.). Moscow: Mashinostroenie (in Russian).

Kopetman, L. N. (1965). Temperature field in slab due to line source of finite width (in wide-layer deposit). *Welding Production, 1*, 2–5 (in Russian).

Korenev, B. G. (1971). *Introduction to the theory of Bessel functions* (288 pp.). Moscow: Nauka (in Russian).

Korenev, B. G. (1980). *Problems of theory of heat conduction and thermoelasticity. Solutions in Bessel functions* (401 pp.). Moscow: Nauka (in Russian).

Lankalapalli, K. N., Tu, J. F., & Gartner, M. (1996). A model for estimating penetration depth of laser welding processes. *Journal of Physics D: Applied Physics, 29*, 1831–1841.

Loewen, E. G., & Shaw, M. C. (1954). On the analysis of cutting-tool temperatures. *Transactions of the ASME, 76*(1), 217–231.

Lopota, V. A., Turichin, G. A., Valdaytseva, E. A., Malkin, P. E., & Gumenyuk, A. V. (2006). Computer system for modelling of electron beam and laser welding. *Automatic Welding, 2*, 18–21 (in Russian).

Makhnenko, V. I. (1967). Calculation of thermal processes in butt welding of dissimilar plates. *Physics and Chemistry of Materials Processing,* (6), 78–82 (in Russian).

Makhnenko, V. I. (1976). *Calculation methods for the investigation of the kinetics of welding stresses and deformations* (320 pp.). Kiev: Naukova Dumka (in Russian).

Makhnenko, V. I., & Yegorova, L. A. (1975). Areas of application of rapidly moving heat source model for calculation of temperature in welding. *Automatic Welding, 1*, 68–69 (in Russian).

Michailov, V. G., & Karkhin, V. A. (1997). Calculation of temperature fields and cycles in welding. In *Proceedings of International Congress "Mechanical Engineering Technologies'97", Welding*, 17–19 September 1997, (pp. 13–18). Sofia, Bulgaria.

Myers, P. S., Uyehara, O. A., & Borman, G. L. (2007). Fundamentals of heat flow in welding. *Welding Research Council Bulletin, 123*, 1–46.

Negoda, E. N. (2008). *Heat processes in welding* (125 pp.). Vladivostok: DVGTU Publishing (in Russian).

Nguyen, N. T. (2004). *Thermal analysis of welds* (334 pp.). Southampton: WIT Press.

Nguyen, N. T., Mai, W.-Y., Simpson, S., & Ohta, A. (2004). Analytical approximate solution for double ellipsoidal heat source in finite thick plate. *Welding Journal, 3*, 82-s–93-s.

Nguyen, N. T., Ohta, A., Matsuoka, K., Suzuki, N., & Maeda, Y. (1999). Analytical solutions for transient temperature of semi-infinite body subjected to 3-D moving heat sources. *Welding Journal, 8*, 265-s–274-s.

Okamoto, T. (1929). Theoretical and experimental researches on electric resistance welding. *Memoirs of the College of Engineering, Kyoto Imperial University, 6*(1), 1–111.

Parshin, S. G. (2016). *Metallurgy of underwater and hyperbaric welding* (402 pp.). St. Petersburg: Polytechnic University Publishing (in Russian).

Piekarska, W. (2002). Mathematical modelling of HAZ metal structure during laser welding. In V. I. Makhnenko (Ed.), *Proceedings of International Conference on Mathematical Modelling and*

Information Technologies in Welding and Related Processes (pp. 114–117). Kiev: Paton Welding Institute Publishing.

Pilipenko, A. (2001). Computer simulation of residual stress and distortion of thick plates in multi-electrode submerged arc welding. Their mitigation techniques (222 pp.). Doctoral thesis. Trondheim: Norwegian University of Science and Technology.

Pilipenko, A. Y., Karkhin, V. A., Grong, O., & Halmoy, E. (1999). Simulation of thermal cycles in multi-electrode submerged arc welding. In J. Martikainen & H. Eskelinen (Eds.), *International Conference on Efficient Welding in Industrial Applications* (pp. 91–98). Finland: Acta Universitatis Lappeenrantaensis 85.

Pittner, A., Karkhin, V., & Rethmeier, M. (2015). Reconstruction of 3D transient temperature fields for fusion welding processes on basis of discrete experimental data. *Welding in the World, 59*(4), 497–512.

Prudnikov, A. P., Brychkov, Y. A., & Marichev, O. I. (1986). *Integrals and series. Elementary functions* (798 pp.). London: Taylor and Francis.

Prudnikov, A. P., Brychkov, Y. A., & Marichev, O. I. (2002). *Integrals and series. Special functions* (750 pp.). London: Taylor and Francis.

Radaj, D. (2003). *Welding residual stresses and distortion. Calculation and measurement* (397 pp.). Duesseldorf: DVS – Verlag.

Ranjbarnodeh, E., Kokabi, A. H., Zerajzadeh, S., Hanke, S., & Fischer, A. (2010). The effect of heat input on residual stresses in dissimilar joints—FE simulation and X-ray validation. In H. Cerjak & N. Enzinger (Eds.), *Mathematical Modelling of Weld Phenomena* (Vol. 9, pp. 637–650). Graz: Verlag der Technischen Universitaet Graz.

Roberts, O. F. T. (1923). The theoretical scattering of smoke in a turbulent atmosphere. In *Proceeding of the Royal Society (London)*, Series A (Vol. 104, pp. 640–655).

Rosenthal, D. (1946). The theory of moving sources of heat and its application to metal treatments. *Transactions of the ASME, 68*(11), 849–865.

Ryabov, V. R., Rabkin, D. M., Kurochko, R. S., & Strizhevskaya, L. G. (1984). *Welding of dissimilar metals and alloys* (239 pp.). Moscow: Mashinostroenie (in Russian).

Rykalin, N. N. (1938a). Temperature distribution in welded joints. *Avtogennoe Delo,* (5), 6–14 (in Russian).

Rykalin, N. N. (1938b). Temperature distribution in welded joints. *Avtogennoe Delo,* (7), 19–24 (in Russian).

Rykalin, N. N. (1941). *Three-dimensional distribution in arc welding* (55 pp.). Moscow: Publication of the USSR Academy of Sciences.

Rykalin, N. N. (1947). *Thermal principles in welding. Part 1. Heat flow in arc welding* (271 pp.). Moscow: Publication of the USSR Academy of Sciences (in Russian).

Rykalin, N. N. (1951). *Calculation of heat flow in welding* (Z. Paley & C. M. Adams, Jr., Trans.) (337 pp.). Moscow.

Rykalin, N. N. (1952). *Die Waermegrundlagen des Schweissvorganges* (265 pp.). Berlin: Verlag Technik (in German).

Rykalin, N. N. (1957). *Berechnung der Waermevorgaenge beim Schweissen* (326 pp.). Berlin: VEB Verlag Technik (in German).

Seyffarth, P. (1978). *Schweiss – ZTU – Schaubilder. Atlas* – Rostock: Wilhelm – Pieck – Universitaet, Band 1 (151 pp.); Band 2 (167 pp.) (in German).

Seyffarth, P., & Kuscher, G. (1982). *Schweiss – ZTU – Schaubilder* (233 pp.). Berlin: VEB Verlag Technik (in German).

Seyffarth, P., Meyer, B., & Scharff, A. (1992). *Grosser Atlas Schweiss – ZTU – Schaubilder* (176 pp.). Duesseldorf: DVS – Verlag (in German).

Sipaylov, V. A. (1978). *Thermal processes in grinding and surface quality control* (167 pp.). Moscow: Mashinostroenie (in Russian).

Sommerfeld, A. (1894). Zur analytischen Theorie der Waermeleitung. *Mathematische Annalen, 45*, 263–277 (in German).

Steen, W. M., Dowden, J., Davis, M., & Kapadia, P. (1988). A point and line source model of laser keyhole welding. *Journal of Physics D: Applied Physics, 21*, 1255–1260.

Swift-Hook, D. T., & Gick, A. E. F. (1973). Penetration welding with lasers. *Welding Journal, 52*(11), 492-s–499-s.

Trigger, K. J., & Chao, B. T. (1951). An analytical evaluation of metal-cutting temperatures. *Transactions of the ASME, 73*(1), 57–68.

Turichin, G., Valdaitseva, E., Pozdeeva, E., Dilthey, U., & Gumeniuk, A. (2008). Theoretical investigation of dynamic behaviour of molten pool in laser and hybrid welding with deep penetration. *The Paton Welding Journal, 7*, 11–15.

Turichin, G. A., Tsibulsky, I. A., Zemlyakov, E. V., & Kharlamov, V. V. (2009). Features of weld formation in laser and hybrid welding of thick metals with fibre lasers. In *Proceedings of St. Petersburg State Polytechnic University, No. 510 "Materials and Chemical Technologies"* (pp. 3–18). St. Petersburg: St. Petersburg State Polytechnic University Publishing (in Russian).

Uglov, A. A., Smurov, I. Y., & Andrianov, S. S. (1981). Calculation of temperature fields due to moving concentrated heat sources. *Physics and Chemistry of Materials Processing, 1*, 35–43 (in Russian).

Watson, G. N. (1995). *A treatise of the theory of Bessel functions* (2nd ed., 814 pp.). Cambridge University Press.

Wilson, H. A. (1904). On convection of heat. *Cambridge Philosophical Society, 12*, 406–423.

Wu, C. S., Wang, H. G., & Zhang, Y. M. (2006). A new heat source model for keyhole plasma arc welding in FEM analysis on the temperature profile. *Welding Journal, 85*(12), 284-s–291-s.

Yampolsky, V. M. (1972). Investigation of features of vacuum arc discharge with hollow cathode of welding type. *Transactions of Institutes of Higher Education. Engineering, 7*, 67–68 (in Russian).

Yazovskikh, V. M., Trushnikov, D. N., & Belenky V. Y. (2012). Thermal processes in electron beam welding of circular welds. *Welding and Diagnostics, 5*, 26–31 (in Russian).

Chapter 6
Temperature Fields in Welding with Pressure

A distinctive feature of pressure welding methods is an additional parameter of welding conditions that is the external mechanical force which is applied to the workpiece during or right after heating. In the former case, the force influences thermal processes and, consequently, should be considered when calculating temperature fields.

Heat sources within welding with pressure are very diverse: electrical resistance of the contact surface of the parts to be welded (spot, seam, resistance butt, flash and induction welding), friction (friction, friction stir welding), electric arc (magnetically impelled arc welding) and others. Upon that, the welding area heats up below or above melting temperature T_m. The terms in the classification of pressure welding methods are common, although they badly (ambiguously) reflect the essence of the welding process.

Further, this chapter considers temperature fields in the main pressure welding methods (resistance spot, flash, friction, friction stir, and magnetically impelled arc welding).

6.1 Resistance Spot Welding

The principle of spot welding was analysed above in Sect. 1.8.

Let us make the following assumptions when formulating the heat conduction problem:

- electrodes and the workpiece are axisymmetric bodies;
- the contact surface of the workpieces is a symmetry plane (electrodes and workpieces are equal by pairs);
- the electromagnetic field is stationary (laws of electrostatics are valid, since the current frequency is relatively low). Then, the current and temperature distributions will be axisymmetric and the corresponding problems will be two-dimensional. The problems of heat conduction and electrostatics are coupled (temperature depends on current and resistance, and current and resistance depend on

V. A. Karkhin, *Thermal Processes in Welding*, Engineering Materials,
https://doi.org/10.1007/978-981-13-5965-1_6

temperature). Therefore, the corresponding differential equations should be solved together.

An axisymmetric temperature field in a body with a volume heat source is described with the following non-linear differential equation in the cylindrical coordinate system

$$c\rho\frac{\partial T}{\partial t} = \frac{\partial}{\partial r}\left(\lambda\frac{\partial T}{\partial r}\right) + \frac{\lambda}{r}\frac{\partial T}{\partial r} + \frac{\partial}{\partial z}\left(\lambda\frac{\partial T}{\partial z}\right) + \rho_e j^2. \quad (6.1.1)$$

Here, temperature T, volumetric heat capacity $c\rho$, thermal conductivity λ, electrical resistivity ρ_e, and current density j are functions of r, z, and t.

Current density j is a vector with radial (j_r) and axial (j_z) components, which are determined by Ohm's law:

$$j_r = \frac{1}{\rho_e}\frac{\partial U}{\partial r};\ j_z = \frac{1}{\rho_e}\frac{\partial U}{\partial z}, \quad (6.1.2)$$

where U is the electrical potential. By the definition of the vector, the current equals:

$$j^2 = j_r^2 + j_z^2. \quad (6.1.3)$$

Potential field U (r, z) is described by Laplace's differential equation (Browne et al. 1995; Cho and Cho 1989):

$$\frac{\partial}{\partial r}\left(\frac{1}{\rho_e}\frac{\partial U}{\partial r}\right) + \frac{1}{\rho_e r}\frac{\partial U}{\partial r} + \frac{\partial}{\partial z}\left(\frac{1}{\rho_e}\frac{\partial U}{\partial z}\right) = 0. \quad (6.1.4)$$

Let us set boundary conditions for a composite axisymmetric body which consists of electrodes and sheets to be welded. Since the body is symmetric about axes r and z, let us consider only the boundary *OABCDEFHKLO* (the boundary of the darkened area in Fig. 6.1a).

For the heat conduction problem, initial and boundary conditions are the following:

$$T(r, z, 0) = T_0 \text{ on } ABCD \quad \text{(water cooling)}; \quad (6.1.5)$$

$$\lambda\frac{\partial T}{\partial n} = \alpha(T - T_0) \text{ on } DEFHK \quad \text{(Newton's law)}; \quad (6.1.6)$$

$$\frac{\partial T}{\partial z} = 0 \text{ on } KLO \quad \text{(symmetry condition)}; \quad (6.1.7)$$

$$\frac{\partial T}{\partial r} = 0 \text{ on } OGA \quad \text{(on } z\text{-axis, symmetry condition)}, \quad (6.1.8)$$

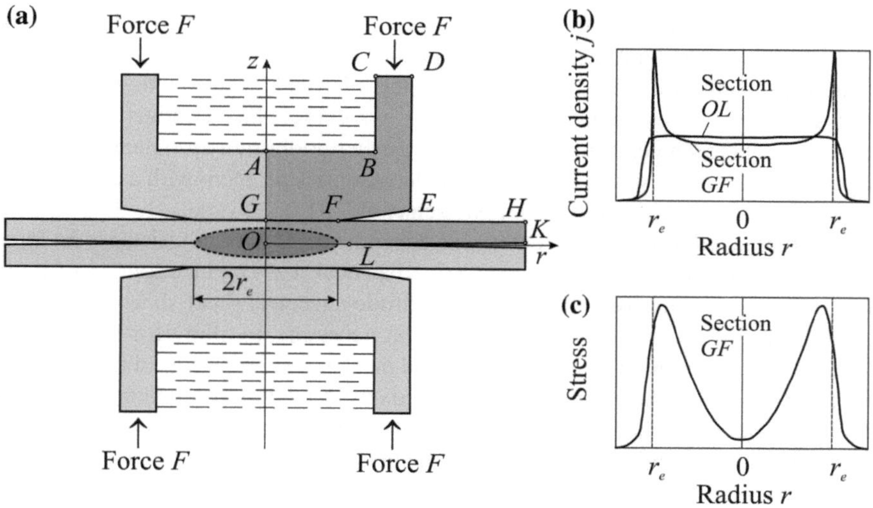

Fig. 6.1 Model for the resistance spot welding (**a**) and distributions of current density (**b**) and axial stress (**c**) in contact sections

where T_0 is the temperature of the medium and cooling water, n is the outward normal to the surface.

Let us consider that the voltage supplied to the electrodes is prescribed (equals U_0 and can be time dependent) and the body's surfaces exposed to water and air are electrically insulated (current does not flow normally to the body's surface, $j_n = 0$). Then, the boundary conditions for the electrical problem will have the form

$$U = U_0/2 \quad \text{on}\, CD; \tag{6.1.9}$$

$$U = 0 \quad \text{on}\, OL; \tag{6.1.10}$$

$$\frac{\partial U}{\partial n} = 0 \quad \text{on}\, ABC \text{ and } DEFHKL. \tag{6.1.11}$$

Naturally, problems of heat conduction and electrostatics are solved with allowance for non-uniformity of the body, which consists of a sheet and an electrode, in its thermophysical and electrical properties.

Let us note that with a gap between the sheets or with a curved working surface of the electrode, the electrode-sheet contact on segment *GF* (Fig. 6.1) can only be partial and changeable during welding. Sliding is possible along the electrode-sheet boundary. The target surface of the electrode-sheet contact can be found by solving a coupled thermomechanical problem with a given cycle of the impressed force *F* (De 2011; Khan et al. 1999).

Since the conditions on the contact surface depend on solving all coupled problems, the following iteration method is applied. The initial contact surface is set and then the problems are solved sequentially in the following cycle: solving an electrostatics problem (determining Joule heating) → solving a heat conduction problem (determining the temperature field) → solving a thermoplasticity problem (determining the contact's new state) → solving an electrostatics problem with a new contact surface, etc. (Murakawa et al. 1995; Nielsen et al. 2013).

A complicated problem is to determine dependence of contact resistance on local temperature and stress (De and Dorn 2005; Khan et al. 1999). Plane sources of Joule heat, caused by contact resistance on the electrode-sheet and sheet-sheet boundaries, are represented as volume heat sources. It is necessary to allow for temperature dependence of thermophysical, electrical and mechanical properties of materials. In the given statement, the problem can be solved only by numerical methods: the finite difference method or the finite element method if the coupled temperature—electrical problem is solved, and the finite element method if the coupled temperature—electrical—mechanical problem is solved.

The main parameters of the resistance spot welding conditions are welding current I (or alternating current amplitude $I_{\max}$), electrode force F, welding time t_w, electrode radius r_e or its diameter d_e (Fig. 1.27).

As an example, let us consider resistance spot welding of the sheets, which are 1.09 mm thick and made of low-carbon steel AISI 1008 (<0.10% C, 0.3–0.5% Mn) (Fig. 6.2) (Khan et al. 1999). The temperature field in the welded sheets is typical for resistance spot welding of steel (Fig. 6.2a). With short welding time, metal does not melt (Fig. 6.2b). The main factor that determines the size of the weld nugget is current (Fig. 6.2c). With the current increase, the nugget increases in both, radius r and axis z. A great influence is exerted by electrode's radius r_e since the contact area between the sheets determines local heating of metal and, consequently, the size of the weld nugget (Fig. 6.2d). During resistance spot welding a small change in the electrode tip curvature leads to large changes in weld nugget development (Babu 2018).

Resistance spot welding of aluminium alloys is characterised by a number of features. In contrast to steel, aluminium alloys have resistance only several times higher than copper alloys for electrodes. The non-conductive layer of aluminium oxide Al_2O_3 on contact surfaces creates additional electrical resistance, which makes the welding process less stable. For welding of aluminium alloys, electrodes are used with a spherical working surface of radius r_f by several times larger than electrode's radius r_e (De and Dorn 2005). The process of spot welding is greatly influenced by the initial state of the electrode-sheet and sheet-sheet contact surfaces, more specifically, contact resistance, which depends on metal's properties and preweld chemical treatment.

Let us consider resistance spot welding of aluminium alloy AlMg0.4Si1.2 (0.25–0.6% Mg, 1.0–1.5% Si, < 0.2% Cu, < 0.2% Mn). The melting temperature range is 848–923 K (Greitmann and Rother 1998). The weld nugget size (isotherm $T_{\max} = 873$ K) depends on welding current I (Fig. 6.3). The nugget size (diameter

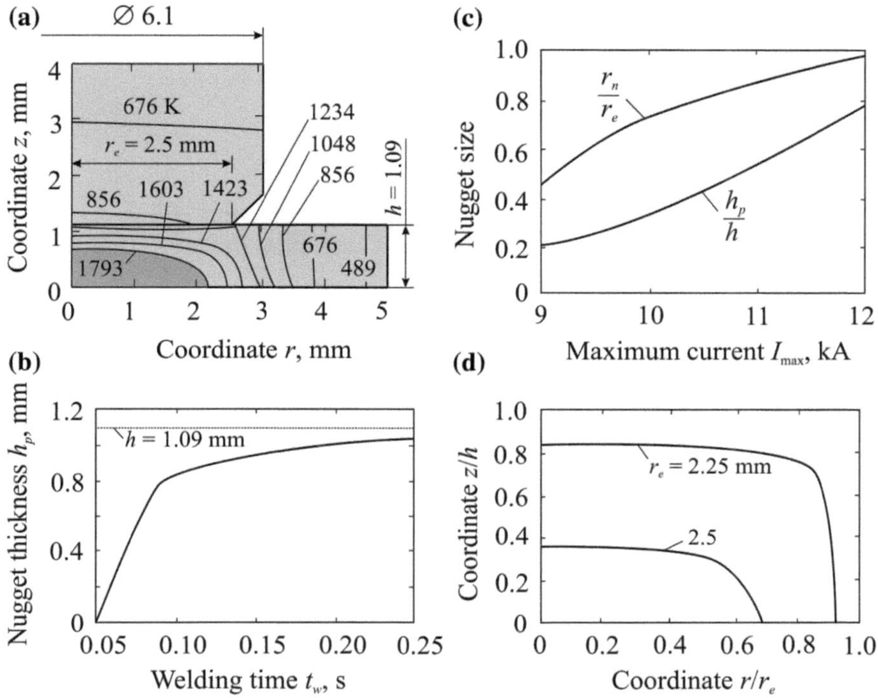

Fig. 6.2 Resistance spot welding of steel sheets of thickness $h = 1.09$ mm (Khan et al. 1999): peak temperature field ($t_w = 0.2$ s, $r_e = 2.5$ mm, $I_{max} = 10$ kA) (**a**); axial nugget thickness h_p as a function of welding time t_w ($r_e = 2.5$ mm, $I_{max} = 10$ kA) (**b**); effect of welding current on the nugget size ($t_w = 0.15$ s, $r_e = 2.5$ mm) (**c**) and effect of electrode tip diameter r_e on the nugget geometry (dimensionless nugget thickness z/h and dimensionless nugget radius r/r_e) ($t_w = 0.2$ s, $I_{max} = 10$ kA) (**d**)

d_n and depth of fusion h_p) is small with current $I = 15$ kA and significantly grows with the current rise up to 30 kA (in the figure, the weld nugget section is darkened).

Figure 6.4a shows that the nugget stops increasing after 0.02 s (i.e. after two cycles of alternating current). Short welding time is a feature of resistance spot welding of aluminium alloys. The nugget size can be significantly increased due to welding current I (Fig. 6.4b). With the growth of electrode force F, the nugget diameter d_n and the depth of fusion h_p slightly decrease (Fig. 6.4c).

Simulation of thermal processes allows determining the effect of each parameter of the resistance spot welding conditions and each factor (e.g. contact resistance) separately and optimising the welding process parameters (Greitmann and Rother 1998; Vukicevic et al. 2010).

Since the temperature dependences of thermal conductivity, enthalpy (specific heat capacity), specific resistance of materials and contact resistance are known only with certain accuracy (generally, with the error not lower than 5%), numerical

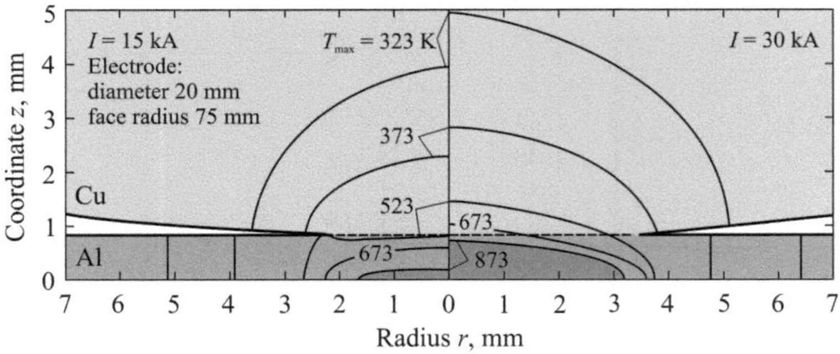

Fig. 6.3 Peak temperature field in a sheet-electrode system in the resistance spot welding of AlMg0.4Si1.2 aluminium alloy sheets 0.8 mm thick (force $F = 2$ kN, welding time $t_w = 0.1$ s) (De and Dorn 2005)

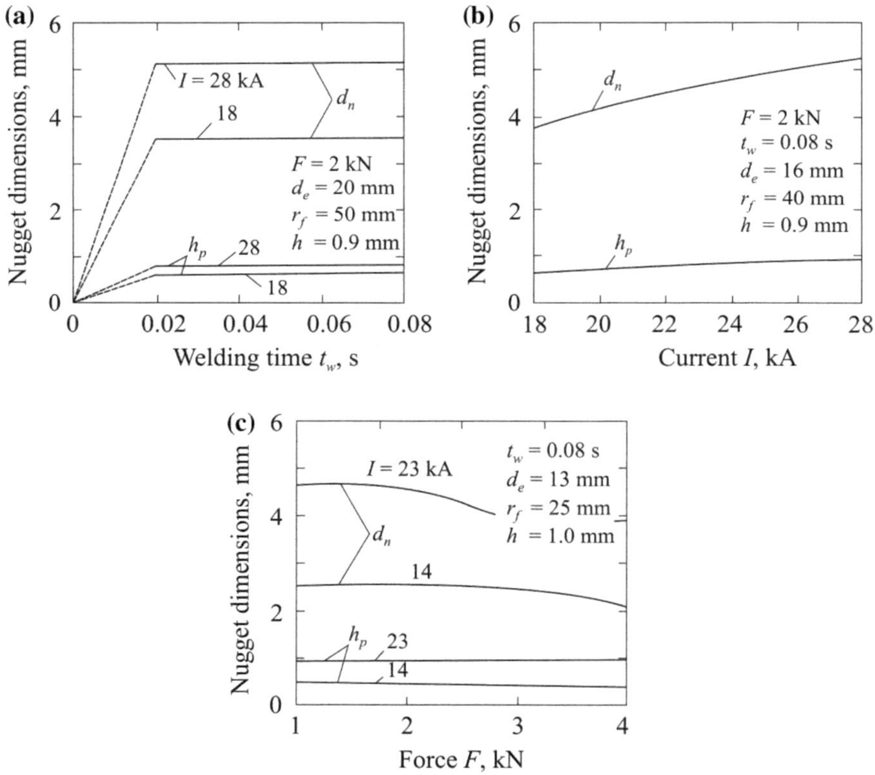

Fig. 6.4 The growth of nugget dimensions with welding time (**a**), with welding current (**b**) and with electrode force (**c**) in resistance spot welding of AlMg0.4Si1.2 aluminium alloy sheets of thickness h (De and Dorn 2005)

simulation is able to determine the influence of specification inaccuracy for each property of the material on thermal processes, including the melting isotherm (the weld nugget size) (Srikunwong et al. 2005). Therefore, simulation leads to better understanding of the welding process and optimising parameters of the welding conditions.

Solution to these coupled temperature—electrical—mechanical problems is time- and effort-consuming. Therefore, it is justified to undertake if high accuracy of the solution is required (Nielsen et al. 2013). If thermal cycles have to be calculated approximately for further estimation of weld and HAZ properties, it is advisable to use more economically efficient analytical solution methods. For example, it is possible to simplify the thermal model for resistance spot welding and consider the heat source as instantaneous and line, uniformly distributed all along the thickness of the plates (formula (4.2.37)). Such a model was used when calculating a cooling rate in the centre of the weld nugget (Grong 1994). A more accurate model can be obtained by solutions for an instantaneous or continuous disk source in a slab (Sects. 5.2.1 and 5.2.2). Other thermal models are also suggested (Kochergin 1987). Let us repeat that complexity of the model is determined by further use of the obtained information.

Similarly, it is possible to frame a solution for resistance seam welding when nuggets are formed at the given spot spacing. The solution for resistance seam welding can be represented as a sum of solutions for resistance spot welding.

6.2 Resistance Butt and Flash Welding

Basic physics of resistance butt and flash welding is described in Sects. 1.9 and 1.10. If cross sections of two workpieces to be welded are equal and their butt ends (faying surfaces) are flat, it is possible to consider the heat source as plane and uniformly distributed, i.e. the temperature field is one-dimensional.

The theory of heating similar and dissimilar bars with current during resistance spot welding was sufficiently developed long ago (Okamoto 1929; Pugin 1959a; Rykalin 1959).

Temperature fields in flash welding were studied in many works (Karkhin et al. 2007, 2008, 2009; Kuchuk-Yatsenko and Lebedev 1976; Nippes et al. 1951, 1953, 1955; Nippes and Chang 1955; Pugin 1959b).

When solving the heat conduction problem, let us make the following assumptions: the bars to be welded are equal, plane $x = 0$ is the symmetry plane (Fig. 6.5a); the extension length l is relatively large and changes in a predetermined way (Fig. 6.5b); there is no heat transfer; the origin of coordinates x is at the bar's butt end (in the centre of the weld), i.e. when heated, it moves relative to the bar's metal at the speed v (Fig. 6.5a); at the fused butt end of the bar ($x = 0$), temperature T_b (problem A) or plane power density q_{2b} (problem B) are prescribed (Fig. 6.5c).

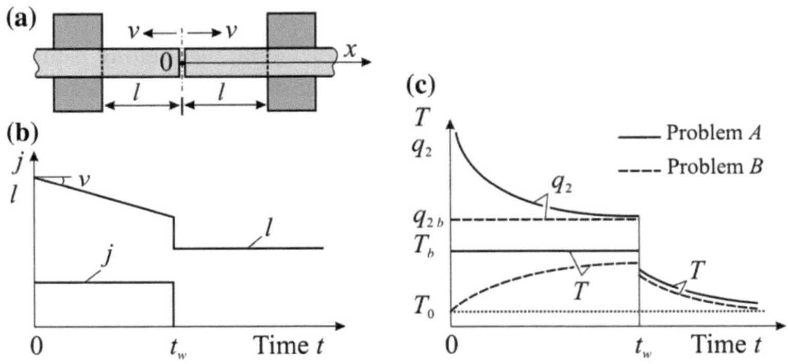

Fig. 6.5 The diagram of flash welding (**a**) and idealised time dependences of current density j, extension length l (**b**), temperature T and the heat flow q_2 at the end of the bar (in the centre of the welded joint) (**c**)

With the made assumptions, the heat conduction problem statement has the form:

1. Linear equation of heat conduction:

$$\frac{\partial T}{\partial t} = a\frac{\partial^2 T}{\partial x^2} + v\frac{\partial T}{\partial x} + \frac{\rho_e j^2}{c\rho}. \tag{6.2.1}$$

2. Initial condition ($t = 0$):

$$T(x, 0) = T_0. \tag{6.2.2}$$

3. Boundary conditions:

$$x = 0,\ t \le t_w:\quad T(0, t) = T_b \quad \text{(problem } A\text{) or} \tag{6.2.3}$$

$$-\lambda\frac{\partial T}{\partial x}(0, t) = q_{2b} \quad \text{(problem } B\text{);} \tag{6.2.4}$$

$$t > t_w:\quad \frac{\partial T}{\partial x}(0, t) = 0; \tag{6.2.5}$$

$$x = \infty, t > 0:\quad T(\infty, t) = T_0, \tag{6.2.6}$$

where t_w is the welding time.

The solution to problem A with a prescribed temperature at the butt end was obtained before by the integral transformation method (see formula (4.2.97)).

In the papers (Karkhin et al. 2007, 2008), it was shown that the influence of Joule heat can be neglected because when welding steel, the heating rate due to flowing current is less than 10 K s^{-1}. Then the solution to the problem is simplified

$$T(x,t) - T_0 = \frac{1}{2}(T_b - T_0)\left[\Phi^*\left(\frac{x+vt}{\sqrt{4at}}\right) + \exp\left(-\frac{vx}{a}\right)\Phi^*\left(\frac{x-vt}{\sqrt{4at}}\right)\right]. \quad (6.2.7)$$

The heat flow at the bar's butt end is determined by the temperature gradient at $x = 0$:

$$\begin{aligned} q_{2b}(0,t) &= -\lambda\frac{\partial T}{\partial x}(0,t) \\ &= \frac{1}{2}\lambda(T_b - T_0)\left[\frac{1}{\sqrt{at}}\exp\left(-\frac{v^2t}{4a}\right) + \frac{v}{a}\Phi^*\left(-\sqrt{\frac{v^2t}{4a}}\right)\right]. \end{aligned} \quad (6.2.8)$$

If time t is relatively large, the thermal state in a moving system is steady (quasi-stationary):

$$T(x,\infty) - T_0 = (T_b - T_0)\exp\left(-\frac{vx}{a}\right); \quad (6.2.9)$$

$$q_{2b}(0,\infty) = (T_b - T_0)c\rho v. \quad (6.2.10)$$

The process of heat saturation is characterised by the function

$$\begin{aligned} \psi_1(\xi,\tau) &= \frac{T(x,t) - T_0}{T(x,\infty) - T_0} \\ &= \frac{1}{2}\left[\exp(2\xi)\Phi^*\left(\frac{\xi}{\sqrt{4\tau}} + \sqrt{\tau}\right) + \Phi^*\left(\frac{\xi}{\sqrt{4\tau}} - \sqrt{\tau}\right)\right]; \\ \xi &= \frac{vx}{2a}; \tau = \frac{v^2t}{4a}, \end{aligned} \quad (6.2.11)$$

where ξ is the dimensionless coordinate; τ is the dimensionless time.

The dependence of function ψ_1 on τ and ξ is given in Fig. 6.6a. The farther the point under study is positioned from the butt end (the larger ξ), the later its temperature approaches the maximum (ψ_1 approaches to 1). The given curves allow determining the time required for heating the predetermined zone up to the predetermined temperature. It is notable that formula (6.2.11) coincides with heat saturation function ψ_3 for the case of a moving point source on the surface of a semi-infinite body (5.1.106).

Problem *B* differs from problem *A* only in the boundary condition at the butt end, where heat flow q_{2b} is prescribed (Fig. 6.5c).

Let us differentiate Eq. (6.2.1) without a constant term with respect to x. As a result, we shall obtain the following boundary problem for flow q_2 ($q_2 = -\lambda\partial T/\partial x$):

$$1.\ \frac{\partial q_2}{\partial t} = a\frac{\partial^2 q_2}{\partial x^2} + v\frac{\partial q_2}{\partial x}; \quad (6.2.12)$$

$$2.\ q_2(x,0) = 0; \quad (6.2.13)$$

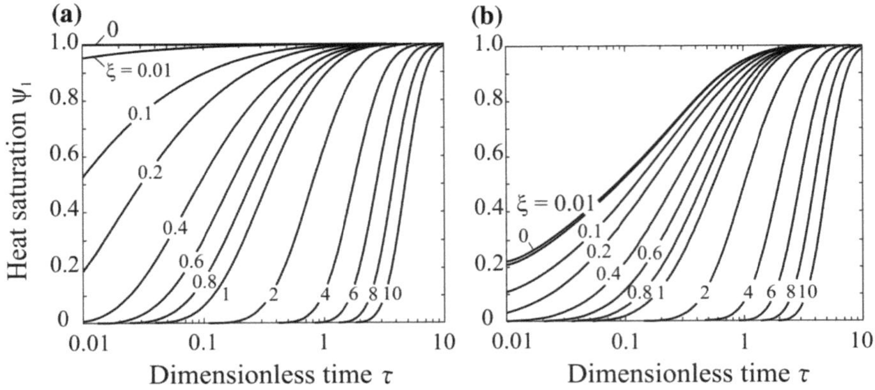

Fig. 6.6 Dependences of the function of heat saturation of the zone in front of the moving plane source on dimensionless time τ and dimensionless coordinate ξ for the cases in which the temperature (**a**) or heat flow (**b**) are prescribed (Karkhin et al. 2008, 2009)

$$3.\ q_2(0, t) = q_{2b}; \tag{6.2.14}$$

$$q_2(\infty, t) = 0. \tag{6.2.15}$$

This problem is in line with the considered problem A and its solution is similar to Eq. (6.2.7):

$$q_2(x, t) = \frac{q_{2b}}{2}\left[\Phi^*\left(\frac{x + vt}{\sqrt{4at}}\right) + \exp\left(-\frac{vx}{a}\right)\Phi^*\left(\frac{x - vt}{\sqrt{4at}}\right)\right]. \tag{6.2.16}$$

With integrating this equation with respect to x we shall find the temperature:

$$T(x, t) - T_0 = -\frac{1}{\lambda}\int\limits_{\infty}^{x} q_2(x, t)dx = -\frac{q_{2b}}{2\lambda}\left\{(x + vt)\Phi^*\left(\frac{x + vt}{\sqrt{4at}}\right) - \sqrt{\frac{4at}{\pi}}\right.$$
$$\left.\times \exp\left(-\frac{(x + vt)^2}{4at}\right) - \frac{a}{v}\left[\exp\left(-\frac{vx}{a}\right)\Phi^*\left(\frac{x - vt}{\sqrt{4at}}\right) - \Phi^*\left(\frac{x - vt}{\sqrt{4at}} + v\sqrt{\frac{t}{a}}\right)\right]\right\}. \tag{6.2.17}$$

In the limiting case ($t \to \infty$), we obtain:

$$T(x, \infty) - T_0 = \frac{q_{2b}}{c\rho v}\exp\left(-\frac{vx}{a}\right). \tag{6.2.18}$$

It follows that:

$$\psi_1(\xi,\tau) = -\frac{1}{2\exp(-2\xi)}\left\{(2\xi+4\tau+1)\Phi^*\left(\frac{\xi}{\sqrt{4\tau}}+\sqrt{\tau}\right)\right.$$
$$\left.-4\sqrt{\frac{\tau}{\pi}}\exp\left[-\frac{(\xi+2\tau)^2}{4\tau}\right]-\exp(-2\xi)\Phi^*\left(\frac{\xi}{\sqrt{4\tau}}-\sqrt{\tau}\right)\right\}. \quad (6.2.19)$$

In Fig. 6.6b, the heat saturation function $\psi_1(\xi, \tau)$ is shown. The curve $\xi = 0$ corresponds to dimensionless temperature in the source plane.

Let us compare the solutions to the considered problems. When prescribing fixed temperature at the butt end $T_b = \text{const}$ (problem A), the heat flow at the butt end at the beginning of heating (6.2.8) exceeds the steady flow (6.2.9). If flow $q_{2b} = \text{const}$ (problem B) is prescribed, the temperature at the butt end (6.2.17) grows monotonically, approaching the limiting value (6.2.18). Time variations in temperature and the flow are schematically shown in Fig. 6.5c.

It should be noted that accurate statement of boundary conditions (T_b or q_{2b}) is problematic. In the initial period of flashing, the average temperature at the butt end is lower than the melting temperature and welding current and, consequently, heat flow q is higher than at the end of flashing (Kuchuk-Yatsenko and Lebedev 1976). This means that problems A and B are natural boundaries of the range, within which the real process runs. In problem A as well as in problem B, the limit distributions of temperature in front of the moving source are equal (formulas (6.2.9) and (6.2.18)), if heat flow q_{2b} is determined by formula (6.2.10), where T_b should be considered as temperature of the expelled metal.

Comparing curves $\psi_1(\xi, \tau)$ in Fig. 6.6a and b it follows that if temperature is prescribed at the butt end, heat saturation attained faster, especially close to the butt end. When using the model of the moving plane source in an infinite rod (Sect. 5.1.3.10), expelling the melted metal is not taken into account. Neglect of expelling of the metal leads to slower heat saturation of the metal in front of the source.

To determine temperature at the cooling stage (after switching off the welding current and upsetting), it is advisable to take temperature distribution at the end of the heating stage as initial and to solve the problem with the method of sources, as it is done in paper (Karkhin et al. 2008).

As an example, let us consider welding the bars of 6.35 × 50.8 mm section made of steel AISI 1020 (0.18–0.23% C, 0.30–0.60% Mn) at the constant speed $v = 0.94\ \text{mm s}^{-1}$ (Nippes et al. 1951). The initial extension $l = 35$ mm. The values were: $T_b = 2500$ K, $\lambda = 0.035\ \text{W mm}^{-1}\ \text{K}^{-1}$, $c\rho = 0.0064\ \text{J mm}^{-3}\ \text{K}^{-1}$, $T_0 = 300$ K. The temperature at the butt end of the bar T_b is close to the average temperature of the metal forcibly expelled from the joint as the workpieces are slowly moved together during flashing (for steel ~2273 K). When prescribing the flow at the butt end q_{2b}, we relied on condition (6.2.18), i.e. we assumed that the limiting temperature $T(0, \infty) = T_b$.

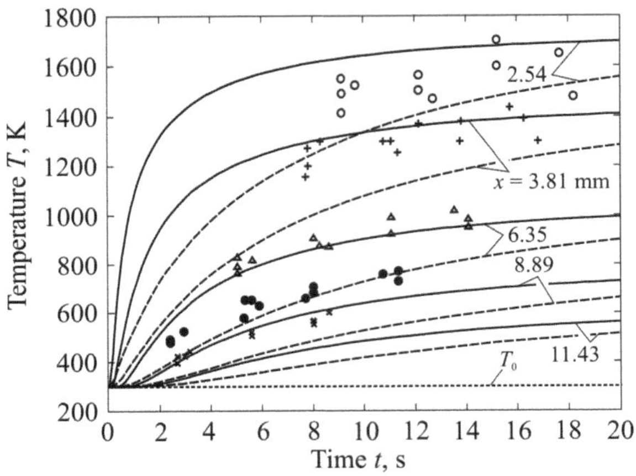

Fig. 6.7 The increase of temperature T at the instantaneous distance x from the interface of the bar at the platen velocity $v = 0.94$ mm s^{-1} (solid line—calculated from Eq. (6.2.7), dashed lines—calculated from Eq. (6.2.17), × •+o—experiments (Karkhin et al. 2008, 2009; Nippes et al. 1951)

Figure 6.7 shows a temperature increase in the moving coordinate system. The temperature increases monotonically, reaching almost limit values in 12 s, if the temperature is prescribed at the butt end. If the flow is prescribed at the butt end, the temperature increases more slowly. The figure presents the results of calculation by formula (6.2.7), when the temperature is prescribed at the butt end, satisfactorily coincides with experimental points. When prescribing the flow at the butt end, the calculation results appear to be underestimated.

It should be noted that with the variable speed v and with the allowance for the temperature dependence of metal properties, the heat conduction problem can be solved numerically (Karkhin et al. 2008).

6.3 Friction Welding

Friction heating of butt ends in round or tubular bars of equal sections and with equal thermophysical properties can be represented as the model of a plane source with power density $q_2 = q/A$. If $q_2 = \text{const}$, the axial distribution of temperature is described by Eq. (5.1.36).

The temperature of the contact section ($x = 0$)

$$T(0, t) - T_0 = \frac{q}{A} \frac{\sqrt{t}}{\sqrt{\pi \lambda c \rho}} \tag{6.3.1}$$

rises with time by the square root law.

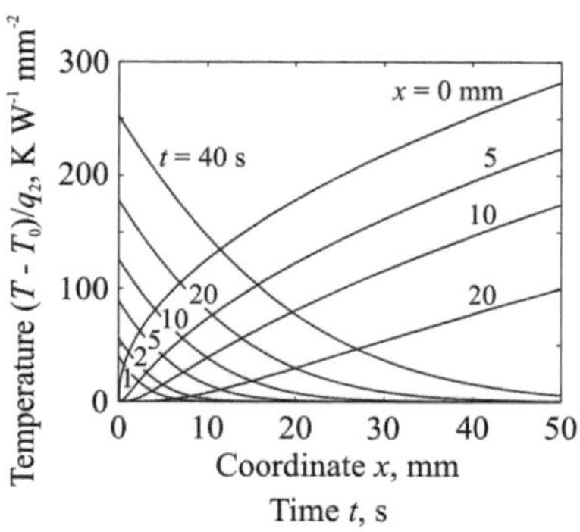

Fig. 6.8 Dependence of relative temperature $(T - T_0)/(q/A)$ in low-carbon steel bar on axial coordinate x and heating time t in friction welding

The heating of bars made of low-carbon steel ($\lambda = 0.04$ W mm^{-1} K^{-1}, $a =$ 8 mm^2 s^{-1}) is shown in Fig. 6.8. With predetermined power density q_2, the curves allow estimating the time of heating up to the prescribed temperature, cooling rate, and the temperature equalisation time after heating.

6.4 Friction Stir Welding

The principle of friction stir welding and calculation models of distributed heat sources were analysed above in Sect. 1.12.

Depending on the required accuracy of solving the heat conduction problem, numerical or functional-analytical methods are used. This may differ in formulation of the problem: geometry of the area under study, boundary conditions, allowance for heat sources, allowance for material properties, etc. If the coupled problems of heat conduction and thermoplasticity are solved, only numerical methods (the finite difference method and the finite element method) can be used. These methods require a lot of input data and a lot of computational time (Colegrove et al. 2007). If it is required to determine only temperature fields and only in the workpiece, the problem formulation can be simplified and solution methods can be both numerical and analytical (Lohwasser and Chen 2010; Mishra and Ma 2005).

Let us consider welding of plates made of aluminium alloy 7075-T6 (5.6–6.1% Zn, 2.1–2.5% Mg, 1.2–1.6% Cu, yield strength $\sigma_{0.2} = 420$ MPa at $T_0 = 298$ K, solidus temperature $T_S = 805$ K) under the following conditions: thickness h is 8 mm, speed v is 0.67 mm s^{-1}, shoulder radius R_{shoul} is 12.5 mm, pin radius R_{pin} is 5 mm, cone pin with a 15° angle, rotation velocity of the tool 535 rpm, a steel tool, and a steel backing plate. The studied body is only welded plates. Coefficients of heat transfer $\alpha = 10{,}000$ W m^{-2} K^{-1}on the plate-tool interface (allowing for a heat loss into the tool), $\alpha = 10$ W m^{-2} K^{-1} for the upper surface outside the tool (allowing for a heat loss into air), and $\alpha = 700$ W m^{-2} K^{-1} for the lower surface

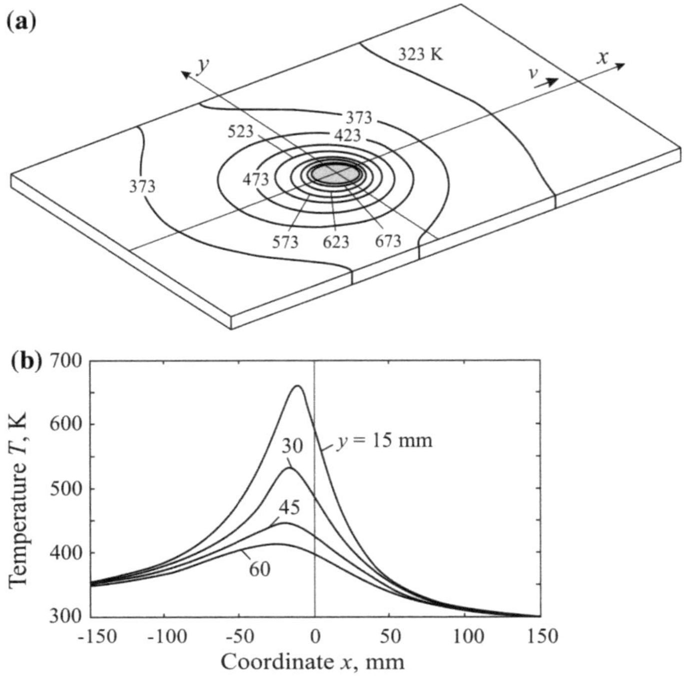

Fig. 6.9 Temperature field (**a**) and temperature profiles along x-axis at distance y from the weld centreline (**b**) in friction stir welding of 7075-T6 aluminium alloy 8 mm thick (Schmidt 2010)

(allowing for the influence of the steel backing plate). The temperature dependence of the alloy's yield strength was given (Schmidt 2010).

As a result of the numerical solution to the problem, an effective power of 1909 W was obtained, moreover, the shoulder takes 83%, the pin side takes 16%, and the pin tip takes 1% of the power (Schmidt 2010). It should be mentioned that this contribution is a characteristic for the considered particular welding conditions. With larger thickness of the plates, the pin radius increases relative to the shoulder radius and, consequently, the heat share that is generated on the pin's surface also increases.

The temperature field on the surface of the welded plates outside the tool is shown in Fig. 6.9a (the contact area of the tool is darkened), and longitudinal temperature distribution in particular sections, located at different distances from the joint plane—in Fig. 6.9b. It is seen that at a distance of 2.5 mm from the tool ($y =$ 15 mm), temperature is quite high. Thermal cycles $T(t)$ for points $y = \text{const}$ are easy to obtain by substituting time t for coordinate x, $t = -x/v$. Since the welding speed $v = 0.67$ mm s^{-1} is low relative to thermal diffusivity of the aluminium alloy $a =$ 48 mm^2 s^{-1}, the heating rate is low and close to the cooling rate.

Let us emphasise that the considered problem can be approximately solved by the analytical method of sources. With such an approach, a workpiece is represented as a

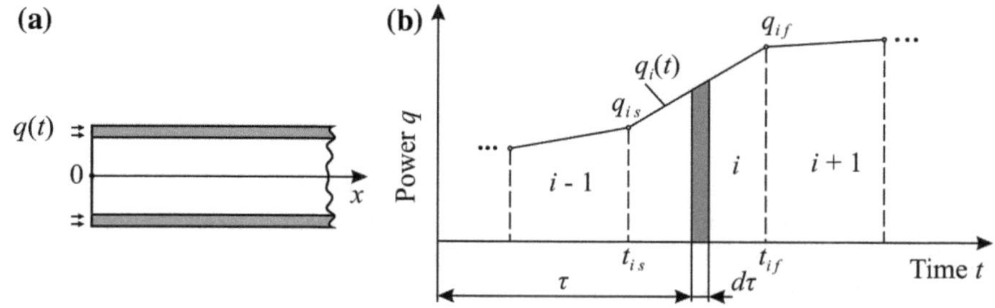

Fig. 6.10 The diagram of heat input in magnetically impelled arc welding (**a**) and time-dependent heat source power (**b**)

slab with the surfaces, the heat transfer coefficients α of which are different. The distributed source is represented as a set of point instantaneous sources distributed both in time and in a predetermined way on the shoulder-workpiece and pin-workpiece contact surfaces. A fundamental solution of the linear problem of heat conduction for the point instantaneous source in the slab is known (Eq. (4.2.30)). Then, the solution to the heat conduction problem is reduced to summation (integration) of temperature increments from the whole set of point instantaneous sources, distributed in time and space. If local generated energy depends on local temperature (e.g. through the material's yield strength), it is easy to synthesise an algorithm to deal with the problem by the iteration method, when the local density of the generated heat at the current iteration is determined by local temperature in the previous iteration. It can be expected that in this case preparing the input data and the required computer time will be less than when using numerical methods.

6.5 Magnetically Impelled Arc Welding

When welding with an arc rotating in a magnetic field (Fig. 1.33), a variable cycle of energy input is used (Abdullaev and Pugin 1969; Gagen and Taran 1970; Kalev et al. 1986).

Let us assume that the speed of the arc along the tube's edge is relatively high. Therefore, it is possible to consider that the heat source is plane, uniformly distributed, and stationary (Fig. 6.10a). Time dependence of source power $q(t)$ is known (Kalev et al. 1986).

Let us present curve $q(t)$ as a piecewise-linear function, each piece of which between boundaries t_{is} and t_{if} is described by linear dependence (Fig. 6.10b):

$$q_i(t) = A_i + B_i t, \tag{6.5.1}$$

where

$$A_i = q_{i\,s} - B_i t_{i\,s};\; B_i = \frac{q_{i\,f} - q_{i\,s}}{t_{i\,f} - t_{i\,s}}.$$

Let us use the method of sources (Green's function method). A temperature increment in point x at time t by the action of the elementary (instantaneous) energy source $q(\tau)d\tau$, which was operating in point $x = 0$ at moment τ during $d\tau$ (Fig. 6.10b), is determined by expression (4.2.42):

$$dT(x,t) = \frac{q(\tau)d\tau}{Ac\rho[4\pi a(t-\tau)]^{1/2}} \exp\left(-\frac{x^2}{4a(t-\tau)} - b(t-\tau)\right), \tag{6.5.2}$$

where A is the area of the body's cross-section. Then the temperature increment due to the ith source will be determined by summing up all temperature increments due to all elementary sources that operated before time t:

$$T_i(x,t) - T_0 = \int\limits_{t_{i\,s}}^{\min(t,\,t_{i\,f})} \frac{A_i + B_i\tau}{Ac\rho[4\pi a(t-\tau)]^{1/2}} \exp\left(-\frac{x^2}{4a(t-\tau)} - b(t-\tau)\right)d\tau. \tag{6.5.3}$$

The integral is expressed as known functions (Mikhailov 1987). In the absence of surface heat transfer ($b = 0$), a solution to the problem is simplified (Kalev et al. 1986). In the case when a power source is constant ($B_i = 0$), the solution coincides with Eq. (5.1.35). The last equation can be used for piecewise-constant function $q(t)$.

Thus, if the time-dependent arc's effective power $q(t)$ is known, piecewise-linear or piecewise-constant approximation is done (a real continuous source is split into N separate sources). The temperature increment in point x at time t is determined by summing up temperature increments of all n separate sources which were active before time t:

$$T(x,t) - T_0 = \sum_{i=1}^{n} [T_i(x,t) - T_0], \quad n \le N. \tag{6.5.4}$$

Figure 6.11 shows time dependence of effective power and thermal cycles of points during tube welding. The influence of current on arc efficiency was taken into account (Gagen and Taran 1970). Effective power $q(t)$ was approximated by a polygonal line (bends are marked with circles in Fig. 6.11). An abrupt increase in power before the end of welding caused a sharp temperature rise near the tube's butt end.

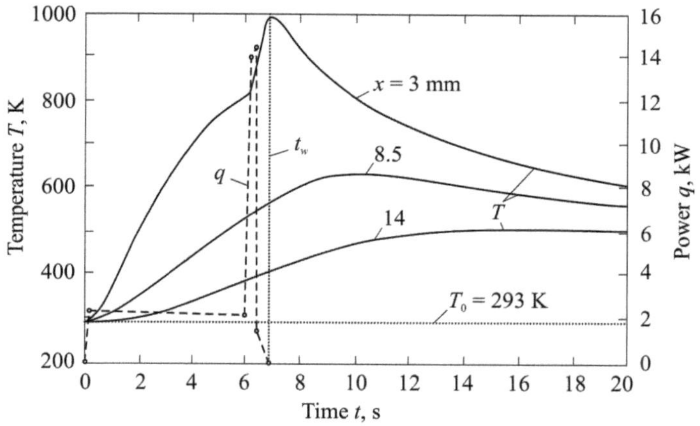

Fig. 6.11 Time-dependent effective power q and thermal cycles of the points at distance x from the tube edge in magnetically impelled arc welding of St2ps steel tube of diameter 33.5 mm and wall thickness 3.2 mm (Kalev et al. 1986)

References

Abdullaev, A. A., & Pugin, A. I. (1969). Heating of pipe edges in magnetically impelled arc butt welding. *Welding Production, 3*, 5–7 (in Russian).

Babu, S. S. (2018). Toward process-based quality through a fundamental understanding of weld microstructural evolution. *Welding Journal, 97*(1), 1-s–18-s.

Browne, D. J., Chandler, H. W., Evans, J. T., & Wen, J. (1995). Computer simulation of resistance spot welding in aluminium: Part 1. *Welding Journal, 10*, 339-s–344-s.

Cho, H. S., & Cho, Y. J. (1989). A study of the thermal behaviour in resistance spot welds. *Welding Journal, 6*, 236-s–244-s.

Colegrove, P., Shercliff, H., Robson, J., Kamp, N., Sullivan, A., & Williams, S. (2007). Integrated process modelling of friction stir welding. In H. Cerjak, H. K. D. H. Bhadeshia, & E. Kozenschnik (Eds.), *Mathematical modelling of weld phenomena* (Vol. 8, pp. 649–679). Graz: Verlag der Technischen Universitaet Graz.

De, A., & Dorn, L. (2005). Computer simulation of resistance spot welding process. In H. Cerjak, H. K. D. H. Bhadeshia, & E. Kozeschnik (Eds.), *Mathematical modelling of weld phenomena* (Vol. 7, pp. 895–923). Graz: Verlag der Technischen Universitaet Graz.

De, A. (2011). Modeling of thermal-electrical-mechanical coupling in fusion welding. In T. J. Lienert, S. S. Babu, T. A. Siewert, & V. L. Acoff (Eds.), *ASM Handbook. Vol. 06A. Welding fundamentals and processes* (pp. 789–796). Materials Park: ASM International.

Gagen, Yu. G., & Taran, V. D. (1970). *Magnetically impelled arc butt welding* (159 pp.). Moscow: Mashinostroenie (in Russian).

Greitmann, M. Y., & Rother, K. (1998). Numerical simulation of the resistance spot welding process using SPOTWELDER. In H. Cerjak (Ed.), *Mathematical modelling of weld phenomena* (Vol. 4, pp. 531–544). London: IOM Communications.

Grong, O. (1994). *Metallurgical modelling of welding* (581 pp.). London: The Institute of Materials.

Kalev, L. Ts., Michailov, V. G., & Vylchev, G. P. (1986). Calculation of temperature fields in magnetically impelled arc butt welding with time-varying power. *Welding Production, 10*, 32–33 (in Russian).

Karkhin, V. A., Khomich, P. N., & Ivanov, S. Yu. (2007). Calculation of temperature distribution during flash welding of steel. In *Proceedings of 2nd International Conference "Computer Technologies in Joining of Materials-2007"* (7 pp.). Tula: Tula State University (in Russian).

Karkhin, V. A., Khomich, P. N., Fedotov, B. V., & Rajamaki, P. (2008). Analysis of thermal cycles in resistance flash welding of steels. *Welding International, 22*(12), 903–908.

Karkhin, V. A., Khomich, P. N., & Ivanov, S. Yu. (2009). Calculation of temperature distribution during flash welding. In *Proceedings of St. Petersburg State Polytechnic University, No. 510 "Materials and Chemical Technologies"* (pp. 193–200). St. Petersburg: St. Petersburg State Polytechnic University Publishing (in Russian).

Khan, J. A., Xu, L., & Chao, Y.-J. (1999). Prediction of nugget development during resistance spot welding using coupled thermal-electrical-mechanical model. *Science and Technology of Welding and Joining, 4*(4), 201–207.

Kochergin, K. A. (1987). *Resistance welding* (240 pp.). Leningrad: Mashinostrenie (in Russian).

Kuchuk-Yatsenko, S. I., & Lebedev, V. K. (1976). *Flash-butt welding* (214 pp.). Kiev: Naukova Dumka (in Russian).

Lohwasser, D., & Chen, Z. (Eds.). (2010). *Friction stir welding. From basics to applications* (424 pp.). Oxford: Woodhead Publishing.

Michailov, V. G. (1987). Heat processes in welding with stationary plane heat source of time-varying power. *Technicheska Misl, 2*, 111–114 (in Bulgarian).

Mishra, R. S., & Ma, Z. Y. (2005). Friction stir welding and processing. *Reports: A Review. Journal Materials Science and Engineering R, 50*, 1–78.

Murakawa, H., Kimura, H., & Ueda, Y. (1995). Weld ability analysis of spot welding on aluminum using FEM. *Transactions of Japan Welding Research Institute, 24*(1), 101–111.

Nielsen, C. V., Zhang, W., Alves, L. M., Bay, N., & Martins, P. A. F. (2013). Modeling of thermo-electro-mechanical manufacturing processes. In *Applications in metal forming and resistance welding* (121 pp.). Springer.

Nippes, E. F., Savage, W. F., McCarthy, J. J., & Smith, S. S. (1951). Temperature distribution during the flash welding of steel. *Welding Journal, 30*(12), 585-s–601-s.

Nippes, E. F., Savage, W. F., Smith, S. S., McCarthy, J. J., & Grotke, G. (1953). Temperature distribution during the flash welding of steel—Part II. *Welding Journal, 32*(3), 113-s–122-s.

Nippes, E. F., & Chang, W. H. (1955). The flash welding of commercial molybdenium—Part II. *Welding Journal, 34*(5), 251-s–261-s.

Nippes, E. F., Savage, W. F., Suzuki, H., & Chang, W. H. (1955). A mathematical analysis of the temperature distribution during flash welding. *Welding Journal, 34*(6), 271-s–285-s.

Okamoto, T. (1929). Theoretical and experimental researches on electric resistance welding. *Memoirs of the College of Engineering. Kyoto Imperial University, 6*(1), 1–111.

Pugin, A. I. (1959a). Heating of bars in resistance butt welding. In N. N. Rykalin (Ed.), *Thermal processes in resistance welding* (pp. 54–133). Moscow: USSR Academy of Sciences Publishing (in Russian).

Pugin, A. I. (1959b). Intermittent heating of carbon steel round bars of big diameter in flash butt welding. In N. N. Rykalin (Ed.), *Thermal processes in resistance welding* (pp. 134–1676). Moscow: USSR Academy of Sciences Publishing (in Russian).

Rykalin, N. N. (1959). Theory of heating the rods by current in upset welding. In N. N. Rykalin (Ed.), *Thermal processes in resistance welding* (pp. 6–53). Moscow: Publication of the USSR Academy of Sciences (in Russian).

Schmidt, H. N. B. (2010). Modelling thermal properties in friction stir welding. In D. Lohwasser, & Z. Chen (Eds.), *Friction stir welding. From basics to applications* (pp. 277–313). Oxford: Woodhead Publishing.

Srikunwong, C., Dupuy, T., & Bienvenu Y. (2005). Influence of electrical–thermal physical properties in resistance spot welding modelling. In H. Cerjak, H. K. D. H. Bhadeshia, & E. Kozeschnik (Eds.), *Mathematical modelling of weld phenomena* (Vol. 7, pp. 999–1022). Graz: Verlag der Technischen Universitaet Graz.

Vukicevic, M., Petrovic, Z., Kolarevic, M., & Bjelic, M. (2010). Simulation model of initial period of spot welding. In H. Cerjak & N. Enzinger (Eds.), *Mathematical modelling of weld phenomena* (Vol. 9, pp. 955–966). Graz: Verlag der Technischen Universitaet Graz.

Chapter 7
Temperature Fields Under Programmed Heat Input

7.1 Overview

Programming the heat input allows a better optimisation of the welding process to obtain a welded joint of the proper quality. The quality of the joint is determined by its geometry as well as by its local properties (mechanical, microstructural, chemical, etc.). In order to predict local properties of the metal, it is required to know its initial state (before welding) and thermal behaviour of the metal during welding (the solution to this problem is described in Chap. 13).

Programmed heat input implies defining a law of heat input in time and space. The former refers to setting energy value and distribution in time, the latter refers to the trajectory of the source centre relative to the weldment. In practice, the input law can be complex and the number of its parameters can be large. In principle, the method of sources allows obtaining solutions to the heat conduction problem with real sources moving along various pathways in bodies of relatively simple geometry.

Let us consider a technique of solving the heat conduction problem by the example of the source, the volumetric power density of which is distributed in space and time according to the known law $q_3(x, y, z, t)$. The form of function q_3 depends on both the distribution law of the source itself and its movement speed and trajectory. Let the centre of the distributed source (where q_3 is maximal) is moving from point B to point E (Fig. 7.1). It is required to find the temperature at point P of the unbounded body at any time t.

Let us use the analytical method of sources (see Sect. 4.2.1). In order to do this, the time interval of source action 0–t_f has to be divided into elementary time increments dt. In other words, a continuous distributed heat source can be represented as a combination of sequent instantaneous distributed heat sources. Besides, every instantaneous distributed heat source can be represented as a set of instantaneous point sources, the energy of which is $dQ(x, y, z, t) = q_3(x, y, z, t)dxdydzdt$. The

V. A. Karkhin, *Thermal Processes in Welding*, Engineering Materials,
https://doi.org/10.1007/978-981-13-5965-1_7

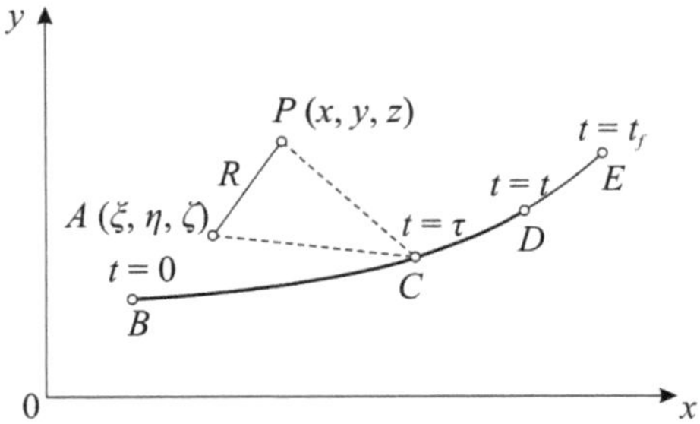

Fig. 7.1 Diagram for calculating the temperature at the point *P* due to a moving distributed heat source with the given trajectory *BCDE* (Ivanov et al. 2015, 2017)

temperature increment at point $P(x, y, z)$ at time t due to the elementary point source in point $A(\xi, \eta, \zeta)$, which operated at time τ and was positioned at a distance R from point P (Fig. 7.1), is calculated by formula (4.2.6):

$$dT(x, y, z, t) = \frac{q_3(\xi, \eta, \zeta, \tau) d\xi d\eta d\zeta d\tau}{c\rho[4\pi a(t - \tau)]^{3/2}} \exp\left(-\frac{R^2}{4a(t - \tau)}\right), \tag{7.1.1}$$

where $R^2 = (x - \xi)^2 + (y - \eta)^2 + (z - \zeta)^2$.

The temperature around a moving distributed heat source is determined by the total temperature increments due to all elementary (instantaneous point) sources:

$$T(x, y, z, t) - T_0 = \int_{-\infty}^{\infty}\int_{-\infty}^{\infty}\int_{-\infty}^{\infty}\int_{0}^{t} \frac{q_3(\xi, \eta, \zeta, \tau)}{c\rho[4\pi a(t - \tau)]^{3/2}}$$
$$\times \exp\left(-\frac{(x - \xi)^2 + (y - \eta)^2 + (z - \zeta)^2}{4a(t - \tau)}\right) d\xi d\eta d\zeta d\tau. \tag{7.1.2}$$

However, if the heat source domain is bounded, integration intervals should be bounded too. If study time t exceeds action time t_f of the source, the upper integration limit should be expressed by the action time of the source, i.e. the upper limit is equal to min (t, t_f).

In general, integrals (7.1.2) can be calculated by quadrature formulas (Simpson's, Gauss's and other rules). In some cases, integrals of (7.1.2) type can be expressed as known functions. This depends on function type q_3, and integration can also be facilitated by the transformation from the fixed coordinate system into the moving coordinate system related to the source (Chap. 5).

It is worth noting that solutions (7.1.2) can be obtained in a similar way for bodies of basic geometric shapes (semi-infinite body, slab, plate, rod, solid cylinder). The difference is in formula (7.1.1), as it is necessary to choose the fundamental solution (Green's function) for a corresponding body. Boundedness of a body with heat-impermeable boundaries can be considered by the method of images.

7.2 Temperature Fields of Pulsed Power Sources

Temperature fields of instantaneous sources and constant power sources were considered earlier in Chap. 5. In this chapter, temperature fields of pulsed power sources will be analysed.

Sources of pulsed power are used for various types of welding (beam methods, resistance seam, pulsed arc welding, intermittent welding, etc.). For example, pulsed power in arc welding can be generated by changing the current and voltage in time or the varying wire feed rate. In the general case, the variable component of the effective arc power is also present when welding with a consumable electrode using DC welding sources because of current pulsation and periodic heat transfer by drops of the molten electrode. Therefore, the analysis of the influence of the shape and frequency of pulses on the geometry of the weld pool, the process of solidification of the liquid metal, the thermal cycles of the weld metal and the HAZ are of particular interest. The distribution of the heat of welding sources of pulsed power has been studied in many works (Akatsevich and Karkhin1998; Dowden et al. 1998; Gurevich 1969; Karkhin 1993; Karkhin and Michailov 1996; Karkhin and Akatsevich 1996; Karkhin et al. 1998; Ravi Vishnu et al. 1991; Ravi Vishnu and Easterling 1993; Ravi Vishnu 1996; Reddy et al. 2001; Saraev 1994; Saraev et al. 1997; Shigaev 1992; Tsai and Hou 1988).

The dependence of the effective power q on time t is determined by the welding power source, the control system, and can be described by different functions. For example, the periodic function $q(t)$ for welding with a pulsating arc is described by a piecewise-exponential function (Fig. 7.2a, b) (Wagner 1980). In the case of a trapezoidal pulse, the dependence $q(t)$ is described by five independent parameters: high pulse (peak) power q_p, low pulse (background) power $q_{b,}$ high pulse time (peak duration) $t_{p,}$ low pulse time (background duration) $t_{b,}$ and the slope-up and slope-down pulse time t_s (Fig. 7.2d). The other parameters (mean value of pulsed power q_0, cycle duration (total pulse time) t_c, frequency f) are expressed in terms of independent parameters:

$$q_0 = [q_p t_p + q_b t_b + (q_p + q_b) t_s] / t_c; \tag{7.2.1}$$

$$t_c = t_p + t_b + 2t_s; \tag{7.2.2}$$

$$f = 1/t_c. \tag{7.2.3}$$

When welding of magnesium alloys with a consumable electrode, the current, voltage and, consequently, power take on complex forms (Fig. 7.2e) (Rethmeier et al. 2001; Street 1990).

When solving a linear problem of heat conduction, we assume that:

- the body is homogeneous and unbounded;
- the material properties (thermal conductivity λ, thermal diffusivity a, volumetric heat capacity $c\rho$) and surface heat transfer coefficient α are independent of temperature;

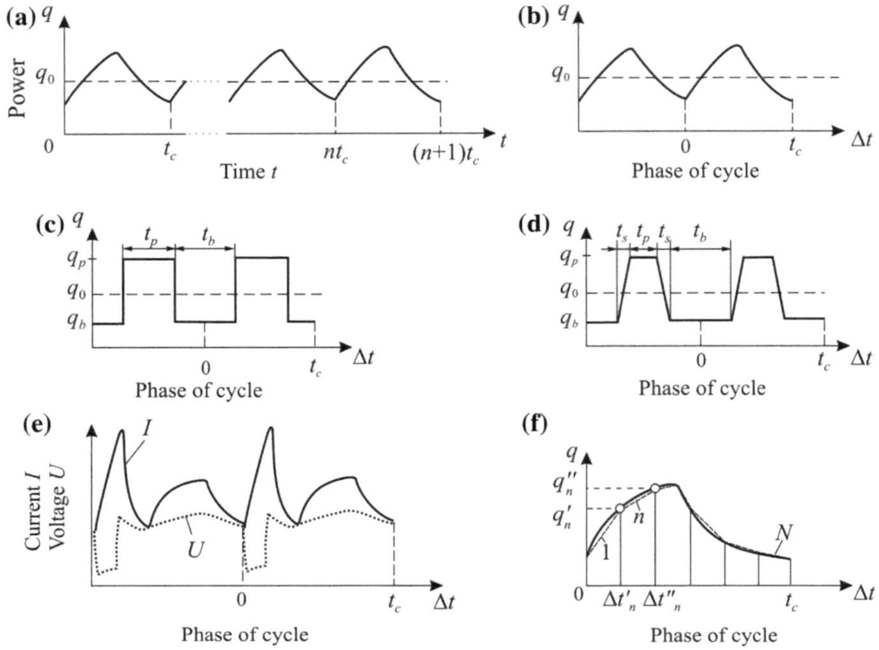

Fig. 7.2 Schematic diagrams of pulsed power

- the thermal state is steady (the number of cycles n and the time from the beginning of welding are large, the source moves along the x-axis rectilinearly with a constant speed v, Fig. 5.1.7);
- the centre of the source coincides with the origin of the moving coordinate system;
- initial temperature T_0 and ambient temperature T_∞ are constant and are taken as temperature reference point ($T_0 = T_\infty = 0$).

Let us consider four basic typical problems for the following source—heated body combinations: a point source on the surface of a semi-infinite body, a point source in an infinite slab, a line source in an infinite plate with surface heat transfer, a plane source in an infinite rod with surface heat transfer. First, we give the solution in the general form, and then for the rectangular and trapezoidal pulses (Fig. 7.2c, d). In case we have the fundamental solutions to these problems, we can take into account the boundedness of the body by means of the method of images, and the non-uniform distribution of the heat source—by means of the Green's function method (the source method).

The sequence of solving typical problems is as follows (Akatsevich and Karkhin1998; Karkhin 1993; Karkhin and Akatsevich 1996; Karkhin et al. 1998):

(1) function $q(t)$ is expanded into a Fourier series (which is possible, since it meets the Dirichlet's conditions):

$$q(t) = q_0 + \sum_{k=1}^{\infty} q_k(t) = \frac{a_0}{2} + \sum_{k=1}^{\infty} [a_k \cos(k\omega t) + b_k \sin(k\omega t)], \tag{7.2.4}$$

where $\omega = 2\pi/t_c$ is the circular frequency; a_k and b_k are the Fourier coefficients:

$$a_k = \frac{2}{t_c}\int_0^{t_c} q(t)\cos(k\omega t)dt;\ b_k = \frac{2}{t_c}\int_0^{t_c} q(t)\sin(k\omega t)dt\,(k = 0,\ 1, 2, \ldots); \tag{7.2.5}$$

(2) temperature increment T_k due to each kth harmonic of the source q_k is determined;
(3) increments T_k are summed over all k;
(4) the limit of the sum is determined with $n \to \infty$.

Thus, the temperature increment due to pulsed source $T(x, y, z, t)$ will be calculated as the sum of temperature increment $T_0(x, y, z, t)$ due to the source of constant power $q_0 = a_0/2$ and temperature increments $T_k(x, y, z, t)$ due to all kth power sources:

$$q_k(t) = a_k \cos(k\omega t) + b_k \sin(k\omega t) \quad (k = 1, 2, \ldots), \tag{7.2.6}$$

that is, the solution to the problem will be sought in the following form:

$$T(x, y, z, t) = T_0(x, y, z, t) + \sum_{k=1}^{\infty} T_k(x, y, z, t). \tag{7.2.7}$$

7.2.1 Point Source on a Semi-infinite Body

Let us assume that there is no surface heat transfer and find the solution to the problem in the form (5.1.107).

The solution to the problem $T_0(x, y, z, t)$ for constant power q_0 was obtained earlier (Sect. 5.1.3.1). The solution to the problem $T_k(x, y, z, t)$ for the kth harmonic q_k in the moving coordinate system can be obtained by replacing q_0 in (5.1.39) with q_k:

$$T_k(x, y, z, t) = \int_0^t \frac{2(a_k \cos(k\omega\tau) + b_k \sin(k\omega\tau))}{c\rho[4\pi a(t-\tau)]^{3/2}} \times \exp\left(-\frac{[x + v(t-\tau)]^2 + y^2 + z^2}{4a(t-\tau)}\right) d\tau. \tag{7.2.8}$$

Let us consider the steady state (time $t \to \infty$). Time will be denoted by Δt, $0 \le \Delta t \le t_c$, and it will be shifted to the beginning of the nth period ($n = \infty$) (Fig. 7.2). Using substitution $u = 4a(t - \tau)$, the expression (7.2.8) is reduced to the form:

$$T_k(x, y, z, \Delta t) = \frac{1}{2\lambda\pi^{3/2}}\exp\left(-\frac{vx}{2a}\right) \times \{[a_k\cos(k\omega\Delta t) + b_k\sin(k\omega\Delta t)]\, I_1 + [a_k\sin(k\omega\Delta t) - b_k\cos(k\omega\Delta t)]I_2\}, \quad (7.2.9)$$

where I_1 and I_2 are integrals (Prudnikov et al. 1986):

$$I_1 = \int_0^\infty u^{-3/2}\exp\left(-\left(\frac{v}{4a}\right)^2 u - \frac{R^2}{u}\right)\cos\left(\frac{k\omega}{4a}u\right)du = \frac{\sqrt{\pi}}{R}\exp\left(-\frac{vR}{2a}\sqrt{A_k + 1/2}\right)\cos\left(\frac{vR}{2a}\sqrt{A_k - 1/2}\right); \quad (7.2.10)$$

$$I_2 = \int_0^\infty u^{-3/2}\exp\left(-\left(\frac{v}{4a}\right)^2 u - \frac{R^2}{u}\right)\sin\left(\frac{k\omega}{4a}u\right)du = \frac{\sqrt{\pi}}{R}\exp\left(-\frac{vR}{2a}\sqrt{A_k + 1/2}\right)\sin\left(\frac{vR}{2a}\sqrt{A_k - 1/2}\right); \quad (7.2.11)$$

$$A_k = \frac{1}{2}\sqrt{1 + \left(\frac{4ka\omega}{v^2}\right)^2}.$$

With I_1 and I_2 in (7.2.9), we obtain:

$$T_k(x, R, \Delta t) = \frac{1}{2\pi\lambda R}\exp\left(-\frac{v}{2a}\left(x + R\sqrt{A_k + 1/2}\right)\right) \times \left[a_k\cos\left(k\omega\Delta t - \frac{vR}{2a}\sqrt{A_k - 1/2}\right) + b_k\sin\left(k\omega\Delta t - \frac{vR}{2a}\sqrt{A_k - 1/2}\right)\right]. \quad (7.2.12)$$

Finally, taking into account (5.1.41) and (7.2.12), we obtain the solution to the heat conduction problem (7.2.7) in the limiting state (Karkhin 1990, 1993; Karkhin et al. 1998; Karkhin and Khomich 2005):

$$T(x, y, z, \Delta t) = T_0(x, y, z)F_3(R, \Delta t), \quad (7.2.13)$$

where

$$T_0(x, y, z) = \frac{q_0}{2\pi\lambda R}\exp\left(-\frac{v(x+R)}{2a}\right); \tag{7.2.14}$$

$$F_3(R, \Delta t) = 1 + \sum_{k=1}^{\infty}\exp\left(-\frac{vR}{2a}\left[\sqrt{A_k + \frac{1}{2}} - 1\right]\right) \times \left[\frac{a_k}{q_0}\cos\left(k\omega\Delta t - \frac{vR}{2a}\sqrt{A_k - \frac{1}{2}}\right) + \frac{b_k}{q_0}\sin\left(k\omega\Delta t - \frac{vR}{2a}\sqrt{A_k - \frac{1}{2}}\right)\right]; \tag{7.2.15}$$

$$R = \sqrt{x^2 + y^2 + z^2}; \qquad A_k = \frac{1}{2}\sqrt{1 + \left(\frac{8\pi ka}{t_c v^2}\right)^2}.$$

As can be seen, the solution (7.2.13) is the product of function T_0, describing the steady temperature field due to a constant power source (formulas (7.2.14) and (5.1.41) are identical), and the function F_3, which takes into account the influence of the source pulsation on the temperature field. Function F_3 depends on the distance between the point under study and the source R and the moment (phase) of observation Δt. The obtained solution makes it possible to determine all the characteristics of the temperature field in a semi-infinite body with a surface point source of pulsed power with any pulse shape at any time (in any phase) Δt.

If the source moves in a massive (infinite) body, the expression (7.2.13) should be divided by 2, then the formula will allow us to obtain solutions for any pulsed power sources distributed in space in unbounded and bounded slabs by means of methods of sources and images.

For trapezoidal and rectangular pulses, it is convenient to select the phase of period Δt so that function $q(t)$ would be even (Fig. 7.2c, d). Then the Fourier coefficients (7.2.5) are as follows:

$$a_0 = 2\frac{q_p t_p + q_b t_b + (q_p + q_b)t_s}{t_c};$$

$$a_k = \frac{-2}{\pi^2 k^2}\frac{(q_p - q_b)t_c}{t_s}\sin\left(k\pi\frac{t_b + t_s}{t_c}\right)\sin\left(k\pi\frac{t_s}{t_c}\right); \qquad b_k = 0 \tag{7.2.16}$$

for trapezoidal pulses and

$$a_0 = 2\frac{q_p t_p + q_b t_b}{t_c}; \quad a_k = (-1)^k\frac{2(q_p - q_b)}{\pi k}\sin\left(k\pi\frac{t_p}{t_c}\right); \qquad b_k = 0 \tag{7.2.17}$$

for rectangular (square) pulses ($t_s = 0$).

The accuracy of calculating the temperature is determined by the number of retained terms of the series, that is, the accuracy of the approximation of the initial pulse shape. The more terms K of the series are taken into account, the more

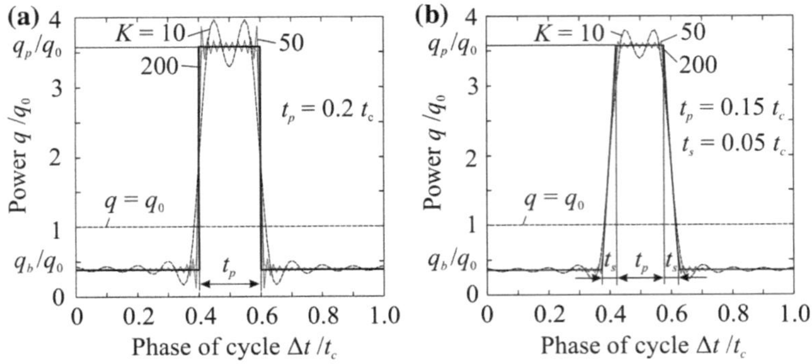

Fig. 7.3 Representation of rectangular (**a**) and trapezoidal (**b**) pulses as a Fourier series (K is the number of terms in the series)

accurately the pulse shape is approximated (Fig. 7.3). With the given accuracy, the number of terms in the series depends on the shape of the pulse. For example, for rectangular pulses, the closer time t_p or t_b to zero, the more terms of the series are required.

Virtually any power distribution $q(\Delta t)$ can be represented in the form of a piecewise linear curve (Fig. 7.2f), each nth piece of which passes through the given points $\Delta t_n', q_n'$ and $\Delta t_n'', q_n''$. The Fourier coefficients (7.2.5) for the nth piece are as follows:

$$
\begin{aligned}
a_{kn} &= \frac{1}{\pi k}\Big[\big(A_n + B_n\Delta t_n''\big)\sin\big(k\omega\Delta t_n''\big) + \frac{B_n}{k\omega}\cos\big(k\omega\Delta t_n''\big) \\
&\quad - \big(A_n + B_n\Delta t_n'\big)\sin\big(k\omega\Delta t_n'\big) - \frac{B_n}{k\omega}\cos\big(k\omega\Delta t_n'\big)\Big]; \\
b_{kn} &= \frac{1}{\pi k}\Big[-\big(A_n + B_n\Delta t_n''\big)\cos\big(k\omega\Delta t_n''\big) + \frac{B_n}{k\omega}\sin\big(k\omega\Delta t_n''\big) \\
&\quad + \big(A_n + B_n\Delta t_n'\big)\cos\big(k\omega\Delta t_n'\big) - \frac{B_n}{k\omega}\sin\big(k\omega\Delta t_n'\big)\Big]; \\
A_n &= \frac{q_n'\Delta t_n'' - q_n''\Delta t_n'}{\Delta t_n'' - \Delta t_n'}; \qquad B_n = \frac{q_n'' - q_n'}{\Delta t_n'' - \Delta t_n'}.
\end{aligned}
\tag{7.2.18}
$$

The Fourier series expansion (7.2.4) is represented as the sum of the Fourier series expansions of the separate linear nth pieces of the function $q(\Delta t)$ with coefficients (7.2.18). Then the temperature is determined by summing the temperature increments, which are calculated according to (7.2.13)–(7.2.15) over all linear pieces of curve $q(\Delta t)$ (Fig. 7.2f).

Taking into account (7.2.17), function F_3 for rectangular pulses takes on the following form

$$F_3(R, \Delta t) = 1 + 2\frac{(q_p - q_b)t_p}{q_p t_p + q_b t_b} \sum_{k=1}^{\infty} (-1)^k \frac{\sin(\pi k t_p / t_c)}{\pi k t_p / t_c}$$
$$\times \exp\left(-\frac{vR}{2a}\left[\sqrt{A_k + \frac{1}{2}} - 1\right]\right) \cos\left(k\omega\Delta t - \frac{vR}{2a}\sqrt{A_k - \frac{1}{2}}\right) \tag{7.2.19}$$

or in the criteria form:

$$F_3(\rho_3, \tau) = 1 + 2\varepsilon \sum_{k=1}^{\infty} (-1)^k \frac{\sin(\pi k \tau_p)}{\pi k \tau_p} \exp\left(-\rho_3\left[\sqrt{A_k + \frac{1}{2}} - 1\right]\right)$$
$$\times \cos\left(2\pi k \tau - \rho_3 \sqrt{A_k - \frac{1}{2}}\right); \quad A_k = \frac{1}{2}\sqrt{1 + (8\pi k \varphi)^2}; \tag{7.2.20}$$

$$\rho_3 = \frac{vR}{2a}; \quad \tau = \frac{\Delta t}{t_c}; \quad \tau_p = \frac{t_p}{t_c}; \quad \varphi = \frac{a}{t_c v^2}; \quad \varepsilon = \frac{(q_p - q_b)t_p}{q_p t_p + q_b t_b}. \tag{7.2.21}$$

Here ρ_3 is the dimensionless volumetric radius vector, τ is the dimensionless phase of the period, τ_p is the dimensionless pulse duration, φ is the dimensionless frequency, and ε is the dimensionless high pulse (peak) power (five parameters in total).

If the source is stationary ($v = 0$), the expression (7.2.19) is less complicated:

$$F_3(R, \Delta t) = 1 + 2\frac{(q_p - q_b)t_p}{q_p t_p + q_b t_b} \sum_{k=1}^{\infty} (-1)^k \frac{\sin(\pi k t_p / t_c)}{\pi k t_p / t_c}$$
$$\times \exp\left(-\sqrt{k\pi}\frac{R}{\sqrt{a t_c}}\right) \cos\left(k\omega\Delta t - \sqrt{k\pi}\frac{R}{\sqrt{a t_c}}\right). \tag{7.2.22}$$

Let us analyse the behaviour of function F_3 (7.2.20), which describes the effect of rectangular pulses on the temperature field (Fig. 7.4). For simplicity, we assume that the low pulse (background) power is $q_b = 0$, that is, the dimensionless high pulse (peak) power $\varepsilon = 1$.

It follows from (7.2.20) that the heat waves are propagated concentrically, and their magnitude depends on the distance to the source (ρ_3). The temperature field pulsates with the frequency of the source, but with the phase shift (time lag), which is as great as the distance from the source to the point under study ρ_3 and the frequency φ (Fig. 7.4a). Due to the time lag of heat waves, the maximum value of function $F_{3\max}$ is reached after the end of the pulse, and minimum value $F_{3\min}$—after the start. In the figure, the beginning of the pulse is shown by the open circle, and the end is shown by the solid circle. In all cases: $F_{3\max} \geq 1$ and $F_{3\min} \leq 1$.

The distance between the heat waves (wavelength) for each kth harmonic is expressed according to (7.2.19) in the explicit form:

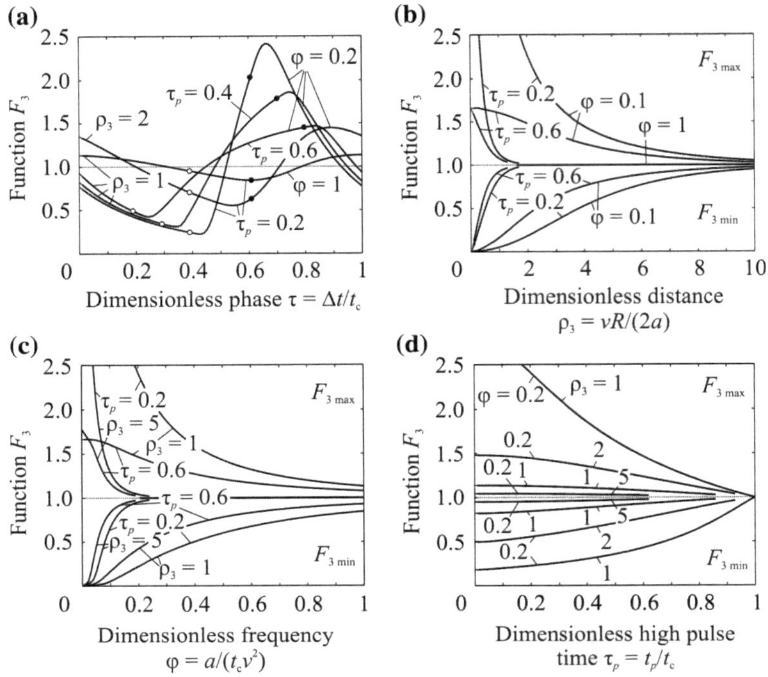

Fig. 7.4 Function accounting for the influence of source power pulsation F_3 over the dimensionless phase of cycle τ (**a**), the dimensionless distance from the point under study to source ρ_3 (**b**), dimensionless frequency φ (**c**) and dimensionless peak duration (**d**) at dimensionless pulse power $\varepsilon = 1$ of the point source on the semi-infinite solid (Karkhin 1993; Karkhin and Michailov 1996; Karkhin et al. 1998)

$$\Delta R = \frac{4\pi a}{v\sqrt{\frac{1}{2}\sqrt{1+\left(\frac{8\pi ka}{t_c v^2}\right)^2}-\frac{1}{2}}}. \quad (7.2.23)$$

If the travel speed is relatively small $\left(v^2 \ll 8\pi ka/t_c\right)$, expression (7.2.23) is simplified:

$$\Delta R = 2\sqrt{\pi a t_c/k} = 2\sqrt{\pi a/(kf)}. \quad (7.2.24)$$

It is obvious that the wavelength is inversely proportional to the square root of the frequency f. At frequency of 1–20 Hz, the wavelength in steel ($a = 8$ mm^2 s^{-1}) is equal to 10–2 mm.

It also follows from (7.2.19) and (7.2.20) that as the distance from the source increases, the pulsation amplitude decreases very rapidly (Fig. 7.4b). Thus, heat waves propagate from the source in concentric circles with a rapidly decaying amplitude, and the entire system moves simultaneously with the source at speed v. When

$(4\pi ka/t_c)^{1/2} \gg v$ (this condition is often satisfied even for the first harmonic), the amplitude of the pulsation decays according to $\exp\left(-R[\pi kf/a]^{1/2}\right)$, that is, the higher frequency f is, the faster the decay is (Fig. 7.4c).

The effect of pulse duration τ_p on the temperature field is of practical interest (Fig. 7.4d). It can be seen that function $F_{3\max}(\tau_p)$ is monotonically falling at any distance ρ_3 and frequency φ and the maximum amplitude for all points corresponds to instantaneous pulses $(\tau_p \to 0)$. We can also see that there is no thermal resonance when a set of points with a certain radius vector in which the maximum temperature oscillations take place corresponds to each mode of operation of the heat source of pulsed power (to each pulse duty factor $\tau_p > 0$) (Gurevich 1969). It should be noted that the apparent resonance mode is obtained if we assume that function F_3 reaches its maximum at the end of the pulse and its minimum—at the end of the pause, which contradicts the graphs in Fig. 7.4a.

The effect of the dimensionless high pulse (peak) power ε is obvious from Eqs. (7.2.20) and (7.2.21): function F_3 is proportional to ε, and ε varies from 0 $(q_p = q_b)$ to 1 $(q_b = 0)$.

The influence of the pulses shape on the temperature field (function F_3) will be considered for the pulses of the rectangular and sinusoidal shape with the same averaged power q_0 and amplitude (Fig. 7.5a). According to the energy conservation law, the duration of the rectangular pulse is $t_p = 2t_c/\pi$ $(\tau_p = 2/\pi)$. For the sinusoidal shape of the pulse, the Fourier coefficients are the following: $a_k/q_0 = 2(1–4k^2)$, $b_k = 0$. Figure 7.5b shows that the total amplitude $\Delta F_3 = F_{3\max} - F_{3\min}$ is larger for the rectangular pulse, and the shape of the pulse has a significant effect only on the temperature of the points located at a relatively small distance from the source. In the source zone $(\rho_3 \to 0)$, $F_{3\max}$ and $F_{3\min}$ are determined only by the maximum and minimum power values. Since the latter are identical for the cases under consideration, then $F_{3\max} \to t_p/t_c = \pi/2$ and $F_{3\min} \to 0$ both for sinusoidal and rectangular pulses. In the hard mode (t_p is small, q_p/q_0 is large), the difference in ΔF_3 values can be significant (Fig. 7.4d).

Frequency is one of the easily adjustable parameters. Let us consider, as an example, the effect of the frequency of the pulsed surface point source on the temperature state of a massive steel body with the following input data: $\lambda = 0.05$ W mm^{-1} K^{-1}, $a = 10$ mm^2 s^{-1}, $T_m = 1793$ K, $q_0 = 500$ W, $v = 5$ mm s^{-1}, $T_0 = 293$ K and $t_c = 0.5$ s, $t_p = 0.05$ s, $q_b = 0$.

Figure 7.6a, b, show the calculated surface "macrosections" when the source moves from left to right. At frequency $f = 2$ Hz, the width of the weld is varying, point A determines the minimum width of the weld, and point B determines the maximum width of the weld. It can be seen from Fig. 7.6c that the thermal cycles of these points are fundamentally different (the time taken to reach the peak temperature at the point under study is taken as the reference point): point A is heated twice to the melting temperature, point B is reheated only to 773 K. The solidification of the welding pool also differs significantly.

Frequency has a very strong effect on the cooling rate at the given temperature (Fig. 7.6c) and, therefore, on the retention time of the metal in a given temperature

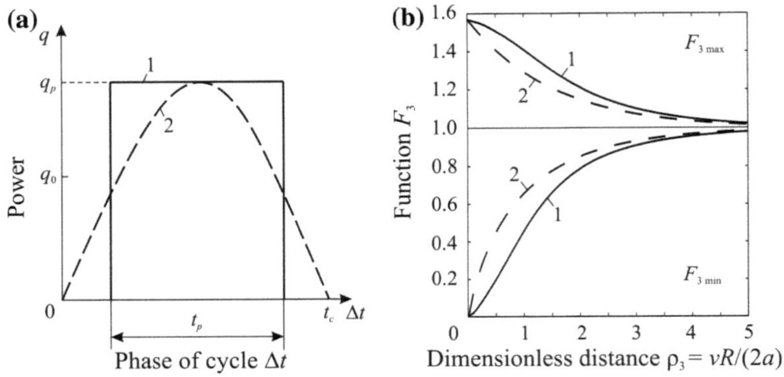

Fig. 7.5 Rectangular (1) and sinusoidal (2) power pulses (**a**) and the effect of the pulse shape on function F_3 (ρ_3) at frequency $\varphi = 0.2$ (**b**) (Karkhin 1993; Karkhin and Akatsevich 1996; Karkhin et al. 1998)

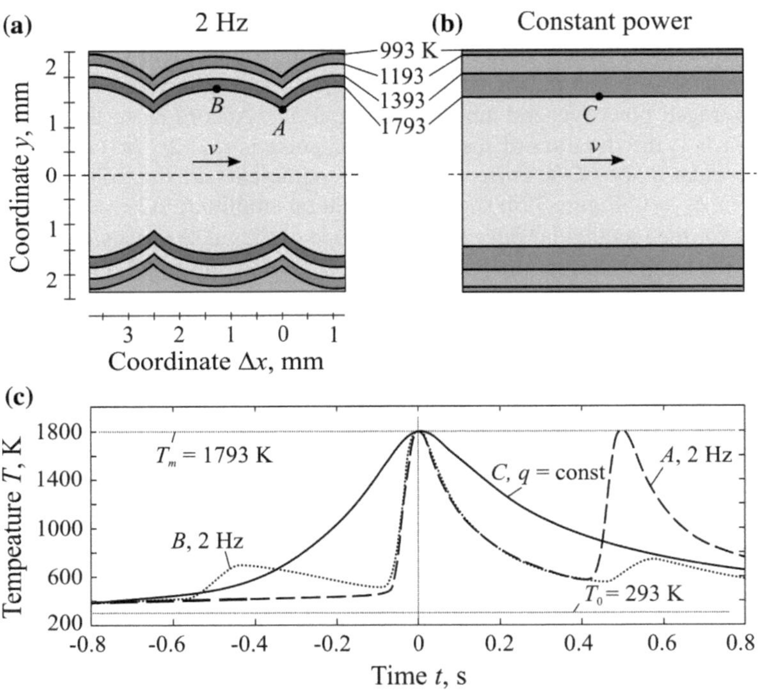

Fig. 7.6 Peak temperature field in the surface of a thick body in welding with periodic (**a**) and time-constant (**b**) power and thermal cycles of metal in the weld interface at points *A*, *B* and *C* (**c**) (Karkhin and Michailov 1996; Karkhin and Akatsevich 1996)

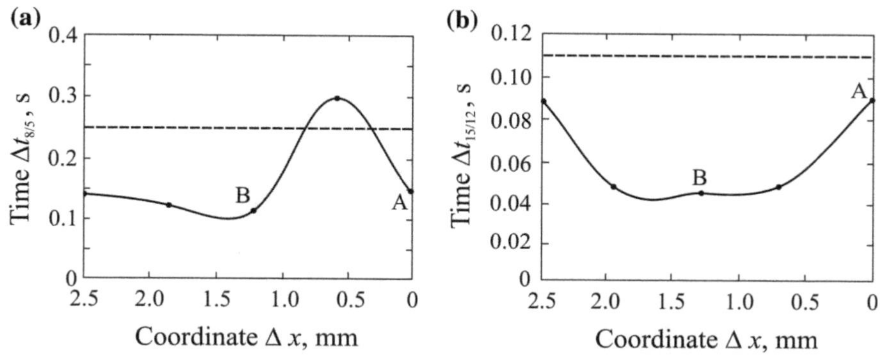

Fig. 7.7 Total time the material along the weld interface spent within temperature range 1073–773 K (800–500 °C) (**a**) and 1773–1473 K (1500–1200 °C) (**b**) (solid lines—periodic power source at frequency $f = 2$ Hz, dashed lines—time-constant power source) (Karkhin and Michailov 1996; Karkhin and Akatsevich 1996)

range. An important characteristic of the thermal cycle of steel is the cooling time in the range of 1073–773 K (800–500 °C) (total cooling time $\Delta t_{8/5}$ determines the microstructure, including hardness, susceptibility of the metal to cold cracking, etc.) and 1773–1473 K (1500–1200 °C) (time $\Delta t_{15/12}$ determines the grain growth, diffusion and other processes in the HAZ, see Chap. 13). At points A and B, time $\Delta t_{8/5} =$ 0.15 s and 0.11 s, time $\Delta t_{15/12} = 0.09$ s and 0.04 s, and at point C (constant power) the time is much higher—0.25 s and 0.11 s, respectively.

The total cooling time of the material in the characteristic temperature ranges varies along the weld interface (Fig. 7.7), so we can expect that local properties of the metal will also be variable.

Thus, using a pulsating source, it is possible to reduce $\Delta t_{12/15}$ (to reduce grain growth, etc.) and, without degrading other characteristics of the microstructure, it is possible to control the solidification process, etc. In other words, additional parameters of the pulsed power source (frequency and shape of the pulse) allow us to optimise the welding processes (see Chaps. 12 and 13).

In general, the cycle and pulse duration can be of any value. Formulas (7.2.13)–(7.2.15) allow us to calculate temperature fields during intermittent welding, including tacking. It can be seen from Fig. 7.8 that the temperature field at pulsed power differs significantly from the temperature field at the constant power source (Fig. 5.8d), even if their power q_0 is the same.

7.2.2 *Point Source in a Slab*

Let us assume that the surfaces of an infinite slab of thickness h are heat-impermeable, the moving point source of pulsed power is at points ξ, η, ζ, and the origin of the moving coordinate system is on the surface of the slab (Fig. 4.5). Using the

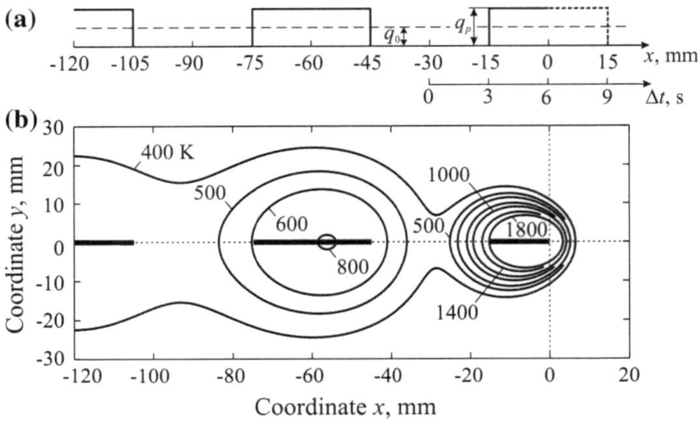

Fig. 7.8 Temperature field in intermittent welding of a thick body ($q_p = 10{,}000$ W, $q_0 = 5000$ W, $v = 5$ mm s^{-1}, $a = 8$ mm^2 s^{-1}, $\lambda = 0.04$ W mm^{-1} K^{-1}, $T_0 = 300$ K): the diagram of pulsed power (**a**) and isotherms in the surface of the body in the middle of pulse ($\Delta t = 6$ s) (**b**)

fundamental solution (7.2.13)–(7.2.15) and the method of images we can obtain the general solution for a slab in the form of the following series (Karkhin 1993; Karkhin and Michailov 1996; Karkhin and Akatsevich 1996; Karkhin et al. 1998):

$$T(x, y, z, \Delta t) = T_0(x, y, z) F_3(R, \Delta t), \tag{7.2.25}$$

where

$$T_0(x, y, z) = \frac{q_0}{4\pi\lambda} \exp\left(-\frac{v(x-\xi)}{2a}\right) \sum_{i=-\infty}^{\infty} \sum_{j=-1,1} \frac{1}{R_{i,j}} \exp\left(-\frac{vR_{i,j}}{2a}\right); \tag{7.2.26}$$

$$\begin{aligned} F_3(R, \Delta t) = 1 + &\sum_{i=-\infty}^{\infty} \sum_{j=-1,1} \frac{1}{R_{i,j}} \sum_{k=1}^{\infty} \exp\left(-\frac{vR_{i,j}}{2a}\sqrt{A_k + \frac{1}{2}}\right) \\ &\times \left[\frac{a_k}{q_0} \cos\left(k\omega\Delta t - \frac{vR_{i,j}}{2a}\sqrt{A_k - \frac{1}{2}}\right)\right. \\ &\left.+ \frac{b_k}{q_0} \sin\left(k\omega\Delta t - \frac{vR_{i,j}}{2a}\sqrt{A_k - \frac{1}{2}}\right)\right] \Big/ \sum_{i=-\infty}^{\infty} \sum_{j=-1,1} \frac{1}{R_{i,j}} \exp\left(-\frac{vR_{i,j}}{2a}\right); \end{aligned}$$

$$R_{i,j} = \sqrt{(x-\xi)^2 + (y-\eta)^2 + (z - j\zeta - 2ih)^2}. \tag{7.2.27}$$

If the pulses are rectangular (Fig. 7.2c), then

$$F_3(R, \Delta t) = 1 + 2\frac{(q_p - q_b)t_p}{q_p t_p + q_b t_b} \sum_{i=-\infty}^{\infty} \sum_{j=-1,1} \frac{1}{R_{i,j}} \sum_{k=1}^{\infty} (-1)^k \frac{\sin(\pi k t_p / t_c)}{\pi k t_p / t_c}$$

$$\times \exp\left(-\frac{vR_{i,j}}{2a}\sqrt{A_k+\frac{1}{2}}\right)\cos\left(k\omega\Delta t-\frac{vR_{i,j}}{2a}\sqrt{A_k-\frac{1}{2}}\right)$$

$$/\sum_{i=-\infty}^{\infty}\sum_{j=-1,1}\frac{1}{R_{i,j}}\exp\left(-\frac{vR_{i,j}}{2a}\right) \tag{7.2.28}$$

or in the criteria form:

$$F_3(\rho_3,\tau)=1+2\varepsilon\sum_{i=-\infty}^{\infty}\sum_{j=-1,1}\frac{1}{\rho_{3\,i,j}}\sum_{k=1}^{\infty}(-1)^k\frac{\sin(\pi k\tau_p)}{\pi k\tau_p}$$

$$\times \exp\left(-\rho_{3\,i,j}\sqrt{A_k+\frac{1}{2}}\right)\cos\left(2\pi k\tau-\rho_{3\,i,j}\sqrt{A_k-\frac{1}{2}}\right)$$

$$/\sum_{i=-\infty}^{\infty}\sum_{j=-1,1}\frac{1}{\rho_{3\,i,j}}\exp(-\rho_{3\,i,j});$$

$$\rho_{3\,i,j}=vR_{i,j}/(2a). \tag{7.2.29}$$

It should be mentioned that the convergence of series with respect to i depends on parameters ρ_3, φ, τ_p and τ, and the number of retained terms of the series must be greater with smaller ρ_3 and with τ_p closer to 0 or 1.

If thickness is relatively large ($h\to\infty$), then we have a point source model in a semi-infinite body (only one term remains for $i=0$ from the series). If we also take $\xi=\eta=\zeta=0$, we obtain a model of the point source at the origin of the coordinates on the surface of a semi-infinite body, and formulas (7.2.25)–(7.2.29) turn into (7.2.13)–(7.2.21).

7.2.3 Line Source in a Plate

The temperature field around a line source of pulsed power in an infinite plate of thickness h with surface heat transfer is described by the formula (Karkhin 1993; Karkhin and Akatsevich 1996; Karkhin et al. 1998):

$$T(x,y,\Delta t)=T_0(x,y)F_2(r,\Delta t), \tag{7.2.30}$$

where

$$T_0(x, y) = \frac{q_0/h}{2\pi\lambda}\exp\Bigl(-\frac{vx}{2a}\Bigr)\mathrm{K}_0\left(\frac{vr}{2a}\sqrt{1+\frac{4ab}{v^2}}\right); \tag{7.2.31}$$

$$F_2(r, \Delta t) = 1 + \sum_{k=1}^{\infty}\left\{\left[\frac{a_k}{q_0}\cos(k\omega\Delta t) + \frac{b_k}{q_0}\sin(k\omega\Delta t)\right]\mathrm{Re}\,\mathrm{K}_0(u)\right.$$
$$\left.-\left[\frac{a_k}{q_0}\sin(k\omega\Delta t) - \frac{b_k}{q_0}\cos(k\omega\Delta t)\right]\mathrm{Im}\,\mathrm{K}_0(u)\right\}/\mathrm{K}_0\left(\frac{vr}{2a}\sqrt{1+\frac{4ab}{v^2}}\right); \tag{7.2.32}$$

$$r = \sqrt{x^2+y^2}; \quad b = \frac{2\alpha}{c\rho h}; \quad u = \frac{vr}{2a}\sqrt{1+\frac{4ab}{v^2}+\mathrm{i}\frac{4k\omega a}{v^2}}; \quad \mathrm{i} = \sqrt{-1}.$$

Here $\mathrm{Re}\,\mathrm{K}_0(u)$ and $\mathrm{Im}\,\mathrm{K}_0(u)$ are real and imaginary components of the cylindrical function of the imaginary argument.

If the pulse is rectangular (Fig. 7.2c), function F_2 in its dimensionless form is described by the expression:

$$F_2(\rho_2, \tau) = 1 + 2\varepsilon\sum_{k=1}^{\infty}(-1)^k\frac{\sin(\pi k\tau_p)}{\pi k\tau_p}$$
$$\times\left[\mathrm{Re}\,\mathrm{K}_0(u)\cos(2\pi k\tau) - \mathrm{Im}\,\mathrm{K}_0(u)\sin(2\pi k\tau)\right]/\mathrm{K}_0(\rho_2, \mu_2); \tag{7.2.33}$$

$$\rho_2 = \frac{vr}{2a}; \quad \mu_2 = \sqrt{1+\frac{4ab}{v^2}}; \quad u = \rho_2\sqrt{\mu_2^2 + \mathrm{i}8\pi k\varphi}.$$

Function F_2 can be expressed in the integral form (Karkhin et al. 1998):

$$F_2(\rho_2, \tau) = 1 + 2\varepsilon\sum_{k=1}^{\infty}(-1)^k\frac{\sin(\pi k\tau_p)}{\pi k\tau_p}\int\limits_0^{\infty}\exp\left(\rho_2\mu_2\left[1-\sqrt{B_k+\frac{1}{2}\mathrm{ch}(t)}\right]\right)$$
$$\times\cos\left(2\pi k\tau - \rho_2\mu_2\sqrt{B_k - \frac{1}{2}\mathrm{ch}(t)}\right)dt/\left[\exp(\rho_2\mu_2)\mathrm{K}_0(\rho_2\mu_2)\right]; \tag{7.2.34}$$

$$B_k = \frac{1}{2}\sqrt{1+\left(\frac{8\pi ka}{t_c v^2\mu_2^2}\right)^2}.$$

The structure of the integral is close to the integral representation of the modified Bessel function K_0, therefore, the algorithm for the numerical determination of K_0 function can be taken as the basis in calculations.

7.2.4 Plane Source in a Rod

One-dimensional temperature field of a moving plane source of pulsed power in an infinite rod with a cross section of area A and perimeter p is described by the formula (Karkhin and Michailov 1996; Karkhin and Akatsevich 1996; Karkhin et al. 1998):

$$T(x, \Delta t) = T_0(x) F_1(x, \Delta t), \tag{7.2.35}$$

where

$$T_0(x) = \frac{q_0/A}{c\rho v \mu_1} \exp\left(-\frac{v(x + |x|\mu_1)}{2a}\right); \tag{7.2.36}$$

$$\begin{aligned} F_1(\rho_1, \tau) = 1 + \sum_{k=1}^{\infty} \frac{1}{2C_k} \exp\left(-\rho_1\mu_1\left[\sqrt{C_k + \frac{1}{2}} - 1\right]\right) \\ \times \left[\left(\frac{a_k}{q_0}\sqrt{C_k + \frac{1}{2}} - \frac{b_k}{q_0}\sqrt{C_k - \frac{1}{2}}\right)\cos\left(2\pi k\tau - \rho_1\mu_1\sqrt{C_k - \frac{1}{2}}\right)\right. \\ \left. + \left(\frac{a_k}{q_0}\sqrt{C_k - \frac{1}{2}} + \frac{b_k}{q_0}\sqrt{C_k + \frac{1}{2}}\right)\sin\left(2\pi k\tau - \rho_1\mu_1\sqrt{C_k - \frac{1}{2}}\right)\right]; \end{aligned}$$

$$\rho_1 = \frac{v|x|}{2a}; \quad \mu_1 = \sqrt{1 + \frac{4a\alpha p}{v^2 c\rho A}}; \quad C_k = \frac{1}{2}\sqrt{1 + \left(\frac{8\pi k a}{t_c v^2 \mu_1^2}\right)^2}. \tag{7.2.37}$$

If the pulses are rectangular (Fig. 7.2c), then

$$\begin{aligned} F_1(\rho_1, \tau) = 1 + 2\varepsilon \sum_{k=1}^{\infty} (-1)^k \frac{\sin(\pi k\tau_p)}{\pi k\tau_p} \exp\left(-\rho_1\mu_1\left[\sqrt{C_k + \frac{1}{2}} - 1\right]\right) \frac{1}{\sqrt{2C_k}} \\ \times \cos\left(2\pi k\tau - \rho_1\mu_1\sqrt{C_k - \frac{1}{2}} - \arctan\sqrt{\frac{2C_k - 1}{2C_k + 1}}\right). \end{aligned} \tag{7.2.38}$$

The solutions for four main calculation schemes (7.2.13), (7.2.25), (7.2.30) and (7.2.35) have the same structure:

$$T(X, \Delta t) = T_0(X) F(R, \Delta t), \tag{7.2.39}$$

where T_0 is the steady temperature field of constant-power source q_0, X is the coordinate vector of the point under study, R is the distance from the point under study to the source, and Δt is the observation phase.

All the characteristics of the periodicity of the source are included in function F, which can be considered as a dimensionless temperature, which is the ratio of the temperatures due to the pulsed power source and the constant power source.

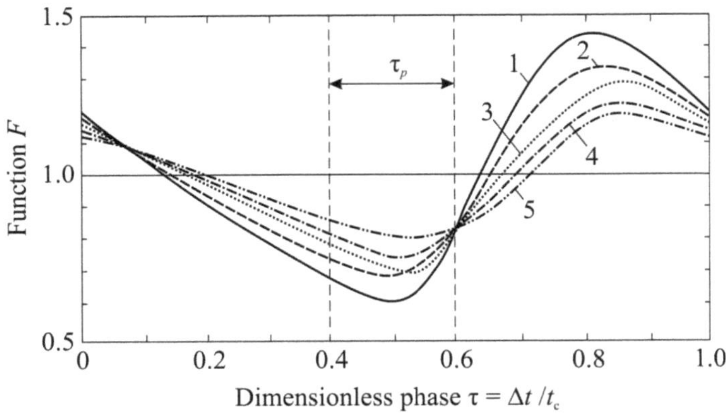

Fig. 7.9 Dependences of the function accounting for the influence of source pulsation on the dimensionless phase of cycle τ at a distance of $\rho = 1.0$ from the source, frequency $\varphi = 0.5$, peak duration $\tau_p = 0.2$ and pulse power $\varepsilon = 1.0$ for (1) semi-infinite solid; (2) slab ($vh/(2a) = 0.75$); (3) plate ($\mu_2 = 1.2$); (4) plate ($\mu_2 = 1$) and (5) rod ($\mu_1 = 1$) (Karkhin and Akatsevich 1996; Karkhin et al. 1998)

Let us analyse function F for rectangular pulses. It can be seen from formulas (7.2.20), (7.2.29), (7.2.34) and (7.2.38) that it depends only on five dimensionless independent parameters: radius vectors ρ ($vR/(2a)$, $vr/(2a)$, $v|x|/(2a)$), pulse duration τ_p, frequency φ, pulse energy ε and observation phase τ.

The dependence of F on phase τ for various combinations of the source—heated body is shown in Fig. 7.9. It can be seen that the more massive the body (more directions of heat removal) and the greater the surface heat transfer we have, the greater the deviation of F from 1 is. The analysis of the cosine argument in formulas (7.2.20), (7.2.29) and (7.2.38) proves that the temperature pulsates with the frequency of source k/t_c, but with a phase shift, which is as great as the distance from the point under study to the source and the frequency. Due to the time lag of heat waves, the minimum value of function F is reached after the pulse start ($\tau = 0.4$), and the maximum value is reached after its termination ($\tau = 0.6$) (Fig. 7.9).

The effect of relative energy ε of the pulse on function F results from its definition (7.2.21), and it varies from 0 (when $q_p = q_b$) to 1 (when $q_b = 0$). The results of the analysis of the effect of ρ, φ, τ_p and the shape of the pulse were obtained earlier in Sect. 7.2.1 for the point source on a semi-infinite body (Fig. 7.4), and they remain qualitatively valid also for other cases under consideration. For example, it is proved that as the distance from the source increases (increasing ρ), the temperature range decreases exponentially. If the pulses are hard ($\tau_p \to 0$), then $\sin(\pi k \, \tau_p)/(\pi k \, \tau_p) \to 1$ and the series are simplified.

Thus, the obtained analytical solutions to the basic typical problems (a point source on the surface of a semi-infinite body, a point source in an infinite slab, a line source in an infinite plate with surface heat transfer, a plane source in an infinite rod with surface heat transfer) allow us to calculate the steady temperature fields taking into consideration real forms and frequencies of pulses. These solutions are fundamental, and using them as the base, we can obtain solutions for spatially distributed sources in bounded bodies by means of methods of sources and images.

7.2.5 Distributed Pulsed Power Sources

Earlier in Sect. 5.2, using the source method, solutions were obtained for moving distributed sources of constant power based on fundamental solutions for concentrated sources. Similarly, the heat conduction problem for moving distributed sources of pulsed power can be solved using fundamental solutions for moving concentrated sources (Sects. 7.2.1–7.2.4). To do this, any spatially distributed source is represented as a set of elementary sources, the increment of temperature due to an elementary source is calculated and the increments are summed up. The resulting sum will be the temperature increment due to the entire distributed source. If the body is bounded, the method of images is additionally used to fulfill the boundary conditions.

Let us consider the sequence of the solution to the problem using the example of a moving source of pulsed power on the surface of a thick body. The radial distribution of power density is described by the law of normal distribution (Akatsevich and Karkhin 1998; Karkhin et al. 1998):

$$q_2(r, \Delta t) = \frac{C}{\pi} q(\Delta t) \exp\left(-Cr^2\right), \tag{7.2.40}$$

where r is the radial distance from the centre of the source, C is the concentration factor and q is the power of the source.

We divide the distributed source (the entire surface of the body) along radius r from 0 to ∞ and along angle φ from 0 to 2π into elementary point sources (Fig. 7.10). The power of an elementary source with the coordinates ρ, φ is equal to

$$dq(\rho, \Delta t) = \frac{C}{\pi} q(\Delta t) \exp\left(-C\rho^2\right) d\varphi \rho d\rho. \tag{7.2.41}$$

The increment of temperature due to this source is described by Eqs. (7.2.13)–(7.2.15). Then the temperature field of the entire normally distributed source is determined by summing the temperature increments of all elementary sources (Akatsevich and Karkhin 1998; Karkhin et al. 1998):

$$\begin{aligned} T(x, y, z, \Delta t) - T_0 = {} & \frac{q_0 C}{2\pi^2 \lambda} \int_0^\infty \int_0^{2\pi} \frac{1}{R} \exp\left(-\frac{v(x - \rho\cos\varphi + R)}{2a} - C\rho^2 \right) \\ & \times F_3(R, \Delta t) d\varphi \rho d\rho; \\ R = {} & \sqrt{(x - \rho\cos\varphi)^2 + (y - \rho\sin\varphi)^2 + z^2}, \end{aligned} \tag{7.2.42}$$

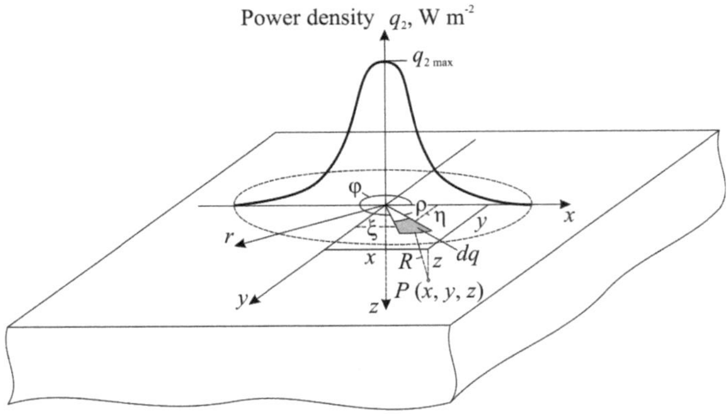

Fig. 7.10 Model for a normally distributed circular source on a semi-infinite solid (Akatsevich and Karkhin 1998)

and function $F_3(R, \Delta t)$ is determined by Eq. (7.2.15).

If the body is limited in thickness, that is, it is a slab, then the formula for calculating the temperature becomes somewhat more complicated (Akatsevich and Karkhin 1998; Karkhin et al. 1998).

Figure 7.11 shows dynamics of melting and solidification of the pool when welding manganese steel of 12 mm in thickness. The results of the experiment (Ravi Vishnu and Easterling 1993) were taken as the basis. Properties of steel are the following: $a = 9.4017$ mm^2 s^{-1}, $\lambda = 0.033$ W mm^{-1} K^{-1}, $T_m = 1773$ K, $m = 0.00067$ K^{-1} (see Eq. (3.5.1)). The gas tungsten arc welding conditions: $q_p = \eta_h UI = 0.8 \times 13.5 \times 175 = 1890$ W, $q_b = 0.8 \times 9.5 \times 25 = 190$ W, $q_0 = 489$ W, $C = 0.133$ mm^{-2}, $t_p = 0.18$ s, $t_b = 1.23$ s, $t_s = 0.105$ s, $t_c = 1.62$ s, $v = 1.1$ mm s^{-1}, $T_0 = 293$ K.

The lower surface of the thick plate has practically no effect, therefore the temperature can be calculated by formulas (7.2.42), (7.2.15) and (7.2.16) for trapezoidal pulses (Fig. 7.2d). Below are the temperature fields and thermal cycles calculated with allowance for the temperature dependence of the thermal conductivity (coefficient m) according to formula (3.5.5).

Figure 7.11b shows the longitudinal section ($y = 0$) in different phases of the Δt of nth cycle. The open circles indicate the current position of the source centre. The beginning of the $n - 2$ cycle is taken as the origin of time t and coordinate x. The axes of some crystal are calculated provided that they grow normally to the solidification front (isotherm T_m). It can be seen that the crystal change the direction of growth after each repeated melting and the growth can be opposite in direction to the welding direction, which is confirmed by the experiments (Ravi Vishnu and Easterling 1993). Unlike welding with a constant-power source, a liquid pool exists for a limited time, and some areas of the weld joint undergo repeated melting.

The rate of solidification of the molten metal of the weld pool is of particular interest. Figure 7.12 shows changes in the position of the length of the pool tail (isotherm T_m) and the rate of solidification of metal v_s on x-axis $\left(v_s = -\partial L / \partial t + v\right)$.

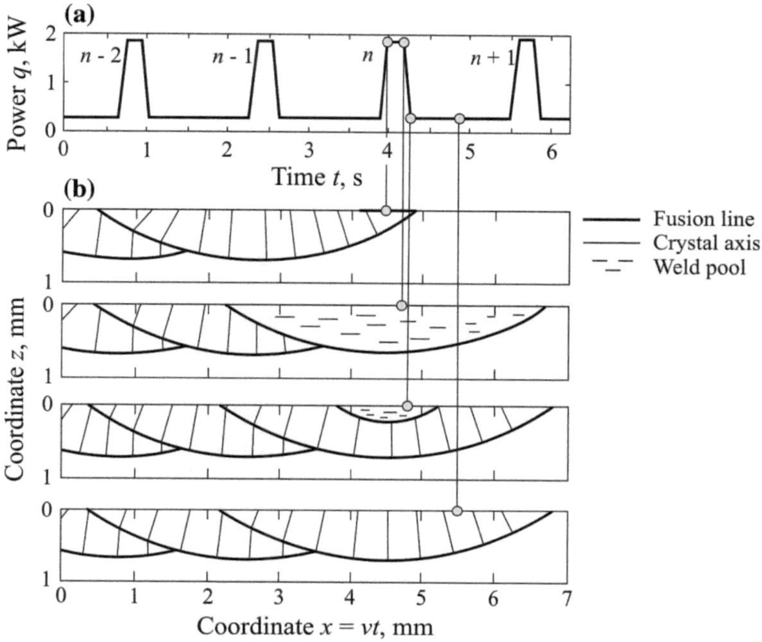

Fig. 7.11 Periodic power of welding arc (**a**) and central longitudinal section of weld and pool ($y = 0$) (**b**) at different instances of time (Akatsevich and Karkhin 1998; Karkhin et al. 1998)

Fig. 7.12 The time-dependent length of weld pool tail L and the solidification rate v_s (Akatsevich and Karkhin 1998; Karkhin et al. 1998)

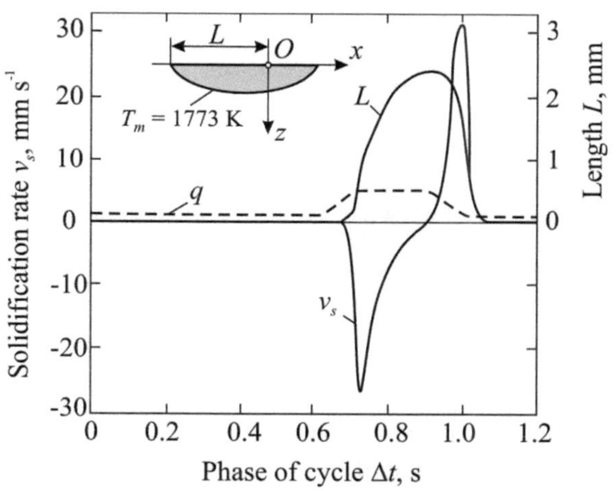

It can be seen that the rate of solidification can be an order of magnitude higher than the welding speed v. This means that when welding with sources of pulsed and constant power, high-temperature processes (grain growth, chemical heterogeneity, etc.) can be quite different.

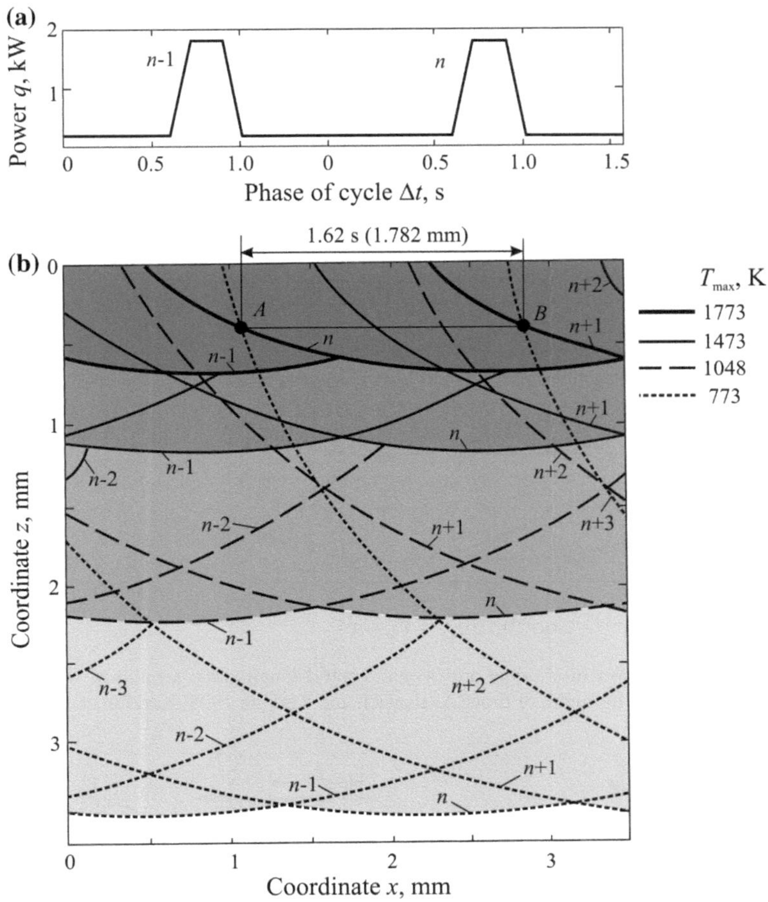

Fig. 7.13 Periodic power of welding arc (**a**) and isotherms of peak temperature T_{max} in the central longitudinal section of weld and HAZ (**b**) (number of isotherm corresponds to the number of forming pulse (Akatsevich and Karkhin 1998; Karkhin et al. 1998)

Figure 7.13 shows the calculated "macrosections" with the position of different isotherms T_{max} from the corresponding pulse. The position of the source at the beginning of the $n-1$ pulse is taken as the origin of x coordinate. The calculated depth of fusion is 0.61–0.71 mm, the experimental depth is 0.61 mm (Ravi Vishnu and Easterling 1993). The width of the zone of coarse grains ($T_{max} = 1473$–1773 K) is 0.86–1.23 mm.

The weld metal and the HAZ metal undergo a complex thermal cycle (Fig. 7.14). The intersection of a transverse plane containing point A or B by the source is taken as the origin of time t. The thermal cycle and, consequently, the microstructure of the weld metal and the HAZ metal are periodically repeated along the weld interface with step $vt_c = 1.782$ mm. For example, points A and B, located at the weld interface at a

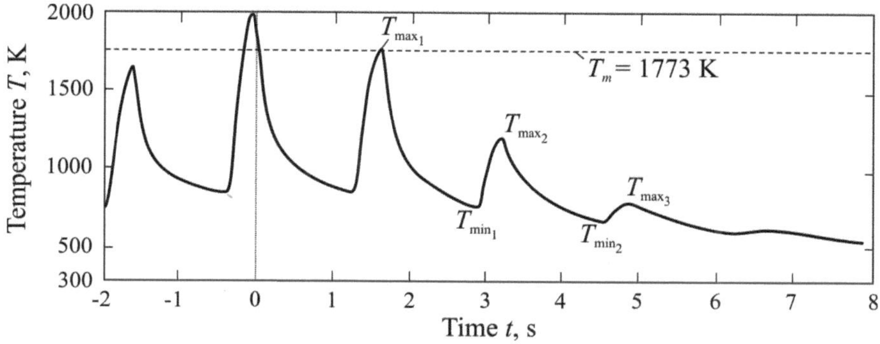

Fig. 7.14 Thermal cycle at the remelting boundary in points *A* and *B* (Akatsevich and Karkhin 1998; Karkhin and Michailov 1996; Karkhin et al. 1998) (see Fig. 7.13)

depth of 0.34 mm, undergo multiple heating and cooling: after the last melting ($T_{\max 1} = T_m = 1773$ K), the metal cools below the temperature of the start of martensite formation M_s ($T_{\min 1} = 733$ K $< M_s = 753$ K), then at the next pulse it is heated up to temperature $T_{\max 2} = 1146$ K (higher than $A_3 = 1133$ K) and cooled down to $T_{\min 2} = 650$ K, etc. Therefore, it can be expected that after cooling the metal at points *A* and *B* has a microstructure of partially tempered martensite (Karkhin et al. 1998).

The analysis of dependence of the thermal cycle on the parameters of the welding by the pulsed source showed that the retention time of the metal at high temperatures ($T > 1500$ K) is determined mainly by pulse and pause duration, the cooling time at low temperatures (in the range 1073–773 K) is determined by the low pulse (background) power, and the size of the fusion zone—by the average power (Akatsevich and Karkhin 1998).

It should be noted that the technique for calculating the steady-state temperature field described here is suitable for any spatially distributed source of pulsed power.

The solution to the heat conduction problem allowing for other complex physical phenomena (convection of the liquid metal of the pool, arc pressure, deflection of the free surface of the pool) requires the use of numerical methods (Reddy et al. 2001; Sarayev 1997).

7.3 Weaving Heat Sources

Welding by weaving heat sources is more flexible than welding with a rectilinear source trajectory, since it is characterised by a large number of independent parameters by which heat input can be regulated in the welded body. The application of this welding technique makes it possible in some cases to exclude or reduce additional heating (Makhnenko and Kravtsov 1976).

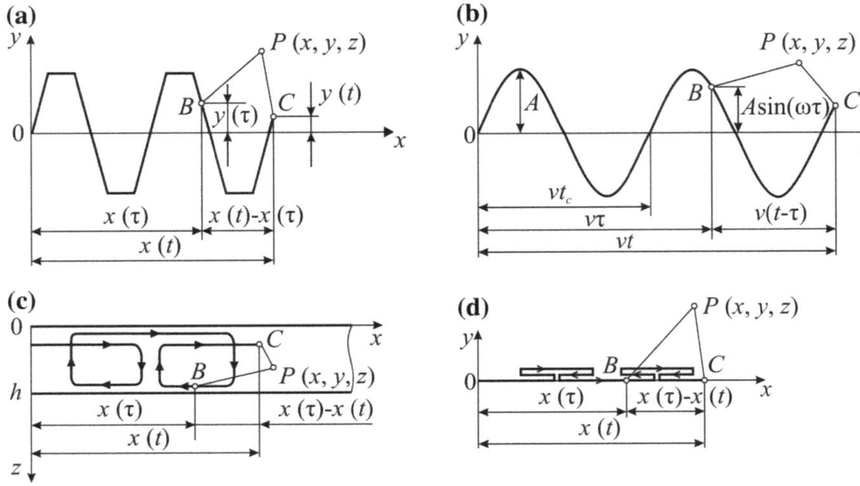

Fig. 7.15 Welding weaving patterns: trapezoidal (**a**), sinusoidal (**b**), back-and-forth square with penetration (**c**) and back-and-forth along weld (**d**)

When welding, the centre of the heat source can periodically change the mechanical trajectory along the weld centreline, and the source parameters (its power and distribution) can also change. Oscillations may be transverse (Fig. 7.15a, b), longitudinal (Fig. 7.15d) or longitudinal-transverse (Fig. 7.15c). In the general case, the trajectory of the source can be quite complicated.

Depending on the shape of the trajectory, the parameters of the heat source and the assumptions, the solution to the heat conduction problem of the type (7.1.2) may take a different form.

Let us first consider the temperature field when welding by a surface heat source with transverse oscillations (Fig. 7.15a, b). The cases with the high frequency of transverse oscillations, when a liquid pool exists over the entire width of the source action and the weld width is constant, were considered above (Sects. 5.2.3.10, 5.2.4.1 and 5.2.4.2). Here, the characteristics of the temperature fields at a relatively low frequency of transverse oscillations of the source will be analysed.

Independent parameters of the surfacing conditions according to the pattern in Fig. 7.16a are effective power q, speed v, length of weld bead L, time of surfacing one bead t_* and surfacing pitch H (Makhnenko 1966).

The temperature at arbitrary point P of a thick body at time t from the start of surfacing by an oscillating electrode is obtained by superimposing the heat equalisation processes during the action of moving concentrated sources on straight lines 0, 1, 2, …, N in the corresponding periods of time (Makhnenko and Kravtsov 1976):

$$T(x, y, z, t) - T_0 = \sum_{n=0}^{N}{}' \frac{q}{2\pi\lambda R_n} \exp\left(-\frac{v(x + R_n)}{2a}\right)$$

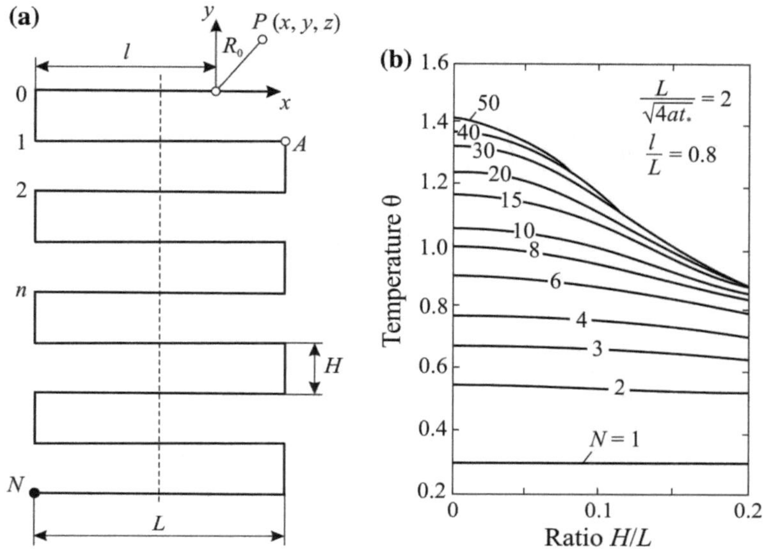

Fig. 7.16 Ladder weave pattern for surface welding of thick plate (**a**) and dimensionless temperature θ at point A as a function of H/L ratio (**b**) (Makhnenko and Kravtsov 1976)

$$\times \left[\psi_3(R_n, t_n + t_*) - \psi_3(R_n, t_n)\right], \tag{7.3.1}$$

where R_n is the distance in space from the centre of the nth source to point P at time t; t_n is the duration of heat diffusion from the nth source, $t_n = (n - 1 + l/L)t_*$; t_* is the time of surfacing one bead, $t_* = (L + H)/v$; ψ_3 is the heat saturation function for a moving point source on a semi-infinite body (5.1.108). The prime for the sum sign means that $\psi(R_n, t_n) = 0$ for $n = 0$.

Figure 7.16b shows the curves of the dimensionless temperature $\theta = 2\pi\lambda L(T - T_0)/q$ at point A as a function of the dimensionless parameters for a different number of oscillations N. It is seen that the temperature can be controlled by changing the parameters of the source trajectory.

Allowance for the surface heat transfer and the body's boundedness in thickness complicates the solution to the problem. For a long surfacing ($N \to \infty$), dimensionless parameters vh/a, H/h and L/h have the most significant effect on the temperature field. The influence of heat transfer (Biot number $\mathrm{Bi} = \alpha h/\lambda$) is insignificant near the source of surfacing (Makhnenko et al. 1971; Makhnenko and Kravtsov 1976).

Technically, sinusoidal transverse oscillations are the simplest (Fig. 7.15b):

$$y(t) = A\sin(\omega t), \quad \omega = 2\pi/t_c. \tag{7.3.2}$$

The temperature field due to the normally distributed source moving rectilinearly over the surface of an h thick slab for time t at constant speed v with transverse oscillations of amplitude A and a period of time t_c is easily obtained by the source method with Eq. (5.2.17) (Karkhin et al. 2011; Yermakov et al. 2006) :

$$T(x, y, z, t) - T_0 = \frac{q/h}{4\pi\lambda}\int\limits_0^t \frac{1}{t + t_0 - \tau}\exp\left(-\frac{(x - v\tau)^2 + [y - A\sin(\omega\tau)]^2}{4a(t + t_0 - \tau)}\right) \times \left[1 + 2\sum_{i=1}^{\infty}\cos\left(\frac{\pi i z}{h}\right)\exp\left(-i^2\pi^2\frac{a(t - \tau)}{h^2}\right)\right]d\tau, \quad (7.3.3)$$

where $t_0 = r_e^2/(4a)$, r_e is the normal radius of the source.

If thickness of the slab is relatively large ($h \to \infty$), the solution to the problem is obtained using formula (5.2.9) (Danilov and Marmur 1980):

$$T(x, y, z, t) - T_0 = \frac{q}{4c\rho(\pi a)^{3/2}}\int\limits_0^t \frac{1}{(t + t_0 - \tau)(t - \tau)^{1/2}} \times \exp\left(-\frac{(x - v\tau)^2 + [y - A\sin(\omega\tau)]^2}{4a(t + t_0 - \tau)} - \frac{z^2}{4a(t - \tau)}\right)d\tau. \quad (7.3.4)$$

Formulas (7.3.3) and (7.3.4) are obtained for the fixed coordinate system. For the steady state ($t \to \infty$), it is more convenient to use a moving system and integrate over the time from t to 0. The physical meaning of this technique is based on the fact that after many oscillations of the moving source, the temperature field around the source depends only on the last oscillations (the effect of the beginning of the weld can be neglected).

Figure 7.17 shows the peak temperature field and the thermal cycles of surface points A and B located at the weld interface during the plasma surfacing of a thick body. It can be seen from Fig. 7.17a that the weld interface is wavy with the frequency of the source. The depth of fusion is variable and asymmetrical relative to the longitudinal weld plane (Fig. 7.17b). Thermal cycles of the points in the weld interface differ significantly in the rate of heating and cooling and retention time at a high temperature (Fig. 7.17c).

If the frequency of the transverse oscillations is relatively high, we can assume that in the cross section the heat is input simultaneously. The distribution of the heat in the cross section can be easily found from the energy balance condition. For example, if the surface source oscillates about longitudinal x-axis by the sinusoid with amplitude A, the power density in the cross section is described by the hyperbolic law:

$$q_1(y) = \frac{2q}{\pi\sqrt{A^2 - y^2}}. \quad (7.3.5)$$

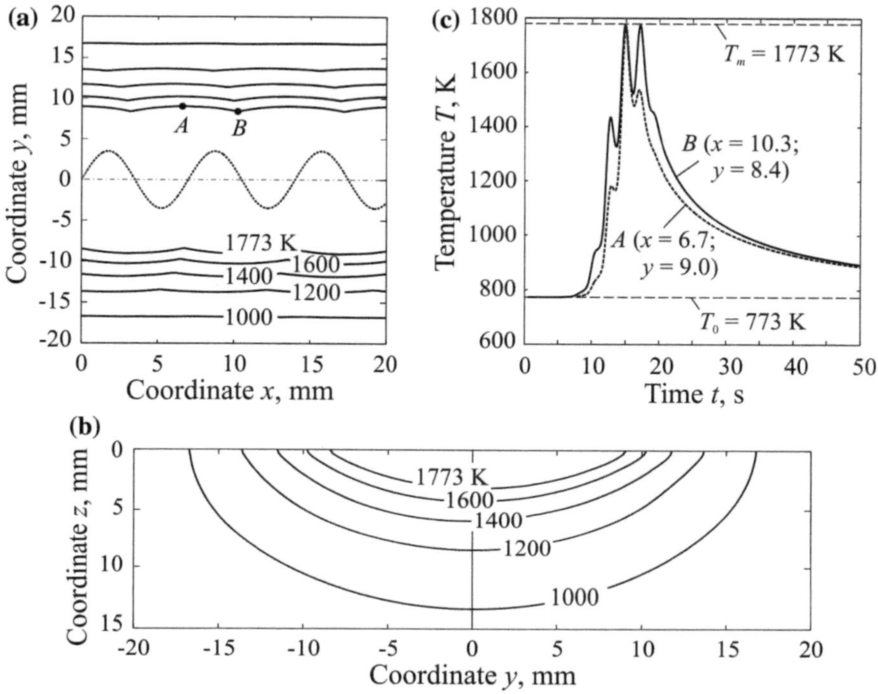

Fig. 7.17 Arc trajectory (dotted line) and peak temperature field in the surface (**a**) and in cross-section with point *A* (**b**) and thermal cycles at points *A* and *B* in weld interface (**c**) in plasma surfacing: $q = 3640$ W, $r_e = 8.15$ mm, $v = 3$ mm s^{-1}, step $vt_c = 7$ mm, amplitude $A = 3.5$ mm, $T_0 = 773$ K, carbon steel ($a = 8$ mm^2 s^{-1}, $\lambda = 0.04$ W mm^{-1} K^{-1}, $T_m = 1773$ K) (Karkhin et al. 2011)

The production of thick-walled steel structures is effective with the use of multi-arc welding. Arcs can have different parameters and, by programming them, it is possible to optimise welding processes more flexibly (thermal cycles, kinetics of strains and stresses, etc.). For example, with two-arc welding of butt joints, the control parameters are: the powers of both arcs, the distance between them, the direction and the speed of the first arc (Fig. 7.18a) (Serenko et al. 1993). The trajectory of the first arc is reciprocating in the plane of the welded edges with the change in the z coordinates, depending on the cyclogram of the current and arc voltage. The second arc moves along the gap on the surface ($z = 0$) at welding speed v.

Temperature at any point of the body at any time t can be found using the method of sources. To do this, trajectories of both arcs in the time period 0–t are divided into elementary sections with instantaneous sources, temperature increments due to instantaneous sources at the point under study are calculated, and all increments of temperature are summed.

Figure 7.18b shows the thermal cycles of surface points *A* and *B* in section I–I, located in the middle between the points of penetration of arc 1, during two-arc

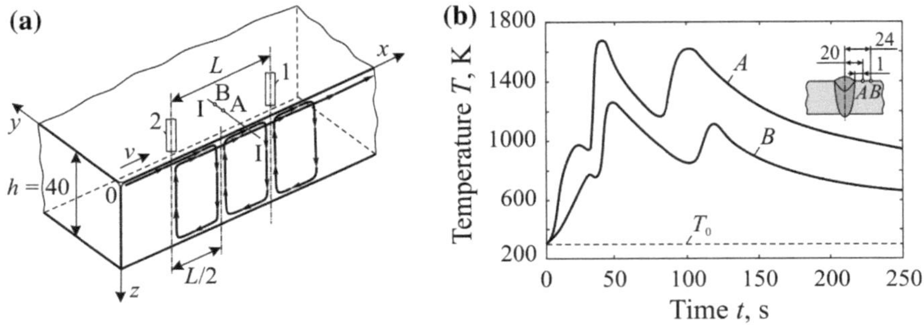

Fig. 7.18 Schematic of trajectory of arcs 1 and 2 (**a**) and thermal cycles at surface points A and B at $L = 150$ mm (**b**) (Serenko et al. 1993)

welding. Cyclic changes in power, speed, and direction of the first arc cause the temperature pulsation. Due to the use of two powerful arcs, the same effect is achieved as with multi-layer welding with short blocks. A peculiar thermal cycle can exclude preheating, as well as postheat treatment when welding thick steel plates (Serenko et al. 1993).

Thus, the calculation of the temperature field during welding by an oscillating source of heat is made by describing the trajectory of the sources in time, choosing the formula for the corresponding instantaneous source and integrating over time. The source trajectory and the time dependences of its parameters can be arbitrary in this case.

It should be noted once again that solving each specific problem by the method of sources it is necessary to find the corresponding solution for the source of unit power (the Green's function) and take into account the spatial and temporal distribution of the welding heat source. For many stationary and non-stationary boundary problems of mathematical physics, Green's functions are already known (Polyanin and Nazaikinskii 2016).

References

Akatsevich, V. D., & Karkhin, V. A. (1998). Application of analytical models for investigation of thermal processes in periodic power source welding. In *Proceedings of St. Petersburg State Technical University, No. 473 "Advanced Materials: Technologies and Researches"* (pp. 106–113). St. Petersburg: St. Petersburg Polytechnic University Publishing (in Russian).

Danilov, A. I., & Marmur, G. Z. (1980). Calculation of temperature in the surface of semi-infinite solid due to arc with sinusoidal path. *Welding Production, 12*, 5–6 (in Russian).

Dowden, J. M., Ducharme, R., & Kapadia, P. (1998). Time-dependent line and point sources: A simple model for time-dependent welding processes. *Laser in Engineering, 7*, 215–228.

Gurevich, V. I. (1969). On calculation of temperature fields due to periodic sources in welding. *Physics and Chemistry of Materials Processing, 2*, 3–10 (in Russian).

Ivanov, S. Yu., Karkhin, V. A., & Michailov, V. G. (2015). Modelling of thermal processes in welding of joints with curvilinear welds. In *Proceedings of Tula University. Technical Sciences* (No. 6, Pt. 2, pp. 62–66).

Ivanov, S. Yu., Karkhin, V. A., & Mikhailov, V. G. (2017). Optimisation of welding conditions for curvilinear joints. *Welding International, 31*(12), 929–933.

Karkhin, V. A. (1990). *Principles of heat transfer in weldments* (100 pp.). Leningrad: Leningrad State Technical University Publishing (in Russian).

Karkhin, V. A. (1993). Calculation of temperature fields in using heat sources with a periodically changing power. *Automatic Welding, 6*, 3–7 (in Russian).

Karkhin, V. A., & Akatsevich, V. D. (1996). Modelling the thermal processes in pulsed power welding. In *Proceedings of St. Petersburg Technical University, No. 463 "Advanced Materials: Science and Technologies"* (pp. 92–99) (in Russian).

Karkhin, V. A., & Khomich, P. N. (2005). Optimisation of pulsed GTA welding. In V. A. Sudnik (Ed.), *Proceedings of International Conference on Computer Technologies in Joining of Materials* (pp. 194–208). Tula: Tula State University (in Russian).

Karkhin, V. A., & Michailov, V. G. (1996). Modelling the thermal behaviour of metals during pulsed power welding. *Journal of Materials Science and Technology, 4*(3), 28–41.

Karkhin, V. A., Michailov, V. G., & Akatsevich, V. D. (1998). Modelling the thermal behaviour of weld and heat-affected zone during pulsed power welding. In H. Cerjak (Ed.), *Mathematical modelling of weld phenomena* (Vol. 4, pp. 411–426). London: IOM Communications.

Karkhin, V. A., Yermakov, S. A., & Topolyansky, P. A. (2011). *Computer-aided design of technological processes in welding. Modelling systems for welding processes* (58 pp.). St. Petersburg: St. Petersburg Polytechnic University Publishing (in Russian).

Makhnenko, V. I. (1966). Calculation of process parameters of deposition on massive workpieces with weaving electrode. *Welding Production, 5*, 4–7 (in Russian).

Makhnenko, V. I., & Kravtsov, T. G. (1976). *Thermal processes in mechanized deposition on circular cylinder-shaped workpieces* (159 pp.). Kiev: Naukova Dumka (in Russian).

Makhnenko, V. I., Shimanovsky, V. P., & Pivtorak, N. I. (1971). Selection of deposition conditions with weaving electrode for workpieces of finite thickness without formation of common pool. *Automatic Welding, 10*, 71–72 (in Russian).

Polyanin, A. D., & Nazaikinskii, V. E. (2016). *Handbook of linear partial differential equations for engineers and scientists* (2nd ed., 1609 pp.). Boca Raton: Chapman and Hall/CRC.

Prudnikov, A. P., Brychkov, Yu. A., & Marichev, O. I. (1986). *Integrals and series. Elementary functions* (798 pp.). London: Taylor and Francis.

Ravi Vishnu, P. (1996). Modelling of pulsed current weld microstructures. *Welding International, 10*(11), 881–889.

Ravi Vishnu, P., & Easterling, K. E. (1993). Phenomenological modelling of heat flow and microstructural changes in pulsed GTA welds in a quenched and tempered steel. In H. Cerjak & K. Easterling (Eds.), *Mathematical modelling of weld phenomena* (pp. 241–299). London: The Institute of Materials.

Ravi Vishnu, P., Li, W., & Easterling, K. E. (1991). Heat flow model for pulsed welding. *Materials Science and Technology, 7*, 649–659.

Reddy, A. A., Guha, B., Achar, D. R. G., & Lalam, S. H. (2001). Finite element analysis of heat affected zone behaviour in pulsed GTA weldments. In H. Cerjak (Ed.), *Mathematical modelling of weld phenomena* (Vol. 5, pp. 393–400). London: IOM Communications.

Rethmeier, M., Wiesner, S., & Wohlfahrt, H. (2001). Einfluesse auf die statische und dynamische Festigkeit von MIG-geschweissten. *DVS, 216*, 327–332 (in German).

Saraev, Yu. N. (1994). *Pulsed welding procedures and surfacing* (107 pp.). Novosibirsk: Nauka (in Russian).

Saraev, Yu. N., Krektuleva, R. A., & Kosyakov, V. A. (1997). Mathematical modelling technological processes of pulsed argon TIG welding. *Welding International, 11*(10), 807–809.

Serenko, A. N., Shaferovsky, V. A., & Pivtorak, N. I. (1993). Temperature fields in two-arc welding with programmed conditions. *Welding International, 7*(10), 802–804.

Shigayev, T. G. (1992). Effect of modulated current welding parameters on weld geometry. *Welding Production, 2*, 10–12 (in Russian).

Street, J. A. (1990). *Pulsed arc welding: An introduction* (57 pp.). An Abington Publishing.

Tsai, C. L., & Hou, C. A. (1988). Theoretical analysis of weld pool behavior in the pulsed current GTAW process. *Transactions of the ASME: Journal of Heat Transfer, 110*(2), 160–165.

Wagner, F. A. (1980). *Equipment and pulsed arc welding techniques* (117 pp.). Moscow: Energia (in Russian).

Yermakov, S. A., Sosnin, N. A., Karkhin, V. A., & Topolyansky, P. A. (2006). Plasma-jet hardfacing with transverse oscillations. In *8th International Conference "Technologies of Repair, Restoration and Hardening of Machine Elements and Tools"*, 11–14 April 2006, (pp. 36–44). St. Petersburg. Part 1. St. Petersburg: Polytechnic University Publishing (in Russian).

Chapter 8
Thermal Cycles of Metal During Welding

8.1 General Information

Heating of the welded joint is applied in almost all welding techniques. The thermal cycle of a metal (the change in the temperature of the metal in time) is determined by dimensions of the welded joint, initial and boundary conditions, thermophysical properties of the metal, parameters of the heating source and position of the metal to the source. Since the welding sources of heat are concentrated (as a rule, their dimensions are much smaller than the dimensions of the welded body), the heating of the metal is characterised by significant non-uniformity both in time and in space. For example, in arc welding of steel, the gradient of the peak temperature near the weld can reach hundreds of degrees per millimeter. Therefore, the thermal cycles of the metal of the adjacent zones of welded joint can be significantly different.

The thermal cycle T (x, y, z, t) of an arbitrary point x, y, z shown in Fig. 8.1 is typical for single-pass welding with simple power distribution in time and space. It is characterised by the following main parameters: peak temperature $T_{\max}$, dwell time of the metal Δt_{T_1} above preset temperature T_1, retention time of the metal $\Delta t_{T_2/T_3}$ in the temperature range T_2 –T_3 and heating and cooling rate $\partial T/\partial t$, which is equal to the tangent of angle α of the inclination of tangent line to curve $T(t)$ at point t. The average rate of heating and cooling $(\partial T/\partial t)_{av} = (T_3 - T_2)/\Delta t_{T_2/T_3}$ is the function of $\Delta t_{T_2/T_3}$ (in the figure it can be seen as the tangent of angle β between the secant and the t-axis). If the function $\partial T/\partial t$ varies insignificantly in the interval T_2 –T_3, rates $\partial T/\partial t$ and $(\partial T/\partial t)_{av}$ are almost equal (angles α and β are almost equal). With the complicated heat input, for example, during multi-pass and pulsed welding, these rates can differ even in sign. The given characteristics of the thermal cycle are extremely important for predicting the properties of the metal in different zones of the welded joint (see Chap. 13).

When choosing a model for calculating the thermal cycle of the metal, we should start with the question: what is the required accuracy of calculating parameters of the thermal cycle? For example, when predicting the microstructure of the welded joint, it does not make sense to use the precise calculation model (which requires

V. A. Karkhin, *Thermal Processes in Welding*, Engineering Materials,
https://doi.org/10.1007/978-981-13-5965-1_8

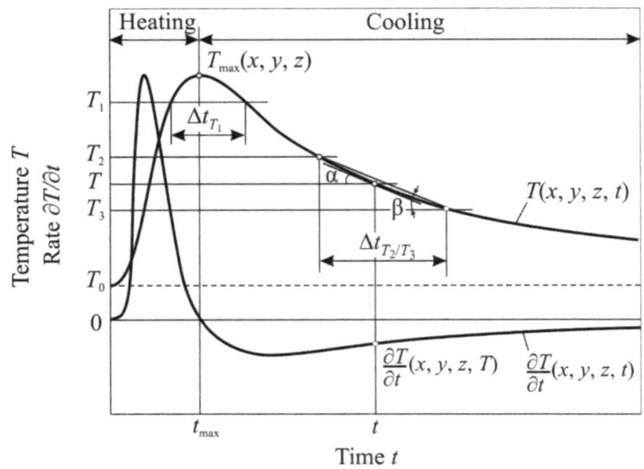

Fig. 8.1 Temperature T and heating/cooling rate $\partial T/\partial t$ as functions of time t for point x, y, z

a lot of time for calculations) if the accuracy level of the metallurgical model used later is much lower. Therefore, it is recommended to choose simple calculation models, which are the simpler, the more assumptions are made. Thus, the assumptions of the body unboundedness in some directions, the heat-insulation of the boundary, constant initial temperature T_0, the absence of heat fluxes in some directions, limiting concentration of the source, constant power, speed and direction of the source movement, homogeneity and temperature independence of material properties simplify formulation and, consequently, the search of the solution to the heat conduction problem.

The thermal cycles of the metal when the body is heated by instantaneous concentrated sources were considered in Sects. 4.2.1 and 5.1.1. In this section we will give the simplest models for calculating the main parameters of the thermal cycle for the following combinations of rapidly moving concentrated sources and heated bodies: (1) a point source on the surface of a massive (semi-infinite) body; (2) a point source on the surface of an infinite slab and (3) a line source in an infinite plate.

8.2 Rapidly Moving Point Source on a Semi-infinite Body

When heating by a rapidly moving point source on the surface of a semi-infinite body, the thermal cycle of a point located at distance r from the axis of movement is described by Eq. (5.1.121), and its cooling rate w is described by Eq. (5.1.128):

$$\begin{aligned} w(r,t) &= -\frac{\partial T}{\partial t}(r,t) = -\frac{q/v}{2\pi\lambda t^2}\exp\left(-\frac{r^2}{4at}\right)\left(\frac{r^2}{4at}-1\right) \\ &= -\frac{T(r,t)-T_0}{t}\left(\frac{r^2}{4at}-1\right). \end{aligned} \tag{8.2.1}$$

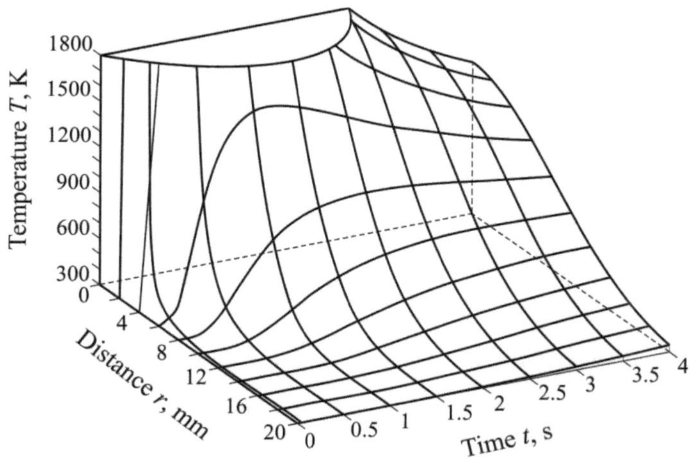

Fig. 8.2 Temperature as a function of time and distance for the rapidly moving point source on the semi-infinite solid ($q = 5000$ W, $v = 5$ mm s^{-1}, $\lambda = 0.04$ W mm^{-1} K^{-1}, $a = 8$ mm^2 s^{-1}, $T_0 =$ 300 K)

As can be seen from Fig. 8.2, the cooling rate w is maximal on the axis of the movement of the source (the weld centreline), $r = 0$. The farther the point under study r is, the longer the peak temperature is reached. The radial distribution of peak temperature $T_{\max}(r)$ is described by expression (5.1.129) and is shown in Fig. 5.1.37a.

The analysis of formula (8.2.1) and Fig. 8.2 shows that at $T \ll T_{\max}$ (that is, for $t \gg r^2/(4a)$), cooling rate w at the given temperature T only slightly depends on coordinate r and in accordance with (5.1.137) at $r = 0$ it is equal to the following:

$$w(0, T) = 2\pi\lambda \frac{(T - T_0)^2}{q/v}. \tag{8.2.2}$$

This formula can be used to calculate the cooling rate of metal throughout the high-heated zone.

The calculation of the cooling rate according to (8.2.2) is practically applicable for a bead on a steel body with thickness of at least 25–30 mm, depending on the preset temperature T and the welding conditions (Rykalin 1951, 1957). With lesser thickness, the calculation model for the slab should be used (Sect. 8.3).

The cooling time of weld centreline $\Delta t_{T_2/T_3}$ in the temperature range from T_2 to T_3 is determined by formula (5.1.138).

The dwell time of metal Δt_{T_1} above preset temperature T_1 can be found using formulas (5.1.121) and (5.1.129). We introduce a dimensionless temperature:

$$\theta_3 = \frac{T_1 - T_0}{T_{\max} - T_0} = \frac{r^2}{4at}\exp\left(1 - \frac{r^2}{4at}\right) = \frac{1}{\tau_3}\exp\left(1 - \frac{1}{\tau_3}\right); \tag{8.2.3}$$

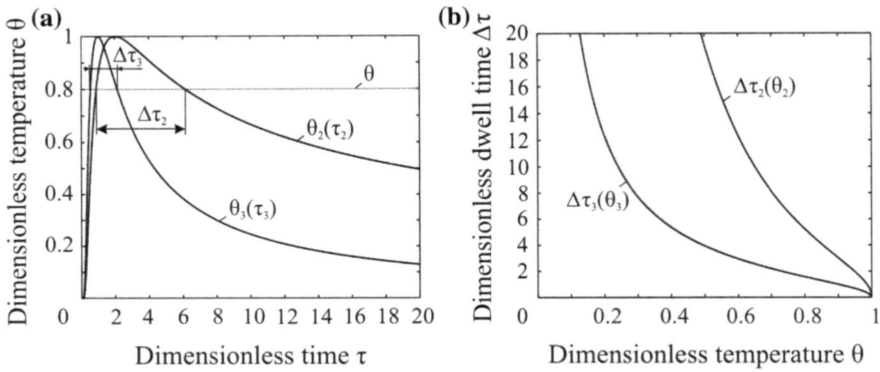

Fig. 8.3 Dimensionless thermal cycles θ (τ) (**a**) and dimensionless dwell time $\Delta\tau$ above certain temperature θ (**b**) for a rapidly moving point source on the semi-infinite solid (θ_3, τ_3, $\Delta\tau_3$) and a rapidly moving line source in the plate (θ_2, τ_2, $\Delta\tau_2$) (Rykalin 1951, 1957)

$$\tau_3 = 4at/r^2, \tag{8.2.4}$$

where τ_3 is the dimensionless time (the Fourier number). Function $\theta_3(\tau_3)$ and corresponding inverse function $\Delta\tau_3(\theta_3)$ are shown in Fig. 8.3.

To determine dwell time Δt_{T_1} above temperature T_1, we express r^2 from (5.1.129) and put into Eq. (8.2.4), from which we obtain the following (Frolov 1988):

$$\Delta t_{T_1} = \frac{1}{2\pi \mathrm{e}} \frac{q/v}{\lambda(T_{\max} - T_0)} \Delta\tau_3. \tag{8.2.5}$$

8.3 Rapidly Moving Point Source on a Slab

When heating by a rapidly moving point source on the surface of an infinite slab, the thermal cycle of the metal at point *y, z* relative to the axis of movement of the source is described by the following equation:

$$T(y, z, t) - T_0 = \frac{q/v}{2\pi\lambda t} \sum_{i=-\infty}^{\infty} \exp\left(-\frac{y^2 + (z - 2ih)^2}{4at}\right). \tag{8.3.1}$$

This follows from Eq. (5.1.144), if the source is at the origin of coordinates.

Let us determine the cooling rate of metal $w = -\,\partial T/\partial t$ at the given temperature T. Differentiating Eq. (8.3.1) with respect to time t, we obtain the following (Rykalin 1948):

$$\frac{\partial T}{\partial t}(y, z, t) = -\frac{T(y, z, t) - T_0}{t}\left(1 - \frac{y^2}{4at} - \frac{A}{B}\right);$$

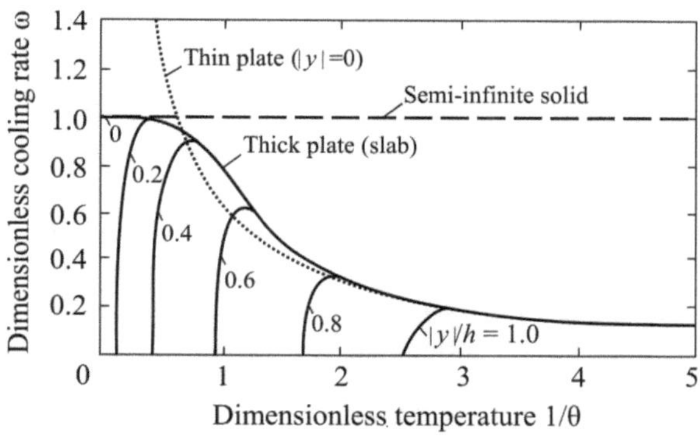

Fig. 8.4 Dimensionless instantaneous cooling rate as a function of dimensionless temperature in bead-on-plate welding (Rykalin 1951, 1957)

$$A = \sum_{i=-\infty}^{\infty} \frac{(z-2ih)^2}{4at} \exp\left(-\frac{(z-2ih)^2}{4at}\right);$$

$$B = \sum_{i=-\infty}^{\infty} \exp\left(-\frac{(z-2ih)^2}{4at}\right). \tag{8.3.2}$$

Formulas (8.3.1) and (8.3.2) allow us to calculate all the characteristics of the thermal cycle at any point y, z.

The solution (8.3.2) can be expressed in terms of dimensionless cooling rate ω and dimensionless temperature θ (Rykalin 1951, 1957):

$$\omega = -\frac{q/v}{2\pi\lambda(T-T_0)^2} \frac{\partial T}{\partial t}; \; \theta = \frac{\pi c\rho h^2}{2q/v}(T-T_0). \tag{8.3.3}$$

The dependence of ω on θ for the surface of a slab is shown in Fig. 8.4. It connects the dependencies for a point source on a semi-infinite body and a line source in a plate. The cooling rates of the points on the slab located on the axis of the movement of the source ($y = z = 0$) are close to the cooling rates of the points on the semi-infinite body, if $1/\theta < 0.4$ (the temperature is high). They are close to the cooling rates of the plate points, if $1/\theta > 2.5$ (the temperature is low).

Cooling rate $\omega(|y|/h)$ of the points on the slab outside the axis of movement of source ($|y|/h > 0$) increases from $\omega = 0$ (when the temperature is maximum), reaches its maximum and becomes equal to the rate for the points on the source movement axis ($|y|/h = 0$).

With a point source on a semi-infinite body, dimensionless cooling rate ω of the bead for any temperature θ is equal to

$$\omega = -\frac{q/v}{2\pi\lambda(T-T_0)^2} \frac{\partial T}{\partial t} = 1, \tag{8.3.4}$$

from which Eq. (5.1.137) is obtained. The curve for the line source in the plate is described by a hyperbola:

$$\omega\frac{1}{\theta} = \frac{2}{\pi} = 0.6366. \tag{8.3.5}$$

Putting (8.3.3) into this expression, we obtain Eq. (5.1.157).

To calculate cooling rate $w = -\,\partial T/\partial t$ on the surface of the slab at temperature T, we calculate dimensionless temperature θ, then from the curves in Fig. 8.4 we find dimensionless cooling rate ω, from which we obtain the following:

$$w = \frac{2\pi\lambda(T - T_0)^2}{q/v}\omega. \tag{8.3.6}$$

The cooling rate of the metal in the slab ($z > 0$) is determined by formulas (8.3.1) and (8.3.2).

8.4 Rapidly Moving Line Source in a Plate

In Sect. 5.1.4.3, formulas for calculating the peak temperature (5.1.152) and (5.1.153), the rate of cooling of the metal at any time (5.1.149), the cooling rate of the metal on the weld centreline at the given temperature (5.1.157) and the cooling time of the metal on the weld axis in the given temperature range (5.1.158) are presented and analysed. Here we give a model for calculating the dwell time of the metal above the given temperature.

Let us assume that there is no surface heat transfer ($b = 0$). Then, using Eqs. (5.1.146) and (5.1.153), the thermal cycle can be represented in a dimensionless form (Rykalin 1951, 1957):

$$\theta_2 = \frac{T - T_0}{T_{\max} - T_0} = \sqrt{2\mathrm{e}}\frac{1}{\sqrt{\tau_2}}\exp\left(-\frac{1}{\tau_2}\right); \tag{8.4.1}$$

$$\tau_2 = \frac{4at}{y^2}, \tag{8.4.2}$$

where τ_2 is the dimensionless time (the Fourier number).

The dependence $\theta_2(\tau_2)$ is shown in Fig. 8.3a. The dwell time of metal $\Delta\tau_2$ at the peak temperature ($\theta = 1$) is zero (Fig. 8.3b). As the temperature decreases, dwell time increases, faster in the plate than in a semi-infinite body.

To determine dwell time Δt_{T_1} above temperature T_1, we express y from (5.1.153) and put into Eq. (8.4.2), from which we obtain (Rykalin 1951, 1957; Frolov 1988):

$$\Delta t_{T_1} = \frac{1}{8\pi\mathrm{e}}\frac{1}{\lambda c\rho}\left[\frac{q/(vh)}{T_{\max} - T_0}\right]^2\Delta\tau_2. \tag{8.4.3}$$

As can be seen from formulas (8.2.5) and (8.4.3), with an increase in heat input q/v and/or in preheating temperature T_0, dwell time Δt_{T_1} above temperature T_1 increases.

It should be noted here that simple formulas for calculating parameters of the thermal cycle were obtained due to assumption of the limiting concentration of the heat sources and the neglect of the longitudinal heat fluxes, which reduced the dimensionality of the problem. This approach is reasonable if the actual source is highly concentrated and the welding speed is relatively high (Peclet number Pe $=$ vx/a is great). The error in calculating the temperature according to the model of rapidly moving concentrated sources was estimated earlier in Sect. 5.1.4.

8.5 Cooling Time

In order to predict the microstructure and mechanical properties of the HAZ metal in welding of steel, the cooling time in the given temperature range, from 1123 K (850 °C) or 1073 K (800 °C) to 773 K (500 °C) is required (see Sects. 13.2 and 13.4.1). Earlier simple formulas (5.1.138) and (5.1.158) were obtained for calculating the cooling time of the weld centreline when a semi-infinite body is heated by a rapidly moving point source and a plate is heated by a rapidly moving line source. These formulas imply the thermal insulation of the body and the absence of longitudinal fluxes. Let us find the approximate formulas for calculating cooling time $\Delta t_{8/5}$ in interval 1073 K (800 °C)–773 K (500 °C) for arc welding of different steel joints (Degenkolbe et al. 1984).

For the point source on the surface of a semi-infinite body, taking into account (1.2.15) and (5.1.138), we obtain the cooling time:

$$\Delta t_{8/5} = K_3 \eta_h Q_{1\,gross}\left(\frac{1}{773 - T_0} - \frac{1}{1073 - T_0}\right);$$
$$K_3 = 1/(2\pi\lambda); \quad Q_{1\,gross} = UI/v. \qquad (8.5.1)$$

It results from (8.5.1) that in the case of a three-dimensional (3D) temperature field the cooling time is proportional to arc energy per unit length of weld $Q_{1\,gross}$.

For a line source in the plate, taking into account (1.2.15) and (5.1.158), we obtain the cooling time:

$$\Delta t_{8/5} = K_2 \eta_h^2 \left(\frac{Q_{1\,gross}}{h}\right)^2 \left[\frac{1}{(773 - T_0)^2} - \frac{1}{(1073 - T_0)^2}\right];$$
$$K_2 = 1/(4\pi c\rho\lambda). \qquad (8.5.2)$$

It can be seen that in the case of a two-dimensional (2D) temperature field, the cooling time is proportional to the square of the heat input per unit thickness of the plate. With increasing initial temperature T_0, the cooling time increases.

Table 8.1 Thermal efficiency of the arc (Degenkolbe et al. 1984; Seyffarth et al. 1992)

Welding technique	Efficiency η_h
Submerged arc welding	1.0
Welding with rutile electrodes	0.9
Welding with basic electrodes	0.8
Welding in CO_2	0.85
Welding with a consumable electrode in Ar or He	0.75
Welding with non-consumable electrode in Ar or He	0.65

The experiments show that when welding structural low-alloy steels, coefficients K_2 and K_3 depend on initial temperature T_0 (Degenkolbe et al. 1984; Seyffarth et al. 1992; Uwer and Degenkolbe 1975):

$$K_3 = 8.06 - 0.005T_0, \text{ mm K W}^{-1}; \tag{8.5.3}$$

$$K_2 = 547 - 0.43T_0, \text{ mm}^4 \text{ K}^{-2} \text{ J}^{-1} \text{ W}^{-1}. \tag{8.5.4}$$

The formulas above are valid for bodies of simple geometric shapes. The influence of the type of the welded joint is taken into account by additional coefficients F_3 (for the 3D case) and F_2 (for the 2D case) (Degenkolbe et al. 1984; Seyffarth et al. 1992).

$$\Delta t_{8/5} = (8.06 - 0.005T_0)\eta_h Q_{1\,gross}\left(\frac{1}{773 - T_0} - \frac{1}{1073 - T_0}\right)F_3; \tag{8.5.5}$$

$$\Delta t_{8/5} = (547 - 0.43T_0)\eta_h^2\left(\frac{Q_{1\,gross}}{h}\right)^2\left[\frac{1}{(773 - T_0)^2} - \frac{1}{(1073 - T_0)^2}\right]F_2. \tag{8.5.6}$$

Here T_0 is measured in K, $Q_{1\,gross}$ is measured in J mm^{-1}. The thermal efficiency of arc η_h is defined with respect to the efficiency for submerged arc welding, which is taken as a unit (Table 8.1).

The values of coefficients F_2 and F_3 are given in Table 8.2. They take into account the effect of all possible heat fluxes in the welded joint. The more directions of heat diffusion there are, the less is cooling time (Fig. 8.5).

Formulas (8.5.5) and (8.5.6) are of the wide practical application, and they are included into international standards (for example, EN 1011-2). However, it should be noted that these formulas were obtained with considerable assumptions. For example, they are not advisable to be used if the welding speed is relatively low (the longitudinal heat fluxes cannot be neglected), and the surface heat transfer is significant. More precisely, the cooling rate in the given temperature range can be found on the basis of functional-analytical and numerical methods, taking into account the geometry of

Table 8.2 Coefficients of weld type influence on cooling time $\Delta t_{8/5}$ (Degenkolbe et al. 1984)

Type of weld		Coefficient	
		F_2	F_3
Submerged arc deposit	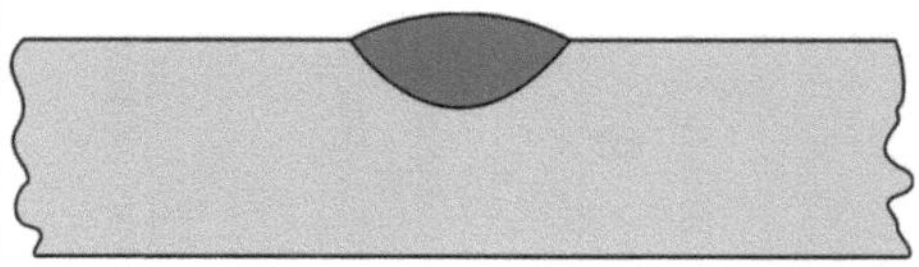	1	1
Multi-pass butt weld	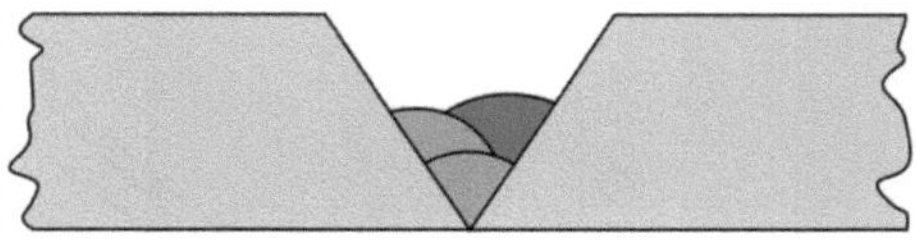	0.9	0.9
Single-pass corner weld	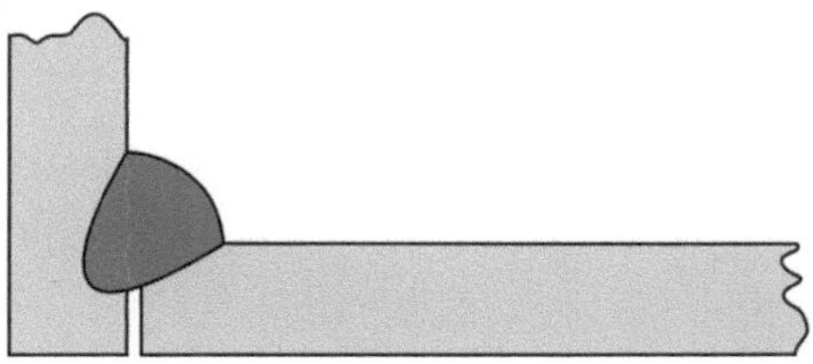	0.9–0.67	0.67
Single-pass fillet weld	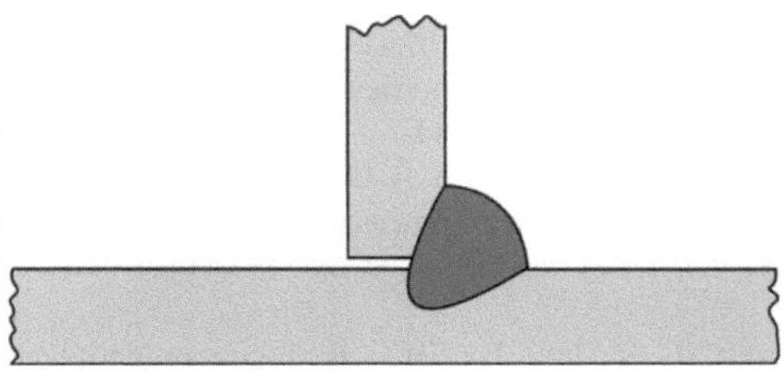	0.45–0.67	0.67

the body and the distribution of the sources. It is recommended to introduce another correction factors in the standard EN 1011-2 with respect to the welding position and the welded joint configuration (Haelsig et al. 2017).

In multi-pass welding, the thermal cycles of the weldment depend on the interpass temperature, that is, the temperature in a multi-pass weld and adjunct base metal immediately prior to the application of the next pass (Dyakov et al. 1984; Rykalin 1950).

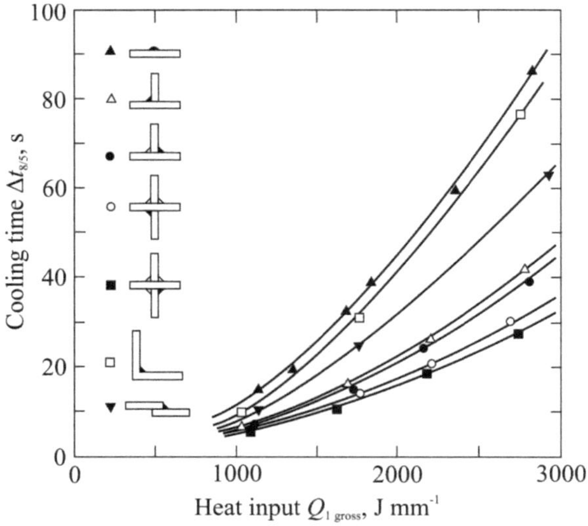

Fig. 8.5 Influence of a welded joint type on cooling time $\Delta t_{8/5}$ in the case of submerged are welding of 10 mm thick steel plates (initial temperature $T_0 =$ 293 K) (Degenkolbe et al. 1984)

References

Degenkoble, J., Uwer, D., & Wegmann, H. (1984). Characterisation of weld thermal cycles with regard to their effect on the mechanical properties of weld joints by the cooling time $t_{8/5}$ and its determination (17 pp.). IIW Doc. IX-1336-84.

Dyakov, Yu. G., Karkhin, V. A., & Anikovsky, A. A. (1984). Kinetics of the stresses and deformations in multi-pass welding of clad metal plates. *Automatic Welding,* (8), 18–21.

Frolov, V. V. (Ed.). (1988). *Theory of welding processes* (559 pp.). Moscow: Vysshaya Shkola (in Russian).

Haelsig, A., Pehle S., Kusch M., & Mayr, P. (2017). Reducing potential errors in the calculation of cooling rates for typical arc welding processes. *Welding in the World, 61,* 745–754.

Rykalin, N. N. (1948). Theory of cooling during arc welding. In *Proceedings of Bauman Moscow Technical Institute: Welding Technique* (Vol. 1, pp. 5–26). Moscow: Mashgiz (in Russian).

Rykalin, N. N. (1950). Calculation of thermal cycles of base metal in multi-pass arc welding. In *Proceedings of the USSR Academy of Sciences. Technical Sciences* (Vol. 2, pp. 233–246) (in Russian).

Rykalin, N. N. (1951). *Calculation of heat flow in welding* (Z. Paley & C. M. Adams, Jr., Trans.) (337 pp.). Moscow.

Rykalin, N. N. (1957). *Berechnung der Waermevorgaenge beim Schweissen* (326 pp.). Berlin: VEB Verlag Technik (in German).

Seyffarth, P., Meyer, B., & Scharff, A. (1992). *Grosser Atlas Schweiss – ZTU – Schaubilder* (176 pp.). Duesseldorf: DVS – Verlag (in German).

Uwer, D., & Degenkolbe, J. (1975). Temperaturzyklen beim Lichtbogenschweissen, Einfluss des Waermebehandlungszustandes and der chemischen Zusammensetzung von Staehlen auf die Abkuehlzeit. *Schweissen und Schneiden, 8,* 303–306 (in German).

Chapter 9
Melting and Solidification of Base Metal

9.1 Weld Pool Dimensions

If the temperature field in the body is known, it is possible to highlight the region, heated above melting point $T_{m,}$ and all its geometric characteristics: length L_m, width $W_{m,}$ depth H_m and volume V_m of the weld pool and its cross-sectional area (weld cross-section) A_m (Fig. 9.1). The shape of the weld cross-section is characterised by relative depth of fusion H_m/W_m and the shape coefficient $\mu = A_m/(H_m W_m)$.

Expressing dimensions of the weld pool in an explicit form as a function of the geometric dimensions of the body, properties of the metal, and parameters of the heat source is possible only with considerable assumptions. The relationship between sizes of the given isothermal surface in the cases of a point source on the surface of a semi-infinite body and a line source in a plate is described by Eqs. (5.1.57)–(5.1.59)

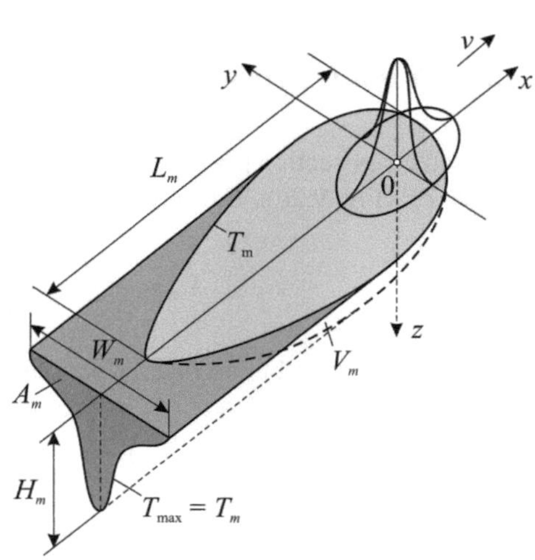

Fig. 9.1 Geometric parameters of the molten pool and weld

V. A. Karkhin, *Thermal Processes in Welding*, Engineering Materials,
https://doi.org/10.1007/978-981-13-5965-1_9

and is shown in Figs. 5.13, 5.14 and 5.24. In the case of rapidly moving sources, Eqs. (5.1.127), (5.1.130), (5.1.148) and (5.1.154) have a simple form.

The shape coefficient μ is usually less than one and depends on the welding technique. For example, with manual arc welding, $\mu = 0.6$–0.7 (Rykalin 1951, 1957).

9.2 Melting Efficiency of Heat Source

Of all the heat put into the heated body, only part of it is spent on heating the metal to melting temperature T_m. The rest is spent on overheating the weld pool above T_m and on heating the body to temperature below T_m.

The ratio of the heat spent on melting the base metal per unit length of the weld to the net heat input is called the thermal efficiency of the base metal melting (Rykalin 1947, 1951, 1952, 1957):

$$\eta_T = \frac{A_m \rho H_m}{q/v} = \frac{A_m c\rho (T_m - T_0)}{q/v}, \tag{9.2.1}$$

where H_m is the specific enthalpy at temperature T_m (more precisely, at liquidus temperature T_L), A_m is the cross-sectional area of the weld metal (Fig. 9.1).

Coefficient η_T characterises efficiency of using the input energy. It is the criterion of technological similarity for welding by the fully penetrating arc (Sosnin and Karkhin 2007). The coefficient depends on the metal properties ($c\rho$, T_m), the welding conditions (q/v, T_0) and the law of heat input, which influences the cross-sectional area of weld A_m. Let us determine thermal efficiency η_T of the point source on a thick body and the line source in the plate.

9.2.1 *Rapidly Moving Point Source on a Semi-infinite Body*

The weld cross-section is a semicircle of radius r_m, bounded by melting temperature $T_{\max} = T_m$ (Fig. 9.2c). The area of semicircle A_m can be found from Eq. (5.1.129):

$$A_m = \frac{\pi r_m^2}{2} = \frac{q/v}{\mathrm{e} c\rho (T_m - T_0)}. \tag{9.2.2}$$

Substituting (9.2.2) into (9.2.1), we obtain

$$\eta_T = \frac{1}{\mathrm{e}} = 0.3679. \tag{9.2.3}$$

Theoretically, only 36.79% of the input energy is spent on melting the metal, the rest is lost due to overheating the liquid metal and heating the solid metal.

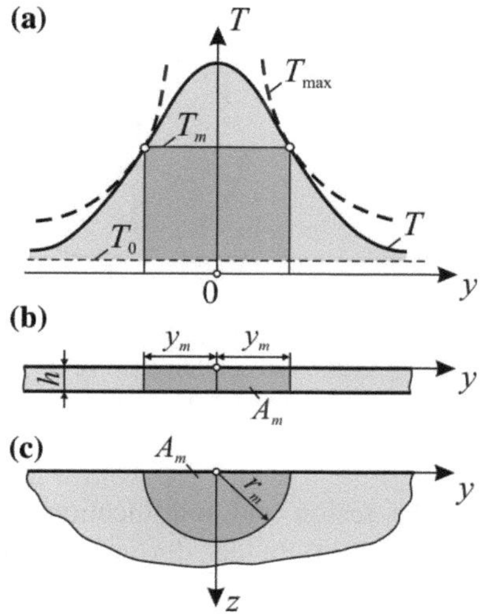

Fig. 9.2 Temperature distribution (**a**) due to rapidly moving sources in a plate (**b**) and on the surface of a semi-infinite solid (**c**) (Rykalin 1947, 1951, 1952, 1957)

9.2.2 Rapidly Moving Line Source in a Plate

The cross-section of the weld is a rectangle (Fig. 9.2b). Weld metal area A_m can be found from Eq. (5.1.154):

$$A_m = 2y_m h = 2\frac{1}{\sqrt{2\pi \mathrm{e}}}\frac{q/(vh)}{c\rho(T_m - T_0)}h. \tag{9.2.4}$$

Substituting (9.2.4) into (9.2.1), we obtain

$$\eta_T = \frac{2}{\sqrt{2\pi \mathrm{e}}} = 0.4839. \tag{9.2.5}$$

Thus, thermal efficiency of a concentrated source in the plate is higher than in the semi-infinite body. It depends on the type of the source and does not depend on properties of the material. Distribution of the source has a large influence on η_T (Berezovsky 1979; Rykalin 1947, 1951, 1952, 1957).

9.3 Solidification of Weld Pool

Movement of the solidification front (isothermal surface T_m of the tail part of the weld pool) determines the rate and direction of crystallisation and, consequently, the primary microstructure of the weld metal.

Usually, the following assumptions are made when analysing the solidification process.

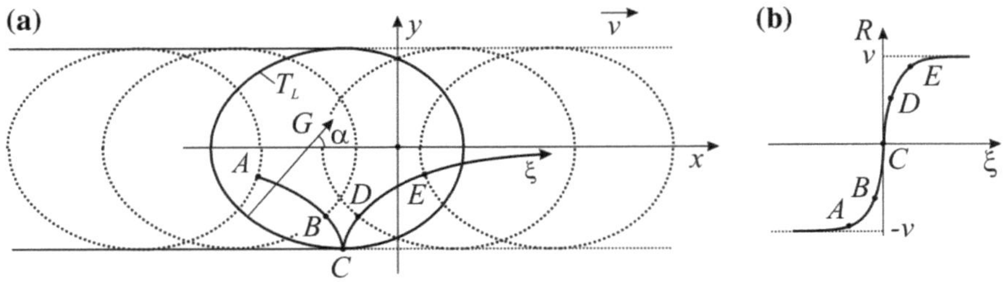

Fig. 9.3 Successive weld pool contours and trajectories of solid/liquid interface of a crystal during melting (*ABC*) and solidification (*CDE*) (**a**) and rate of melting and solidification *R* (**b**) (Karkhin et al. 2010; Rajamaki et al. 2010)

- The fusion and solidification fronts are set by liquidus isotherm T_L = const (Fig. 9.3a). In fact, during solidification (as the impurity concentration rises ahead of the solidification front), value T_L decreases and, consequently, the tail part of the pool is extended (Karkhin et al. 2010; Messler 1999).
- Direction of melting and growth of crystal coincides with the direction of temperature gradient *G* (Fig. 9.3a), which means that the influence of crystallographic orientation and competitive growth (grain selection) is not taken into account (Karkhin et al. 2002a, b, c). Then, on the given boundary surface of the weld pool, melting and solidification rates are determined as follows:

$$R = \frac{d\xi_{SL}(t)}{dt} = v\cos\alpha, \tag{9.3.1}$$

where ξ_{SL} is the coordinate of the interphase boundary of the crystal growth trajectory ξ, α is the angle between the welding direction (*x*-axis) and the direction of temperature gradient *G* at point ξ_{SL} on the pool boundary (Fig. 9.3a). The solidification rate is zero at the weld interface (at point *C*), and it is maximal on the weld axis where $R = v$ (Fig. 9.3b). In practice, the local growth rate of the crystals never will exceed the welding speed *v* (Grong 1994).

The primary microstructure of the weld metal is connected with the morphology of solidification, which is determined by the state of the supercooled liquid metal (impurity and temperature distributions) ahead of the solidification front. Four limiting cases of solidification of alloys are usually considered (Chalmers 1962; Kou 2003; Lancaster 1999; Messler 1999). These cases are conveniently described in D_S/R-D_L/R plane, where D_S and D_L are the impurity diffusion coefficients in the solid and liquid phases, respectively (Fig. 9.4) (Karkhin et al. 2009, 2010). Parameter D_L/R characterises the width of the zone of increased concentration of the impurity ahead of the solidification front during the steady state of this zone.

Case 1 (*C*1). Distribution of the impurity concentration in each phase is equilibrium (uniform) due to intense diffusion ($D_S \to \infty$ and $D_L \to \infty$) and/or long diffusion time ($R \to 0$). This corresponds to the phase diagrams, for example, the Fe-C diagram.

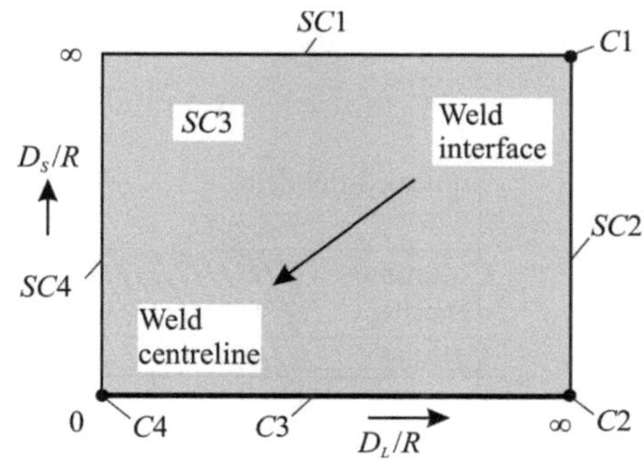

Fig. 9.4 The limiting cases ($C1$, $C2$, $C3$ and $C4$) and the intermediate subcases ($SC1$, $SC2$, $SC3$ and $SC4$) of solidification on the parameter plane: D_S and D_L are the diffusion coefficients in solid and liquid, respectively; R is the solidification rate (Karkhin et al. 2009, 2010)

Case 2 ($C2$). There is no diffusion of the impurity in the solid phase ($D_S = 0$). The distribution of impurities in the liquid phase is uniform due to ideal mechanical mixing and intensive diffusion ($D_L \to \infty$). This is the well-known calculation model of Gulliver-Scheil (Messler 1999).
Case 3 ($C3$). There is no diffusion of the impurity in the solid phase ($D_S = 0$). Diffusion occurs only in liquid ($0 < D_L < \infty$).
Case 4 ($C4$). The concentration of the impurity in the liquid does not change over time due to the absence of mechanical mixing and diffusion ($D_L = 0$).

All four solidification cases are limiting (D_S/R and / or D_L/R take discrete values 0 or ∞). With the additional removal of one of these constraints, intermediate subcases $SC1$, $SC2$, $SC3$, and $SC4$ are obtained (Fig. 9.4). Then the range of possible values of the parameters is continuous. Intermediate subcase $SC3$ best takes into account welding conditions. As the crystals grow, the solidification rate R varies from zero at the weld interface to the welding speed at the centre of the weld. In this case, the point characterizing the case of solidification moves on the parameter plane within the area of $SC3$ (in Fig. 9.4, this movement is indicated by an arrow).

Morphology of solidification is determined by the ratio of the local temperature gradient G to solidification rate R. The values of G and R can be obtained by solving the heat conduction problem. Since ratio G/R varies greatly in the process of solidification from the boundary to the centre of the weld, different types of the primary microstructure of the weld metal are formed in the cross section: planar, cellular, columnar-dendritic and equiaxed-dendritic (Fig. 9.5). For quantitative prediction, it is necessary to know the concentration field of the impurity at the solidification front, which can be obtained by solving the diffusion problem for the non-homogeneous body, taking into account the different solubility of the impurity in the phases (Karkhin et al. 2009, 2010).

The trajectory of the crystal growth is determined by the shape of the boundary surface of the weld pool. Explicit models describing the equations of the crystal axes were proposed earlier (Prokhorov 1968, 1979). The approach, which allows for the growth of grains along the crystallographic directions (which, in general, do not coincide with the direction of gradient G) and the competitive grain growth during the solidification is more accurate but more complex (Karkhin et al. 2002a,

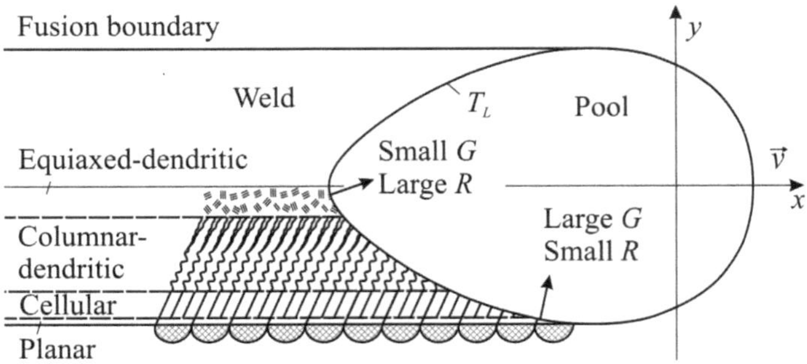

Fig. 9.5 Schematic of primary microstructure formation of the weld metal

b, c; Ploshikhin and Bergmann 1998). With this approach, the following physical assumptions are accepted:

- the solidification front is set by liquidus isotherm T_L;
- size distribution and crystallographic orientation of grains are known;
- the solidification problem is two-dimensional, which is true for the case of welding thin plates;
- there is a selection of grains according to the following criterion: the growth rate of the grain whose crystallographic direction is closer to the direction of the temperature gradient is higher than that of the neighbouring grain.

The solution to the crystallographic problem is the set of grain boundaries, that is, the calculated "macrosection" of the weld in the plane of the plate, the statistical processing of which gives the distribution of the grain sizes and directions of the crystallographic orientation of the grains at any distance from the weld interface.

As an example, Fig. 9.6 shows the calculated "macrosection" of the weld during melting the edge of an AC120 (1.15% Si, 0.4% Mg) aluminium alloy plate 1.15 mm thick with a laser beam of 0.6 mm in diameter and a power of $q_{gross} = 1700$ W. It was assumed that the distribution of the size and crystallographic orientation of the grains of the base metal is random and uniform, and the initial diameter of the grain lies within the range 20–30 μm. The temperature problem is solved by inverse modelling (Chap. 11).

The welding conditions have a significant effect on the shape and texture of the weld metal (Fig. 9.6). The shape of the weld pool determines the preferential crystallographic orientation of grains (the most probable angle γ of the crystallographic orientation of the grains). The more elongated is the shape of the pool (the greater the speed v), the greater is angle γ (Fig. 9.6a). The wider is the weld pool relative to the size of the base metal grains, the more intensive is the selection of grains, which leads to the difference in the texture and, therefore, in the properties between the weld metal and the base metal. The difference increases with the distance from the weld interface (Fig. 9.6b).

The solidification front of the weld pool is of a complex shape, which depends on a heat source, dimensions and properties of the weld metal. Knowing the weld pool

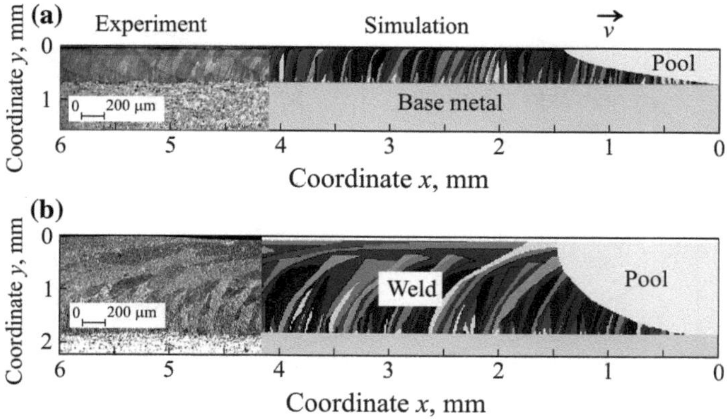

Fig. 9.6 Experimental and simulated grain structure in the longitudinal section of a 1.15 mm thick AC120 aluminium alloy plate after remelting the plate edge by a laser beam at power q_{gross} = 1700 W and speed $v = 50$ mm s^{-1} (**a**) and 16.7 mm s^{-1} (**b**) (only the tail part of the weld pool is shown) (Karkhin et al. 2002a, c)

geometry, we can determine a trajectory and a growth rate of crystal by numerical methods (Karkhin et al. 2009, 2010; Prokhorov 1968; Rajamaki et al. 2010). For a generalised analysis of the trajectory and the growth rate it is advisable to find the solution in the form of equations.

Let us find the equation of the columnar crystal axis for a moving weld pool under assumption that the solidification front retains its constant shape, moves at constant speed v along the x-axis, is approximated by the elliptic arc in the plane containing the x-axis and the crystal grows along the normal to the front (orthogonal trajectory).

Movement of the solidification front can be presented by the family of ellipses (Fig. 9.7a)

$$\varphi(x, y, x_0) = \left(\frac{x - x_0}{L_r}\right)^2 + \left(\frac{y}{w}\right)^2 - 1 = 0 \tag{9.3.2}$$

or

$$x = L_r\sqrt{1 - (y/w)^2} + x_0. \tag{9.3.3}$$

Generally, orthogonal trajectories are described by the equation (Korn and Korn 1968):

$$\frac{\partial \varphi}{\partial y}dx = \frac{\partial \varphi}{\partial x}dy. \tag{9.3.4}$$

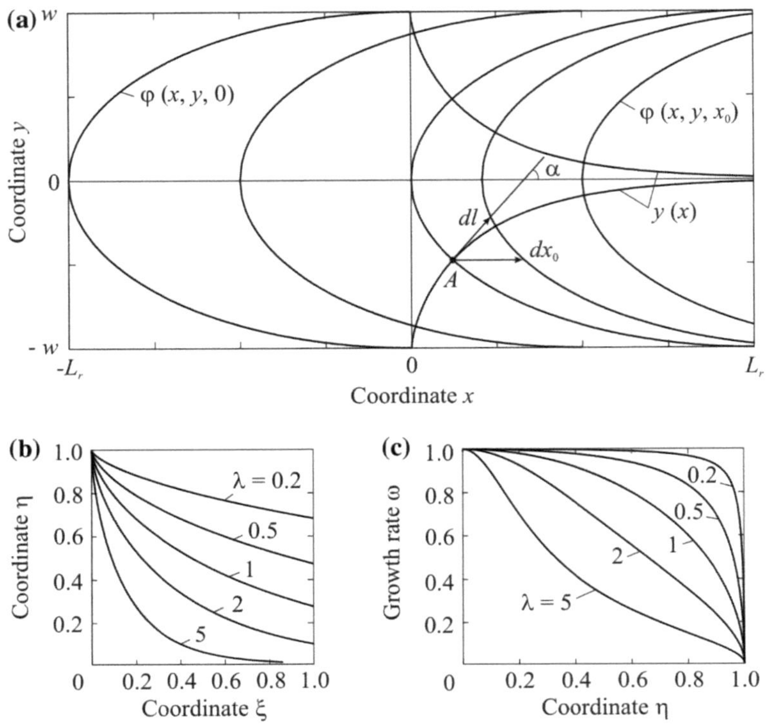

Fig. 9.7 Growth of crystal under solidification of a moving ellipse-shaped weld pool: family of ellipses φ (x, y, x_0) and axes of growing crystals y (x) (**a**), views of crystal axes in dimensionless coordinates (**b**), dependence of a dimensionless growth rate of crystals $\omega = v_s/v$ on dimensionless distance to the weld interface $\eta = y/w$ (**c**)

With allowance for (9.3.2) from (9.3.4) we obtain:

$$dx = \left(\frac{w}{L_r}\right)^2 \frac{x - x_0}{y} dy. \tag{9.3.5}$$

From (9.3.2) we determine the parameter of family x_0 and insert it into (9.3.5), resulting into:

$$x = \frac{w^2}{L_r} \int \frac{\sqrt{1 - (y/w)^2}}{y} dy. \tag{9.3.6}$$

Using the following integral (Prudnikov et al. 1986)

$$\int \frac{1}{x}\left(a^2 - x^2\right)^{1/2} dx = \left(a^2 - x^2\right)^{1/2} - a \ln\left|\frac{a + \left(a^2 - x^2\right)^{1/2}}{x}\right| \tag{9.3.7}$$

we obtain the equation of the axis

$$x = \frac{w^2}{L_r}\left[\sqrt{1-(y/w)^2} - \ln\left|\frac{1+\sqrt{1-(y/w)^2}}{y/w}\right|\right] + C. \tag{9.3.8}$$

Taking the coordinate of the growth start for the upper crystal as $x = 0$ and $y = w$, we find the constant of integration $C = 0$.

Figure 9.7a shows that crystal $y(x)$ starts growing perpendicularly to the weld interface, then its axis gradually bends asymptotically approaching the x-axis. The trajectory depends on ellipse axes L_r and w, and their ratio L_r/w.

Equation (9.3.8) can be expressed in the dimensionless form:

$$\xi = \frac{1}{\lambda}\left(\sqrt{1-\eta^2} - \ln\left|\frac{1+\sqrt{1-\eta^2}}{\eta}\right|\right);$$

$$\xi = \frac{x}{w}; \eta = \frac{y}{w}; \lambda = \frac{L_r}{w}. \tag{9.3.9}$$

Axes of crystals (9.3.9) are shown in Fig. 9.7b, proving that the more elongated the tail of the weld is (the higher the value of λ is), the more curved the crystal axis will be.

The rate of weld solidification determines distribution of impurities ahead of the solidification front. Therefore, it is an essential feature that characterises microstructure and properties of the weld metal (Karkhin et al. 2009, 2010; Rajamaki et al. 2010).

When the solidification front is shifted by dx_0 value, the crystal at point A will increase by dl value in the direction of the normal to the front, which is determined by the relation:

$$\frac{dl}{dx_0} = \cos\alpha, \tag{9.3.10}$$

where α is the angle between the normal and the x-axis (Fig. 9.7a). On the other hand, the value of cosα is defined by the equation of ellipse x (y) (formula (9.3.3)):

$$\cos\alpha = \frac{1}{\sqrt{1+(dx(y)/dy)^2}} = \left(1+\lambda^2\frac{\eta^2}{1-\eta^2}\right)^{-1/2}, \tag{9.3.11}$$

as Eq. (9.3.3) results into

$$\frac{dx}{dy} = -\frac{L_r}{w}\frac{y/w}{\sqrt{1-(y/w)^2}} \tag{9.3.12}$$

Expression (9.3.10) can be interpreted as a dimensionless crystal growth rate

$$\omega = \frac{dl}{dx_0} = \frac{1}{v}\frac{dl}{dt} = \frac{v_s}{v}, \tag{9.3.13}$$

where v is the speed of weld pool motion, $v_s = dl/dt = R$ is the solidification rate (crystal growth rate), t is the time elapsed from the crystal growth start ($x_0 = vt$).

Figure 9.7c presents a crystal growth rate with a different ratio of the weld pool tail to its half-width λ. It is visible that the elongated tail of the weld pool (λ is high in values) shows a maximum growth acceleration of solidification $\partial\omega/\partial\eta$ with η low in values (near the weld centreline), whereas relatively wide weld pools (λ is low in values) show a maximum growth acceleration of solidification with η high in values (near the weld interface). Variability of a solidification rate leads to the non-uniform distribution of liquating elements along the crystal axis (Karkhin et al. 2009, 2010; Rajamaki et al. 2010).

Equations (9.3.8)–(9.3.13) enable calculating the shape of the crystal axis and its growth rate. The change from dimensionless values to dimensional ones is made using formulas (9.3.9) and (9.3.13). It is easy to determine changes in the width of columnar crystal in the plane (two-dimensional) and volume (three-dimensional) crystallisation with the assumption that crystal grows continuously and along the temperature gradient (Prokhorov 1968).

It is significant that the obtained results refer to the solidification front of the elliptical shape. The two geometrical parameters (L_r and w) of the solidification front depend on welding conditions and metal properties; they can be determined analytically in an explicit form using models for rapidly moving heat sources (Sect. 5.1.4), or empirically. If the solidification front is different in shape from the ellipse, for example, is tear- or egg-shaped (Karkhin et al. 2016; Prokhorov 1968; Yadaiah and Bag 2014), then crystal's trajectory and growth rate can be calculated by numerical methods.

9.4 Effects of Latent Heat

While the amount of the heat of phase transitions in the solid state is relatively small and can often be neglected, the amount of the heat of melting in the specific enthalpy of the liquid metal can reach 40% or more (for example, in aluminium alloys). Therefore, when analysing segregation of chemical elements, the formation of hot cracks, microstructural and other phenomena near the weld pool, the effects of the heat of melting and solidification should be taken into account.

Allowing for the amount of heat of phase transformations leads to a non-linear heat conduction problem (see Eq. (3.6.4)). Introducing the area-specific heat source power density q_2 instead of volumetric power density q_3 ($q_3 = q_2/h$), we obtain the non-linear heat equation for the plate in the following form (Karkhin et al. 2005c):

$$\nabla^2 T + \frac{v}{a}\frac{\partial T}{\partial x} - \frac{2\alpha}{\lambda h}(T - T_\infty) + \frac{q_2}{\lambda h} + \frac{v}{ac}\frac{\partial(\Delta H)}{\partial x} = 0 \tag{9.4.1}$$

Non-linearity refers to the last term of Eq. (9.4.1) with phase transition heat ΔH, which depends non-linearly on temperature. This term can be formally presented as an additional distributed source and sink in the area of phase transformations (Fig. 3.6). Then non-linear Eq. (9.4.1) can be changed into an integral equation, which is solved by the iteration method with the required accuracy (Karkhin et al. 2004, 2005a, b, c):

$$T^{(i)}(x, y) - T_0 = \int\int_{D_{hs}} q_2(\xi, \eta) G(x, y, \xi, \eta) d\xi d\eta + v\rho h \int\int_{D_{mz}} \frac{\partial(\Delta H^{(i)})}{\partial x}(\xi, \eta) G(x, y, \xi, \eta) d\xi d\eta, \quad (9.4.2)$$

where

$$G(x, y, \xi, \eta) = \frac{1}{2\pi\lambda h} \exp\left(-\frac{v(x-\xi)}{2a}\right) \mathrm{K}_0\left(\mu \frac{v\sqrt{(x-\xi)^2 + (y-\eta)^2}}{2a}\right);$$
$$\mu = \sqrt{1 + \frac{8a\alpha}{c\rho h v^2}}. \quad (9.4.3)$$

Here D_{hs} is the coverage area of the real heat source, where $q_2 \neq 0$; D_{mz} is the mushy zone, where $\partial(\Delta H)/\partial x \neq 0$ (Fig. 3.6); i is the iteration number; G is the Green's function (temperature at the point x, y due to the moving line source of the unit power acting at the point ξ, η). The Bessel function K_0 is infinite at point $x = \xi$, $y = \eta$, therefore it is recommended to integrate near this singular point by the Gaussian quadrature in the polar coordinate system (Ramachandran 1994).

The iteration process is as follows (Karkhin et al. 2004, 2005a, b, c; Prakash et al. 1987):

(1) successively at the ith iteration, $\partial(\Delta H^{(i)})/\partial x$ is determined by the known $\Delta H^{(i)}$ distribution and temperature $T^{(i)}$ is calculated from Eq. (9.4.2);
(2) $\Delta H^{(i+1)}$ is calculated at the $i + 1$ iteration taking into account the following constraints:

$$\Delta H^{(i+1)} = \Delta H^{(i)} + \omega c[T^{(i)} - T(\Delta H^{(i)})];$$
$$\Delta H^{(i+1)} = \max[0, \; \Delta H^{(i+1)}];$$
$$\Delta H^{(i+1)} = \min[L, \; \Delta H^{(i+1)}], \quad (9.4.4)$$

where ω is the iteration parameter, $\omega > 0$; $T(\Delta H)$ is the inverse function (T is the function of the heat of melting ΔH). From the convergence of the iteration process ($\Delta H^{(i+1)} \approx \Delta H^{(i)}$) we obtain the phase transformation condition $T = T(\Delta H)$.

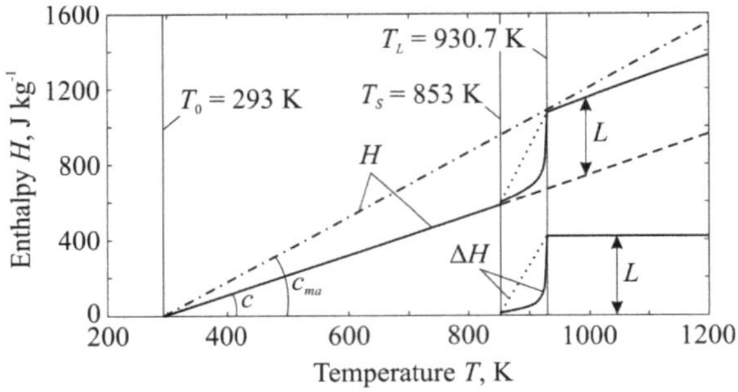

Fig. 9.8 Temperature dependences of enthalpy H and heat of melting ΔH for AlSi0.5 aluminium alloy (dotted lines are simplified relationships) (Karkhin et al. 2004, 2005a, b)

Equation (9.4.1) can be solved with any given accuracy. The solution error is determined by the accuracy of integration (9.4.2) and can be made as small as is wished.

It should be noted that the heat of solidification can differ from the heat of melting, for example, when the pool is doped with a filler material. This can be taken into account in Eqs. (9.4.2) and (9.4.4). The influence of the plate's limited width can be taken into account with the method of images.

The presented method allows us to account for the heat of any phase transitions, for example, for $\alpha \rightarrow \gamma$ transformations in steel. Similarly, the solution for the three-dimensional case can be obtained, with the only difference in Eqs. (9.4.1)–(9.4.3) (since the Green's function is also different) (Karkhin et al. 2005a).

Let us analyse the influence of the heat of melting and solidification on the thermal processes during the laser welding of plates of aluminium alloy AlSi0.5 containing 0.5% Si, under the following conditions: $h = 1.5$ mm, $\lambda = 200$ W m^{-1} K^{-1}, $\rho = 2710$ kg m^{-3}, $c = 1.048$ J g^{-1} K^{-1}, $a = 70.4$ mm^2 s^{-1}, $T_S = 853$ K, $T_L = 930.7$ K, $L = 419.3$ J g^{-1}, $\alpha = 50$ W m^{-2} K^{-1}, $q = 1350$ W, with the radius of the uniformly distributed source $r_0 = 0.25$ mm, $v = 46.67$ mm s^{-1}, $T_0 = T_\infty = 293$ K. The power of the heat source (the effective power of the laser beam) q was determined by inverse modelling (Chap. 11). In the numerical solution, the properties shown in Fig. 9.8 were used.

When calculating the temperature field, the following assumptions for the temperature dependence of enthalpy $H(T)$ and heat capacity $c(T)$ can be made according to equation $c = dH/dT$: non-linear dependence of the enthalpy of phase transition $\Delta H(T)$ (solid curve in Fig. 9.8), simplified linear dependence $\Delta H(T)$ (dotted curve), neglect of $\Delta H(T)$ and constant heat capacity c (dashed line) and neglect of $\Delta H(T)$ and constant heat capacity c_{ma} (dot-dashed line). The liquidus (T_L) and solidus (T_S) isotherms in the quasi-stationary state are shown in Fig. 9.9. It follows from the figure that if we do not take into account the heat of melting and solidification ($\Delta H(T) = 0$) it gives us an overestimated weld width at $c =$ const and an underestimated weld width at $c_{ma} =$ const.

The type of the temperature dependence on the heat of melting and solidification $\Delta H(T)$ has a strong influence on the shape of the tail part of the weld pool (T_L isotherm). The release of heat of solidification mainly near liquidus temperature

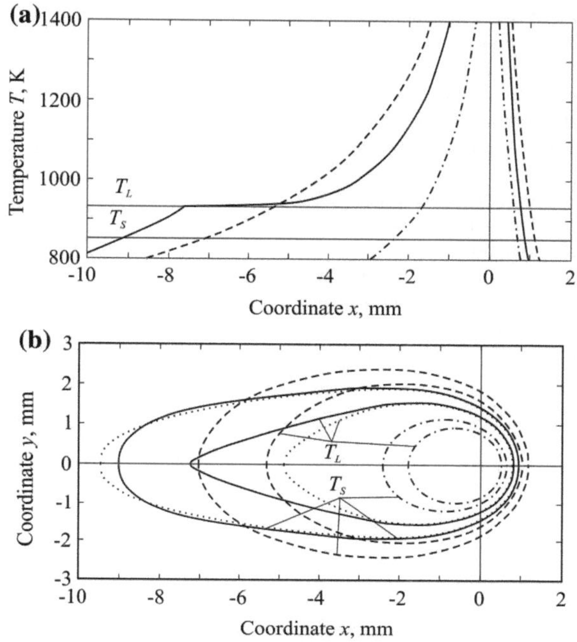

Fig. 9.9 Dependence of the distribution of temperature along x-axis ($y = 0$) (**a**) and weld pool shape and size (**b**) on the heat of melting and solidification in laser welding of AlSi0.5 aluminium alloy: —— non-linear dependence ΔH (T); ······· linear dependence ΔH (T); – – – ΔH $(T) = 0$, heat capacity $c = \text{const}$; –·– ΔH $(T) = 0$, heat capacity $c_{ma} = \text{const}$ (Karkhin et al. 2005a)

T_L leads to the sharpened tail of the weld pool (solid curve T_L in Fig. 9.9b) and a very small temperature gradient just ahead of the solidification front (solid curve in Fig. 9.9a). The sharp form of the tail part of the liquid pool in the form of ripples on the weld surface can be observed experimentally (Karkhin et al. 2004, 2005a, b). If $\Delta H(T)$ dependence is assumed to be simplified linear, the width of the mushy zone increases significantly, the tail part of the pool turns out to be less sharp, and the width of the pool practically does not change (the dashed curve in Fig. 9.9b).

The dimensionless heat of melting of AlSi0.5 aluminium alloy is as follows:

$$\Lambda = \frac{L}{c(T_L - T_0)} \tag{9.4.5}$$

and is equal to 0.6648, which corresponds to 40% of the heat of melting in the specific enthalpy of the metal at liquidus temperature T_L. For austenitic stainless steel AISI 304 (0.1% C, 18% Cr, 8% Ni), the dimensionless heat of melting is much smaller: $\Lambda = 0.2294$ (Karkhin et al. 2005a).

For the plate and the assumed $H(T)$ dependence (Fig. 9.8), the thermal efficiency of the base metal melting is as follows:

$$\eta_T = \frac{W_m h\rho[c(T_L - T_0) + L]}{q/v}. \tag{9.4.6}$$

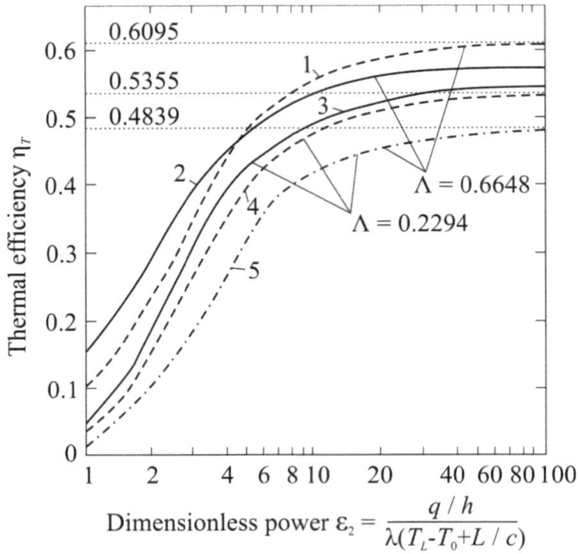

Fig. 9.10 Dependence of the thermal efficiency on latent heat of melting in welding of heat-impermeable aluminium alloy plate ($\Lambda = 0.6648$) and stainless steel plate ($\Lambda = 0.2294$): 1, 4—solution to the linear problem with the changed effective power; 2, 3—solution to the non-linear problem; 5—solution to the linear problem with changed properties (Karkhin et al. 2005a, b)

Let us analyse the effect of the heat of melting L on thermal efficiency η_T when welding heat-insulated plates (Fig. 9.10). Curves 2 and 3 were obtained by solving a non-linear problem (Eq. (9.4.1)). It can be seen that the values of η_T can far exceed 0.4839 (see Eq. (9.2.5). The greater the heat of melting is, the higher is the value η_T (for an aluminium alloy it is higher than for steel).

The solution to differential Eq. (9.4.1) is followed by certain difficulties because of the last non-linear term $\partial(\Delta H)/\partial x$ that takes into account the heat of melting. The question arises: whether this term can be excluded, and, in the remaining linear equation, whether the heat of melting can be approximately taken into account in one of the ways: (1) by changing the coefficients in the differential equation (material properties) or (2) by changing input energy q_2. Let us estimate the error of these two approximate methods of considering the heat of melting.

In the first way, we replace in (9.4.1) the heat capacity c with $c_{ma} = c + L/(T_L - T_0)$ and the thermal conductivity λ with $\lambda_{ma} = \lambda + a\rho L/(T_L - T_0)$ so that the thermal diffusivity a and, consequently, the Peclet number Pe would not change ($a = \lambda/(c\rho) = \lambda_{ma}/(c_{ma}\rho)$ and $\mathrm{Pe} = vx/a = \text{const}$). It should be emphasised that such a replacement is equivalent to allowing for the heat of melting in the specific enthalpy of the metal at liquidus temperature T_L. In other words, $c_{ma}(T_L - T_0) = H(T_L)$, where $H(T_L)$ is the given enthalpy at the liquidus temperature (Fig. 9.8). This approach is used in the work (Rykalin 1951, 1957). The corresponding dependence is shown in Fig. 9.10 (curve 5). It does not depend on speed and is general for all materials with different values of the heat of melting.

In the second way, the energy spent for melting of the metal and distributed along the melting front is attributed to the centre of the source, that is, effective power q is formally replaced by $q - vW_m h\rho L$ (Sosnin and Fedotov 1989). Then we define the width of the weld W by the iteration method and the thermal efficiency by formula (9.4.6) (Fig. 9.10, curves 1 and 4). The values of the asymptotes in this approach can be determined analytically using the model for a rapidly moving line source. In the absence of surface heat transfer:

$$W_m = \frac{q/(vh)}{\sqrt{\pi e/2}c\rho(T_L - T_0) + \rho L}. \tag{9.4.7}$$

Substituting (9.4.7) into (9.4.6), we obtain:

$$\eta_T = \frac{1 + \Lambda}{\sqrt{\pi e/2} + \Lambda}. \tag{9.4.8}$$

It results from this formula that the thermal efficiency increases with increasing Λ (for example, due to preheating temperature T_0). The asymptote values are 0.6095 for the aluminium alloy and 0.5355 for the austenitic steel (Fig. 9.10).

From comparison of curve 2 with 1 and 5 (aluminium alloy) and curve 3 with 4 and 5 (steel) in Fig. 9.10 it follows that in the general case they do not coincide. In general, the accuracy of option 2 (curves 1 and 4) is much higher. Option 1 gives underestimated values (curve 5).

The dependence of the thermal efficiency of the base metal melting on the effective energy has a maximum, which is the lowest when the surface heat transfer is the highest (Karkhin et al. 2005a, b, c).

The heat of melting and solidification influences the shape and size of the weld pool and, consequently, the characteristics of the solidification process (the shape of the "ripples" of the weld, the rate of solidification, the shape of the crystal axes, the segregation of the elements, etc.). These conclusions are confirmed by experiments (Karkhin et al. 2005a, c; Kou 1981; Kou et al. 1982).

References

Berezovsky, B. M. (1979). Thermal efficiency of base metal melting by surface welding arc. *Automatic Welding*, (10), 18–21 (in Russian).

Chalmers, B. (1962). *Physical metallurgy* (2nd ed., 468 pp.). Wiley.

Grong, O. (1994). *Metallurgical modelling of welding* (581 pp.). London: The Institute of Materials.

Karkhin, V. A., Ivanov, S. Yu., Homich, P. N., & Rajamaki, P. (2010). Analysis of solute macro- and microsegregation in fusion welding. In H. Cerjak & N. Enzinger (Eds.), *Mathematical modelling of weld phenomena* (Vol. 9, pp. 155–175). Graz: Verlag der Technischen Universitaet Graz.

Karkhin, V. A., Il'in, A. S., Ploshikhin, V. V., & Prikhodovskii, A. A. (2005b). Effect of the heat of melting and solidification on the thermal efficiency of the melting process. *Welding International*, *19*(3), 212–216.

Karkhin, V. A., Ilin, A. S., Ploshikhin, V. V., & Prikhodovsky, A. A. (2004). Effect of heat of aluminium alloy melting on shape and sizes of weld pool. *Paton Welding Journal*, (1), 2–7.

Karkhin, V. A., Ilyn, A. S., Pesch, H. J., Prikhodovsky, A. A., Plochikhine, V. V., Makhutin, M. V., & Zoch, H.-W. (2005c). Effects of latent heat of fusion on thermal processes in laser welding of aluminium alloys. *Science and Technology of Welding and Joining, 10*(5), 597–603.

Karkhin, V. A., Khomich, P. N., & Ivanov, S. Y. (2016). Calculation of thermal processes around moving weld pool by boundary element method. *Welding and Diagnostics,* (4), 32–35 (in Russian).

Karkhin, V. A., Khomich, P. N., & Rajamaki, P. (2009). Modelling of macrosegregation in fusion welding. In *Proceedings of St. Petersburg State Polytechnic University, "Materials and Chemical Technologies"* (Vol. 510, pp. 164–179). St. Petersburg: St. Petersburg State Polytechnic University Publishing (in Russian).

Karkhin, V. A., Pesch, H. J., Ilin, A. S., Prikhodovsky, A. A., Plochikhine, V. V., Makhutin, M. V., & Zoch, H.-W. (2005a). Effects of latent heat of fusion on thermal processes during welding. In H. Cerjak, H. K. D. H. Bhadeshia, & E. Kozeschnik (Eds.), *Mathematical modelling of weld phenomena* (Vol. 7, pp. 39–62). Graz: Verlag der Technischen Universitaet Graz.

Karkhin, V. A., Plochikhine, V. V., & Bergmann, H. W. (2002a). Solution of inverse heat conduction problem for determining heat input, weld shape, and grain structure during laser welding. *Science and Technology of Welding and Joining, 7*(4), 224–231.

Karkhin, V. A., Plochikhine, V. V., & Bergmann, H. W. (2002c). Simulation of thermal and solidification processes in laser welding of aluminium plates. *The Paton Welding Journal, 54*(8), 10–14.

Karkhin, V. A., Plochikhine, V. V., Ilyin, A. S., & Bergmann, H. W. (2002b). Inverse modelling of fusion welding processes. In H. Cerjak (Ed.), *Mathematical modelling of weld phenomena* (Vol. 6, pp. 1017–1042). London: Maney Publishing.

Korn, G. A., & Korn, T. M. (1968). *Mathematical handbook for scientists and engineers. Definitions, theorems and formulas for reference and review* (2nd ed., 1130 pp.). New York: McGraw-Hill.

Kou, S. (1981). Simulation of heat flow during the welding of thin plates. *Metallurgical Transactions A, 12A*(12), 2025–2030.

Kou, S. (2003). *Welding metallurgy* (2nd ed., 461 pp.). Hoboken: Wiley.

Kou, S., Kanevsky, T., & Fyfitch, S. (1982). Welding thin plates of aluminum alloys—A quantitative heat-flow analysis. *Welding Journal,* (6), 175-s–181-s.

Lancaster, J. F. (1999). *Metallurgy of welding* (6th ed., 446 pp.). Cambridge: Abington Publishing.

Messler, R. W., Jr. (1999). *Principles of welding: Processes, physics, chemistry, and metallurgy* (662 pp.). New York: Wiley.

Ploshikhin, V. V., & Bergmann, H. W. (1998). Simulation of grain structures in laser beam welds undergoing the planar solidification mode. In H. Cerjak (Ed.), *Mathematical modelling of weld phenomena* (Vol. 4, pp. 150–165). London: IOM Communications.

Prakash, C., Samonds, M., & Singhal, A. K. (1987). A fixed grid numerical methodology for phase change problems involving a moving heat source. *International Journal of Heat Mass Transfer, 30*(12), 2690–2694.

Prokhorov, N. N. (1968). *Physical processes in metals during welding. Vol. 1: Physics of metals and crystallisation process* (696 pp.). Moscow: Metallurgiya (in Russian).

Prokhorov, N. N. (1979). *Technological strength of welds during crystallisation* (248 pp.). Moscow: Metallurgiya (in Russian).

Prudnikov, A. P., Brychkov, Yu. A., & Marichev, O. I. (1986). *Integrals and series. Elementary functions* (798 pp.). London: Taylor and Francis.

Rajamaki, P., Karkhin, V. A., & Homich, P. N. (2010). Analysis of macrosegregation near fusion boundary in fusion welding. *Science and Technology of Welding and Joining, 15*(1), 31–39.

Ramachandran, P. A. (1994). *Boundary element method in transport phenomena* (406 pp.). Southampton: Computation Mechanics Publication.

Rykalin, N. N. (1947). *Thermal principles in welding. Part 1. Heat flow in arc welding* (271 pp.). Moscow: Publication of the USSR Academy of Sciences (in Russian).

Rykalin, N. N. (1951). *Calculation of heat flow in welding* (Z. Paley & C. M. Adams, Jr., Trans.) (337 pp.). Moscow.

Rykalin, N. N. (1952). *Die Waermegrundlagen des Schweissvorganges* (265 pp.). Berlin: Verlag Technik (in German).
Rykalin, N. N. (1957). *Berechnung der Waermevorgaenge beim Schweissen*. Berlin: VEB Verlag Technik (in German).
Sosnin, N. A., & Fedotov, B. V. (1989). Formation of the pool and thermal conditions in plasma welding with a penetrating arc. *Welding Production,* (9), 19–20 (in Russian).
Sosnin, N. A., & Karkhin, V. A. (2007). Melting efficiency is criterion of technological similarity in keyhole plasma welding. In J. Martikainen (Ed.), *3rd JOIN International Conference on Total Welding Management in Industrial Applications*, Lappeenranta, Finland, 21–24 August 2007, Acta Universitatis Lappeenrantaensis (Vol. 274, pp. 288–294).
Yadaiah, N., & Bag, S. (2014). Development of egg-configuration heat source model in numerical simulation of autogenous fusion welding process. *International Journal of Thermal Sciences, 86*, 125–138.

Chapter 10
Heating and Melting of Filler Metal

Fusion welding relies on filler metal in the form of a covered electhrode (in manual metal arc welding), electrode wire (in mechanised arc welding and electroslag welding), wire, rods and bars (in gas tungsten arc welding, laser, electron beam and gas welding).

10.1 Heating and Melting of Covered Electrodes

The electric current that flows along the electrode of finite length L heats it up. The end of the electrode is also heated by the arc and melts (Fig. 10.1).

When analysing electrode heating, it is possible to define two sources of heat: a conventional heat source, which is distributed uniformly along electrode length L and which allows for the effect of the flowing current, and a moving plane source, located on the end of the electrode and which allows for the effect of the arc (Fig. 10.1). The conventional source causes a steady rise of temperature along the length of the

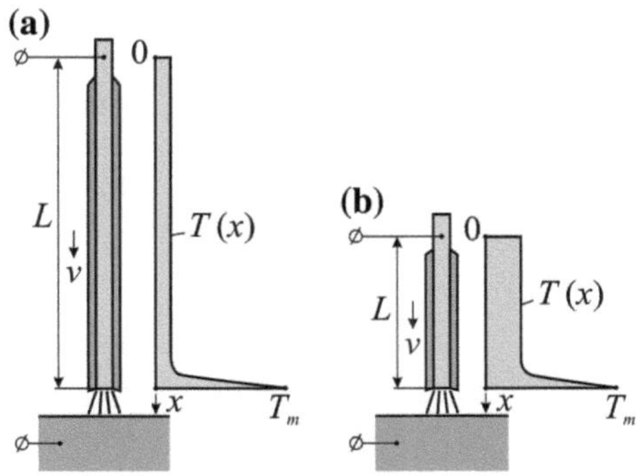

Fig. 10.1 Temperature distribution along the covered electrode at short-time heating (**a**) and long-time heating (**b**) (Frolov 1988; Nerovny 2016)

V. A. Karkhin, *Thermal Processes in Welding*, Engineering Materials,
https://doi.org/10.1007/978-981-13-5965-1_10

electrode, whereas the moving plane source causes a rise of temperature only near the arc. The latter can be determined from Eqs. (5.1.103) and (5.1.104).

However, it is complicated to calculate heating and melting of the electrode without determining the effect of the cover, temperature dependences of the electrode properties and an arc power contribution to electrode melting. These problems are specified in the books (Rykalin 1951, 1957).

10.2 Heating of Electrode Wire

Many research papers focus on heating an electrode wire (Halmoy and Karkhin 1999; Karkhin et al. 2001, 2002, 2005; Kim et al. 1991; Paton 1948; Vasilyev et al. 1974; Waszink and Heuvel 1982). In addition, voltage drop in the electrode extension was considered (Jonsson et al. 1995; Karkhin et al. 2002, 2005; Lancaster 1986; Waszink and van den Heuvel 1979).

Assume that radial temperature distribution across the electrode is uniform, which was shown in paper (Paton 1948). Select a segment of the electrode wire with diameter d_e and dx in length that moves at speed v in the environment at temperature T_∞ (Fig. 10.2b). Heat balance of the segment for dt period of time, which allows for heat conduction, surface heat transfer and Joule heating, is described by the following equation (Vasilyev et al. 1974)

$$\lambda \frac{\partial^2 T}{\partial x^2} \frac{\pi}{4} d_e^2 dx dt - \alpha (T - T_\infty) \pi d_e dx dt + \rho_e j^2 \frac{\pi}{4} d_e^2 dx dt - c\rho \frac{\pi}{4} d_e^2 dx \frac{\partial T}{\partial t} dt = 0, \tag{10.2.1}$$

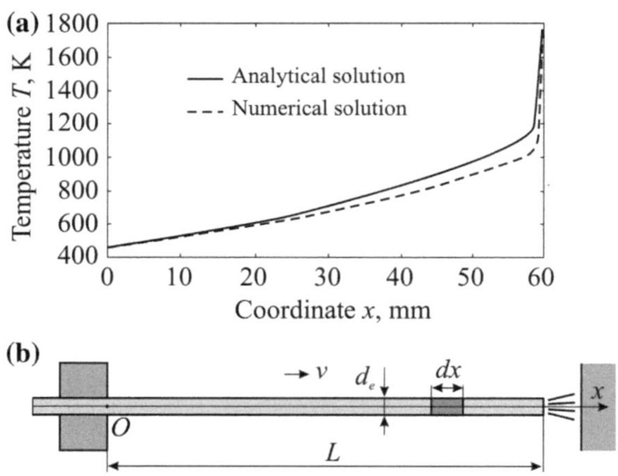

Fig. 10.2 Temperature distribution in the electrode extension (**a**) and the model for calculation (**b**) (Vasilyev et al. 1974)

where j is the current density, ρ_e is the specific electrical resistance.

Consider the steady-state process, $\partial T/\partial t = 0$. Turning to the coordinate system, stationary relative to the arc, we obtain the ordinary differential equation

$$\frac{d^2T}{dx^2} - \frac{v}{a}\frac{dT}{dx} - \frac{4\alpha}{d_e\lambda}(T - T_\infty) + \frac{\rho_e}{\lambda}j^2 = 0. \tag{10.2.2}$$

Generally, Eq. (10.2.2) is non-linear, as all properties of the material depend on temperature. Assume, a, λ, α are constant and resistance ρ_e is described by linear dependence $\rho_e = \rho_0 + \rho_1 T$. Consequently, Eq. (10.2.2) results into a linear differential equation

$$\frac{d^2T}{dx^2} - \frac{v}{a}\frac{dT}{dx} + \left(\frac{\rho_1}{\lambda}j^2 - \frac{4\alpha}{d_e\lambda}\right)T + \frac{\rho_0}{\lambda}j^2 + \frac{4\alpha}{d_e\lambda}T_\infty = 0 \tag{10.2.3}$$

under the following boundary conditions:

$$\begin{aligned} T &= T_0 \quad \text{for } x = 0; \\ T &= T_m \quad \text{for } x = L; \end{aligned} \tag{10.2.4}$$

where L is the electrode extension (Fig. 10.2).

The solution of problem (10.2.3)–(10.2.4) is represented by the following expression (Karkhin et al. 2002, 2005; Vasilyev et al. 1974):

$$\begin{aligned} T &= T_0\frac{\exp(Ax) - \exp(Bx + [A-B]L)}{1 - \exp([A-B]L)} \\ &+ T_m\frac{\exp(B[x-L]) - \exp(Ax - BL)}{1 - \exp([A-B]L)} + \frac{\rho_0 j^2 + \frac{4}{d_e}\alpha T_\infty}{\rho_1 j^2 - \frac{4}{d_e}\alpha} \\ &\times \left\{\frac{\exp(Ax) - \exp(Bx + [A-B]L) + \exp(B[x-L]) - \exp(Ax - BL)}{1 - \exp([A-B]L)} - 1\right\}, \end{aligned} \tag{10.2.5}$$

where

$$A = \frac{v}{2a} - \sqrt{\left(\frac{v}{2a}\right)^2 - \frac{\rho_1}{\lambda}j^2 + \frac{4\alpha}{d_e\lambda}}; \; B = \frac{v}{2a} + \sqrt{\left(\frac{v}{2a}\right)^2 - \frac{\rho_1}{\lambda}j^2 + \frac{4\alpha}{d_e\lambda}}.$$

The first term allows for the effect of heat release in the electrode—tube contact, the second—the effect of the arc, the third—the effect of the flowing current.

Figure 10.2a presents distribution of temperature in the Sv-08A steel electrode wire extension with diameter $d_e = 4$ mm during direct current reverse polarity welding at $I = 600$ A and $v = 21.2$ mm s^{-1}. The curves are plotted according to analytical solution (10.2.5) and numerical solution of Eq. (10.2.2) with boundary conditions (10.2.4) considering temperature dependences of a, λ, α and ρ_e.

In a particular case if there is no surface heat transfer ($\alpha = 0$) and specific resistance ρ_e is constant, the solution to the problem is simplified (Karkhin et al. 2002, 2005):

$$T(x) = \left(T_m - T_0 - \frac{\rho_e j^2 L}{c\rho v}\right)\frac{\exp(vx/a) - 1}{\exp(vL/a) - 1} + \frac{\rho_e j^2}{c\rho v}x + T_0 \tag{10.2.6}$$

or in a dimensionless form

$$\theta(\xi) = \left(1 - \frac{g}{\text{Pe}}\right)\frac{\exp(\text{Pe}\xi) - 1}{\exp(\text{Pe}) - 1} + \frac{g}{\text{Pe}}\xi, \tag{10.2.7}$$

where dimensionless temperature θ, coordinate ξ, feed rate Pe (Peclet number) and energy g are expressed as it follows:

$$\theta = \frac{T - T_0}{T_m - T_0}; \ \xi = \frac{x}{L}; \ \text{Pe} = \frac{vL}{a}; \ g = \frac{\rho_e j^2 L^2}{\lambda(T_m - T_0)}.$$

Equation (10.2.7) shows that the current has an impact on temperature growth proportional to the distance from the contact tube (second term), and the temperature increases exponentially as the arc is approached (first term). Figure 10.3 illustrates temperature distributions along the electrode extension when welding steel, aluminium and titanium under a feasible range of values for current density j, feed rate v and electrode extension L. The higher values of electrical and thermal conductivities represent, the lower electrode temperature and the larger arc affected zone will be. The fraction of Joule heat in electrode melting amounts to 5–25% for aluminium, 8–90% for steel, 25–98% for titanium. The analysis of the effect of welding parameters on temperature distributions along the electrode extension is provided in the papers (Karkhin et al. 2002, 2005).

Fig. 10.3 Dimensionless temperature distributions along the electrode extension of steel (St), aluminium (Al) and titanium (Ti) wires under a feasible range of welding conditions (Karkhin et al. 2002, 2005)

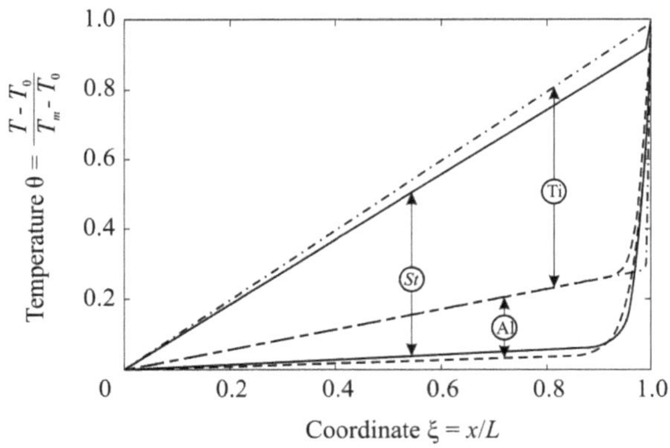

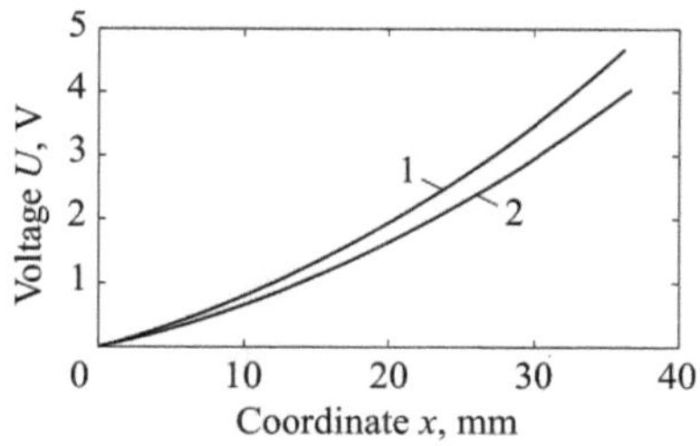

Fig. 10.4 Voltage drop distribution in the electrode extension of C-Mn steel: 1–$d_e = 1.6$ mm, $I = 308$ A, $v = 96$ mm s^{-1}, $L =$ 36.6 mm; 2–$d_e = 1.2$ mm, I = 150 A, $v = 74$ mm s^{-1}, L = 38.7 mm (Karkhin et al. 2002)

The analytical solution of the problem is considerably complicated in pulsed power (pulsed current) welding (Karkhin et al. 2001, 2002, 2005). This approach is based on the fundamental solution (7.2.35).

Knowing temperature distributions along the electrode extension $T(x)$, it is possible to find a voltage drop in the electrode extension

$$U(x) = \int_0^x \rho_e(T(x))jdx. \tag{10.2.8}$$

The publications (Karkhin et al. 2002, 2005) provide the analytical solution of this problem, showing that the voltage drop in the electrode extension can be significant. For instance, in steel wire welding voltage falls to 5 V (Fig. 10.4). Notably, similar curves result from experiments (Naidu et al. 2003; Waszink and van den Heuvel 1979).

10.3 Melting of Electrode Wire

Electrode melting is caused by both Joule heat, when the electrode is heated by the flowing current, and the arc (anode heating if polarity is reverse). Under the steady conditions the rate of electrode melting is equal to the feed rate; it depends on current I and the length of extension L (Welding 1991)

$$v = c_1 I + c_2 L I^2, \tag{10.3.1}$$

where c_1 and c_2 are constants, allowing for effects of anode or cathode heating and Joule heating respectively. If current is low and extension is short, the first term is more significant. If wire diameter is small, extension is long, current is high, the second term is more significant. For instance, during gas metal arc welding (reverse polarity) with a low-carbon electrode of 1.6 mm in diameter coefficients c_1 = 0.152 mm s^{-1}A^{-1} and $c_2 = 4.42 \times 10^{-4}$ s^{-1} A^{-2} (Welding 1991). Let us find coefficients c_1 and c_2 analytically.

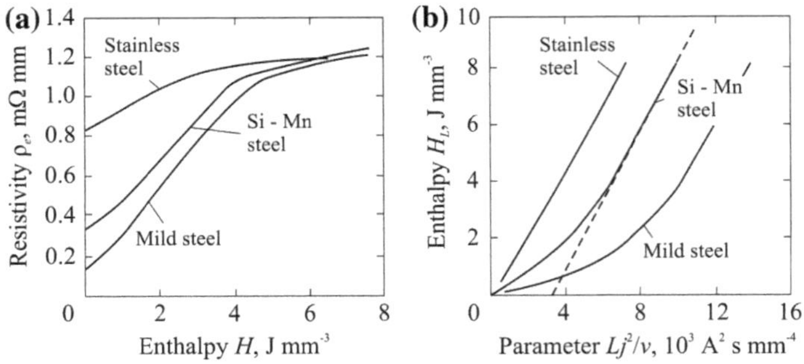

Fig. 10.5 Resistivity of some steel wires as functions of specific heat content (**a**) and the resistive heat content as a function of parameter Lj^2/v (**b**) (Halmoy 1986)

Under the steady conditions, in accordance with Ohm's law, the heat content of a moving electrode element (Fig. 10.2b) is changed at the following rate (Halmoy 1991):

$$\frac{dH}{dt} = v\frac{dH}{dx} = \rho_e j^2. \tag{10.3.2}$$

Note, resistivity ρ_e depends considerably on temperature (its enthalpy H) (Fig. 10.5a).

Assume that enthalpy H starts in point $x = 0$. Integrating Eq. (10.3.2), we get

$$\frac{j^2}{v}\int_0^L dx = \int_0^{H_L} \frac{dH}{\rho_e} = f(H_L) \tag{10.3.3}$$

or

$$\frac{Lj^2}{v} = f(H_L). \tag{10.3.4}$$

Here H_L is the heat content of metal at the end of the extension (at $x = L$). Usually, conditions of gas metal arc welding are in compliance with linear sections of curves H_L (Lj^2/v) (Fig. 10.5b):

$$H_L = \rho_L \frac{Lj^2}{v} - b, \tag{10.3.5}$$

where ρ_L and b are known constants.

Power required for electrode melting and droplet detachment is equal to:

$$H_m Av = \varphi I + H_L Av \tag{10.3.6}$$

or

$$H_L + \varphi \frac{j}{v} = H_m, \tag{10.3.7}$$

where H_m is an empirical constant (total energy per unit of electrode's volume), A is the cross-sectional area of the wire, φ is an empirical constant (potential drop at the anode).

Inserting (10.3.5) into (10.3.7), we obtain analytical dependence of a wire feed rate on current and extension length (Halmoy 1986):

$$v = \frac{1}{H_m + b}\left(\varphi j + \rho_L L j^2\right). \tag{10.3.8}$$

The form of this equation is agrees with empirical formula (10.3.1).

Si–Mn steel has the following constants in Eq. (10.3.8): $H_m = 11.1$ J mm^{-3} (a little higher than liquidus temperature), $b = 4.0$ J mm^{-3} (dependent on steel resistivity), $\varphi = 3.50$ V (equal to the work function of steel), $\rho_L = 1.2$ mΩ mm (equal to the resistivity of hot steel). Thus, the wire feed rate (melting rate) is the following (Halmoy 1991)

$$v = 0.23j + 8.0 \times 10^{-5} L j^2, \tag{10.3.9}$$

where dimensions: v—mm s^{-1}, j—A mm^{-2} and L—mm. This dependence is shown in Fig. 10.6.

Based on a similar physical model, a calculation method for melting of electrode wire during pulsed power welding was developed (Halmoy and Karkhin 1999).

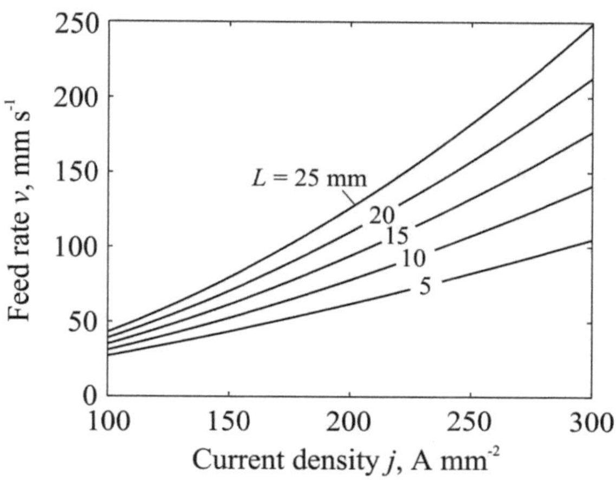

Fig. 10.6 Electrode feed rate as a function of current density j and electrode extension L

10.4 Heating and Melting of Filler Wire

The solution to the problem is obtained for a moving point source on the surface of a solid cylinder (Sect. 5.1.3.11); it ensures solution for any surface distributed source by using the method of sources. Generally, the solution can be presented in the form of a double integral taken over the filler wire surface exposed to heating (Carlson et al. 2013).

Solving a heat conduction problem can be considerably simplified with the following assumptions: along the wire cross-section the temperature is constant, the temperature field in a moving coordinate system is steady and the surface heat transfer is unavailable (Fig. 10.7). Therefore, distribution of enthalpy increase H along wire axis x is expressed in cylindrical coordinates

$$H(x) = \frac{4}{\pi d_e^2 \rho v} \int_x^L \int_0^\pi q_2(\theta, x) \cos \theta d_e d\theta dx \tag{10.4.1}$$

or in Cartesian coordinates

$$H(x) = \frac{4}{\pi d_e^2 \rho v} \int_x^L \int_{-d_e/2}^{d_e/2} q_2(x, y, z) dx dy; z = d_e/2 \cos \theta. \tag{10.4.2}$$

Assume that a normally distributed source of r_e radius does not depend on θ angle (z coordinate), its axis is tilted to x-axis at β angle and shifted from x-axis by y_0. Then the heat flux on the wire surface is described with the equation similar to (1.2.8):

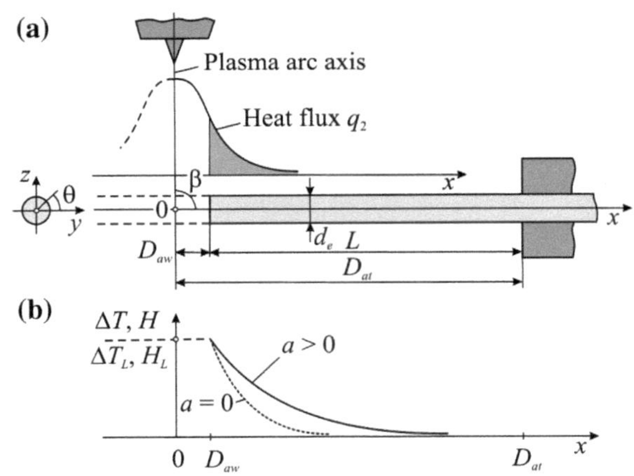

Fig. 10.7 Schematic diagram of filler wire heating and melting: heat flux q_2 distribution (**a**); temperature ΔT (enthalpy H) distribution along wire axis x (**b**) (Carlson et al. 2013)

$$q_2(x, y) = \frac{q}{\pi r_e^2} \sin\beta \exp\left(-\frac{(y-y_0)^2}{r_e^2} - \frac{x^2}{(r_e/\sin\beta)^2}\right). \qquad (10.4.3)$$

Substituting (10.4.3) into (10.4.2), we obtain

$$H(x) = \frac{q}{\pi d_e^2 \rho v}\left[\Phi\left(\frac{d_e/2 + y_0}{r_e}\right) + \Phi\left(\frac{d_e/2 - y_0}{r_e}\right)\right] \times \left[\Phi\left(\frac{\sin\beta}{r_e} D_{at}\right) - \Phi\left(\frac{\sin\beta}{r_e} x\right)\right]. \qquad (10.4.4)$$

Extension length L can result from (10.4.4) provided that enthalpy conforms to liquidus temperature H_L at the wire tip:

$$L = \frac{r_e}{\sin\beta}\Phi^{-1}\left(\Phi\left(\frac{\sin\beta}{r_e} D_{at}\right) - \frac{\pi d_e^2 \rho v H_L}{q} \times \left[\Phi\left(\frac{d_e/2 + y_0}{r_e}\right) - \Phi\left(\frac{d_e/2 - y_0}{r_e}\right)\right]^{-1}\right), \qquad (10.4.5)$$

where $\Phi^{-1}(u)$ is an inverse error function.

Extension length L is very sensitive to effective heat source power q. The higher power q is, the shorter length L will be. Thermal diffusivity of material a does not affect the extension length, it affects only temperature distribution in the wire (Fig. 10.7b).

Figure 10.8 exemplifies the dependence of the distance between arc axis and wire tip D_{aw} on v feed rate of bronze wire AWS 5.7 (2.8–4.0% Si, ≤1.0% Sn, ≤1.0% Zn, ≤1.5% Mn) under the following conditions: $q = 462$ W, $r_e = 1.0$ mm, $d_e = 1.2$ mm, $D_{at} = 10$ mm, $y_0 = 0$, $\beta = 90°$, the distance between the non-consumable electrode and the wire equals 1.9 mm (Carlson et al. 2013).

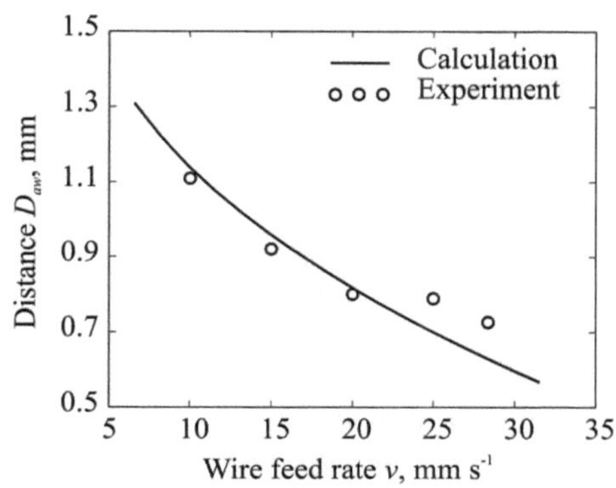

Fig. 10.8 Effect of a feed rate for a bronze wire of 1.2 mm in diameter on the distance between the arc axis and wire tip D_{aw} (Carlson et al. 2013)

References

Carlson, B. E., Wang, H.-P., Karkhin, V. A., Ivanov, S. Yu., Turichin, G. A., & Valdaitseva, Y. A. (2013). Mathematical model of filler metal heating and melting for plasma arc brazing. In C. Sommitsch & N. Enzinger (Eds.), *Mathematical modelling of weld phenomena* (Vol. 10, pp. 267–286). Graz: Varlag der Technischen Universitaet Graz.

Frolov, V. V. (Ed.). (1988). *Theory of welding processes* (559 pp.). Moscow: Vysshaya Shkola (in Russian).

Halmoy, E. (1986). Electrode wire heating in terms of welding parameters. In J. F. Lancaster (Ed.), *The physics of welding* (pp. 330–337).

Halmoy, E. (1991). Current-voltage process characteristics in gas metal arc welding. In *PED-51, Welding and Joining Processes, ASME* (pp. 17–27).

Halmoy, E., & Karkhin, V. A. (1999). Dynamic simulation of aluminium and steel electrode melting in pulsed GMAW. In *Welding Conference LUT JOIN'99,* Lappeenranta, Finland (pp. 106–117).

Jonsson, P. G., Szekely, J., Madigan, R. B., & Quinn, T. P. (1995). Power characteristics in GMAW: Experimental and numerical investigation. *Welding Journal*, (3), 93-s–102-s.

Karkhin, V. A., Fedotov, B. V., Babkin, I. N., & Subramaniam, S. (2001). Analysis of temperature distributions in the consumable wire extension in pulsed current welding. In *Proceedings of 3rd All-Russian Conference on Computer Technologies in Joining of Materials*, Tula (pp. 95–99) (in Russian).

Karkhin, V. A., Fedotov, B. V., Babkin, I. N., & Subramaniam, S. (2002). Analysis of temperature and voltage drop distributions in the wire extension in continuous and pulsed arc welding. In H. Cerjak (Ed.), *Mathematical modelling of weld phenomena* (Vol. 6, pp. 121–141). London: The Institute of Materials, Minerals and Mining.

Karkhin, V. A., Fedotov, B. V., Babkin, I. N., & Subramaniam, S. (2005). Distribution of temperature and voltage drop in the extension of a consumable electrode in welding with continuous and pulsed current. *Welding International*, *19*(1), 52–61.

Kim, Y.-S., McEligot, D. M., & Eagar, T. W. (1991). Analyses of electrode heat transfer in gas metal arc welding. *Welding Journal*, (1), 20-s–31-s.

Lancaster, J. F. (1986). The electric arc in welding. In J. F. Lancaster (Ed.), *The physics of welding* (2nd ed., pp. 146–227). Oxford: Pergamon Press.

Naidu, D. S., Ozcelik, S., & Moore, K. L. (2003). *Modeling, sensing and control of gas metal arc welding* (351 pp.). Amsterdam: Elsevier.

Nerovny, V. M. (Ed.). (2016). *Theory of welding processes* (2nd ed., 702 pp.). Moscow: MVTU Publishing (in Russian).

Paton, B. E. (1948). Investigation of the process of electrode heating during automatic submerged arc welding. In *Proceedings on Automatic Submerged Arc Welding* (Vol. 3, pp. 13–28) (in Russian).

Rykalin, N. N. (1951). *Calculation of heat flow in welding* (Z. Paley & C. M. Adams, Jr., Trans.) (337 pp.). Moscow.

Rykalin, N. N. (1957). *Berechnung der Waermevorgaenge beim Schweissen* (326 pp.). Berlin : VEB Verlag Technik (in German).

Vasilyev, E. M., Demyantsevich, V. P., & Karkhin, V. A. (1974). Analysis of temperature distribution in electrode extension. In *Proceedings of Leningrad Polytechnic Institute, Welding Production* (Vol. 334, pp. 67–70) (in Russian).

Waszink, J. H., & van den Heuvel, G. J. P. M. (1979). Measurements and calculations of the resistance of the wire extension in arc welding. In *International Conference on Arc Physics and Weld Pool Behaviour* (pp. 227–239), London, 8–10 May 1979.

Waszink, J. H., & Heuvel, G. J. P. M. (1982). Heat generation and heat flow in the filler metal in GMA welding. *Welding Journal*, *61*(8), 269-s–282-s.

Welding (1991). *Handbook* (8th ed., Vol. 1, 638 pp.). Miami: American Welding Society.

Chapter 11
Inverse Heat Conduction Problems in Welding

11.1 Formulation of an Inverse Heat Conduction Problem

In the previous chapters, we addressed a direct heat conduction problem: the heat conduction problem is solved with allowance for input data (parameters of the welding heat source, body geometry and material properties). As a result, temperature field and all its characteristics are obtained. This procedure can be presented in a schematic form: independent input parameters $p_1, p_2, \ldots, p_K$ of vector $\mathbf{p}$ are given, and a dependent vector (response function) $\mathbf{T}$ is obtained with the following parameters (output data): $T_1, T_2, \ldots, T_J$ (Fig. 11.1a):

$$\mathbf{T} = f(\mathbf{p}). \tag{11.1.1}$$

Equation (11.1.1) is a mathematical model of the process.

Another problem often arises in the welding practice: finding such parameters of the heat source that would provide the required characteristics of the temperature field (weld size, cooling rate at the specified temperature, peak temperature at the specified points, etc.). To put it differently, the components of vector $\mathbf{T}$ $\{T_1, T_2, \ldots, T_J\}$ serve as input parameters, while the components of vector $\mathbf{p}$ $\{p_1, p_2, \ldots, p_K\}$ serve as output parameters (Fig. 11.1b):

$$\mathbf{p} = f^{-1}(\mathbf{T}). \tag{11.1.2}$$

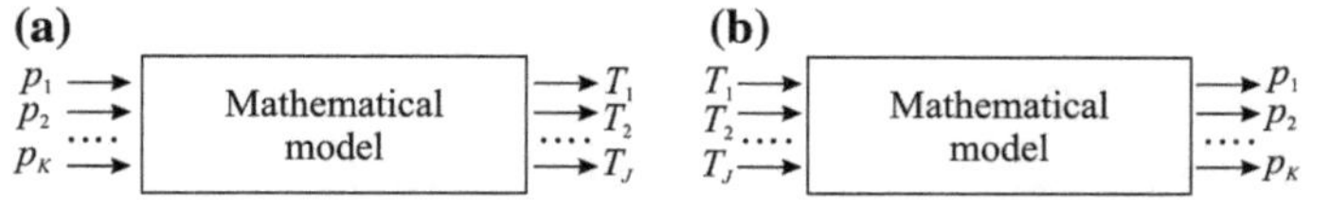

Fig. 11.1 Schemes of direct (**a**) and inverse (**b**) problems of heat conduction

V. A. Karkhin, *Thermal Processes in Welding*, Engineering Materials,
https://doi.org/10.1007/978-981-13-5965-1_11

Such problems are inverse, since welding conditions (heat source, welding speed, etc.) are determined according to the result (temperature field). Solving the inverse problem makes it possible to restore the entire temperature field and find all of its characteristics. A formulation of the inverse heat conduction problem, methods of its solution and example solutions of practical problems are provided below. Particular attention is drawn to implementation of the methods. Inverse heat conduction problems and methods for their solution are considered in numerous publications (Beck et al. 1985; Beck 1991; Karkhin et al. 2001a, c; Kurpisz and Nowak 1995; Tikhonov et al. 1990; Zabaras 1991).

Usually, it proves difficult to achieve the required result, and all that is left is to be content with the best possible (optimum) result. This poses the question of what result can be considered "the best", what changes can be beneficial and what the limits of such changes are.

Answering the first question requires formulation of an optimality criterion for comparative assessment of feasible solutions and selection of the best one. Other questions can be answered using a mathematical model for the optimisation object, i.e. a solution to the direct heat conduction problem. Variables that change during value optimisation are called optimisation parameters, or design variables, while conditions or restrictions that define acceptable values of the variables are called constraints.

The optimality criterion is called an objective function. Its definition is the most important and difficult stage in formulating the inverse problem. It reflects the researcher's view of the problem and is hereby subjective. With regard to welding, it can be formulated based on the following augmented sum of squares function (Karkhin et al. 2001a, c, 2002a, b, c, d, e, 2003a, b, c, 2007e, 2010):

$$F(\mathbf{p}) = \sum_{j=1}^{J} w_j^f [f_j^m - f_j(\mathbf{p})]^2 + \sum_{k=1}^{K} w_k^p [p_k^0 - p_k]^2 + R.T. \to \min, \qquad (11.1.3)$$

where $\mathbf{p}$ is the vector of the sought parameters, $\mathbf{p} = \{p_1, p_2, \ldots, p_K\}$; J is the number of observations; K is the number of the sought parameters; f_j and f_j^m are the calculated and prescribed (measured) characteristics of the temperature field at jth point; p_k and p_k^0 are the kth sought parameter and its value specified a priori; w_j^f and w_k^p are weighting factors for the characteristic of the temperature field at jth point and the kth sought parameter; $R.T.$ is the regularisation term:

$$R.T. = w_0 \sum_{n=1}^{N} [p_n]^2 + w_1 \sum_{n=1}^{N-1} [p_{n+1} - p_n]^2 + w_2 [p_{n+2} - 2p_{n+1} - p_n]^2. \quad (11.1.4)$$

Here, $N \le K$. Sought parameter p_k can be a characteristic of the heat source, cooling rate, etc. (value dimensions can be different) (Fig. 11.2).

Response function f (the prescribed and calculated values at jth point, f_j^m and f_j) can be a thermal, metallurgical, chemical or mechanical characteristic (Fig. 11.2).

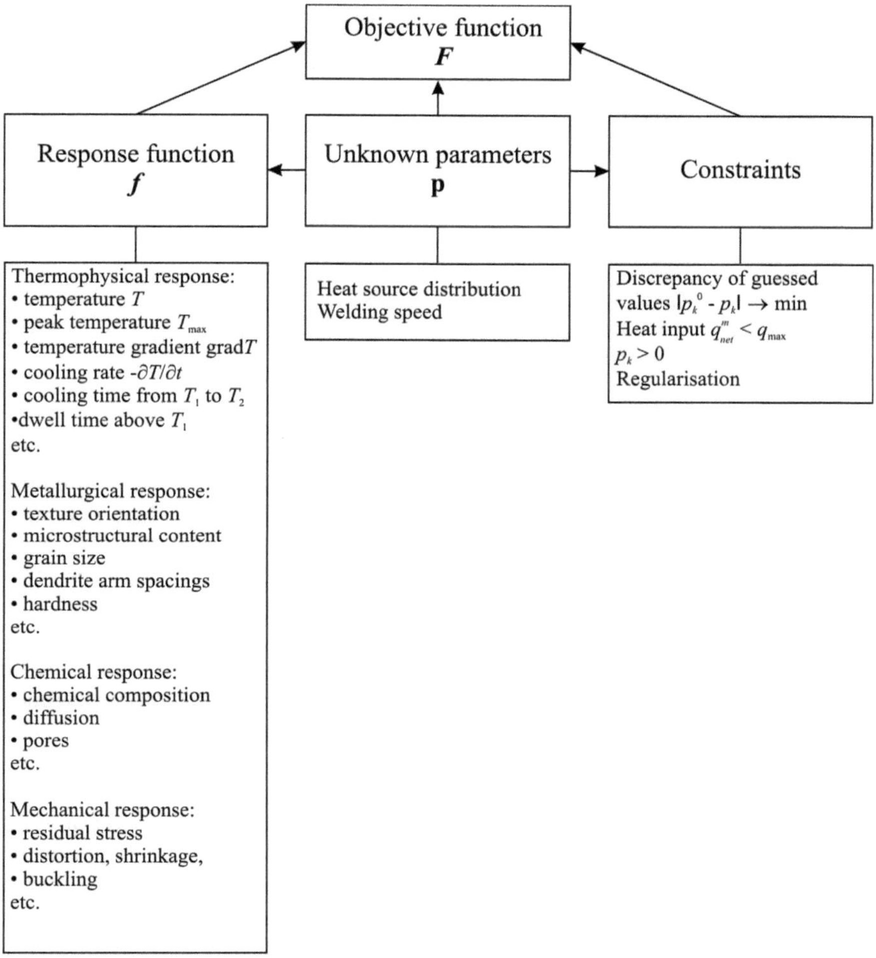

Fig. 11.2 Master chart of the objective function for fusion welding (Karkhin et al. 2001c, 2002a, b, e, 2015)

Weld dimensions in the cross-section (isotherm $T_{\max} = T_L$), HAZ width, geometry of a ripple on the surface of the weld (geometry of the tail part of the weld pool interface) can serve as a thermal response function. If the response function is not thermal, it is assumed that there are models that associate thermoactivation metallurgical, chemical and mechanical phenomena with the characteristics of the temperature field.

Constraints can contain the value of the kth parameter p_k^0 specified a priori, constraints on its sign and magnitude, the value of effective power q_{net} ($\eta_h < 1$), etc. (Fig. 11.2).

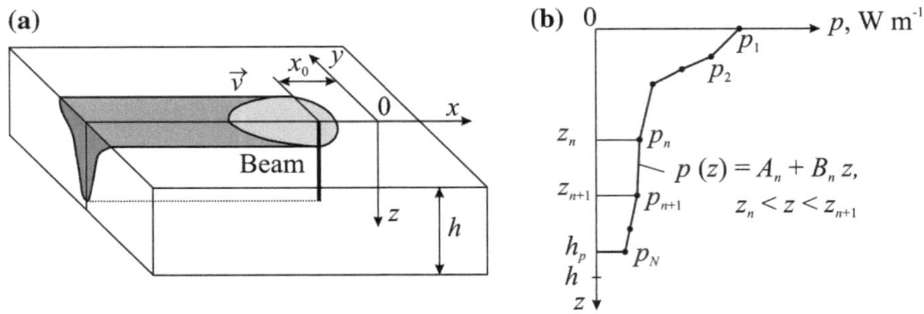

Fig. 11.3 Moving reference frame (**a**) and an approximation of the line source power distribution by piecewise-linear function (**b**) (Karkhin et al. 2001c, 2002d)

It is recommendable to take weighting factors w_j^f inversely proportional to measurement error variance (mean-square error) f_j^m (Beck et al. 1985) and weighting factors w_k^p inversely proportional to the squared interval within which parameter p_k can change relative to prescribed value p_k^0 (Tikhonov and Arsenin 1977). Since response function f_j and parameters p_k have different dimensions, the dimensions of all weighting factors are chosen so as to ensure that all terms of the objective function (11.1.3) have the same dimension.

Regularisation makes it possible to smooth the sought power distribution function and assign its form if it is specified, for example, by a continuous piecewise linear function (Karkhin et al. 2001a, c, 2002b, d). As an example, Fig. 11.3 shows a piecewise linear approximation of the power distribution of a beam with N nodes. By increasing w_0 in (11.1.4), it is possible to reduce the maximum magnitude of estimated values p_n; by increasing w_1, it is possible to reduce the first derivatives of the power distribution function and bring all values p_n closer to a constant value; by increasing w_2, it is possible to reduce the second derivatives of the distribution function. If the step between nodes is not constant, the second derivative (the last term in (11.1.4)) is written differently. Regularisation reduces ill-posedness of the inverse problem. The ill-posedness implies that minor changes in the initial conditions bring about significant changes in problem solution.

As can be seen from Sect. 11.2, the solution to the inverse heat conduction problem in relation to unknown heat source parameters comes down to a repeated solution to the direct problem. In turn, the direct problem can be solved using a numerical method in sufficiently comprehensive formulation (with allowance for surface and electromagnetic phenomena, convection of the liquid metal of the weld pool) (Kim et al. 2003; Kumar et al. 2005; Kumar and DebRoy 2007; Zhang et al. 2004) or using an analytical method in simplified formulation based on the concept of a volumetric apparent source (Karkhin et al. 2002b, 2006, 2007b, e, 2010).

The concept of a volumetric apparent source revolves around taking into account effects of liquid metal convection and the heat of phase transformations, apart from the real source (Fig. 11.4) (Karkhin et al. 2002b, 2006, 2007b, e, 2010; Radaj 2003). It should be emphasised that dimensionality of the power density of the apparent

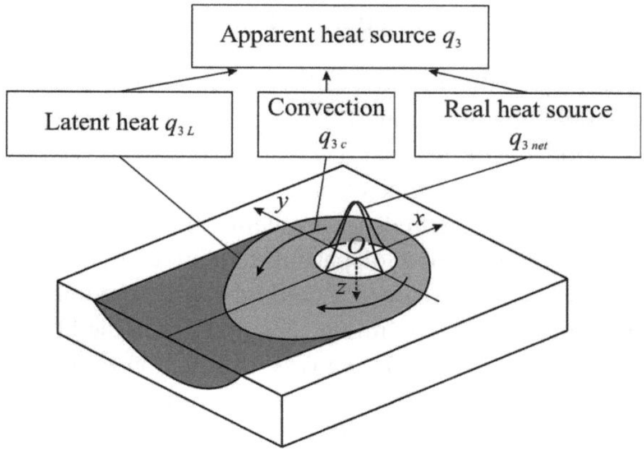

Fig. 11.4 Schematic diagram illustrating a moving reference frame and apparent heat source (Karkhin et al. 2002b, e, 2003b, 2015)

source equals dimensions of the temperature field even if the real source has lesser dimensionality. For example, if a thick plate is heated by a surface source during welding (the real source being two-dimensional), the temperature field and, consequently, the apparent source are three-dimensional. It is proved that the effective power of the real source equals the power of the apparent source (Karkhin et al. 2002b). This makes it easy to determine heat source efficiency of the real source (arc, beam, etc.) according to the obtained value of apparent source power (Karkhin et al. 2002b).

11.2 Solution of an Inverse Heat Conduction Problem

Solving the inverse problem involves finding the minimum of function F (11.1.3) with allowance for constraints. This problem has a number of mathematically unpleasant features. Firstly, it is non-linear, i.e. unknown parameter p_k appears in the equation in a non-linear way. Secondly, there can be non-unique solutions (function F can have multiple local minima in the parameter space).

Objective function minimisation pertains to optimisation methods, which have been properly developed (Alifanov 1994; Attetkov et al. 2001; Gill 1984; Moiseyev et al. 1978; Panteleyev and Letova 2002; Vanderplaats 1984). The vast majority of numerical methods are iterative, i.e. they generate a sequence of points in accordance with the prescribed rule that includes an end criterion. With specified initial vector $\mathbf{p}^0$, such methods generate sequence $\mathbf{p}^0, \mathbf{p}^1, \mathbf{p}^2, \ldots$, with sequence $\{\mathbf{p}^m\}$ having the following property:

$$F(\mathbf{p}^{m+1}) < F(\mathbf{p}^m), \quad m = 0, 1, 2, \ldots \tag{11.2.1}$$

General sequence formation rule $\{\mathbf{p}^m\}$ has the following form:

$$\mathbf{p}^{m+1} = \mathbf{p}^m + \ell_m d^m, \quad m = 0, 1, 2, \ldots \tag{11.2.2}$$

where ℓ_m is the step size, d^m is the direction in which function $F(\mathbf{p})$ decreases, for example, the direction of the negative of the gradient of the objective function.

Initial vector $\mathbf{p}^0$ is set according to physical considerations and a priori information about the minimum position.

Numerical methods of problem solution can be divided into three groups based on the highest order of partial derivatives of function F used in the formation of d^m and ℓ_m: (1) zero-order methods, which require only values of function $F(\mathbf{p})$ in searching for the optimum; (2) first-order methods, which use information about the gradient, or first derivatives, of function $F(\mathbf{p})$; (3) second-order methods, which require knowledge of the second derivatives of function $F(\mathbf{p})$ (Attetkov et al. 2001; Panteleyev and Letova 2002). Let us examine these methods in sequence and illustrate them through the example of the following methodological welding problem.

Let us assume that plates with thickness $h = 5$ mm are welded under the following conditions: gross power $q_{gross} = 7000$ W, speed $v = 10$ mm s^{-1}, initial temperature $T_0 = 300$ K. Thermal properties of the material: $a = 10$ mm^2 s^{-1}, $\lambda = 0.05$ W mm^{-1} K^{-1}, $T_m = 1773$ K. The surface source is bounded, normally distributed: source efficiency $\eta_h = 0.85$, normal source radius $x_e = y_e = r_e = 4$ mm, source boundaries $x' = y' = -r_c = -6$ mm and $x'' = y'' = r_c = 6$ mm (Sect. 5.2.1.7). The length of the weld pool on top side $L_m = 21.06$ mm. The width of the weld pool on top side $W_m = 10.432$ mm, on bottom side $W_m = 4.566$ mm (Fig. 11.5). The temperature obtained using Green's function method (by integrating instantaneous sources (5.2.49) over the acting time of the real source in moving coordinates) is as follows:

$$T(x, y, z) - T_0 = \frac{q}{c\rho} \int_0^{t_w} \theta(x, t)\theta(y, t)\theta(z, t)dt, \tag{11.2.3}$$

where

$$\theta(x, t) = \frac{1}{[\Phi(x''/x_e) - \Phi(x'/x_e)]\sqrt{\pi(4at + x_e^2)}} \exp\left(-\frac{(x + vt)^2}{4at + x_e^2}\right)$$
$$\times \left\{\Phi\left(\frac{-(x + vt)x_e}{\sqrt{4at(4at + x_e^2)}} + \frac{x''}{x_e}\sqrt{1 + \frac{x_e^2}{4at}}\right) - \Phi\left(\frac{-(x + vt)x_e}{\sqrt{4at(4at + x_e^2)}} + \frac{x'}{x_e}\sqrt{1 + \frac{x_e^2}{4at}}\right)\right\};$$

$$\theta(y, t) = \frac{1}{[\Phi(y''/y_e) - \Phi(y'/y_e)]\sqrt{\pi(4at + y_e^2)}} \exp\left(-\frac{y^2}{4at + y_e^2}\right)$$
$$\times \left\{\Phi\left(\frac{-yy_e}{\sqrt{4at(4at + y_e^2)}} + \frac{y''}{y_e}\sqrt{1 + \frac{y_e^2}{4at}}\right) - \Phi\left(\frac{-yy_e}{\sqrt{4at(4at + y_e^2)}} + \frac{y'}{y_e}\sqrt{1 + \frac{y_e^2}{4at}}\right)\right\};$$

$$\theta(z, t) = \frac{1}{h}\left[1 + 2\sum_{n=1}^{\infty} \cos\left(\frac{\pi nz}{h}\right) \exp\left(-\pi^2 n^2 \frac{at}{h^2}\right)\right].$$

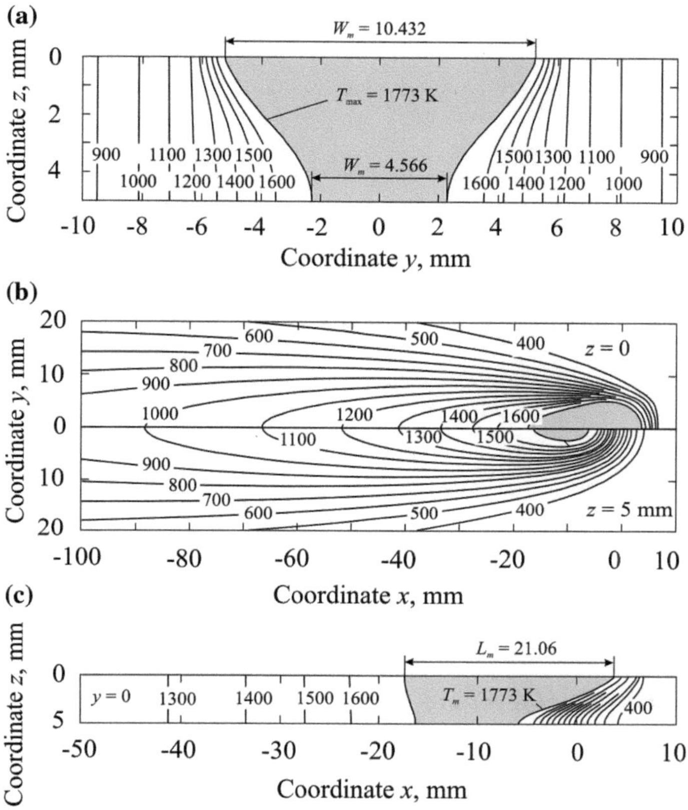

Fig. 11.5 Peak temperature field around the moving surface bounded normally distributed source (**a**), steady temperature field on top side and bottom side (**b**) and in the longitudinal section of the slab (**c**) ($q_{gross} = 7000$ W, $\eta_h = 0.85$, $r_e = 4$ mm, source boundary $r_c = 6$ mm, $v = 10$ mm s^{-1}, $h = 5$ mm, $a = 10$ mm^2 s^{-1}, $T_0 = 300$ K, $\lambda = 0.05$ W mm^{-1} K^{-1}, $c\rho = 0.005$ J mm^{-3} K^{-1}; the weld pool is shaded)

11.2.1 Zero-Order Method

In zero-order methods, what is only needed is to be able to calculate the function at any point of its domain. The major advantage of this method is its simple algorithm with allowance for constraints, the major drawback is a large number of calculations with the increasing number of sought parameters.

Let two parameters of the source be unknown, namely heat source efficiency η_h and radius r_e ($x_e = y_e = r_e$). Let us take as prescribed parameters (1) weld pool length L_m, (2) peak temperature at the weld interface on top side (T_{max} ($W_m/2$, 0) = T_m) and (3) peak temperature at the weld interface on

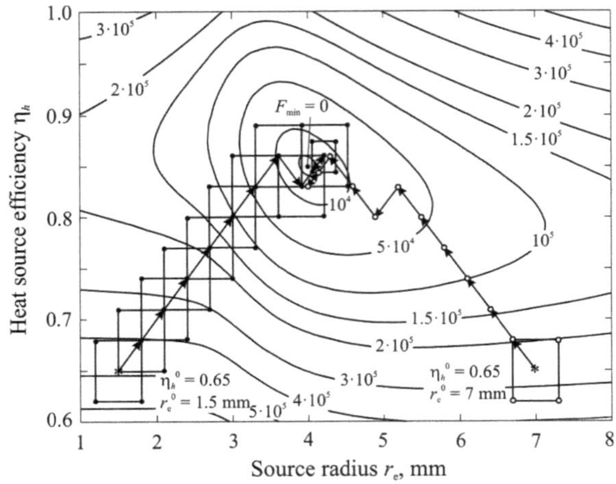

Fig. 11.6 Contour map of the objective function and direct search for unknown source parameters η_h and r_e at different initial iterations of η_h^0 and r_e^0

bottom side (T_{max} $(W_m/2, h) = T_m$). Let us assign weights $w_1^f = 1000\ \text{mm}^{-2}$, $w_2^f = w_3^f = 1\ \text{K}^{-2}$, other weights in Eqs. (11.1.3) and (11.1.4) equal zero.

Thus, objective function (11.1.3) for the examined example takes the following form:

$$F(\eta_h, r_e) = w_1^f[L_m - L_m(\eta_h, r_e)_{\substack{y=0\\z=0}}]^2 + w_2^f[T_m - T\max(\eta_h, r_e)_{\substack{y=W/2\\z=0}}]^2 + w_3^f[T_m - T\max(\eta_h, r_e)_{\substack{y=W/2\\z=h}}]^2. \tag{11.2.4}$$

The objective function field with two sought parameters of the source (heat source efficiency η_h and radius r_e) is shown in Fig. 11.6. Let us find values η_h and r_e using the zero-order method. The strategy for determining them using the direct search algorithm begins with assigning the initial point ($\eta_h^0 = 0.65$ and $r_e^0 = 1.5$ mm) and calculating the objective function at the centrepoint (the base point indicated by an asterisk) and at the vertices of a rectangle with sides of the initial length of $\Delta\eta_h^0 = 0.06$ and $\Delta r_e^0 = 0.6$ mm. The sides are parallel to the axes. The set of rectangle vertices (in general, corners of a hypercuboid) along with the base point constitute a pattern. After calculating function values at the vertices, let us select the vertex at which the function value is lower than at the base point as the new base point. Then let us repeat the above procedure of building a pattern and select the next base point, etc. If there is no such vertex (i.e. all function values are higher), the base point remains the same and a pattern is built with the reduced (for example, in half) rectangle side length. The search is complete once the rectangle side length becomes lower than the prescribed value.

The objective function field in Fig. 11.6 is the simplest case of objective function form (with only one minimum and simple contour line topography). If another point is taken, ($\eta_h^0 = 0.65$ and $r_e^0 = 7$ mm), the solution also converges to minimum F_{min}.

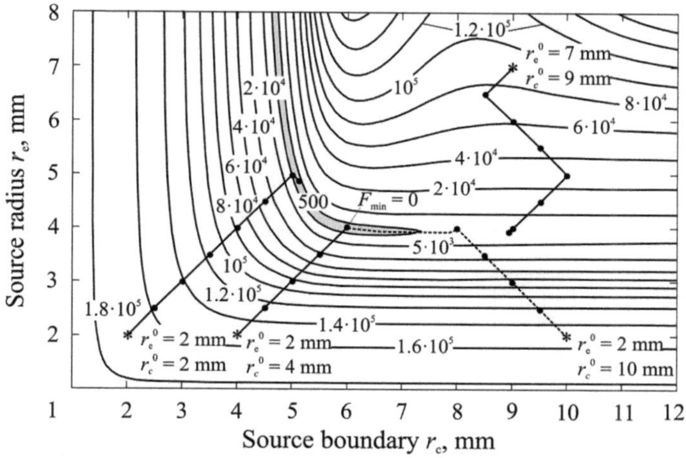

Fig. 11.7 Contour map of the objective function and direct search for unknown source parameters r_e and r_c at different initial iterations of r_e^0 and r_c^0

Let us now consider the same example where only source radius r_e and source boundary r_c are sought (heat source η_h is known). Figure 11.7 shows that the contour lines of $F = \text{const}$ are significantly prolonged near the minimum. The decrease surface in one direction is significantly larger than in the other (function F looks like the banana function because of its distinctive geometry). The search may come to an end after reaching this area (shaded), since the parameter increment is insignificant. In such cases, it is recommendable to set different initial approximations and use additional techniques such as rotating the rectangle around the base point.

Another faster way of finding the objective function is the simplex search algorithm (Attetkov et al. 2001). In the case when two parameters are unknown, a triangle is used instead of a rectangle (in general, a multi-dimensional polyhedron). The basic concept of the algorithm is as follows: the objective function value at the vertices of the triangle is calculated, then the worst vertex at which the function value is higher is replaced by a new vertex, forming a new triangle. The new vertex is a mirror image of the worst vertex. This procedure is then repeated.

A geometric illustration of the simplex search for the minimum of the objective function is shown in Figs. 11.8 and 11.9. If surface F has a banana structure, it is possible to find F_{min} by setting different initial approximations. Figure 11.9 shows four cases, only one of which converges to the minimum.

Many versions of the direct search algorithm have been developed in order to streamline the process of finding the minimum of the objective function (Attetkov et al. 2001).

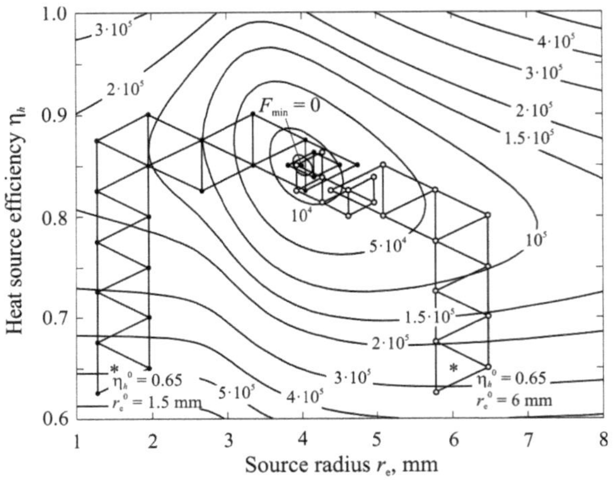

Fig. 11.8 Contour map of the objective function and simplex search for unknown source parameters η_h and r_e at different initial iterations of η_h^0 and r_e^0

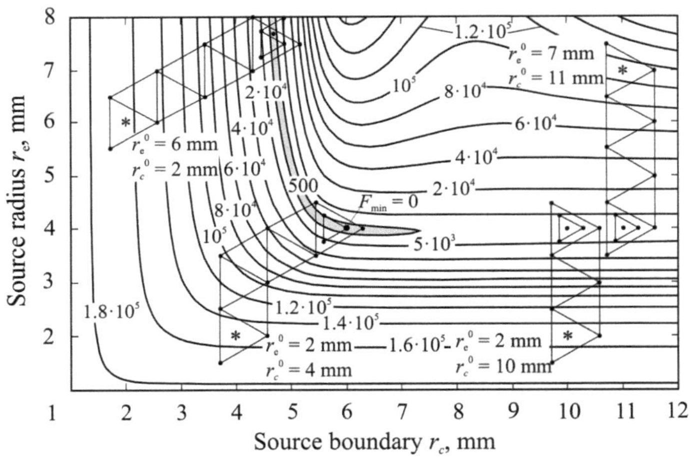

Fig. 11.9 Contour map of the objective function and simplex search for unknown source parameters r_e and r_c at different initial iterations of r_e and r_c

11.2.2 First-Order Method

If objective function $F(\mathbf{p})$ is differentiable, an algorithm of finding its minimum can be developed using the information about the gradient of the function. The search direction in the gradient descent method is taken as the negative of the gradient of the objective function—grad $F(\mathbf{p})$, while relaxation sequence elements $\{\mathbf{p}^m\}$ are built using the following recurrence relation (Attetkov et al. 2001):

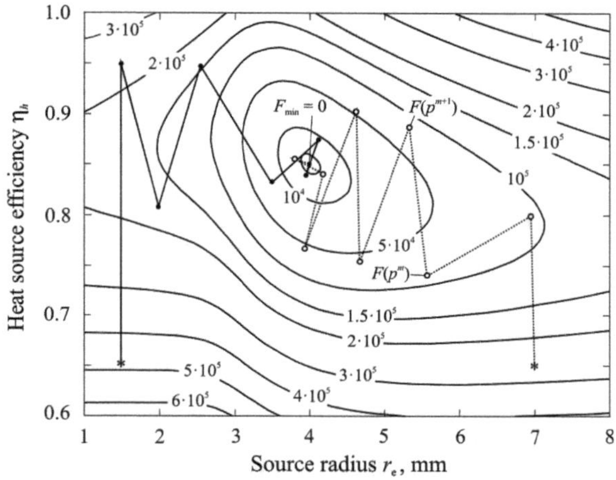

Fig. 11.10 Contour map of the objective function and search for unknown source parameters η_h and r_e by a steepest-descent algorithm at different initial iterations

$$\mathbf{p}^{m+1} = \mathbf{p}^m - \ell_m \text{grad}\, F(\mathbf{p}^m), \quad m = 0, 1, 2, \ldots \tag{11.2.5}$$

Step size ℓ_m is specified and remains constant as long as the function decreases at the points of the sequence. Otherwise, the step value is reduced (for example, in half) until the condition $\ell_m < \ell_{\min}$ is met.

A geometric illustration of the described algorithm in a two-dimensional case is shown in Fig. 11.10 (the initial conditions are the same as those in Sect. 11.2.1). When the surface of the objective function is more complex, it is recommendable to increase search precision (i.e. reduce value $\ell_{\min}$) and set different initial approximations (Fig. 11.11).

There are other first-order methods, which are more complex, but require fewer iterations in order to achieve the same search precision with a comparable amount of calculation during each iteration (Attetkov et al. 2001).

11.2.3 Second-Order Method

If objective function $F(\mathbf{p})$ is twice differentiable, the efficiency of the process of finding the optimum point can be increased by using information not only about function gradient grad $F(\mathbf{p})$, but also about second-order derivatives (the so-called Hessian matrix). During each $(i + 1)$ iteration of the algorithm in its simplest form, the objective function is approximated in the neighborhood of point $\mathbf{p}^i$ by a quadratic function, and then point $\mathbf{p}^{i+1}$ of the minimum of this quadratic function is determined. During the next iteration, a new quadratic approximation is built, this time in the neighbourhood of point $\mathbf{p}^{i+1}$ (Attetkov et al. 2001).

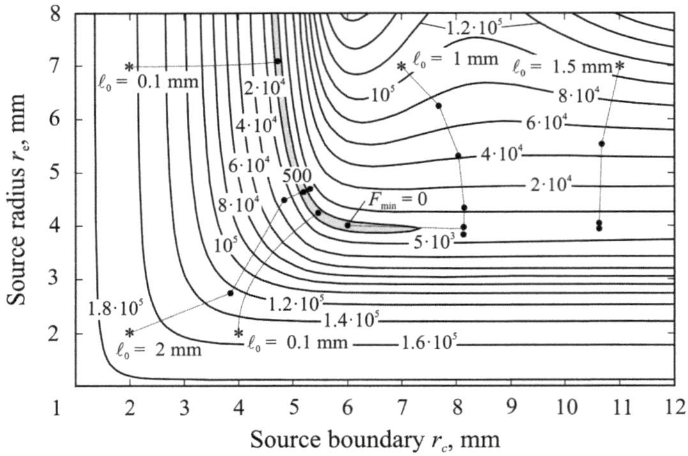

Fig. 11.11 Contour map of the objective function and search for unknown source parameters r_e and r_c by a steepest-descent algorithm at different initial iterations and step l_0

The minimum condition of function F corresponds to the condition under which derivative F equals zero with respect to each parameter p_ℓ ($\ell = 1, \ldots, K$), which leads Eq. (11.1.3) to the following system of equations (Karkhin et al. 2002b):

$$\frac{\partial F}{\partial p_\ell} = -2\sum_{j=1}^{J} w_j^f [f_j^m - f_j(\mathbf{p})] S_{j\ell} - 2w_i^p [p_\ell^0 - p_\ell] = 0, \tag{11.2.6}$$

where $S_{ij} = \partial f_j\,(\mathbf{p})/\partial p_\ell$ is the sensitivity coefficient. Here, a regularisation term is omitted as taking it into account makes the formula more complex (Karkhin et al. 2002b).

Equation (11.2.6) is non-linear and can be solved using any iteration method such as the Gauss–Newton algorithm. For this purpose, let us express unknown parameter p_k^{i+1} and function $f_j(\mathbf{p}^{i+1})$ during the next $(i + 1)$ iteration in terms of known values p_k^i and $f_j(\mathbf{p}^i)$ during the current iteration i (i.e. linearise function f_j):

$$p_k^{i+1} = p_k^i + \Delta p_k; \tag{11.2.7}$$

$$f_j(\mathbf{p}^{i+1}) = f_j(\mathbf{p}^i) + \sum_{k=1}^{K} S_{jk}^i \Delta p_k. \tag{11.2.8}$$

By substituting (11.2.8) into (11.2.5), we shall get K linear equations with respect to K unknown increments of parameters Δp_k:

$$\mathbf{A}\Delta\mathbf{p} = \mathbf{B}. \tag{11.2.9}$$

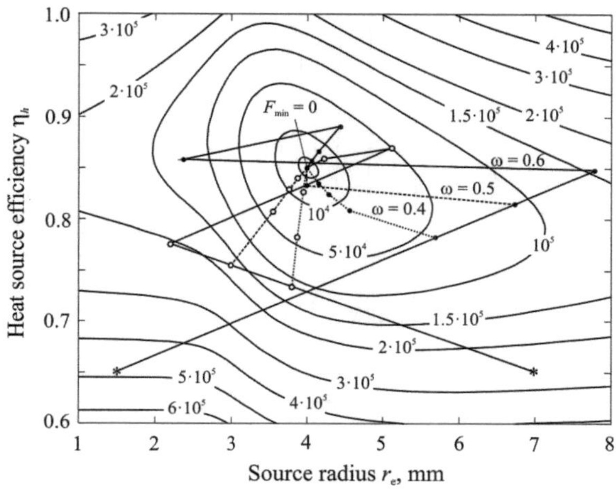

Fig. 11.12 Contour map of the objective function and search for unknown source parameters η_h and r_e by a second-order method at different initial iterations and iteration parameter ω

Each ℓth equation ($\ell = 1, \ldots, K$) takes the following form:

$$\sum_{k=1}^{K}\left(\sum_{j=1}^{J} w_j^f S_{jk}^i S_{j\ell}^i\right)\Delta p_k + w_\ell^p \Delta p_\ell$$

$$= \sum_{j=1}^{J} w_j^f [f_j^m - f_j(\mathbf{p}^i)] S_{j\ell}^i + w_\ell^p [p_\ell^0 - p_\ell^i]. \quad (11.2.10)$$

In order to obtain Eq. (11.2.9), it is necessary to solve a number of direct problems for determining f_j and S_{jk} during each iteration (the minimum number of direct problems equals $K + 1$).

The value of the unknown parameter during the next ($i + 1$) iteration is calculated according to the following formula:

$$p_k^{i+1} = p_k^i + \omega_k \Delta p_k, \quad (11.2.11)$$

where ω_k is the iteration parameter that determines a convergence rate, $\omega_k > 0$; Δp_k is the solution to Eq. (11.2.9). The iteration process ends once the increment of each parameter Δp_k or the value of function F is sufficiently insignificant. The obtained solution is, by definition, a solution to the inverse problem (Karkhin et al. 2002b).

Figure 11.12 shows a graphic illustration of the method in a two-dimensional case (the initial conditions are the same as those in Sects. 11.2.1 and 11.2.2). It can be seen that relaxation parameter ω_k is the major parameter that affects the convergence rate. With some values of ω_k ($\omega_1 = \omega_2 > 0.6$ in this example) the process diverges or converges extremely slowly (in the case of small values of ω_k). The optimum value of ω_k depends on the initial approximation.

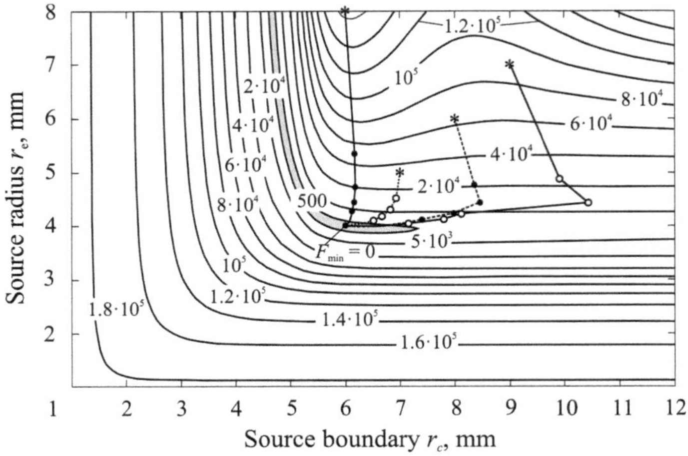

Fig. 11.13 Contour map of the objective function and search for unknown source parameters r_e and r_c by a second-order method at different initial iterations and relaxation parameter $\omega = 0.3$

If the objective function has a banana structure, the initial parameter value becomes important (Fig. 11.13). Thus, the examined problem has a solution if the initial parameters have the following values: $r_e = 4.5$–8.0 mm and $r_c = 6$–10 mm, otherwise the process diverges.

One advantage of the examined method is that it takes significantly fewer iterations and, consequently, less time to solve the inverse problem. However, this method requires the objective function to be relatively smooth. This method has numerous modifications (Attetkov et al. 2001).

Inverse modelling methods make it possible to restore the temperature field according to the individual characteristics of the welded joint in

- laser welding of aluminium alloys (Karkhin et al. 2001a, b, 2002a, b), C-Mn steel (Karkhin et al. 2001c, 2002b, 2007a, e, 2010), austenitic steel (Karkhin et al. 2001c, 2002b), ferritic chromium steel (Karkhin et al. 2002d, 2003c), stainless steel (Lambrakos and Michopoulos 2007), nickel alloy (Pittner et al. 2010);
- electron beam welding of low-carbon steel (Karkhin et al. 2001a, 2002b), austenitic Cr-Ni steel (Karkhin et al. 2001c, 2002b), aluminium alloy (Lambrakos and Milewski 2010; Zervaki et al. 2007);
- arc welding of copper (Rajamaki et al. 2007);
- hybrid laser-arc welding of medium-carbon steel (Karkhin et al. 2013a);
- hybrid laser-GMA welding of steel (Pittner 2012; Pittner et al. 2015);
- friction welding of steel (Lopera et al. 2013; Maalekian et al. 2007);
- friction stir welding (De Vuyst et al. 2004; Fonda and Lambrakos 2002).

Besides, the required solutions to the direct head conduction problem can be obtained using analytical and numerical methods. A preliminary sensitivity analysis (for example, temperature change with respect to the change in the source parameter)

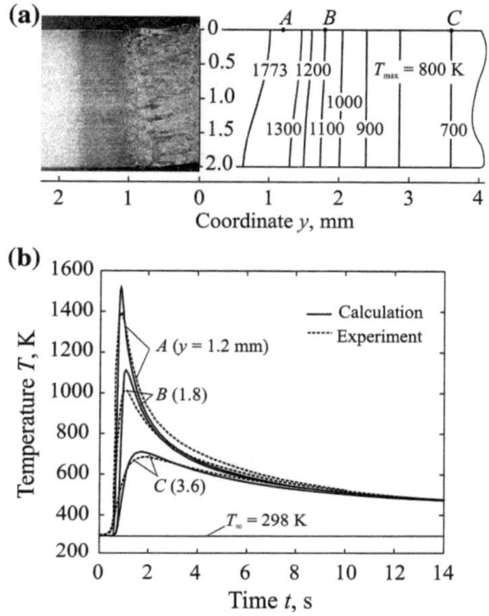

Fig. 11.14 The experimental (left) and calculated (right) cross section of the welded joint (**a**) and the thermal cycles of the points on the surface of the sheet (**b**) in laser welding of low-alloy steel 2 mm thick (Karkhin et al. 2007a, e)

makes it possible to better select a number of unknown parameters, and identify the best position of points when a thermal cycle is taken as the response function (Gabriel et al. 2007).

Let us consider laser welding of low-carbon steel plates with the thickness of 2 mm ($C \leq 0.20\%$, 1.6% Mn). Welding conditions: gross beam power $q_{gross} = 1800$ W, speed $v = 16.67$ mm s^{-1}, initial temperature $T_0 = 293$ K (Ossenbrink et al. 2003).

The coordinates of the points of the weld interface in the cross-section and the measured thermal cycles of the three surface points are taken as input data for the inverse problem (Fig. 11.14). An apparent volumetric source continuously distributed according to the normal law in the plane of the plate and exponential law along the entire thickness (Fig. 3.13). A solution to the inverse problem for 8 unknown source parameters makes it possible to restore the entire three-dimensional temperature field (Fig. 11.15).

It should be remarked that methods for inverse problem solution can be used not only to determine the characteristics of a heat source, but also to calculate properties of liquid metal of the weld pool (for example, thermal conductivity and viscosity) (De and DebRoy 2004, 2005; Kumar et al. 2005) and temperature dependences of the thermal and physical properties of metal (Beck et al. 1985; Beck 1991). The methods make it possible to obtain information about welding processes that cannot be measured such as measurement of hydrogen evolved from the test piece during welding (Karkhin et al. 2007c, d, 2009; Karkhin and Levchenko 2008).

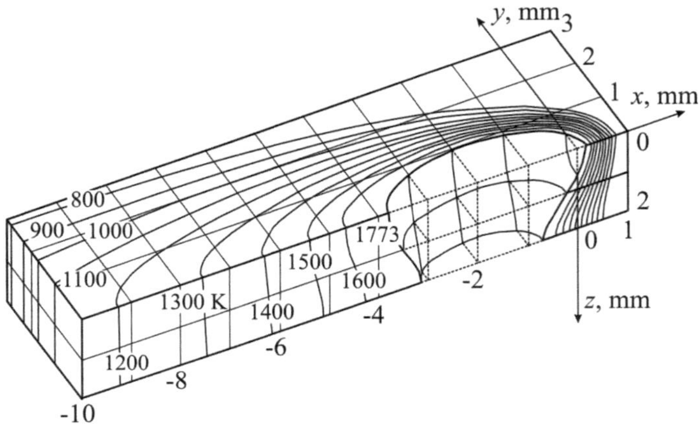

Fig. 11.15 The steady 3D temperature field in laser welding of low-alloy steel 2 mm thick (half of the welded joint without the liquid weld pool is shown) (Karkhin et al. 2007a, e)

References

Alifanov, O. M. (1994). *Inverse heat transfer problems* (348 pp.). Berlin: Springer.

Attetkov, A. V., Galkin, S. V., & Zarubin, V. S. (2001). *Methods of optimisation* (440 pp.). Moscow: Moscow State Technical University Publishing (in Russian).

Beck, J. V. (1991). Inverse problems in heat transfer with application to solidification and welding. In *Modelling of casting, welding and advanced solidification processes* (Vol. V, pp. 503–515). Warrendale, PA: TMS.

Beck, J. V., Blackwell, B., & St. Clair, C. R., Jr. (1985). *Inverse heat conduction: Ill posed problems* (308 pp.). Wiley-Interscience.

De, A., & DebRoy, T. (2004). Probing unknown welding parameters from convective heat transfer calculation and multivariable optimization. *Journal of Physics D: Applied Physics, 37*, 140–150.

De, A., & DebRoy, T. (2005). Reliable calculations of heat and fluid flow during conduction mode laser welding through optimization of uncertain parameters. *Welding Journal,* (7), 101-s–112-s.

De Vuyst, T., D'Alvise, L., Simar, A., de Meester, B., & Pierret, S. (2004). Inverse analysis using a genetic algorithm for the finite element modelling of friction stir welding. In *5th International Friction Stir Welding Symposium,* Metz, France, 14–16 September 2004 (16 pp.).

Fonda, R. W., & Lambrakos, S. G. (2002). Analysis of friction stir weld using an inverse problem approach. *Science and Technology of Welding and Joining, 7*(3), 177–181.

Gabriel, F., Ayrault, D., Fontes, A., Roatta, J. L., & Raynaud, M. (2007). Global method for estimation of heat source parameters dedicated to narrow gap GTA welding. In H. Cerjak, H. K. D. H. Bhadeshia, & E. Kozenschnik (Eds.), *Mathematical modelling of weld phenomena* (Vol. 8, pp. 891–906). Graz: Verlag der Technischen Universitaet Graz.

Gill, P. E. (1984). *Practical optimization* (401 pp.). London: Academic Press.

Karkhin, V. A., Homich, P. N., Ivanov, S. Yu., & Karimi J. (2013a). Prediction of microstructure and mechanical properties of weld metal in hybrid laser-arc welding. In *Proceedings of the 7th International Scientific and Technical Conference on Beam Technologies and Laser Application,* 18–21 September 2013, (pp. 38–45). St. Petersburg, Russia: St. Petersburg Polytechnic University Publishing.

Karkhin, V. A., Homich, P. H., & Michailov, V. G. (2007a). Analytical-experimental technique for calculating the temperature fields in laser welding. In *11th NOLAMP Conference in Laser Processing of Materials,* Lappeenranta, Finland (pp. 21–27).

Karkhin, V. A., Il'in, A. S., & Ploshikhin, V. V. (2003c). Solving an inverse problem of heat conductivity, taking the heat of melting and solidification into account. *Welding International, 17*(12), 971–974.

Karkhin, V. A., Ilin, A. S., & Ploshikhin, V. V. (2003a). Inverse modelling of deep penetration welding. In IIW Doc. XII-1738-03 (9 pp.).

Karkhin, V. A., Ilin, A. S., Ploshikhin, V. V., & Bergmann, H. W. (2001a). Inverse modelling of the heat input distribution during deep penetration laser and electron beam welding. In F. O. Olsen & J. K. Kristensen (Eds.), *Proceedings of 8th Conference on Laser Materials Processing in the Nordic Countries*, Copenhagen, 13–15 August 2001 (pp. 289–300).

Karkhin, V. A., Ilin, A. S., Ploshikhin, V. V., & Prikhodovsky, A. A. (2003b). Effect of latent heat on thermal efficiency of base metal fusing. In IIW Doc. XII-1739-03 (10 pp.).

Karkhin, V. A., Ilyin, A. S., Plochikhine, V. V., & Bergmann, H. W. (2001c). Solution of inverse heat conduction problem for determining heat input distributions during laser and electron beam welding. In *Proceedings of 3rd All-Russian Conference on Computer Technologies in Joining of Materials*, Tula (pp. 119–127) (in Russian).

Karkhin, V. A., Khomich, P. N., & Michailov, V. G. (2006). Prediction of microstructure and mechanical properties of weld metal with consideration for real weld geometry. In W. Lucas & V. I. Makhnenko (Eds.), *Proceedings of Joint International Conference "Computer Technology in Welding and Manufacturing (16th International Conference) and Information Technologies in Welding and Related Processes (3rd International Conference)"*, Kiev (pp. 162–166).

Karkhin, V. A., Khomich, P. N., & Michailov, V. G. (2007b). Models for volume heat sources and functional-analytical technique for calculating the temperature fields in butt welding. In H. Cerjak, H. K. D. H. Bhadeshia, & E. Kozeschnik (Eds.), *Mathematical modelling of weld phenomena* (Vol. 8, pp. 819–834). Graz: Verlag der Technischen Universitaet Graz.

Karkhin, V. A., Khomich, P. N., Michailov, V. G., & Ivanov, S. Yu. (2015). Solution of inverse heat conduction problem in fusion welding. *Welding and Diagnostics,* (3), 20–23 (in Russian).

Karkhin, V. A., Khomich, P. N., Ossenbrink, R., & Michailov, V. G. (2007e). Calculation-experimental method for the determination of the temperature field in laser welding. *Welding International, 21*(5), 391–395.

Karkhin, V. A., & Levchenko, A. M. (2008). Computer-aided determination of diffusible hydrogen in deposited weld metal. *Welding in the World, 52*(3/4), 3–11.

Karkhin, V. A., Levchenko, A. M., & Homich, P. N. (2009). Express determination of diffusible hydrogen in deposited weld metal. In *Proceedings of St. Petersburg State Polytechnic University, "Materials and Chemical Technologies"* (Vol. 510, pp. 180–192). St. Petersburg: St. Petersburg State Polytechnic University Publishing (in Russian).

Karkhin, V. A., Levchenko, A. M., & Khomich, P. N. (2007d). A method for the rapid determination of the concentration of diffusible hydrogen in deposited metal. *Welding International, 21*(6), 466–470.

Karkhin, V. A., Levchenko, A. M., Khomich, P. N., & Michailov, V. G. (2007c). Computer-aided determination of diffusible hydrogen in deposited weld metal. In H. Cerjak, H. K. D. H. Bhadeshia, & E. Kozeschnik (Eds.), *Mathematical modelling of weld phenomena* (Vol. 8, pp. 785–798). Graz: Verlag der Technischen Universitaet Graz.

Karkhin, V. A., Pittner, A., Schwenk, C., & Rethmeier, M. (2010). Heat source models in simulation of heat flow in fusion welding. In *Proceedings of 5th International Conference "Mathematical Modelling and Information Technologies in Welding and Related Processes"*, Crimea, 25–28 May 2010 (pp. 56–60).

Karkhin, V. A., Plochikhine, V. V., & Bergmann, H. W. (2001b). Reconstruction of the heat distribution in the molten pool during laser beam welding. In *Lasers in Manufacturing. Proceedings of the 1st International WLT-Conference on Lasers in Manufacturing*, Munich, June 2001 (pp. 269–272).

Karkhin, V. A., Plochikhine, V. V., & Bergmann, H. W. (2002a). Solution of inverse heat conduction problem for determining heat input, weld shape, and grain structure during laser welding. *Science and Technology of Welding and Joining, 7*(4), 224–231.

Karkhin, V. A., Plochikhine, V. V., & Bergmann, H. W. (2002c). Simulation of thermal and solidification processes in laser welding of aluminium plates. *The Paton Welding Journal, 54*(8), 10–14.
Karkhin, V. A., Plochikhine, V. V., & Ilyin, A. S. (2002d). Inverse computer modelling of thermal processes during welding. In *Proceedings of 3rd International Conference "Computer Modelling 2002"*, 6–8 June 2002, St. Petersburg (pp. 54–59). St. Petersburg: St. Petersburg State Polytechnic University Publishing (in Russian).
Karkhin, V. A., Plochikhine, V. V., Ilyin, A. S., & Bergmann, H. W. (2002b). Inverse modelling of fusion welding processes. In H. Cerjak (Ed.), *Mathematical modelling of weld phenomena* (Vol. 6, pp. 1017–1042). London: Maney Publishing.
Karkhin, V. A., Plochikhine, V. V., Ilyin, A. S., & Bergmann, H. W. (2002e) Inverse modelling of fusion welding processes. *Welding in the World, 46*(11/12), 2–13.
Kim, C.-H., Zhang, W., & DebRoy, T. (2003). Modeling of temperature field and solidified surface profile during gas-metal arc fillet welding. *Journal of Applied Physics, 94*(4), 2667–2679.
Kumar, A., & DebRoy, T. (2007). Heat transfer and fluid flow during gas-metal-arc fillet welding for various joint configurations and welding positions. *Metallurgical and Materials Transactions A, 38*(3), 506–519.
Kumar, A., Zhang, W., Kim, C.-H., & DebRoy, T. (2005). A smart bi-directional model of heat transfer and free surface flow in gas metal arc fillet welding for practicing engineers. In H. Cerjak, H. K. D. H. Bhadeshia, & E. Kozeschnik (Eds.), *Mathematical modelling of weld phenomena* (Vol. 7, pp. 3–37). Graz: Verlag der Technischen Universitaet Graz.
Kurpisz, K., & Nowak, A. J. (1995). *Inverse thermal problems* (258 pp.). Southampton: Computational Mechanics Publications.
Lambrakos, S. G., & Michopoulos, J. G. (2007). Algorithms for inverse analysis of heat deposition processes. In H. Cerjak, H. K. D. H. Bhadeshia, & E. Kozenschnik (Eds.), *Mathematical modelling of weld phenomena* (Vol. 8, pp. 847–879). Graz: Verlag der Technischen Universitaet Graz.
Lambrakos, S. G., & Milewski, J. O. (2010). Analysis of welding and heat deposition processes using an inverse-problem approach. In H. Cerjak, & N. Enzinger (Eds.), *Mathematical modelling of weld phenomena* (Vol. 9, pp. 1025–1056). Graz: Verlag der Technischen Universitaet Graz.
Lopera, J., Mucic, K., Fuchs, F., & Enzinger, N. (2013). Linear friction welding of high strength chains: Modelling and validation. In C. Sommitsch, & N. Enzinger (Eds.), *Mathematical modelling of weld phenomena* (Vol. 10, pp. 203–218). Graz: Verlag der Technischen Universitaet Graz.
Maalekian, M., Kozeschnik, E., Brantner, H. P., & Cerjak, H. (2007). Inverse modelling of heat generation in friction welding. In H. Cerjak, H. K. D. H. Bhadeshia, & E. Kozenschnik (Eds.), *Mathematical modelling of weld phenomena* (Vol. 8, pp. 881–890). Graz: Verlag der Technischen Universitaet Graz.
Moiseyev, N. N., Ivanov Y. P., & Stolyarova, E. M. (1978). *Optimisation methods* (352 pp.). Moscow: Nauka (in Russian).
Ossenbrink, R., Brand, M., Michailov, V., Wiebe, J., & Wohlfahrt, H. (2003). Numerische Simulation der Eigenspannungen und des Umwandlungsver haltens. In *DVS Berichte* (Vol. 225, pp. 279–283). Duesseldorf: DVS – Verlag (in German).
Panteleyev, A. V., & Letova, T. A. (2002). *Optimisation methods in examples and problems* (544 pp.). Moscow: Vysshaya Shkola (in Russian).
Pittner, A. (2012). *A contribution to the solution of the inverse heat conduction problem in welding simulation* (201 pp.). Berlin: BAM-Dissertationsreihe, Band 85.
Pittner, A., Karkhin, V., & Rethmeier, M. (2015). Reconstruction of 3D transient temperature fields for fusion welding processes on basis of discrete experimental data. *Welding in the World, 59*(4), 497–512.
Pittner, A., Schwenk, C., Weiss, D., & Rethmeier, M. (2010). An efficient solution of the inverse heat conduction problem for welding simulation. In H. Cerjak & N. Enzinger (Eds.), *Mathematical modelling of weld phenomena* (Vol. 9, pp. 761–791). Graz: Verlag der Technischen Universitaet Graz.

Radaj, D. (2003). *Welding residual stresses and distortion. Calculation and measurement* (397 pp.). Duesseldorf: DVS-Verlag.

Rajamaki, P., Karkhin, V. A., & Khomich, P. N. (2007). Determination of the main characteristics of the temperature field for the evaluation of the type of solidification of weld metal in fusion welding. *Welding International, 21*(8), 600–604.

Tikhonov, A. N., & Arsenin, A. Ya. (1977). *Solutions of ill-posed problems* (258 pp.). Winston.

Tikhonov, A. N., Kalner, V. D., & Glasko, V. B. (1990). *Mathematical modelling of technological processes and inverse problems in mechanical engineering* (264 pp.). Moscow: Mashinostroienie (in Russian).

Vanderplaats, G. N. (1984). *Numerical optimization techniques for engineering design* (333 pp.). New York: Hill Publishing Company.

Zabaras, N. (1991). Inverse modeling of solidification and welding processes. In *Modeling of casting, welding and advanced solidification processes* (Vol. 5, pp. 523–530).

Zervaki, A. D., Haidemenopoulos, G. N., & Lambrakos, S. G. (2007). Analysis of heat affected zone using direct and inverse modelling in 6XXX aluminium alloys. In H. Cerjak, H. K. D. H. Bhadeshia, & E. Kozenschnik (Eds.), *Mathematical modelling of weld phenomena* (Vol. 8, pp. 907–923). Graz: Verlag der Technischen Universitaet Graz.

Zhang, W., Kim, C. H., & DebRoy, T. (2004). Heat and fluid flow in complex joints during gas metal arc welding—Part 1: Numerical model of fillet welding. *Journal of Applied Physics, 95*, 5210–5219.

Chapter 12
Optimisation of Welding Conditions

Chapter 11 provided insights into the inverse problem of heat conduction and methods of its solution. From certain effects of the welding source, its unknown parameters were obtained, and the entire temperature field was reconstructed. In principle, the inverse problem is an optimisation problem that can be solved using optimisation methods.

Another type of optimisation problem is also of practical importance: finding the optimum welding conditions according to the required parameters of the weld joint (constant weld width along its length, maximum depth of fusion with the required heat input per unit length of the weld, minimum depth of the fused base metal during deposition welding, minimum residual stresses, etc.). Optimisation problems are examined below.

12.1 Optimum Heat Input in Girth Welding of Thin-Walled Small Diameter Pipes

In girth welding of small diameter pipes, the following optimisation problem arises: finding such a law of heat input along the butt joint so that the size of the weld along its entire length would be constant and equal to the prescribed size. Such a law will naturally depend on the diameter and thickness of the pipe as well as material properties.

As shown in Sect. 5.1.3.3, calculation of the temperature field in an infinitely long thin-walled pipe can be reduced to the calculation of the temperature field in an infinite plate and slab with real and fictitious heat sources. Therefore, let us formulate the optimisation problem for the plate (more specifically, for the slab, since the temperature field is three-dimensional).

Let the source move straight along x-axis at the constant speed v on a plate with thickness h from point A to point B (Fig. 12.1). Let us divide down length AB into K equal sections and assume that at each kth section source power is constant, q_k

V. A. Karkhin, *Thermal Processes in Welding*, Engineering Materials,
https://doi.org/10.1007/978-981-13-5965-1_12

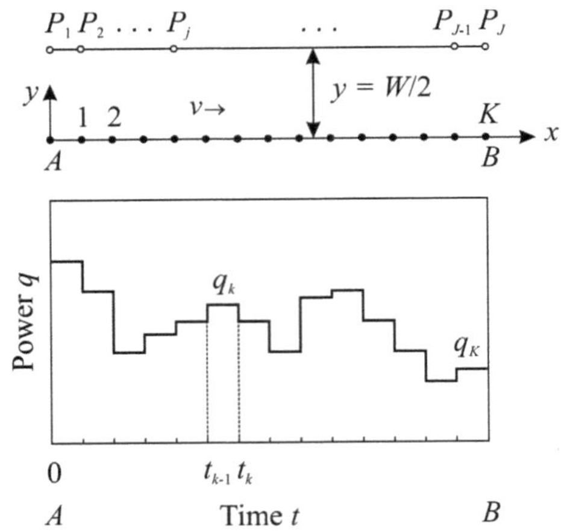

Fig. 12.1 Optimisation problem for heat input in arc welding (Kondo et al. 1998; Kondoh and Ohji 1998)

= const. Let us select points P_j ($j = 1, ..., J$) at distance $y = W/2$ from the weld centreline (W being weld width) and require that the peak temperature at these points be equal to melting temperature, $T_{\max j} = T_m$. Then, similarly to (11.1.3), objective function F can be defined as follows (Kondo et al. 1998; Kondoh and Ohji 1998):

$$F(\mathbf{q}) = \sum_{j=1}^{J} [T_m - T_{\max j}]^2 \rightarrow \min; \tag{12.1.1}$$

$$q_{\min} \leq q_k \leq q_{\max}. \tag{12.1.2}$$

The latter constraints of the sought vector $\mathbf{q} = \{q_1, \ldots, q_k, \ldots, q_K\}$ reflect the physical requirements, $q_{\min}$ is determined by the stability of the welding process, and $q_{\max}$ is determined by the capabilities of welding equipment. In any case, quantity q_k is non-negative.

Let the source be a point and surface one and the surfaces be heat-impermeable. In this case, the general analytical solution to the heat conduction problem is known, Eq. (5.1.78). Then the temperature field at any time t is the sum of temperature increments due to each kth constant power source q_k:

$$T(x, y, z, t) - T_0 = \sum_{k=1}^{K} q_k \cdot g_k(x, y, z, t), \tag{12.1.3}$$

where

$$g_k(x, y, z, t) = \frac{1}{4c\rho(\pi a)^{3/2}} \int_{t_{k-1}}^{t_k} \frac{1}{(t - \tau)^{3/2}}$$

$$\times \sum_{i=-\infty}^{\infty} \sum_{n=-\infty}^{\infty} \exp\left(-\frac{(x - v\tau - n2\pi R)^2 + y^2 + (z - 2ih)^2}{4a(t-\tau)}\right) d\tau \text{ for } \tau < t;$$

$$g_k(x, y, z, t) = 0 \text{ for } \tau \geq t. \qquad (12.1.4)$$

The physical meaning of influence function g_k is temperature increment due to the kth unit power source. The integral in (12.1.4) is expressed in terms of known functions (see formula (5.1.39)).

Objective function (12.1.1) can be expressed in terms of unknown components q_k of vector $\mathbf{q}$ (Kondoh and Ohji 1998):

$$F(\mathbf{q}) = \sum_{j=1}^{J} \left[T_m - \max t \left(\sum_{k=1}^{K} q_k \cdot g_{k,j}(t) \right) \right]^2 \to \min. \qquad (12.1.5)$$

Here, $g_{k,j}(t)$ is the function accounting for the influence of the kth unit power source on the temperature at target point P_j at time t.

It should be remarked that target points P_j can generally be set anywhere in the welded joint with any temperature $T_{\max}$ at each point. Point coordinates and temperature are selected from physical considerations.

Thus, solving the optimisation problem involves finding a sequence of optimum values q_k that would minimise objective function (12.1.5) with regard to constraint (12.1.2). It is advisable to solve this problem using the method of steepest descent (Kondo et al. 1998; Kondoh and Ohji 1998) (see Sect. 11.2.2).

Example 12.1.1 Two aluminium pipes of thickness $h = 3$ mm and outer diameter $D = 30$ mm are welded at the speed $v = 3.5$ mm s^{-1}. Peak temperature equal to melting temperature, $T_{\max j} = T_m = 894$ K, is set at points P_j ($j = 1, 2, \ldots, J$) located on the inner surface of the pipe at a distance $y = 3$ mm. The number of unknown power values K equals the number of points J, $K = J = 24$ (Kondoh and Ohji 1998).

Figure 12.2 shows that with optimum distribution of power along the joint length, weld width (isotherm T_m position) is constant. If arc power is constant, the temperature is lower than melting temperature T_m at the beginning of the weld and much higher at its end. This means that the weld will be narrower than average at the beginning and much wider at the end, which has been experimentally verified.

A similar optimisation problem arises in arc welding of thin structured panels with the curvilinear weld joint. At the optimum distribution of arc power along the weld, the weld width is constant (Ivanov et al. 2015, 2017).

12.2 Optimisation of Pulsed Power Welding Conditions

Pulsed power welding has become widespread due to the known advantages over constant power welding: increased efficiency of base metal fusion and joint penetration, capability of welding in various spatial positions, reduced distortion of welded

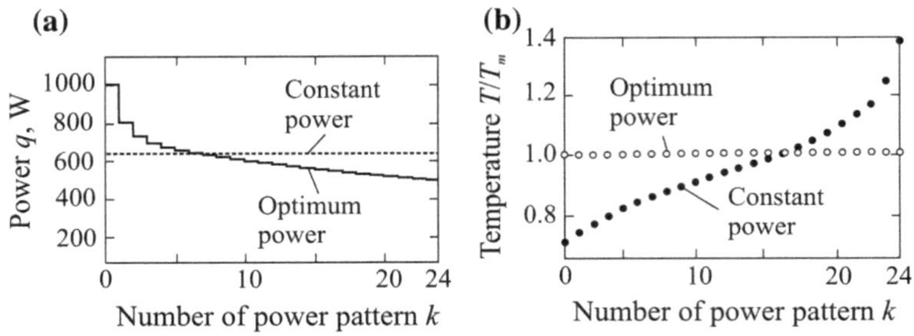

Fig. 12.2 Calculated optimum heat input along the weld (**a**) and relative temperature at a distance of 3 mm from the weld centreline (**b**) in girth welding of aluminium pipe 3 mm thick and 30 mm in diameter (Kondo et al. 1998)

structures, increased chemical homogeneity of weld metal and decreased susceptibility to hot cracking, etc. (Dudko et al. 1987; Karkhin and Khomich 2005, 2007; Saraev 1994; Street 1990).

A practical issue arises in the development of welding equipment and technology: under what conditions can the heat input per unit length of the weld at a required depth of fusion be reduced or the depth of fusion with the given heat input be increased? The answers to these questions are the solutions to the corresponding optimisation problems. Additional variables (pulse shape and frequency) make it possible to better optimise welding conditions compared with constant power welding (Street 1990). The solution to this optimisation problem is based on dealing with the corresponding direct heat conduction problem previously discussed in Sect. 7.2.

Let average heat input per unit length of weld q_{av}/v serve as the objective function. It is required to obtain a minimum depth of fusion $H_{\min}$ (in longitudinal plane $y = 0$ minimum coordinate z of the point with peak temperature $T_{\max}$ equalling melting temperature T_m) under the specified constraints. In this case, the optimisation problem (of minimising the objective function) with constraints of the type of equality and inequality can be presented as follows:

$$\begin{aligned} &q_{av}/v \to \min; \\ &z(y = 0, T_{\max} = T_m) = H_{\min}; \\ &0 < v < v_{\max};\ 0 < q < q_{\max}, \text{ etc.,} \end{aligned} \tag{12.2.1}$$

where parameters of welding conditions (speed v, frequency f, etc.) are optimisation parameters (design variables).

Example 12.2.1 Let us consider the optimisation problem with the following parameters: $H_{\min} = 2$ mm, $v_{\max} = 10$ mm s^{-1}, $q_{\max} = \infty$, $t_p = t_s = 0$ (pulse being as narrow as possible, see Fig. 7.2d), $T_0 = 293$ K, material being thick low-carbon steel ($a = 9.4$ mm^2 s^{-1}, $c\rho = 0.0035$ J mm^{-3} K^{-1}, $T_m = 1723$ K).

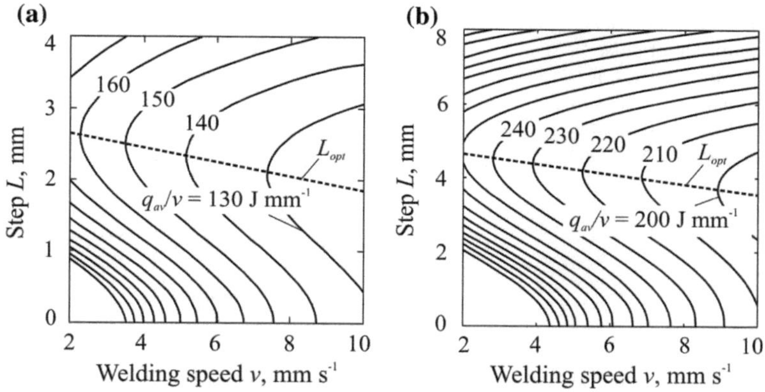

Fig. 12.3 The required effective heat input per unit length of weld q_{av}/v for the depth of fusion $H_{\min} = 2$ mm in pulsed power welding of low-carbon steel with a heat source with radius $r_e = 0$ (**a**) and $r_e = 2.5$ mm (**b**) (Karkhin and Khomich 2007)

Let the source be a surface circular source, normally distributed, with power density depending on the phase of period Δt,

$$q_2(r, \Delta t) = \frac{q(\Delta t)}{\pi r_e^2} \exp\left(-\left[\frac{r}{r_e}\right]^2\right); r^2 = x^2 + y^2, \qquad (12.2.2)$$

where r_e is the normal radius of the heat source. The source centre and origin of coordinates x, y, z move along the x-axis (Fig. 7.10). If thickness has no effect on the high-temperature area, then steady temperature field in this area can be obtained using Green's function method in its integral form (Karkhin and Khomich 2007):

$$T(x, y, z, \Delta t) - T_0 = \int_0^{\infty} \frac{2q(r, \Delta t)}{c\rho(4a\tau + r_e^2)(\pi a\tau)^{1/2}} \times \exp\left(-\frac{(x + v\tau)^2 + y^2}{4a\tau + r_e^2} - \frac{z^2}{4a\tau}\right) d\tau. \qquad (12.2.3)$$

In the general case, source radius r_e may depend on the phase of period Δt (of time τ).

The objective function field with $r_e = 0$ (point source) is shown in Fig. 12.3a, where L is the length between adjacent pulses (step $L = vt_c$). The figure shows that as speed v increases, the required heat input is reduced due to the reduction of losses on account of longitudinal heat fluxes (Karkhin and Khomich 2005, 2007).

At a constant speed (v = const), there is length L_{opt} where value q_{av}/v is minimum (represented as a dotted line in Fig. 12.3a). For example, with $v = 6$ mm s^{-1} and $r_e =$ 0, required heat input per unit length $q_{av}/v = 136$ J mm^{-1}, if $L = L_{opt} = 2.2$ mm, and

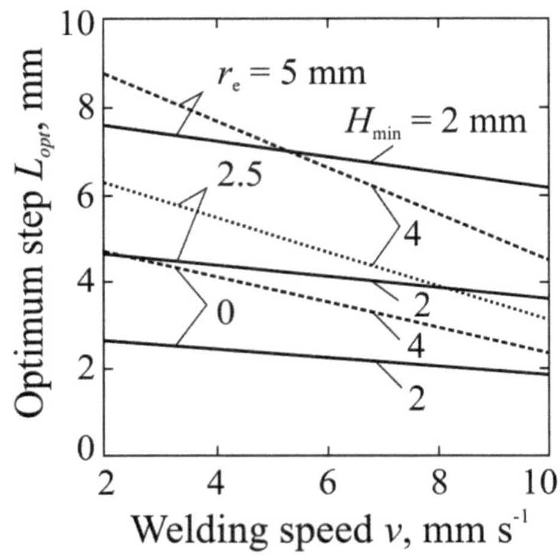

Fig. 12.4 Dependence of the optimum step of pulses L_{opt} on welding speed v at various heat source radius r_e and required depth of fusion H_{min} (Karkhin and Khomich 2007)

$q_{av}/v = 170$ J mm^{-1}, if $L = 0$ (the source power is constant). The global minimum of heat input per unit length of the weld corresponds to speed $v = \infty$ and length $L = 0$ (with no longitudinal fluxes) and can be calculated as a rapidly moving heat source (Rykalin 1951, 1957):

$$q/v = \pi e c\rho (T_m - T_0) H^2/2. \quad (12.2.4)$$

Optimum length L_{opt} (optimum frequency) depends on source radius r_e, resulting from the comparison of Fig. 12.3a, b. The bigger the radius r_e is, the longer the length L_{opt} (less optimum frequency) will be. Value L_{opt} also depends on the required depth H_{min} (Fig. 12.4).

Short pulses have the greatest effect on the penetration of the material (Fig. 12.5). As relative pulse duration t_p/t_c increases, so does the required heat input per unit length of the weld, approaching that of the constant source power. Besides, the optimum conditions shift towards higher frequency (shorter step L).

The depth of fusion reversed in sign with fixed heat input per unit length can be taken as the objective function, which means seeking a maximum depth of fusion:

$$-z(y = 0, T_{max} = T_L) = -H_{min} \rightarrow \min;$$
$$q_{av}/v = \text{const}; 0 < v < v_{max};\ 0 < q < q_{max}, \text{ etc.} \quad (12.2.5)$$

Example 12.2.2 Let us study the effect of distance between pulses L (pulse frequency f) in pulsed gas tungsten arc welding of austenitic steel 304 (0.03% C, 18% Cr, 10% Ni) with the thickness of 6.35 mm in argon: $v = 2.54$ mm s^{-1}, current $I_p = 125$ A, $I_b = 25$ A, $t_p/t_c = 0.25$, $t_s = 0$ (square pulses). Arc radius r_e is found by solving the inverse problem provided that the calculation and experiment in near-

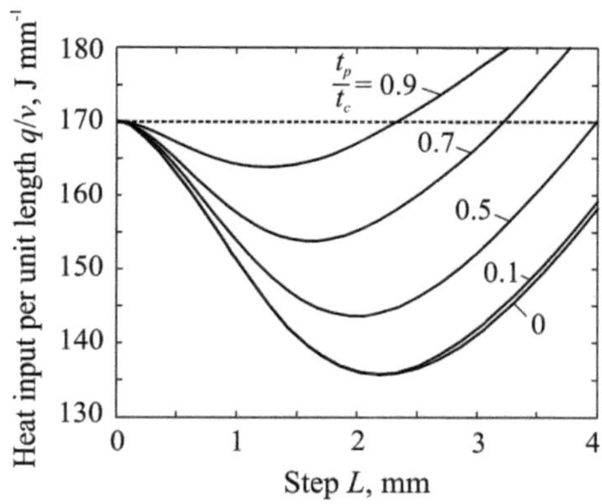

Fig. 12.5 Effect of relative high pulse time t_p/t_c on the effective heat input per unit length of weld q/v required for the depth of fusion $H_{\min} = 2$ mm in pulsed power welding of low-carbon steel with a concentrated heat source ($r_e = 0$) moving at the speed of 6 mm s^{-1} (Karkhin and Khomich 2007)

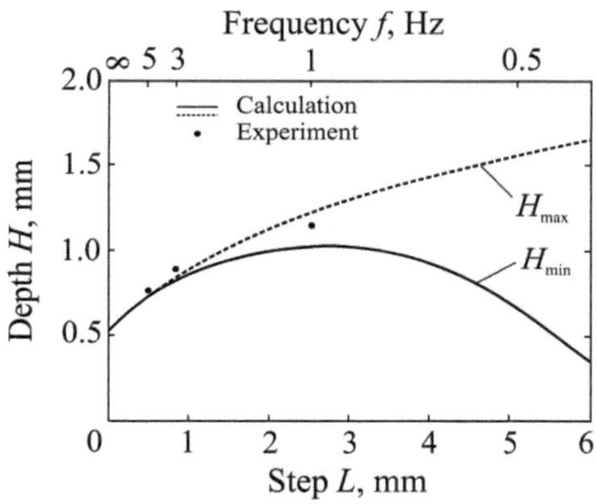

Fig. 12.6 The dependence of the depth of fusion H on the distance between the pulses L (frequency f) in pulsed power welding of 304 austenitic steel (Karkhin and Khomich 2007; Leitner et al. 1973)

constant current welding are consistent ($I_b = 100$ A, $I_p = 125$ A, $f = 1$ Hz) (Leitner et al. 1973).

Figure 12.6 shows that in the beginning as the distance between pulses L increases (frequency f reduces), there is a significant increase in the depth of fusion. At frequencies $f < 3$ Hz (interval $L > 1$ mm), variation in the depth of fusion (difference between $H_{\max}$ and $H_{\min}$) grows increasingly, maximum depth $H_{\max}$ constantly increases, while minimum depth $H_{\min}$ smoothly drops to zero. Hence, it can be concluded that with constant average current (heat input per unit length of the weld) there is a pulse frequency that provides the maximum value of the minimum depth of fusion.

The assessment of solidification of weld pool metal is of practical interest as it determines properties of the weld metal. The rate of liquid metal solidification on the weld centreline ($y = z = 0$) is determined by the tail part of molten pool L_m (Fig. 12.7):

$$v_s = v - \partial L_m / \partial t. \tag{12.2.6}$$

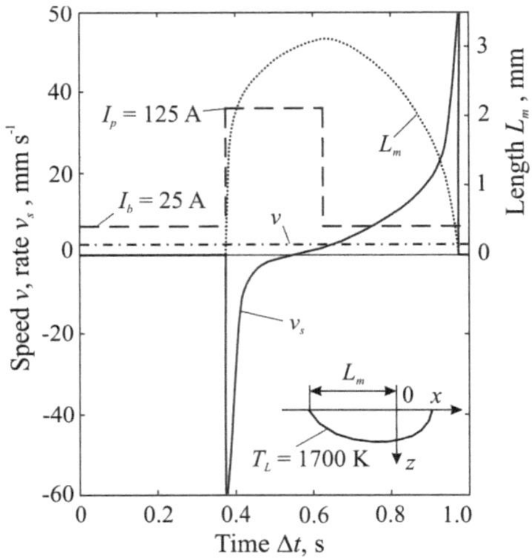

Fig. 12.7 Time dependence of the pool tail length L_m and the rate of liquid metal solidification at weld centreline v_s in pulsed-arc welding of 304 austenitic steel (Karkhin and Khomich 2007)

Figure 12.7 shows second melting at the beginning of the pulse ($v_s < 0$), while after its end solidification rate v_s can be an order of magnitude higher than welding speed v. This means that high-temperature processes in pulsed power welding (grain growth, liquation of elements, etc.) are significantly different from those in constant power welding.

Peak temperature field $T_{\max}$ is shown in Fig. 12.8a. The depth of fusion (isotherms $T_{\max} = T_m$) changes from minimum value $H_{\min}$ (point A) to maximum value $H_{\max}$ (point B). The depth of fusion with constant current equal to average current I_{av} is calculated with the same values r_e and q_{av} (point C). One t can see that penetration efficiency with periodic power is much higher than with constant power.

The thermal cycles of the HAZ metal are significantly different (Fig. 12.8b). The moment when the source centre crosses the cross-section containing the point under study is taken as reference time t ($t = 0$). For example, the dwell time of HAZ metal at high temperature in pulsed power welding is significantly lower than in constant power welding. At a relatively low temperature, cooling rates are almost equal.

12.3 Optimisation of Strip Electrode Shape for Cladding

Previously in Sect. 5.2.3.10, we examined thermal processes in plane strip electrode cladding. At the weld toe, depth of fusion is reduced due to heavy heat dissipation into the base metal (Fig. 5.72d). In these toe zones (where adjacent beads overlap in wide-bead cladding) defects such as slag trap may occur. In order to avoid slag trap, it is necessary to keep metal in a liquid state for a sufficient period of time (Kravtsov 1988). This problem can be formulated as an optimisation problem of obtaining the

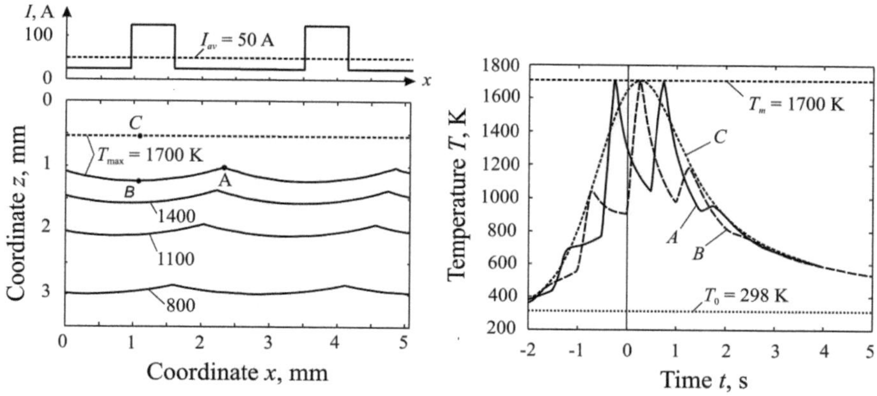

Fig. 12.8 Time dependence of current and the peak temperature field in the longitudinal section of the weld ($y = 0$) (**a**); and the thermal cycles of points *A*, *B* and *C* at the weld interface (**b**) in pulsed arc welding (solid and dashed lines) and constant-current welding (dotted lines) (Karkhin and Khomich 2007)

specified shape and size of weld bead toes by varying the law of heat input in the cross-section of the workpiece.

Arc power is uniformly distributed along the end of the strip with uniform thickness and heat input along the width of the weld pool can be controlled by changing the shape of the strip section. The strip section can be of any shape: straight, semi-ring, C-shaped, Π-shaped, etc. (Fig. 12.9). In these cases, heat input per unit length q_1 along the y-axis is significantly different: from uniform distribution in the case of a straight strip end (Fig. 12.9a) to extremely non-uniform distribution in the case of a Π-shaped strip (Fig. 12.9d). In the latter case, the depth of fusion at the toes can be much larger than in the middle. Varying the shape (curvature) of the strip, it is possible to fulfill the optimisation condition. The objective function should reflect practical requirements such as uniformity of the depth of fusion across the bead and dwell time of metal in a liquid state.

In the case of a plane strip (Fig. 12.9a), the steady-state temperature field in the semi-infinite solid is described by Eq. (5.2.147). The field is shown in Fig. 5.72.

Let us find a solution to the heat conduction problem for a moving semi-ring source (strip in the shape of a half-pipe, Fig. 12.9b) similar to the solution (5.2.105):

$$T(x, y, z, \infty) - T_0 =$$

$$= \frac{q}{2\pi^2\lambda} \int_{-\pi/2}^{\pi/2} \frac{\exp\left(-\frac{v\left[x - r_0\cos\varphi + \sqrt{(x - r_0\cos\varphi)^2 + (y - r_0\sin\varphi)^2 + z^2}\right]}{2a}\right)}{\sqrt{(x - r_0\cos\varphi)^2 + (y - r_0\sin\varphi)^2 + z^2}} d\varphi. \quad (12.3.1)$$

If the strip has convexity with the opposite sign, the formula will change.

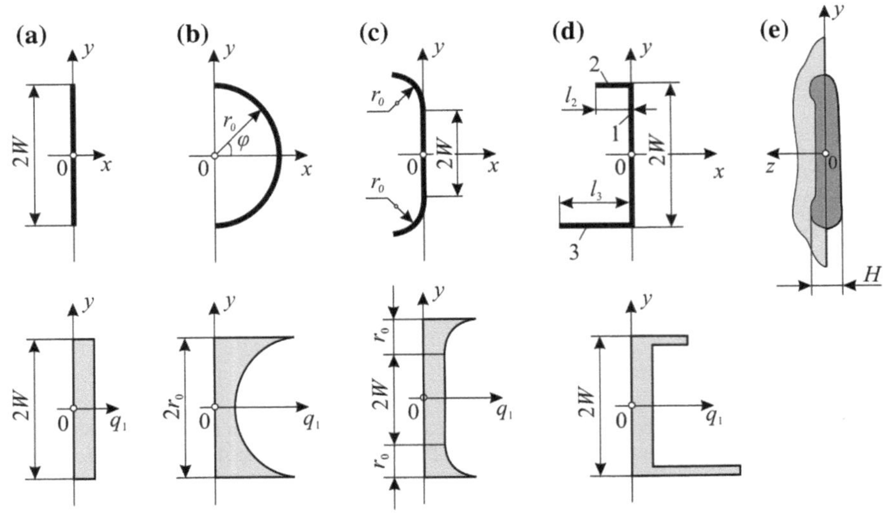

Fig. 12.9 Patterns and corresponding dependences of line power density q_1 on the distance from the weld centreline for rectilinear (**a**), semi-ring (**b**), C-shaped (**c**) and Π-shaped (**d**) line heat sources, and the cross-section shape of deposited layer (**e**)

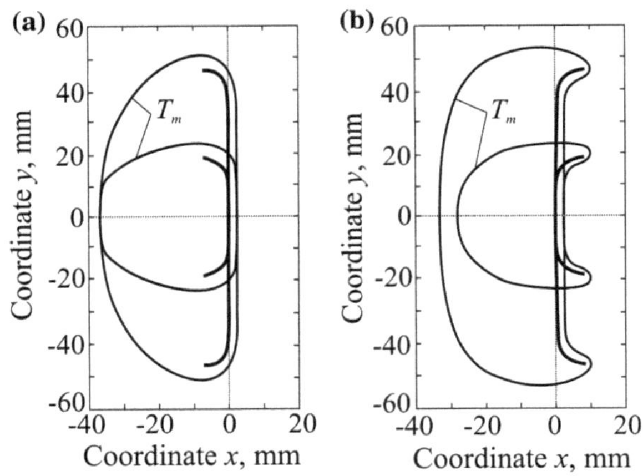

Fig. 12.10 Melting isotherms in the surface of a semi-infinite solid during fusion cladding using steel strips of cross-section 5 mm × 45 mm ($q = 18{,}960$ W, $v = 4.2$ mm s^{-1}) and 5 mm × 100 mm ($q = 32{,}500$ W, $v = 3$ mm s^{-1}) (Belousov et al. 1976)

Figure 12.10 shows contours of the weld pool on the body surface during cladding using a strip of a complex shape (see Fig. 12.9c). One can see that the width and length of the weld pool depend on the orientation of the curved edges, i.e. on the law of heat input.

These problems of heat conduction can be solved for a strip of any shape using Green's function method based on a moving point source by calculating the line integral over a heat curve (Fig. 12.9).

Assuming the model for a rapidly moving source (i.e. ignoring longitudinal heat fluxes) makes the solution much simpler. Let us demonstrate this through the example of cladding using a strip with edges bent at 90° (Fig. 12.9d).

Using Eqs. (5.2.172) and (5.1.31), we shall get

$$\begin{aligned} T(y,z,t)-T_0 = {} & \frac{q/(2W+l_2+l_3)}{2\pi\lambda}\left\{\sqrt{\frac{\pi a}{v^2 t}}\right. \\ & \times \exp\left(-\frac{z^2}{4at}\right)\left[\Phi\left(\frac{y+W}{\sqrt{4at}}\right)-\Phi\left(\frac{y-W}{\sqrt{4at}}\right)\right] \\ & + \mathrm{H}(t-l_2/v)\mathrm{Ei}\left(-\frac{(y-W)^2+z^2}{4a(t-l_2/v)}\right)-\mathrm{Ei}\left(-\frac{(y-W)^2+z^2}{4at}\right) \\ & \left. + \mathrm{H}(t-l_3/v)\mathrm{Ei}\left(-\frac{(y+W)^2+z^2}{4a(t-l_3/v)}\right)-\mathrm{Ei}\left(-\frac{(y+W)^2+z^2}{4at}\right)\right\}, \end{aligned} \tag{12.3.2}$$

where H is the Heaviside unit step function:

$$\mathrm{H}(u)=\begin{cases} 0 & \text{for } u<0; \\ 1 & \text{for } u\geq 0. \end{cases} \tag{12.3.3}$$

The first pair of summands in (12.3.2) allows for the contribution of strip part 1 with width 2 W, the second—of part 2 with width l_2 and the third—of part 3 with width l_3 (Fig. 12.9d). Here, the first moment when the strip crosses the cross-section of the body containing the point under study is taken as time $t = 0$.

It is notable that the proposed method for temperature field calculation can also be applied to other types of surfacing and welding. For example, high-frequency weaving welding with electron beam deceleration in toe zones makes it possible to improve the quality of the weld (by preventing pore formation).

12.4 Optimisation of Welding Conditions with Restrictions for Peak Temperature

During welding, the workpiece is subject to significant heating. Peak temperature is often limited in specified zones of the welded structure in order to prevent undesirable consequences. Therefore, a special welding technique that would allow for such requirements needs to be developed. The law of heat input and boundary conditions are variable parameters of the technique.

Let us assume that two massive bodies in the form of a cuboid need to be welded (Fig. 12.11a). The weld comprises N layers (Fig. 12.11b) with each nth layer consisting of M sections (Fig. 12.11c). Given: permissible temperature on the control (reference) line (point P, Fig. 12.11b), initial temperature T_0, depth of each nth layer

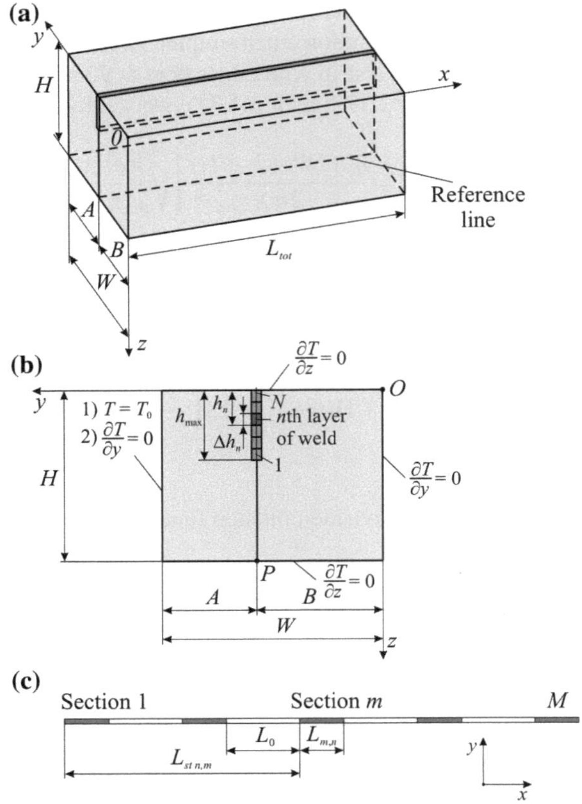

Fig. 12.11 Calculation diagram of the model: parallelepiped with a longitudinal weld (**a**); cross-section of the body (**b**); and section of the weld in the nth layer (**c**)

h_n and its height Δh_n (Fig. 12.11b), effective source power $q_{n,m}$ and welding speed v, possible forced effective cooling of the surface of one of the body's sides. It is required to find such length of each section $L_{n,m}$, time $t_{st\ n,m}$ and longitudinal coordinate $L_{st\ n,m}$ of initiation for each section while minimising the total time after the start of welding (Karkhin et al. 2011, 2016).

Let us adopt a body model with all adiabatic boundaries or with one boundary being isothermal and others being adiabatic (Fig. 12.11b). Technically, it is possible to find an analytical solution to the problem with the boundary conditions of third kind (heat transfer by Newton's law). However, such an approach will make the solution much more difficult.

Let all solid boundaries be heat-insulated, $\partial T/\partial n = 0$, where n is the outward normal, apart from side $y = W$, the temperature of which is maintained constant and equal to the initial T_0.

If the distance from the weld to reference point P is considerably larger than the size of the cross-section of the weld layer, the welding heat source can be represented as a moving point source in a bounded body.

Let the mth section of the nth weld layer in a cuboid run at the speed $v_{n,m}$, start from point $x = L_{st\,n,m}$, $y = B$, $z = h_n$ at time $t_{st\,n,m}$ and end at time $t_{fin\,n,m}$ (Fig. 12.11c). Let us take the solution for a moving point source in an unbounded body (5.1.43) as a basis, taking boundary conditions into account using the method of images. In this case, rise in temperature ΔT ($\Delta T = T - T_0$) at any point x, y, z in fixed coordinates associated with the body at time t is determined by the sum of temperature increments due to all the sources (all weld sections):

$$\Delta T(x, y, z, t) = \frac{1}{8\pi\lambda}\sum_{n=1}^{N}\sum_{m=1}^{M_n} q_{n,m} \sum_{i=-\infty}^{\infty}\sum_{j=-\infty}^{\infty} p'_j \sum_{r=-1,1} p''_r$$

$$\sum_{k=-\infty}^{\infty}\sum_{s=-1,1} [f(x - \xi_{n,m,i},\ y - \eta_{j,r},\ z - \zeta_{n,k,s},\ t - t_{st\,n,m}, v_{n,m})$$

$$- f(x - \xi_{n,m,i},\ y - \eta_{j,r},\ z - \zeta_{n,k,s},\ t - t_{fin\,n,m}, v_{n,m})], \quad (12.4.1)$$

where

$$p'_j = \begin{cases} -1 \text{ if } j \text{ is even;} \\ +1 \text{ if } j \text{ is odd;} \end{cases}$$

$$p''_r = r, \quad \xi_{n,m,i,l} = 2iL_{tot} + l[L_{stn,m} + v(t - t_{stn,m})];$$
$$\eta_{j,r} = (2j + 1)W + rA; \quad \zeta_{n,k,s} = 2kH + sh_n;$$

$$f(x, y, z, t, v) = \begin{cases} \left\{\Phi^*\left(\frac{1}{2}\left[\frac{R}{\sqrt{at}} - v\sqrt{\frac{t}{a}}\right]\right) + \exp\left(\frac{vR}{a}\right)\Phi^*\left(\frac{1}{2}\left[\frac{R}{\sqrt{at}} + v\sqrt{\frac{t}{a}}\right]\right)\right\} \\ \times\frac{1}{R}\exp\left(-\frac{v(x+R)}{2a}\right) \quad \text{for} \quad t > 0; \\ 0 \text{ for } t \le 0; \end{cases}$$

$R = (x^2 + y^2 + z^2)^{1/2}$.

If all boundaries of the body are heat-insulated, the solution takes the form (12.4.1), but with $p_j' = p_r'' = 1$.

It is worth noting that the higher the values of geometric parameters L_{tot}, W and H and less time $t - t_{st\,n,m}$ are, the faster series with respect to i, j, k converge. In practice, the finite number of series terms is determined according to the relative series increment. If the sum of increments of two adjacent series terms is no more than $10^{-7} \cdot 8\pi\lambda/(q_{n,m} R)$ (K), the series can be truncated (Karkhin et al. 2011, 2016).

Example 12.4.1 Let us consider manual arc welding of a stainless maraging steel bar under the following conditions: $W = 85$ mm, $H = 51$ mm, $L_{tot} = 300$ mm, $h_{\max} = 22$ mm, $A = 40$ mm, $c\rho = 0.0037$ J mm^{-3} K^{-1}, $\lambda = 0.0213$ W mm^{-1} K^{-1}, $q = 1820$ W, $v = 2.5$ mm s^{-1}. Let us find the length of a section of the first weld layer where the maximum temperature increment in the reference area (point P) does not exceed the permissible value (Karkhin et al. 2011, 2016).

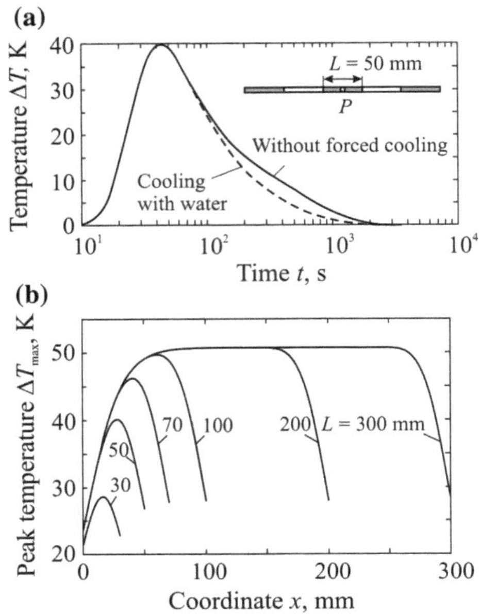

Fig. 12.12 Heating of the reference zone with the length L of the first layer of the weld: the thermal cycles of the reference point P (**a**); and the distribution of the maximum increase in temperature $\Delta T_{\max}$ at the reference line (without forced cooling of the surface) (**b**)

The effect of forced cooling is shown in Fig. 12.12a. There is little effect on peak temperature, but the effect on the cooling process is great. With section length over 100 mm, peak temperature reaches its maximum value (Fig. 12.12b). If the temperature increase only by 40 K is allowed, the maximum permissible section length L is 50 mm. As the groove is filled (the layer moves away from the reference line), the permissible section length increases. The obtained solution allows finding the optimum layout and sequence of weld sections (Karkhin et al. 2011, 2016).

12.5 Optimisation of Plate Edge Preheating Conditions in Butt Welding

Local preheating is often used in order to obtain the required properties of weld metal and HAZ. For example, for thick pipes and plates, induction heating is recommendable. Heating conditions (source power, time and area of heating) depend on the size of the body, properties of the metal, initial temperature and other factors (Ivanov and Karkhin 2009).

Determining preheating conditions can be formulated as the following optimisation problem:

$$t_p(q) \to \min;$$
$$T(x_1, t_p) \leq T_{\max};$$

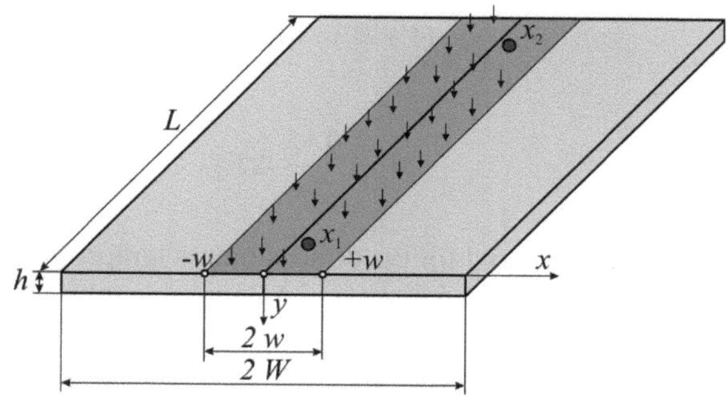

Fig. 12.13 Calculation diagram of the model for preheating of the plates

$$T(x_2, t_p + t_w) \geq T_{\min}, \text{ etc.,} \tag{12.5.1}$$

where t_p is the heating time, q is the effective source power, $T_{\max}$ is the maximum permissible temperature at point x_1 at the time of heating end t_p, $T_{\min}$ is minimum permissible temperature at point x_2 at the time of welding end $t_p + t_w$ (t_w is the time after the welding start).

Let us assume that (1) the body consists of two plates with thickness h, width W and length L (Fig. 12.13); (2) effective power q is constant and uniformly distributed along the width $2w$ and length L; (3) temperature difference across the thickness is relatively small ($\partial T/\partial y = 0$), and temperature changes only along the x-axis; (4) there is no surface heat transfer. It should be noted that the pipe with radius R can be represented as a plate with limited width W ($2\,W = 2\pi R$) in longitudinal welding or with limited length L in circumferential welding ($L = 2\pi R$).

Temperature increment due to a continuous volume source uniformly distributed along longitudinal section A ($A = hL$) and width $2w$ can be found using Green's function method based on the solution for a stationary continuous plane source in an infinite rod (5.1.35):

$$T(x,t) - T_0 = \Delta T(x,t) = \frac{q}{4wA\lambda}\left\{\sum_{i=-1,1}\left[\left(\frac{1}{2}(w+ix)^2 + at\right)\Phi\left(\frac{w+ix}{\sqrt{4at}}\right)\right.\right.$$
$$\left.\left. + \sqrt{\frac{at}{\pi}}(w+ix)\exp\left(-\frac{(w+ix)^2}{4at}\right) - \frac{1}{2}(w+ix)|w+ix|\right]\right\}. \tag{12.5.2}$$

In this case, it is also assumed that plate width W is infinite. The finiteness of width can be easily accounted for using the method of images.

Formula (12.5.2) can be expressed in a dimensionless form:

$$\theta(\xi,\tau) = \sum_{i=-1,1}\left[\left(\frac{1}{2}(1+i\xi)^2 + \tau\right)\Phi\left(\frac{1+i\xi}{\sqrt{4\tau}}\right)\right.$$

$$+\sqrt{\frac{\tau}{\pi}}(1+i\xi)\exp\left(-\frac{(1+i\xi)^2}{4\tau}\right)-\frac{1}{2}(1+i\xi)|1+i\xi|\Bigg]; \tag{12.5.3}$$

$$\theta=\frac{4A\lambda}{qw}\Delta T;\quad \xi=\frac{x}{w};\quad \tau=\frac{at}{w^2}, \tag{12.5.4}$$

where θ, ξ, τ stand for the dimensionless temperature, coordinate and time, respectively. Formulas (12.5.2) and (12.5.3) provide solutions for sources of any distribution by combining sources and sinks.

Let us introduce dimensionless temperature:

$$\overline{\theta}=\frac{T_{\max}(x_1)-T_0}{T_{\min}(x_2)-T_0}=\frac{\Delta T(x_1,t_p)}{\Delta T(x_2,t_p+t_w)-\Delta T(x_2,t_w)}=$$
$$=\frac{\theta(\xi_1,\tau_p)}{\theta(\xi_2,\tau_p+\tau_w)-\theta(\xi_2,\tau_w)} \tag{12.5.5}$$

where ξ_1 and ξ_2 are the dimensionless point coordinates, τ_p and τ_w are the dimensionless heating and welding times.

It should be noted that temperature $\overline{\theta}$ is known, since $T_{\max}(x_1)$, $T_{\min}(x_2)$ and T_0 are given. With known values $\overline{\theta}$, ξ_1, ξ_2 and τ_w it is possible to find heating time τ_p.

Figure 12.14 (top half) shows the dependence of τ_p on $\overline{\theta}$ for edge ($\xi_1=\xi_2=0$) with different values τ_w. From Eq. (12.5.3) and according to the found value τ_p, it is possible to determine dimensionless temperature θ and, consequently, dimensionless effective power κ, $\kappa=\theta^{-1}$ (Fig. 12.14, bottom half). Then effective power q and heating time t_p are obtained from Eqs. (12.5.4):

$$q=4Aw^{-1}\lambda(T_{\max}(x_1)-T_0)\kappa;\quad t_p=w^2a^{-1}\tau_p. \tag{12.5.6}$$

Similarly, it is possible to determine the preheating conditions for any given points x_1 and x_2.

Example 12.5.1 The edge of a steel pipe with the diameter of 1000 mm and the thickness of 20 mm is heated for the width of $2w=2\cdot 75=150$ mm (Fig. 12.15). Pipe cross-sectional area is $A=\pi\cdot 1000\cdot 20=62{,}832$ mm^2. Peak temperature of the edges should not exceed 523 K, $T_{\max}(0)=523$ K. Welding preparation after heating and the first pass takes $t_w=500$ s. The temperature of the edges after welding should not be less than 423 K, $T_{\min}(0)=423$ K. Initial temperature $T_0=253$ K. Find: effective power q and heating time t_p. Let us assume that for steel: $\lambda=0.04$ J mm^{-1} K^{-1} and $a=8$ mm^2 s^{-1}. Then dimensionless time after welding start $\tau_w=8\cdot 500/75^2=0.711$. According to (12.5.5), dimensionless temperature $\overline{\theta}=(523-253)/(423-253)=1.59$. Figure 12.14 shows (in straight lines with arrows) that dimensionless heating time $\tau_p=0.523$ and dimensionless effective power $\kappa=1.136$, which corresponds to the power and heating time (see Eqs. (12.5.4)):

$$q=4\cdot 62,832\cdot 75^{-1}\cdot 0.04\cdot(523-253)\cdot 1.136=41,113\text{ W};$$

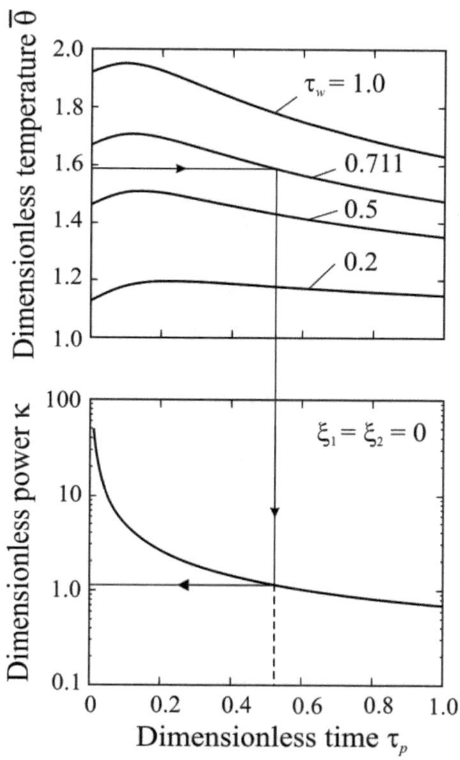

Fig. 12.14 A nomogram for calculating the dimensionless time of preheating τ_p and dimensionless effective power κ (Ivanov and Karkhin 2009)

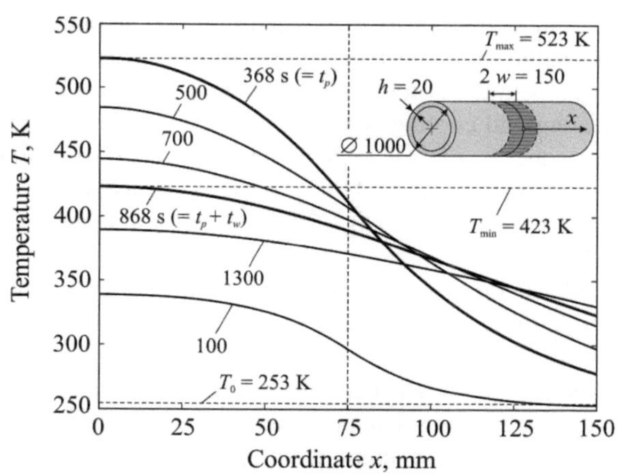

Fig. 12.15 Temperature distribution in the pipe at different times after heating start (Ivanov and Karkhin 2009)

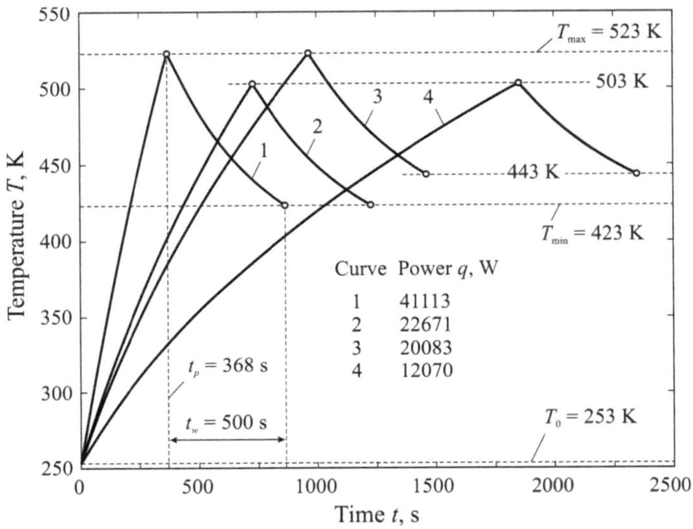

Fig. 12.16 Thermal cycles of the points in the centre of the heating zone at various power of preheating

$$t_p = 75^2 \cdot 8^{-1} \cdot 0.523 = 368\,\text{s}.$$

From known heat source efficiency η_h, it is possible to find total source power equal to q/η_h.

Temperature distributions at different times are shown in Fig. 12.15. It can be seen that at the end of heating, temperature does not exceed 523 K, while at the edge ($x = 0$) temperature is not less than 423 K at the end of welding. The thermal cycle of edge heating and cooling is shown in Fig. 12.16 (curve *1*). Heating time increases with the reduction of peak temperature ($T_{\max} = 503$ K, curve *2*) and increase in minimum temperature ($T_{\min} = 443$ K, curve *3*). Curves *1*–*4* stand for effective power $q =$ 41,113, 22,671, 20,083, 12,070 W and effective energy $q \cdot t_p = 4.20$, 4.59, 5.38, 6.20 kW h.

Consequently, the higher peak permissible heating temperature $T_{\max}$ with specified minimum temperature $T_{\min}$ is, the lower the heating time and energy and higher the source power are.

Since the body is being heated primarily on the side of the source, rapid heating may result in significant temperature difference across the thickness. This may cause plastic strains and, consequently, residual stress in the heating area. Let us assume that a one-dimensional heat flux is directed from the heating surface towards the opposite surface and the body is completely rigid. In this case, residual stresses can be avoided provided that $E\alpha\Delta T_y < \sigma_Y$, where E is the elasticity modulus, α is the linear expansion coefficient, σ_Y is the yield strength. Maximum temperature difference across thickness ΔT_y can be found using the following formula (Luikov 1968):

$$\Delta T_y = \frac{qh}{4wL\lambda}. \quad (12.5.7)$$

If calculated power q is greater than maximum power $q_{\max}$ that equipment can provide, it should be assumed that $q = q_{\max}$ and the inverse problem with respect to the unknown parameter should be solved using Fig. 12.14.

The problem of finding function $q(t)$ upon condition that heating time t_p is minimised also qualifies as an optimisation problem.

12.6 Minimisation of Transient and Residual Stresses

Welding brings about distortion and stresses that have an adverse effect on the behaviour of the welded structure. Welding of certain alloys may result in hot cracking due to the high concentration of strains in the weld behind the weld pool. After cooling, tensile welding stresses may reach the yield strength of the material in the weld and HAZ, which would significantly reduce the structure's fatigue life under cyclic load. Outside the weld and HAZ area, there are residual compressive stresses that may cause the plates to lose stability.

Nowadays, there are many methods to mitigate welding distortion and stresses (Gatovsky and Karkhin 1980; Goldak and Akhlaghi 2005; Okerblom et al. 1963; Radaj 1992, 2003; Sagalevich 1974). Changes in the welding technology (welding conditions, weld sequencing, welding consumables, additional heating) make it possible to improve significantly the technological strength (i.e. reduce the probability of hot and cold cracking) and reduce residual welding stresses and distortion of welded structures. Next, we shall examine methods for optimising local heating conditions in order to minimise transverse strains in the centre of the weld (preventing longitudinal solidification cracks) and to minimise longitudinal residual stresses of the welded structure.

12.6.1 Optimisation of Local Heating Conditions During Welding to Prevent Hot Cracking

In fusion welding, some aluminium, nickel alloys and austenitic steels have become prone to hot cracking, including the emergence of longitudinal solidification cracks (Coniglio and Cross 2009; Dantzig and Rappaz 2009; Katgerman and Eskin 2008; Lippold 2015; Makarov and Yakushin 2014; Pellini 1952; Plochikhine et al. 2003; Ploshikhin et al. 2005, 2007; Prokhorov 1952; Rappaz et al. 1999, etc.). It is believed that the cause of solidification cracks is the mechanical rupture of the narrow liquid

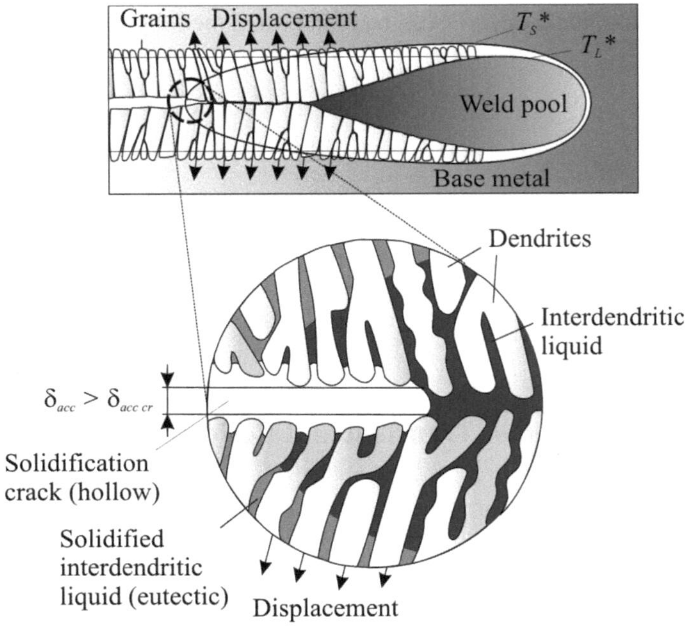

Fig. 12.17 Microscopic nature of the solidification cracking mechanism: schematic representation on the macro- and microscopic scale (Ploshikhin et al. 2007)

film between the dendrite tips at the tail part of the weld pool (Fig. 12.17). This film is enriched in liquating elements due to the segregation effect and therefore solidifies at a relatively low temperature (at local solidus temperature T_S^*, which corresponds to eutectic temperature T_E and which is significantly lower than weld metal solidus temperature T_S at the boundary with the base material). Back-feeding of the liquid metal from the weld pool is constrained.

The mechanical condition of the liquid film can be conveniently characterised by the relative displacement of the sides of the film δ_{acc}, which accumulates as the film is cooled from the actual liquidus temperature T_L^* to the actual solidus temperature T_S^*. If accumulated displacement δ_{acc} reaches critical value $\delta_{acc\ cr}$, the liquid film fails (Fig. 12.17). Displacement δ_{acc} depends on the geometry of the welded joint, welding conditions, properties of the base material and weld metal, and film parameters. Its value is calculated. Critical displacement (crack-opening of film sides) $\delta_{acc\ cr}$ characterises the liquid film's capability of accumulating tensile strains without destruction and depends on the thickness and morphology of the film (Ploshikhin et al. 2005).

Solidification cracking can be prevented by reducing total transverse tensile strains and creating total transverse shortening strains at the tail part of the weld pool during welding (Fig. 12.18b). This can be achieved through additional heating of the metal outside the welding area (Fig. 12.18c). Under heating, the expanding metal at the

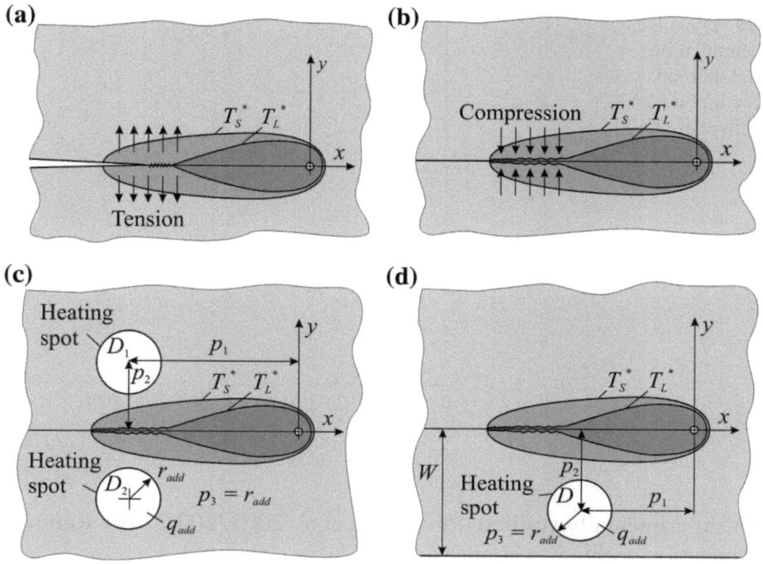

Fig. 12.18 Solidification cracking in welding of plates: generation of a longitudinal solidification crack under transverse tension of a mushy zone (**a**); suppression of longitudinal solidification cracking under compression of a mushy zone (**b**); application of additional heating when the weld is located far away from the plate edge (**c**) and close to the plate edge (**d**)

heating spots prevents displacement accumulation (mutual separation of film sides). This technique is known (Shumilin et al. 1984) and practically used in the laser and electron beam welding of aluminium alloys (von Dobeneck 2005).

Finding optimum local heating conditions can be formulated as the following optimisation problem:

$$
\begin{aligned}
&q_{add}(p_1, p_2, p_3) \rightarrow \min;\\
&k\delta_{acc}\max < \delta_{acc\,cr};\, p_2 > p_3;\, T(x, y) < T_{\max},\, x, y \in D_1 \text{ and } D_2,
\end{aligned}
\tag{12.6.1}
$$

where q_{add} is the power of the additional source; p_i is the heat source parameters (Fig. 12.18c); k is the safety factor; $T_{\max}$ is the peak temperature at the heating spot (assigned from metallurgical considerations), D_1 and D_2 are heating areas.

Figure 12.18d shows another technology option for preventing solidification cracking. Cracking is possible in the welding of a narrow strip with width W, whereas welding a wide plate under the same welding conditions has no cracking (Bergmann and Hilbinger 1998; Ploshikhin et al. 2005, 2007). A non-uniformly-heated narrow plate bends in its plane due to the lack of stiffness, which results in additional transverse tensile strains at the tail part of the weld pool. Additional heating of the edge of the narrow plate prevents bending and, consequently, solidification cracking, which

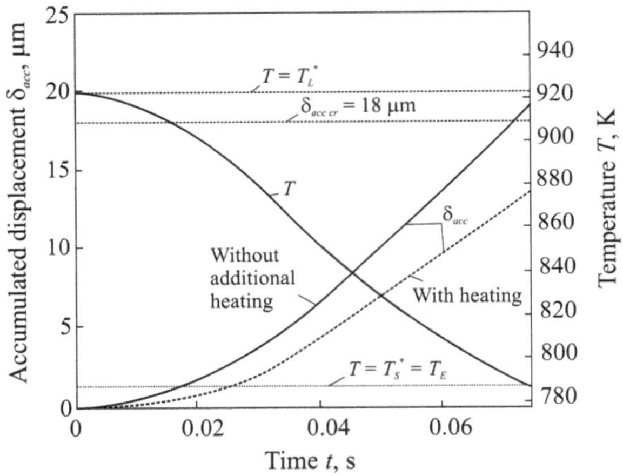

Fig. 12.19 Dynamics of strain accumulation $\delta_{acc}(t)$ and thermal cycles $T(t)$ of the liquid film (Ploshikhin et al. 2007)

has been experimentally verified (Ploshikhin et al. 2005, 2007). The formulation of this optimisation problem is similar to (12.6.1):

$$\begin{aligned} &q_{add}(p_1, p_2, p_3) \to \min; \\ &k\delta_{acc}\max < \delta_{acc\,cr};\ p_2 > p_3;\ p_2 + p_3 \leq W; \\ &T(x, y) < T\max, \quad x, y \in D. \end{aligned} \tag{12.6.2}$$

Example 12.6.1 AA6056 (0.93% Si, 0.75% Mg, 0.79% Cu, 0.618% Mn) aluminium alloy plates with the thickness of 1.5 mm are welded. Input data: narrow plate width $W = 25$ mm, main laser beam power = 3000 W, welding speed $v = 66.7$ mm s^{-1}, critical accumulated displacement $\delta_{acc\,cr} = 18$ µm, safety factor $k = 1.3$, maximum permissible heating temperature $T_{\max} < T_S$.

The thermomechanical problem is solved using the numerical finite element method with provision for element segregation (Ploshikhin et al. 2007). The following solution to the optimisation problem is obtained: $q_{add} = 750$ W, $p_1 = 0$, $p_2 =$ 20 mm. As can be seen from Fig. 12.19, the accumulated displacement of the sides of the liquid film in the interval from the actual liquidus temperature T_L^* to the actual solidus temperature T_S^* is lower than critical value $\delta_{acc\,cr}$ if additional heating is used and higher than $\delta_{acc\,cr}$ without heating (Ploshikhin et al. 2007).

It should be mentioned that there are other solidification cracking criteria such as curvature of the tail part of the weld pool (Tsarkov and Trukhanov 2013), which determines the rate and direction of crystal growth. The curvature itself depends on the latent heat of liquid metal solidification (Figs. 9.4.2 and 9.9) and actual liquidus and solidus temperature (local chemical composition determined by the diffusion of liquating elements). Therefore, the curvature of the solidification front at the tail part of the weld pool can reverse sign (Karkhin et al. 2010), as outlined in Fig. 12.17.

12.6.2 Minimisation of Longitudinal Residual Stresses Using Additional Local Heating During Welding

In the welding of steel, residual stresses in the area of plastic strains, which includes the weld, HAZ and the adjacent base metal, can reach yield strength. Longitudinal residual stresses can be reduced using concurrent or post-welding additional longitudinal tensioning of the welding area by applying an external force or local heating (Burak et al. 1977, 1979; Gatovsky and Karkhin 1980; Michaleris and Sun 1997; Michaleris et al. 1999; Okerblom et al. 1963).

The concept of the transient thermal tensioning involves creating additional longitudinal tensile strains using local heating of the area located outside the welding area and moving along with the welding heat source (Fig. 12.20).

When formulating the optimisation problem, it should be defined which residual stresses should be minimised in which part of the welded structure. These requirements depend on the operating conditions of the structure. For example, in order to increase fatigue strength of the welded structure, residual tensile stresses in the weld and HAZ should be minimised, whereas in order to enhance mechanical stability of plate structures, residual compressive stresses should be minimised.

Let us consider the cross-section of the beam structure consisting of a plate (bottom flange) with a welded web and a top flange (Fig. 12.21). The material is AH36 steel (0.18% C, 0.9–1.6% Mn, 0.1–0.5% Si) (Michaleris and Sun 1997). Rectangular heating bands $d_1 \times d_2$ in the area are positioned symmetrically on both sides of the web, with the band being displaced from the source by d_3 and d_4. The heating bands move at welding speed v. The parameters of the welding source and the heating source are prescribed. It is required to find such sizes of the heating bands so that residual longitudinal stresses σ_x in the defined areas of the bottom flange are minimum.

Let us take the quadratic sum of residual longitudinal stresses σ_x in the reference area with width B as the objective function:

$$F(q_2, d_1, d_2, \ldots, d_m) = \int_B [\sigma_x(y)]^2 dy \to \min; \qquad (12.6.3)$$

$$d_{i\min} \le d_i \le d_{i\max}. \qquad (12.6.4)$$

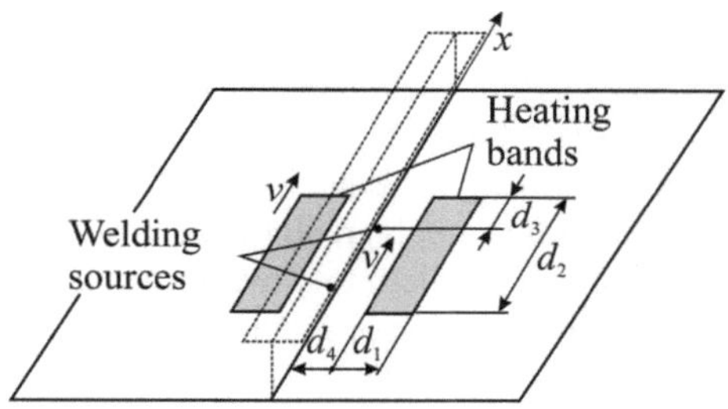

Fig. 12.20 Welding with additional heating (thermal tensioning) (Michaleris et al. 1999)

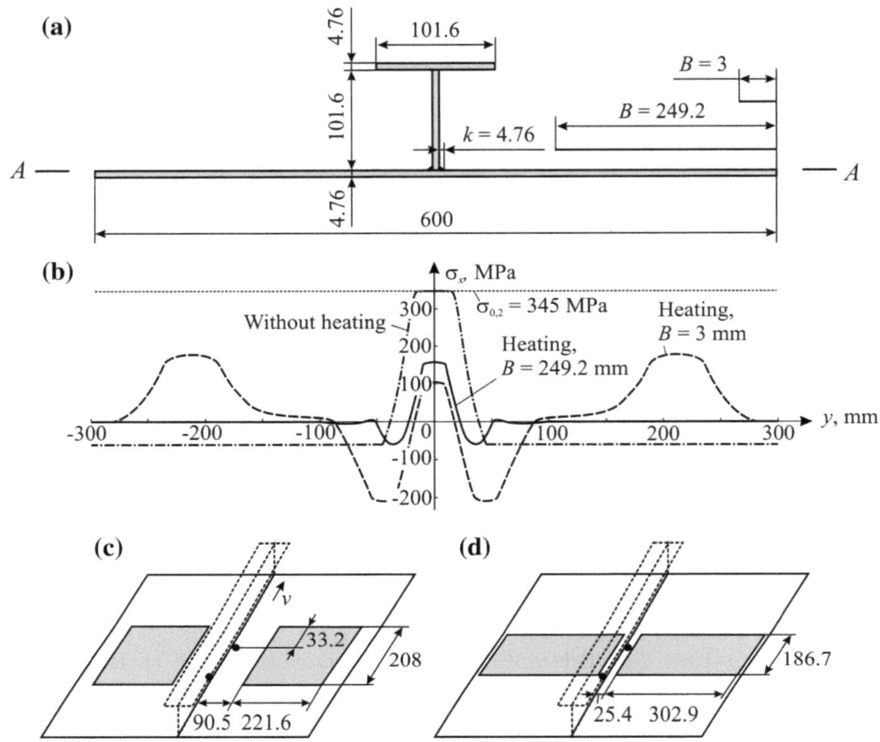

Fig. 12.21 Cross-section of the beam (**a**), distribution of longitudinal residual stresses σ_x in the upper surface of the plate A-A (**b**); configuration of heating bands for minimum stress on plate's edge of width $B = 3$ mm (**c**) and on the plate band of width $B = 249.2$ mm (**d**) (Michaleris and Sun 1997; Michaleris et al. 1999)

where $d_{i\,\min}$ and $d_{i\,\max}$ are the minimum and maximum permissible values of the size (position) of heating source d_i (Fig. 12.20). Other constraints can also be added (for example, on the peak temperature in the heating area).

Dimensions of the beam cross-section are shown in Fig. 12.21a. The power of the first arc is 6682 W, the power of the second arc is 6180.5 W, welding speed $v =$ 10.5 mm s^{-1}. The maximum power density of normally distributed heating flame is 0.5 W mm^{-2} (Michaleris et al. 1999).

The problems of heat conduction and thermoplasticity are solved using the finite element method in the case of a generalised plane-strain condition, which corresponds to the hypothesis of plane sections (total longitudinal deformations ε_x (y, z) are described using the equation of plane). This hypothesis is true so long as the length of the beam is much greater than its width and makes it possible to reduce the three-dimensional problem to a two-dimensional one (Michaleris et al. 1999).

The optimisation problem (12.6.3) is solved using the method of steepest descent (Michaleris et al. 1999).

Minimum residual longitudinal stresses at the edge of the bottom panel ($B = 3$ mm) are produced by the form of heating shown in Fig. 12.21c. Longitudinal stresses are minor at the edge but high in the heating area due to softening the metal at peak temperature in the long heating bands and formation of plastic strains (Fig. 12.21b, curve $B = 3$ mm). These stresses can be reduced by adding constraints on the peak temperature in the heating area when formulating the optimisation problem or by increasing the width of reference area B (Michaleris et al. 1999).

We shall choose the latter option. With the specified value $B = 249.2$ mm, optimum dimensions of the heating band are shown in Fig. 12.21d, and the corresponding stresses are shown in Fig. 12.21b (curve $B = 249.2$ mm). It can be seen that there are virtually no stresses in the reference area with width B. Peak temperature in this area does not exceed 803 K—hence no microstructural transformations take place in the heating area. The residual longitudinal stresses in the weld and HAZ reach only a half of the metal yield strength at room temperature $\sigma_{0.2}$ (Michaleris et al. 1999). It should be emphasised that in the case of welding without heating, stresses reach yield strength $\sigma_{0.2}$ (Fig. 12.21b) (Michaleris and Sun 1997).

The main cause of longitudinal stresses is the non-uniformity of temperature distribution and, consequently, of temperature strains in the cross-section of the welded structure. Local heating of the unrestrained structure outside the welding heating area provides a more uniform distribution of temperature strains, reduces longitudinal elastic strains and prevents formation of longitudinal plastic strains at a relatively low temperature (Gatovsky and Karkhin 1980). The efficiency of residual stress reduction is determined by power density of the heat source and the size and position of the heating area.

12.6.3 Optimisation of Welding Conditions for Obtaining Required Residual Stresses

Another method to obtain minimum or required residual stresses in steel welded joints is based on the optimisation of welding conditions (more specifically, a cooling rate) and the chemical composition of the filler material. The objective function is (1) minimum stress or (2) minimum deviation of residual stress σ in a specific area of the welded joint from the required stress σ_{req}:

$$(1)\ \sigma(\mathbf{p}) \rightarrow \min; \tag{12.6.5}$$

$$(2)\ [\sigma(\mathbf{p}) - \sigma_{req}]^2 \rightarrow \min. \tag{12.6.6}$$

Parameters of welding conditions within the required temperature interval and the chemical composition of the metal serve as vector components of variable parameters **p**. Constraints are selected for practical reasons.

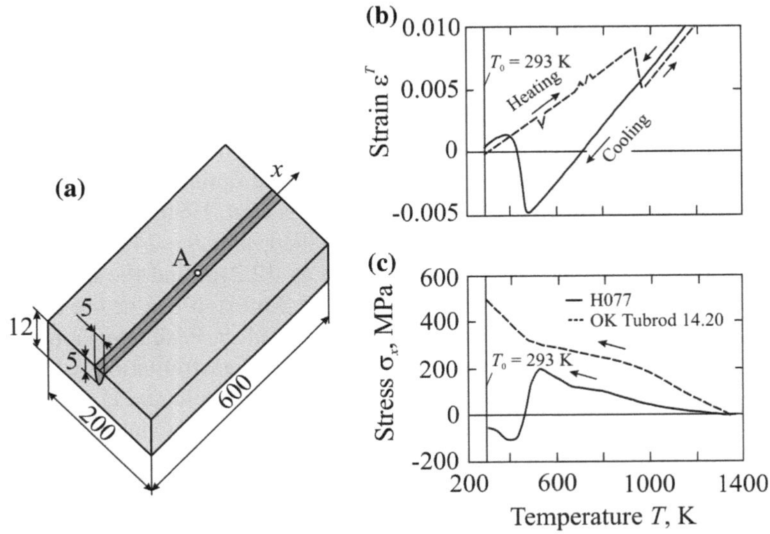

Fig. 12.22 Welded plate (**a**); dilatometric diagram ε^T for H077 steel heating and subsequent cooling at rate 10 K s^{-1} (**b**) and transient longitudinal stress in surface point A plotted against temperature during cooling in welding of 304 steel with electrodes H077 and OK Tubrod 14.20 (**c**) (Dai et al. 2010)

Residual stresses in steel depend on welding conditions, structure rigidity and mechanical properties of the metal (yield strength, elasticity modulus, Poisson's ratio) and phase transformations if they occur at low temperature (Karkhin et al. 1999, 2001, 2003). Under cooling, the austenite with a denser face-centred lattice turns into pearlitic ferrite with a less dense body-centred lattice, i.e. metal expands under cooling (negative linear expansion coefficient, density reduction) (Fig. 12.22b). Temperature strain ε^T (free strain under heating and cooling) is determined by linear expansion coefficient α:

$$\varepsilon^T(T) = \int_{T_0}^{T} \alpha(T)dT. \tag{12.6.7}$$

The temperature interval of phase transformations depends on the cooling rate and chemical composition of steel. Generally, the higher the degree of alloying is, the lower the interval will be. For example, at cooling rate 10 K s^{-1} martensitic transformation in H077 steel (0.01% C, 12.5% Cr, 6.8% Ni, 1.5% Mo, 1.2% Mn, yield strength 840 MPa) occurs in the range 487–387 K (Fig. 12.22b). The higher the cooling rate is, the lower the temperature transformation interval will be. The cooling rate can be adjusted by changing welding conditions (source power, welding speed, heating temperature). It should be remarked that welding conditions determine the percentage of the base metal in the weld metal (dilution) and, consequently, the chemical composition and dilatogram $\varepsilon^T(T)$ of the weld metal.

Let us investigate a method for minimising longitudinal stresses σ_x at point A positioned on the surface of the weld (Fig. 12.22a), in gas metal arc welding of 304L austenitic steel (<0.035% C, 18–20% Cr, 8–13% Ni, <2% Mn, <0.75% Si, yield strength at room temperature 240 MPa) using H077 and OK Tubrod 14.20 electrodes (0.003% C, 19.5% Cr, 10.0% Ni, 1.2% Mn, yield strength 390 MPa).

The kinetics of the stresses is calculated using the finite element method with consideration for the material superplasticity effect during phase transformations. Longitudinal residual stresses are measured using the neutron diffraction method (Dai et al. 2010).

The following conditions for single-pass welding are selected: current $I = 283$ A, arc voltage $U = 26$ V, welding speed $v = 7.7$ mm s^{-1}, initial temperature $T_0 =$ 293 K (Dai et al. 2010). Figure 12.22c shows the kinetics of longitudinal stresses σ_x at point A. It is obvious that in welding using OK Tubrod 14.20, when there are no low-temperature phase transformations in the weld metal, stresses continuously grow, reaching their maximum after complete cooling. In welding using H077, tensile stresses relax with the beginning of austenite decomposition and martensite formation, while after complete cooling they are close to zero. The calculations have been experimentally verified (Dai et al. 2010). Thus, residual stresses in the weld metal can be minimised by varying welding conditions and the chemical composition of the electrode.

There is a similar solution to the second problem: obtaining required residual stresses in a particular area of the welded joint (Eq. (12.6.6)). For example, residual compressive stresses in dangerous areas are desirable as they improve fatigue strength of the welded joint under cyclic load and reduce sensitivity of the metal to stress concentration, including discontinuities such as pores and cracks.

Thus, residual stresses can be controlled through welding conditions and chemical composition of the filler metal.

It should be noted that the formulation and solution of the optimisation problem can be simplified if it is required to find parameters of the heat source with specified temperature conditions, size of the heating area and observation time. In some cases, it is possible to obtain a solution in an analytical form (Melyukov 1996, 2003; Melyukov and Grigoryants 2002).

To date, many examples of the optimisation problem solution have been published. Optimum conditions for scan plasma arc welding make it possible to obtain a temperature field with a lower gradient and, consequently, a more uniform weld along the length, more uniform distribution of microstructure at the weld interface, lower tendency to defect formation (pores, inclusions, cracks, lack of fusion) and lower cooling rate compared to conventional arc welding (Doumanidis and Fourligkas 1995). The same effect is achieved with optimum scan laser welding conditions (Doumanidis 2001). Numerical optimisation is a key tool for the improvement of advanced multi-beam laser welding techniques (Petzet et al. 2005). Optimisation of induction welding makes it possible to obtain uniform distribution of the temperature of the body's work surface that allows high-quality deposition of a powdered hard alloy (Shably et al. 2003). Additional moving heat sinks (for example, weld cooling with carbon dioxide snow) make it possible to reduce cooling time $\Delta t_{8/5}$ and hereby change microstructure and properties of the weld metal and HAZ as well as reduce welding distortion (Van der Aa et al. 2007).

As can be seen from the examples, formulating optimisation problems (objective functions and constraints) can vary as they serve different practical purposes. The objective function can comprise characteristics of different nature (geometrical, metallurgical, economic, etc.), with weighting factors being set by the researcher and reflect his view of the optimisation problem. Therefore, researchers play a decisive role in the formulation of the optimisation problem.

References

Belousov, Yu. V., Leshchinsky, L. K., & Sologub, B. B. (1976). Selection of optimum shape of strip electrode for wide-layer deposit. *Automatic Welding*, (12), 24–28 (in Russian).

Bergmann, H. W., & Hilbinger, R. M. (1998). Numerical simulation of centre line hot cracks in laser beam welding of aluminium close to the sheet edge. In H. Cerjak (Ed.), *Mathematical modelling of weld phenomena* (Vol. 4, pp. 658–668). London: IOM Communications.

Burak, Ya. I., Besedina, L. P., Romanchuk, Ya. P., Kazimirov, A. A., & Morgun, V. P. (1977). Control of longitudinal plastic shrinkage in welding. *Automatic Welding*, (3), 27–29 (in Russian).

Burak, Ya. I., Romanchuk, Ya. P., Kazimirov, A. A., & Morgun, V. P. (1979). Selection of optimum field of preheating in welding of plates. *Automatic Welding*, (5), 15–19 (in Russian).

Coniglio, N., & Cross, C. E. (2009). Mechanisms for solidification crack initiation and growth in aluminium welding. *Metallurgical and Materials Transactions*, *40A*, 2718–2728.

Dai, H., Mark, A. F., Moat, R., Shirzadi, A. A., Bhadeshia, H. K. D. H., Karlsson, L., & Withers, P. J. (2010). Modelling of residual stress minimization through martensitic transformation in stainless steel welds. In H. Cerjak & N. Enzinger (Eds.), *Mathematical modelling of weld phenomena* (Vol. 9, pp. 239–252). Graz: Verlag der Technischen Universitaet Graz.

Dantzig, J. A., & Rappaz M. (2009). *Solidification* (621 pp.). EPFL Press.

Doumanidis, C. C. (2001). Scan welding: Thermal modelling for microstructure control. In H. Cerjak (Ed.), *Mathematical modelling of weld phenomena* (Vol. 5, pp. 863–883). London: IOM Communications Ltd.

Doumanidis, C. C., & Fourligkas, N. (1995). Scan welding: A parametric analysis of thermal characteristics. In *Proceedings of the 4th International Conference on Trends in Welding Research*, 5–8 June 1995, Gatlinburg, Tennessee (pp. 745–750).

Dudko, D. A., Zatserkovny, S. A., Sidoruk, V. S., Taraborkin, L. A., & Makhlin, N. M. (1987). Effect of process parameters of manual pulsed current arc welding on weld shape. *Automatic Welding,* (6), 19–22 (in Russian).

Gatovsky, K. M., & Karkhin, V. A. (1980). *Theory of welding stresses and deformations* (331 pp.). Leningrad: Leningrad Shipbuilding Institute Publishing (in Russian).

Goldak, J. A., & Akhlaghi, M. (2005). *Computational welding mechanics* (321 pp.). Springer.

Ivanov, S. Yu., & Karkhin V. A. (2009). Evaluation of edge preheating conditions in butt welding. In *Proceedings of International Conference on Welding Materials*, 18–22 May 2009, St. Petersburg, Russia (pp. 177–182) (in Russian).

Ivanov, S., Karkhin, V., Panchenko, O., & Michailov, V. (2015). Optimisation technique for arc welding of structured sheet. In *Proceedings of International Conference "Metal 2015"*, 3–5 June 2015, Brno, Czech Republic (6 pp.).

Ivanov, S. Yu., Karkhin, V. A., & Mikhailov, V. G. (2017). Optimisation of welding conditions for curvilinear joints. *Welding International, 31*(12), 929–933.

Karkhin, V. A., Ivanov, S. Yu., Homich, P. N., & Rajamaki, P. (2010). Analysis of solute macro- and microsegregation in fusion welding. In H. Cerjak & N. Enzinger (Eds.), *Mathematical modelling of weld phenomena* (Vol. 9, pp. 155–175). Graz: Verlag der Technischen Universitaet Graz.

Karkhin, V. A., Ivanov, S. Yu., & Vikhman, V. B. (2016). The method for calculating the welding conditions taking into account the restrictions for the maximum temperature in heating. *Welding International, 30*(8), 614–620.

Karkhin, V. A., & Khomich, P. N. (2005). Optimisation of pulsed GTA welding. In V. A. Sudnik (Ed.), *Proceedings of International Conference on Computer Technologies in Joining of Materials* (pp. 194–208). Tula: Tula State University (in Russian).

Karkhin, V. A., & Khomich, P. N. (2007). Minimisation of heat input in pulsed welding. *Welding International, 21*(3), 227–230.

Karkhin, V. A., Kreutz, E. W., Pavlova, N. O., & Schulz, W. (1999). Analysis of residual stress distributions in laser welded joints allowing for low-temperature phase transformations. In V. Kujanpaa & J. Ion (Eds.), *7th Nordic Conference on Laser Processing of Materials*, Lappeenranta, Finland (Vol. 1, pp. 70–80).

Karkhin, V. A., Kreutz, E. W., Pavlova, N. O., & Schulz, W. (2001). Effect of low-temperature phase transformations on residual stress distributions in laser welded joint. In H. Cerjak (Ed.), *Mathematical modelling of weld phenomena* (Vol. 5, pp. 597–614). London: The Institute of Materials.

Karkhin, V. A., Lopota, V. A., & Pavlova, N. O. (2003). Effect of phase transformations on residual stresses in laser welding 09G2 steel. *Welding International, 17*(8), 645–649.

Karkhin, V. A., Ivanov, S. Yu., & Vikhman, V. B. (2011). Computational method for determination of welding conditions with constraints on peak temperature. In *Proceedings of International Conference "Technologies and Equipment EBW-2011"*, 23–26 May 2011, St. Petersburg (pp. 103–115) (in Russian).

Katgerman, L., & Eskin, D. G. (2008). In search of the prediction of hot cracking in aluminium alloys. In H. Herold, C. Cross, & J. Lippold (Eds.), *Hot cracking phenomena in welds* (Vol. 2, pp. 3–18). Springer.

Kondo, K., Ueda, K., & Oji, T. (1998). Optimum heat input control in girth welding of small diameter pipe. *Welding International, 12*(6), 440–447.

Kondoh, K., & Ohji, T. (1998). Algorithm based on non-linear programming method for optimum heat input control in arc welding. *Science and Technology of Welding and Joining, 3*(3), 127–134.

Kravtsov, T. G. (1988). Calculation of thermal processes in deposition with profiled strip electrode. *Welding Production,* (4), 33–35 (in Russian).

Leitner, R. E., McElhinney, G. H., & Pruitt, E. L. (1973). An investigation of pulsed GTA welding variables. *Welding Journal,* (9), 405-s–410-s.

Lippold, J. C. (2015). *Welding metallurgy and weldability* (400 pp.). Hoboken, NJ: Wiley.

Luikov, A. (1968). *Analytical heat conduction theory* (702 pp.). Elsevier.

Makarov, E. L., & Yakushin, B. F. (2014). *Theory of weldability of steels and alloys* (487 pp.). Moscow: Moscow State Technical University Publishing (in Russian).

Melyukov, V. V. (1996). Optimisation of heat state in welding process. *Welding Production,* (1), 9–11 (in Russian).

Melyukov, V. V. (2003). *Optimisation of materials processing with high-concentrated heat flow* (212 pp.). Kirov: VGU Publishing (in Russian).

Melyukov, V. V., & Grigoryants, A. G. (2002). Determination of rapidly moving line heat source power in heating of thin plate. *Welding Production,* (3), 8–9 (in Russian).

Michaleris, P., Dantzing, J., & Tortorelli, D. (1999). Minimization of welding residual stress and distortion in large structures. *Welding Journal, 78*(11), 361-s–366-s.

Michaleris, P., & Sun, X. (1997). Finite element analysis of thermal tensioning techniques mitigating weld buckling distortion. *Welding Journal,* (11), 451-s–457-s.

Okerblom, N. O., Demyantsevich, V. P., & Baykova, I. P. (1963). *Design of fabrication technology of welded structures* (602 pp.). Leningrad: Sudpromgiz (in Russian).

Pellini, W. S. (1952, November). Strain theory of hot tearing. *Foundry*, 125–199.

Petzet, V., Bueskens, C., Pesch, H. J., Karkhin, V., Makhutin, M., Prikhodovsky, A., & Ploshikin, V. (2005). Numerical optimization as a key tool for the improvement of advanced multi-beam laser welding techniques. In A. Bode & F. Durst (Eds.), *High performance computing in science and engineering* (pp. 153–166). Berlin: Springer.

Plochikhine, V., Prikhodovsky, A., & Zoch, H.-W. (2003). Modern approach to modelling of solidification cracking in welds. In B. E. Paton & V. S. Kovalenko (Eds.), *Proceedings of International Conference "Laser Technologies in Welding and Materials Processing"*, 19–23 May 2003, Katsiveli (pp. 137–141). Kiev: Paton Electric Welding Institute Publishing.

Ploshikhin, V., Prikhodovsky, A., Ilin, A., Heimerdinger, C., & Palm, F. (2007). Mechanical–metallurgical approach for prediction of solidification cracking in welds. In H. Cerjak, H. K. D. H. Bhadeshia, & E. Kozeschnik (Eds.), *Mathematical modelling of weld phenomena* (Vol. 8, pp. 87–104). Graz: Verlag der Technischen Universitaet Graz.

Ploshikhin, V., Prikhodovsky, A., Makhutin, M., Ilin, A., & Zoch, H.-W. (2005). Integrated mechanical-metallurgical approach to modelling of solidification cracking in welds. In *Hot cracking phenomena in welds* (Vol. 2, pp. 223–244). Springer.

Prokhorov, N. N. (1952). *Hot cracking during welding* (224 pp.). Moscow: Mashgiz (in Russian).

Radaj, D. (1992). *Heat effects of welding. Temperature field, residual stress, distortion* (348 pp.). Berlin: Springer.

Radaj, D. (2003). *Welding residual stresses and distortion. Calculation and measurement* (397 pp.). Duesseldorf: DVS-Verlag.

Rappaz, M., Drezet, Y.-M., & Gremaud, M. (1999). A new hot-tearing criterion. *Metallurgical and Materials Transactions A, 30A*(2), 449–455.

Rykalin, N. N. (1951). *Calculation of heat flow in welding* (Z. Paley & C. M. Adams, Jr., Trans.) (337 pp.). Moscow.

Rykalin, N. N. (1957). *Berechnung der Waermevorgaenge beim Schweissen* (326 pp.). Berlin: VEB Verlag Technik (in German).

Sagalevich, V. M. (1974). *Methods for eliminating welding deformations and stresses* (248 pp.). Moscow: Mashinostroenie (in Russian).

Saraev, Yu. N. (1994). *Pulsed welding procedures and surfacing* (107 pp.). Novosibirsk: Nauka (in Russian).

Shably, O. N., Pulka, C. V., & Pismenny, A. S. (2003). Optimization of induction hardfacing of thin discs, allowing for thermal and electromagnetic shielding. *Paton Welding Journal,* (9), 20–24.

Shumilin, V. G., Karkhin, V. A., Rakhman, M. I., & Gatovsky, K. M. (1984). Patent 1109280, USSR, MKI B 23K. Method for arc welding. N3224505/25-27 of 23.08.1984.

Street, J. A. (1990). *Pulsed arc welding: An introduction* (57 pp.). An Abington Publishing.

Tsarkov, A. V., & Trukhanov, K. Yu. (2013). Pool curvature as a parameter for optimisation of welding conditions. In *Proceedings of International Conference "Welding and Inspection—2013"* (pp. 371–382). Perm: Perm University Publishing (in Russian).

Van der Aa, E. M., Thiessen, R. G., Murugaiyan, A., Hermans, M. J. M., Sietsma, J., & Richardson, I. M. (2007). Influence of a trailing heat sink on the microstructure and stress distribution in DP600 welds. In H. Cerjak, H. K. D. H. Bhadeshia, & E. Kozenschnik (Eds.), *Mathematical modelling of weld phenomena* (Vol. 8, pp. 387–408). Graz: Verlag der Technischen Universitaet Graz.

Von Dobeneck, D. (2005). *Electron beam welding* (72 pp.). Pro-beam AG.

Chapter 13
Prediction of Local Microstructure and Mechanical Properties of Welded Joint Metal with Allowance for Its Thermal Cycle

13.1 Microstructural Zones of Welded Joint

The behaviour of a welded joint under external conditions (load, temperature, hostile environment, etc.) depends on the local microstructure and local mechanical properties of all welded joint zones (of the weld, HAZ and base metal). In order to predict the microstructure and properties, it is necessary to know the thermal processes in the welded joint, i.e. to solve the heat conduction problem with allowance for body geometry, boundary conditions, welding conditions, and the thermophysical properties of the metal (Fig. 13.1). It is further assumed that the thermal cycles of all welded joint zones are known.

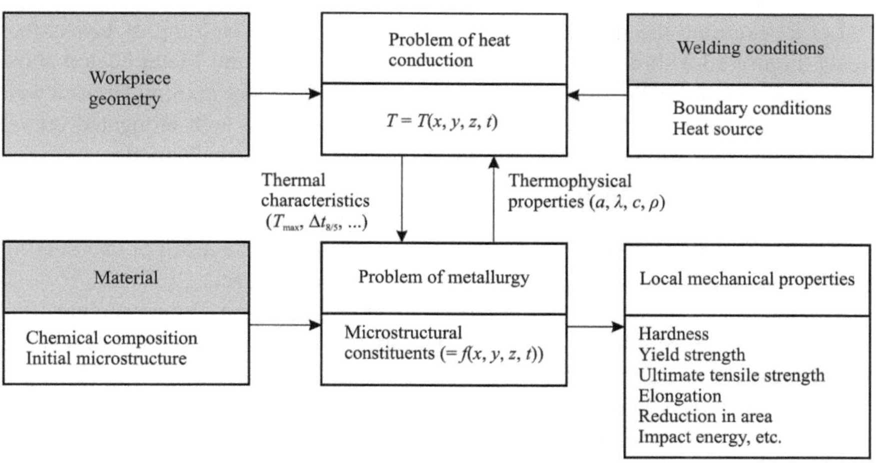

Fig. 13.1 Flow chart for the calculation of the transient temperature field, microstructural transformations and mechanical properties of HAZ metal

V. A. Karkhin, *Thermal Processes in Welding*, Engineering Materials,
https://doi.org/10.1007/978-981-13-5965-1_13

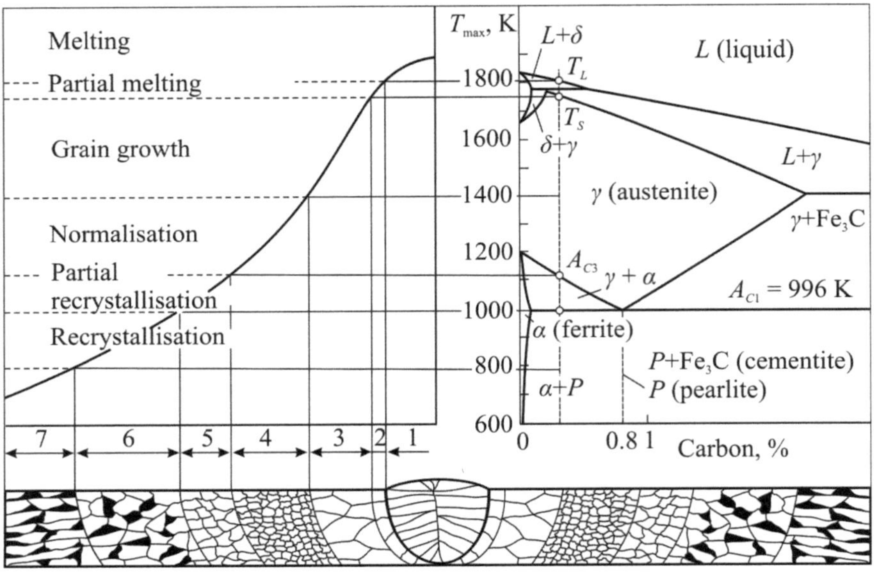

Fig. 13.2 Microstructural zones in the welded joint of low-alloy steel, peak temperature profile and Fe–C diagram (Petrov and Tumarev 1977)

Different welded joint zones are subjected to a different thermal cycle during welding. Metal in the welding zone is heated faster and to a higher temperature than metal positioned at a distance from the weld. As a result, the zones have different local metallurgical and mechanical properties.

Let us examine the microstructural zones in the fusion welding of low-carbon steels. Figure 13.2 shows an equilibrium Fe–C phase diagram. Metal heated above liquidus temperature T_L forms a weld pool, while metal under cooling forms a weld (zone 1 in Fig. 13.2). The weld metal has a cast structure with elongated curved columnar crystals growing on the grains of the base metal (see Chap. 9).

Metal in zone 2 is heated above solidus temperature T_S but below liquidus temperature T_L. In this zone, the base metal is partially melted.

The final coarse grain microstructure of metal in zone 3 is a result of overheating. In this zone, the grain size increases towards the weld interface.

Zone 4, heated slightly above temperature A_{C3}, has a fine grain structure (normalisation structure). Metal heated above temperature A_{C3} (zones 3 and 4) makes a total transition into an austenitic state, it forms a region of complete recrystallization. Austenite has a face-centred lattice.

In the zone of partial recrystallisation 5 ($A_{C1} < T_{max} < A_{C3}$), pearlite colonies fully or partially transform to austenite, while ferrite grains remain almost unchanged. Ferrite has a body-centred lattice. Pearlite is a two-phase layered structure consisting of ferrite and cementite Fe_3C.

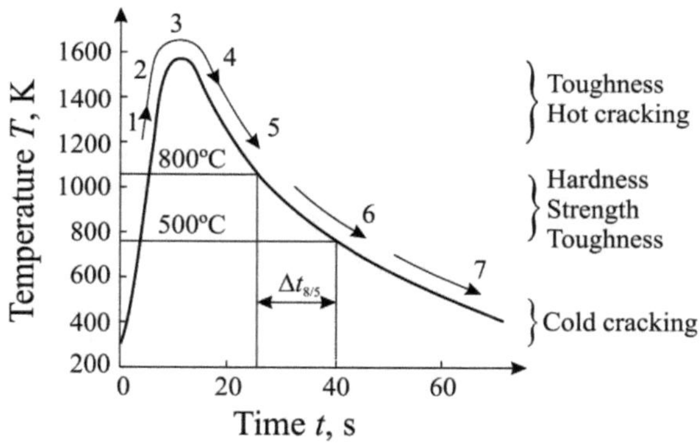

Fig. 13.3 Microstructural change in HAZ of low-alloy steel and its effect on mechanical properties (Buchmayr and Cerjak 1988; Radaj 1992): 1—austenitisation, 2—carbide dissolution, 3—grain growth, 4—carbide coarsening, 5—carbide reprecipitation, 6—phase transformation, 7—martensite formation

The microstructure in zone 6 ($T_{max} < A_{C1}$) remains unchanged if metal was in an annealed condition prior to welding. If metal was work-hardened as a result of cold rolling or bending (i.e. it was in an unstable state), it develops recrystallisation processes. The minimum recrystallisation temperature of pure metals is $0.4T_m$, where T_m is the melting temperature (K). In welded joints, recrystallisation is observed in the area heated above 800 K.

Zone 7 contains base metal not undergoing microstructural changes. Zones 2–6 form a common heat affected zone (HAZ).

The zones of partial melting and coarse grains are critical in a welded joint of low-alloy steels. The thermal cycle is accompanied by physical and chemical phenomena that determine the strength of metal during and after welding such as susceptibility to hot and cold cracking, hardness, strength, toughness (Fig. 13.3). High-alloy steels and non-ferrous alloys have different phase diagrams and HAZ areas.

A sufficiently accurate description of the thermal history of each HAZ and weld point is the basic requirement for the modelling of metallurgical reactions.

13.2 Microstructure of Heat Affected Zone Metal in Single-Pass Welding

The thermal cycles of metal during welding and normal heat treatment are very different in terms of duration, peak temperature and austenitising temperature (Fig. 13.4). The cooling rates of metal can be different by orders of magnitude, hence the microstructure and mechanical properties of metal can be substantially different. Therefore, in order to be able to predict microstructure during welding, it is necessary to use anisothermal time-temperature transformation diagrams. These are constructed on the basis of experimental data obtained by using dilatometric and thermal analysis methods.

A distinction is made between an isothermal time-temperature-transformation diagram (TTT diagram), continuous cooling time-temperature-transformation dia-

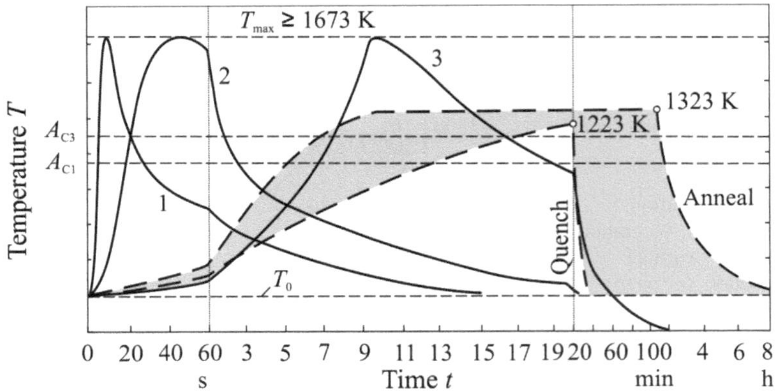

Fig. 13.4 Thermal cycles of HAZ in single-pass fusion welding of low-alloy steel compared to those during heat treatment (shaded) (Shorshorov and Belov 1972): 1—arc welding of 1 mm thick sheets; 2—submerged arc welding of 25 mm thick plates; 3—electroslag welding of 220 mm thick plates

gram (CCT diagram) and welding time-temperature-transformation diagram (welding TTT diagram) (Radaj 1992).

In the case of an isothermal TTT diagram, a steel specimen is heated and held at ~1173 K (~900 °C), then fast cooled to a temperature below A_{C1} and held until the end of transformations. These diagrams cannot be used in welding engineering practices (Radaj 1992).

In the case of a CCT diagram, a steel specimen is held at ~1173 K (~900 °C) and cooled at a constant rate. This case also does not conform to the welding conditions (Radaj 1992).

In the case of a welding TTT diagram, a steel specimen is heated and cooled according to the welding thermal cycle with a peak temperature of 1623 K (dashed curve in Fig. 13.5a). The point in time when the specimen is cooled to 1123 K (850 °C) is taken as the origin of the horizontal axis, $\Delta t = t - t_{850} = 0$. The thermal cycle in logarithmic coordinates $T - \log\Delta t$ is represented as a solid curve.

The decomposition of supercooled austenite of 12CrMo4.3 (0.11% C, 1.02% Cr, 0.29% Mo, 0.19% V) steel begins at the point of intersection of the cooling curve and the first transformation curve (Fig. 13.5b). The corresponding transformation ends at the point of intersection of the cooling curve and the next transformation curve (the percentage of the microstructural constituents is specified at the point of intersection). The products of decomposition of austenite (*A*) include austenite (*M*), bainite (*B*) and ferrite-pearlite (*FP*). The values of metal hardness after cooling are specified at the end of the curves.

The final microstructure and, consequently, properties are determined by metal cooling time $\Delta t_{8.5/5}$ in the range 1123 K (850 °C) − 773 K (500 °C). Figure 13.5, *c* also shows the same welding TTT diagram in coordinates $T - \Delta t_{8.5/5}$, while Fig. 13.6 shows the mechanical properties of steel.

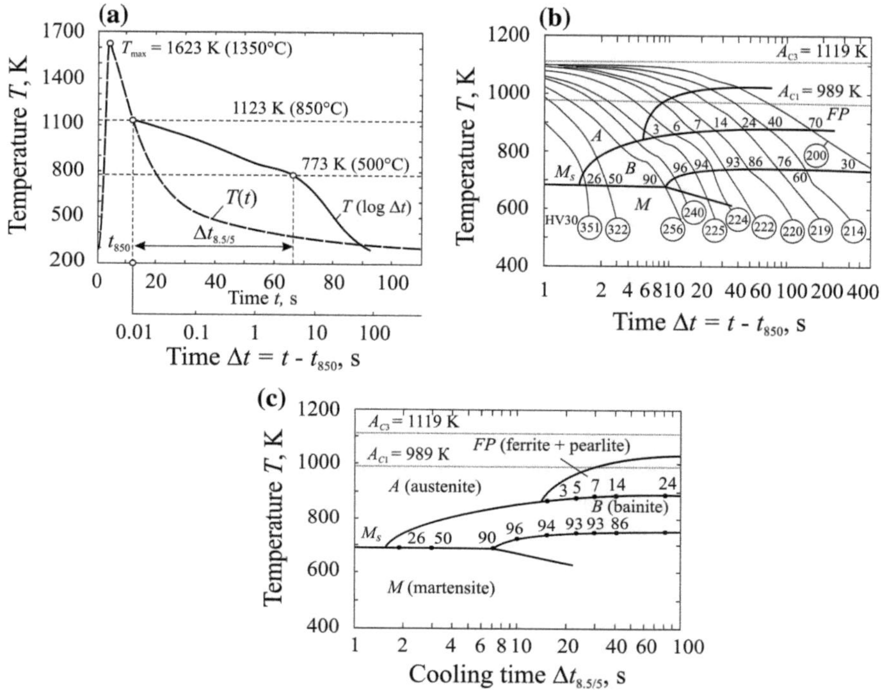

Fig. 13.5 Welding thermal cycles of HAZ metal (**a**), welding time—temperature transformation diagram of 12CrMo4.3 steel (**b**) and dependency of microstructure on cooling time from 1123 K (850 °C) to 773 K (500 °C) (**c**) (Seyffarth et al. 1992)

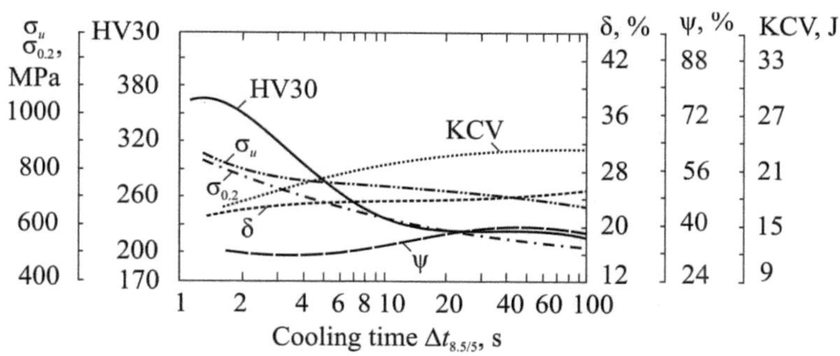

Fig. 13.6 Mechanical properties (ultimate tensile strength σ_u, 0.2% offset yield strength $\sigma_{0.2}$, hardness HV30, elongation δ, reduction in area ψ, impact energy KCV) of 12CrMo4.3 steel versus cooling time from 1123 K (850 °C) to 773 K (500 °C) (Seyffarth et al. 1992)

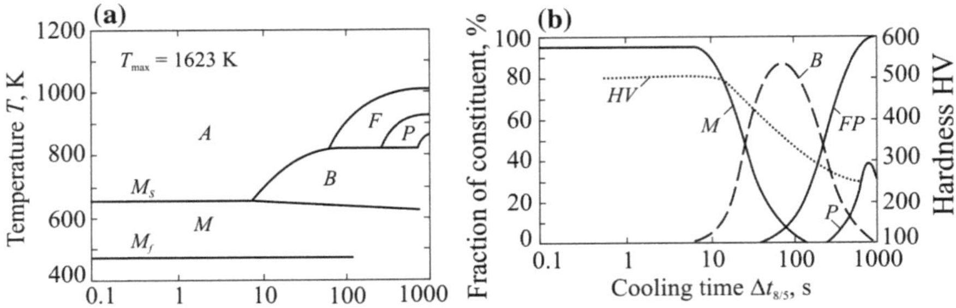

Fig. 13.7 Continuous cooling—transformation diagram of 20KhGSA steel (**a**) and final microstructural constituents and hardness after cooling down (Frolov 1988)

Cooling time $\Delta t_{8/5}$ in the range 1073 K (800 °C)–773 K (500 °C) proposed in paper (Degenkoble et al. 1984) is often used as an argument. Theoretically, it is possible to use the average cooling rate in the given temperature range or the instantaneous cooling rate at the given temperature $T(-\partial T/\partial t\,(T))$ instead of cooling time. However, in the latter case, rate error can be significant, since a derivative error is higher than the error of function $T(t)$. This can be seen in the bends of curves $T(t)$ at the moment of phase transformations accompanied by the liberation of phase transformation heat (Fig. 13.5b).

The anisothermal time-temperature-transformation diagram is supplemented by data on the composition of microstructure (Fig. 13.7). Typical HAZ phase composition curves in the welding of low-alloy steels are shown in Fig. 13.8. Under fast metal cooling conditions ($\Delta t_{8.5/5} < 1$ s), the effect of carbon diffusion is weak, and austenite transformation occurs in a martensite form. The role of diffusion grows with the decrease of the cooling rate, bainite transformation occurs, and ferrite and pearlite are produced. At $\Delta t_{8.5/5} < 60$ s, the martensite phase is virtually extinct, whereas at $\Delta t_{8.5/5} < 200$ s austenite transformation occurs completely in a ferritic-pearlitic form. Thus, it is possible to calculate the thermal cycle (cooling time $\Delta t_{8.5/5}$) and determine HAZ microstructure after single-pass welding from the geometry of the welded joint, metal properties and welding conditions.

The diagram of austenite decomposition depends on the peak temperature (Fig. 13.9). Austenite stability increases as temperature rises. Precise assessment of metal behaviour for different HAZ areas requires a diagram with different peak austenitising temperature. These can be summarised in the form of PTCT-diagrams (peak temperature versus cooling time diagram) (Fig. 13.10). The diagram indicates the type of microstructure resulting from different combinations of peak austenitising temperature $T_{a\,\max}$ and cooling time $\Delta t_{8/5}$. At specified time $\Delta t_{8/5}$ it is possible to find microstructure, hardness and grain size depending on temperature $T_{a\,\max}$ (grain size indicated according to ASTM). Lines with similar hardness HV30 are shown in the dashed lines. A similar shift of the time-temperature transformation diagram is observed when high-strength Ni-Cr-Mo HY-100 steel is cooled from $T_{\max} = 1623$ and 1153 K (Odanovic and Nedeljkovic 2001).

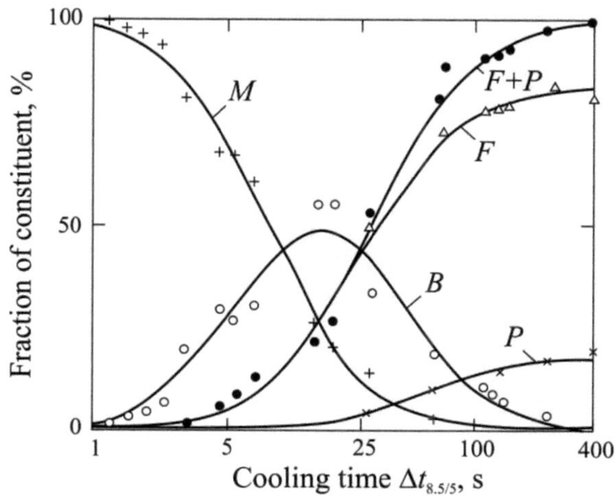

Fig. 13.8 Microstructural constituents of HAZ metal after welding of low-alloy steel (0.09% C, 1.34% Mn, 0.54% Si, 0.96% Cr, 0.44% Ni, 0.32% Cu) as a function of cooling time (Kasatkin 1984c)

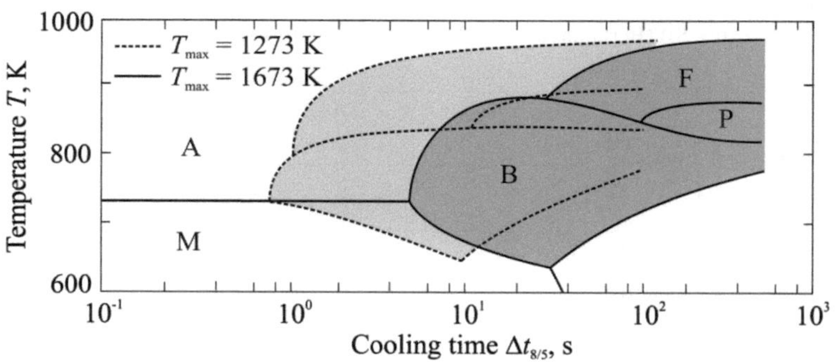

Fig. 13.9 Continuous cooling—transformation diagram for different peak temperature T_{max} for St 52 steel (0.16% C, 1.50% Mn, 0.40% Si) (Berkhout and van Lent 1968)

The diagrams in Figs. 13.9 and 13.10 do not fully reflect the degree of austenitisation during welding. They cover cooling with no allowance for heating conditions, which means that different microstructures can form at the same T_{max} and $\Delta t_{8/5}$, but different heating rates, austenitising time and initial grain size (Michailov et al. 2016; Ossenbrink and Michailov 2007).

There are many empirical formulas to determine the characteristic temperatures of phase diagrams with allowance for the chemical composition of steels (Adams 1958; Buchmayr 1991; Kasatkin 1984c, 1990; Seyffarth and Kassatkin 1979, 2002; Seyffarth and Kuscher 1982). The monograph (Buchmayr 1991) collects formulas of various researchers for calculating melting temperature, temperatures A_{C1} and A_{C3}, temperature for pearlite formation start, start and finish of bainite transformation, start and finish of martensite transformation and other characteristic parameters of the time-temperature transformation diagram.

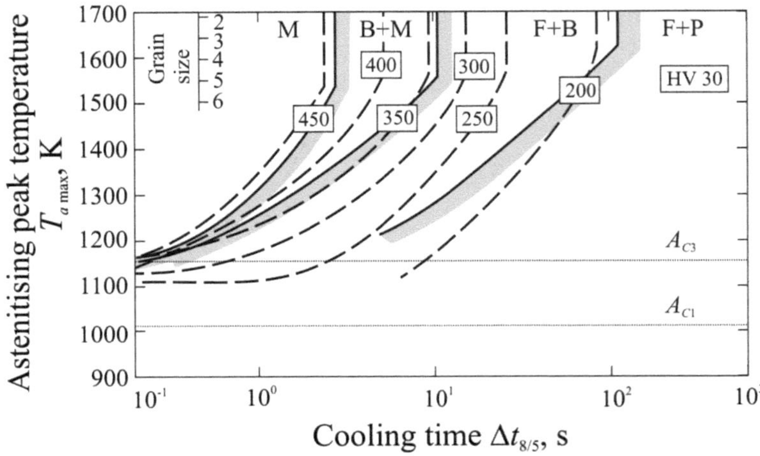

Fig. 13.10 Peak temperature—cooling time diagram of St 52 normalised structural steel (0.16% C, 1.50% Mn, 0.40% Si) (Berkhout and van Lent 1968; Radaj 1992)

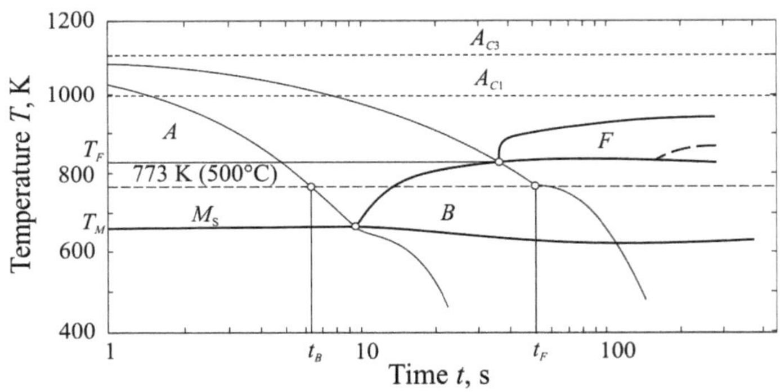

Fig. 13.11 Characteristic temperature (T_F and T_M) and time (t_B and t_F) of the continuous cooling—transformation diagram for austenite decomposion (Seyffarth and Kuscher 1982)

There are also regression equations for temperatures A_{C1}, A_{C3}, T_M and T_F and times t_B and t_F with allowance for the cross-effect of alloying elements (Fig. 13.11) (Kasatkin 1984c; Seyffarth and Kuscher 1982):

$$\begin{aligned}\ln A_{C1} = {} & 6.5792 - 0.038058\mathrm{C} + 0.052317\mathrm{Si} + 0.011872\mathrm{Ni} - 0.045575\mathrm{V} \\ & + 0.18057\mathrm{Al} + 0.011442\mathrm{W} - 0.013403\mathrm{Cu} + 5.5207\mathrm{B} + 0.91209\mathrm{S} - 1.1002\mathrm{P} \\ & + 0.060014\mathrm{Mn}\cdot\mathrm{C} - 0.096628\mathrm{Cr}\cdot\mathrm{C} + 0.050625\mathrm{Cr}\cdot\mathrm{Si} + 0.39802\mathrm{Mo}\cdot\mathrm{C} \\ & - 0.34782\mathrm{Mo}\cdot\mathrm{Mn} + 0.40986\mathrm{Mo}\cdot\mathrm{Si} - 0.12959\mathrm{Mo}\cdot\mathrm{Cr} - 0.048128\mathrm{Ni}\cdot\mathrm{C} \\ & - 0.011090\mathrm{Mn}^2 - 0.03550\mathrm{Si}^2 + 0.010207\mathrm{Cr}^2 + 0.36074\mathrm{Mo}^2 - 0.0030705\mathrm{Ni}^2;\end{aligned} \tag{13.2.1}$$

$$\begin{aligned}\ln A_{C3} = {} & 6.8165-0.47132\text{C} -0.057321\text{Mn}+0.066026\text{Si}-0.050211\text{Cr} \\ & -0.094455\text{Ni}+0.10593\text{Ti}-0.014847\text{W}+2.0272\text{N}+1.0536\text{S} \\ & -0.12024\text{Si}\cdot\text{C}+0.11629\text{Cr}\cdot\text{C}-0.30451\text{Mo}\cdot\text{Mn}+0.68229\text{Mo}\cdot\text{Si} \\ & -0.21210\text{Mo}\cdot\text{Cr}+0.12470\text{Ni}\cdot\text{C}+0.069960\text{Ni}\cdot\text{Mn}+0.014003\text{Ni}\cdot\text{Cr} \\ & +0.29225\text{C}^2+0.015660\text{Mn}^2+0.017315\text{Cr}^2+0.46894\text{Mo}^2+0.0027897\text{Ni}^2;\end{aligned} \tag{13.2.2}$$

$$\begin{aligned}\ln T_M = {} & 6.1705-0.27898\text{C}-0.20939\text{Cr}+0.20985\text{Ni}+0.51078\text{Ti} \\ & +0.32997\text{Al}-0.087418\text{Mn}\cdot\text{C}+0.15897\text{Cr}\cdot\text{Mn}+0.78628\text{Mo}\cdot\text{C} \\ & -0.31837\text{Mo}\cdot\text{Mn}+0.62347\text{Mo}\cdot\text{Si}-0.22623\text{Mo}\cdot\text{Cr}-0.57669\text{Ni}\cdot\text{C} \\ & -0.31989\text{Ni}\cdot\text{Si}+0.081581\text{Ni}\cdot\text{Cr}-0.91909\text{C}^2-0.021565\text{Mn}^2+0.068359\text{Cr}^2 \\ & -0.057869\text{Ni}^2;\end{aligned} \tag{13.2.3}$$

$$\begin{aligned}\ln T_F = {} & 6.3518-1.0588\text{C}+0.19345\text{Mn}+0.28463\text{Si}-0.64127\text{Cr} \\ & +5.3356\text{Mo}+1.7081\text{Ni}-0.76398\text{V}-0.62657\text{Ti}+2.2319\text{Nb}-0.14124\text{Cu} \\ & -625.88\text{B}+3.0092\text{N}+3.1049\text{S}-6.1159\text{P}-0.31011\text{Mn}\cdot\text{C}-0.87325\text{Si}\cdot\text{C} \\ & -0.084200\text{Mn}\cdot\text{Si}+0.94081\text{Cr}\cdot\text{C}+0.31078\text{Mn}\cdot\text{Cr}-1.66627\text{Mn}\cdot\text{Mo} \\ & -35.643\text{Mo}\cdot\text{Si}-2.3338\text{Ni}\cdot\text{C}-0.84595\text{Mn}\cdot\text{Ni}+0.86544\text{Si}\cdot\text{Ni} \\ & +1.0769\text{Mo}\cdot\text{Ni}+2.8027\text{C}^2-0.043512\text{Mn}^2+22.549\text{Mo}^2-0.55888\text{Ni}^2;\end{aligned} \tag{13.2.4}$$

$$\begin{aligned}\ln t_B = {} & 0.4176-2.1913\text{Ni}-5.1502\text{Al}+19.164\text{Nb}+0.71165\text{Cu} \\ & +108.88\text{B}-22.284\text{N}+1.9409\text{Mn}\cdot\text{C}+7.3573\text{Si}\cdot\text{C}-3.5189\text{Cr}\cdot\text{C} \\ & -3.6421\text{Mo}\cdot\text{Mn}+9.0380\text{Mo}\cdot\text{Si}+9.9735\text{Ni}\cdot\text{C}+0.74690\text{Ni}\cdot\text{Mn} \\ & +0.97021\text{Ni}\cdot\text{Cr}-0.94725\text{Si}^2+0.25388\text{Cr}^2;\end{aligned} \tag{13.2.5}$$

$$\begin{aligned}\ln t_F = {} & -2.5139+7.5022\text{C}+1.8910\text{Mn}+5.3040\text{Si}-3.6403\text{Cr} \\ & -259.72\text{Mo}+15.443\text{Ni}-10.901\text{V}+15.296\text{Cr}\cdot\text{C}+1.3923\text{Cr}\cdot\text{Mn} \\ & +243.0\text{Mo}\cdot\text{C}+515.29\text{Mo}\cdot\text{Si}-51.849\text{Mo}\cdot\text{Cr}-86.201\text{Ni}\cdot\text{C}-3.8702\text{Si}^2 \\ & +250.34\text{Mo}^2.\end{aligned} \tag{13.2.6}$$

Here, temperature is expressed in °C.

Phase composition of the HAZ metal in welding steel depends on the thermal cycle. Building calculation models based on the examined physical mechanisms of transformations in steel is currently difficult. Therefore, regression models of phase formation with allowance for cooling rate can be used for practical purposes. Based on the analysis of the known CCT diagrams for low-alloy steels (Seyffarth and Kuscher 1982; Shorshorov and Belov 1972), models for assessing the proportion of martensite M, ferrite and pearlite $F+P$ and bainite B in the HAZ structure (Kasatkin 1984c, 1990) were suggested:

$$M(\Delta t_{8.5/5}) = 1 - \Phi\left(\frac{\ln \Delta t_{8.5/5} - \ln \Delta t_{8.5/5\,M}}{\ln S_M}\right); \tag{13.2.7}$$

$$(F + P)(\Delta t_{8.5/5}) = 1 - \Phi\left(\frac{\ln \Delta t_{8.5/5} - \ln \Delta t_{8.5/5\,F+P}}{\ln S_{F+M}}\right); \tag{13.2.8}$$

$$B(\Delta t_{8.5/5}) = 1 - M(\Delta t_{8.5/5}) - (F + P)(\Delta t_{8.5/5}), \tag{13.2.9}$$

where Φ is the error function (3.7.4), $\Delta t_{8.5/5\,M}$ and $\Delta t_{8.5/5\,F+P}$ are the cooling times at which 50% of martensite or ferritic-pearlitic microstructure are formed, S_M и S_{F+P} are constants that characterize the steepness of composition history curves M and $F + P$ as $\Delta t_{8.5/5}$ increases. Figure 13.8 shows the example of use of such models with a good fit of models to experimental data (Kasatkin 1984c).

For low-alloy steels containing C $\leq 0.3\%$, Mn $\leq 2\%$, Si $\leq 0.8\%$, Cr $\leq 2\%$, Mo $\leq 1\%$, Ni $\leq 2\%$, V $\leq 0.3\%$, Ti $\leq 0.06\%$, Al $\leq 0.06\%$, Nb $\leq 0.1\%$, W $\leq 0.5\%$, Cu $\leq 0.5\%$, parameters $\Delta t_{8.5/5\,M}$, $\Delta t_{8.5/5\,F+P}$, S_M and S_{F+P} can be assessed as follows (Kasatkin 1990):

$$\begin{aligned}\ln \Delta t_{8.5/5M} &= -2.1 + 15.5\mathrm{C} + 0.96\mathrm{Mn} + 0.84\mathrm{Si} + 0.65\mathrm{Cr} + 0.74\mathrm{Mo} \\ &\quad + 0.55\mathrm{Ni} + 0.30\mathrm{V} + 4.0\mathrm{Al} + 0.50\mathrm{W} + 0.80\mathrm{Cu} \\ &\quad - 0.25\mathrm{Nb} - 13.5\mathrm{C}^2 - 0.55\mathrm{Si}^2;\end{aligned} \tag{13.2.10}$$

$$\begin{aligned}\ln S_M &= 0.56 - 0.41\mathrm{C} + 0.10\mathrm{Mn} + 0.14\mathrm{Cr} - 0.30\mathrm{Mo} + 2.7\mathrm{Ti} \\ &\quad + 0.50\mathrm{Cu} - 1.1\mathrm{Nb} + 1.7\mathrm{C}\cdot\mathrm{Mo};\end{aligned} \tag{13.2.11}$$

$$\begin{aligned}\ln \Delta t_{8.5/5F+P} &= 0.34 + 5.2\mathrm{C} + 1.8\mathrm{Mn} + 0.53\mathrm{Si} + 0.33\mathrm{Cr} + 2.9\mathrm{Mo} \\ &\quad + 0.86\mathrm{Ni} + 1.5\mathrm{W} + 1.0\mathrm{Cu} - 6.0\mathrm{C}\cdot\mathrm{V} - 5.1\mathrm{C}^2 + 0.50\mathrm{Si}^2;\end{aligned} \tag{13.2.12}$$

$$\begin{aligned}\ln S_{F+P} &= 0.91 - 0.90\mathrm{C} + 0.09\mathrm{Mn} + 0.08\mathrm{Cr} + 0.34\mathrm{Mo} + 0.15\mathrm{Ni} \\ &\quad + 0.85\mathrm{V} + 2.2\mathrm{Ti} + 0.43\mathrm{W}.\end{aligned} \tag{13.2.13}$$

An analysis of the obtained models shows that under conditions of a welding thermal cycle all alloying elements increase austenite stability (Kasatkin 1984c).

13.3 Microstructure of Heat Affected Zone Metal in Multi-Pass Welding

Low-alloy steels are sensitive to the thermal cycle of welding. In multi-pass welding, HAZ metal is reheated, which determines its final microstructure and, consequently, mechanical properties.

In the welding of low-alloy high-strength tempered martensitic-bainitic steel (0.09% C, 2.63% Ni, 1.05% Cr, 0.27% Si, 0.30% Mn), a coarse grain area is formed

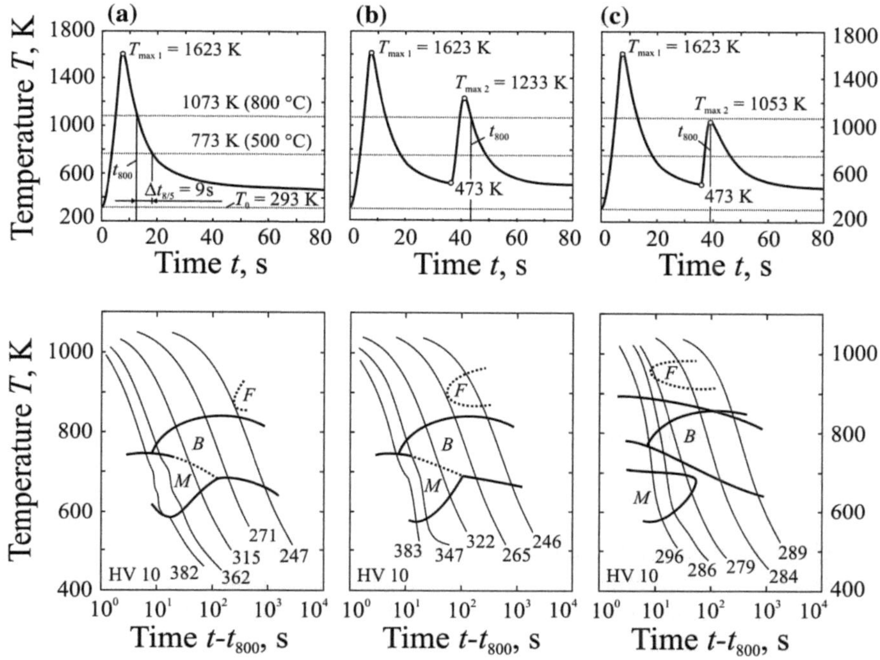

Fig. 13.12 Thermal cycles and continuous cooling transformation diagrams for high-strength low-alloy steel (0.09% C, 2.63% Ni, 1.05% Cr) in a coarse-grained heat affected zone during first pass (**a**) and second pass (**b**, **c**) welding (Gliha 2005)

near the weld interface ($T_{max1} = 1623$ K). Under cooling, austenite decomposes into martensite (if cooling time $\Delta t_{8/5} < 9$ s), bainite and martensite (if 9 s < $\Delta t_{8/5} < 50$ s), bainite (if 50 s < $\Delta t_{8/5} < 400$ s) or bainite with ferrite (if $\Delta t_{8/5} > 400$ s) (Fig. 13.12a). In this case, hardness HV 10 of this area depends heavily on cooling time $\Delta t_{8/5}$. If the final microstructure is fully martensitic, hardness is constant (Gliha 2005).

During the second pass, the metal in the coarse grain area can become hot and turn into austenite completely (if peak temperature T_{max2} is higher than characteristic temperature A_{C3}, $T_{max2} = 1233$ K > A_{C3}) or partially ($A_{C1} < T_{max2} = 1053$ K < A_{C3}) (Fig. 13.12b, c). In the former case, the products of austenite decomposition will include martensite, bainite and ferrite with the same hardness as before the first pass (Fig. 13.12b). In the latter case, the microstructure only partially (up to 70%) turns into austenite while reheating to $T_{max2} = 1053$ K (Fig. 13.12c). Therefore, the CCT diagram is true only for the decomposition of this austenite. A martensite-bainite microstructure is formed even under extremely fast cooling. The martensite start temperature is slightly lower than in the first pass (cf. Fig. 13.12a, c) (Gliha 2005). The dependence of hardness HV10 on the cooling rate is very weak.

The kinetics of HAZ metal microstructure becomes more complex if the number of passes is more than two as in the thick V butt weld. In this case, the welding

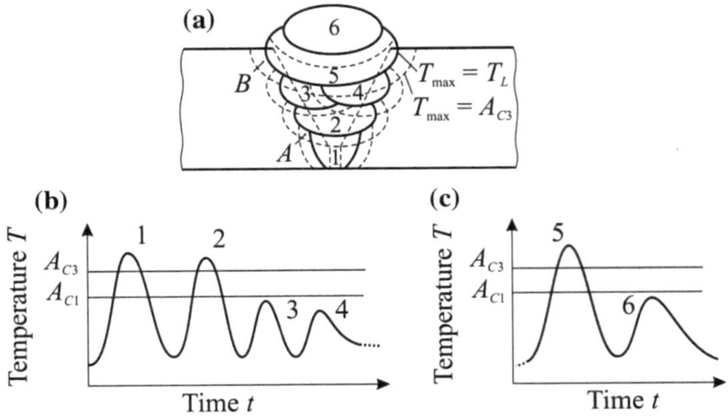

Fig. 13.13 Sketch of multi-pass weld (**a**) and thermal cycles of point *A* (**b**) and *B* (**c**) (Petrov and Tumarev 1977)

of each subsequent bead results in the additional thermal effect both on the metal of previous weld passes and on the HAZ metal. Consequently, the metal changes its microstructure, i.e. its microstructure becomes more equilibrium. The degree of microstructure change is determined by the thermal cycle of metal (peak temperature, cooling rate, etc.), which can be calculated or measured.

As an example, let us select point *A* in the first-pass HAZ (Fig. 13.13a), the peak temperature of which is higher than critical temperature A_{C3} (Fig. 13.13b). During passes 1 and 2, metal transforms into austenite, which later decomposes under further cooling. If the welded metal is quenched steel, then after passes 1 and 2 the metal at point *A* has a quenched microstructure. If under heating by pass 3 the peak temperature is sufficiently high but below A_{C1}, after cooling the metal has a tempered microstructure. The effects of the subsequent passes are minor if peak temperatures are relatively low.

The thermal effect of welding on point *B* is less intricate. Metal is heated above A_{C3} during pass 5. If pass 5 is final, metal remains quenched. It is possible to weld another bead 6 so that the peak temperature in area *B* would reach the tempering temperature. This technological process of improving the HAZ microstructure is called temper bead technique (Petrov and Tumarev 1977).

In actual practice, HAZ in the multi-pass welding of steels has a complex non-homogeneous structure (Fig. 13.14). The boundaries of the HAZ areas of two adjacent passes are determined by peak temperatures. In an HAZ, there are areas 1–4, which are not affected by the subsequent beads, and areas 5–20, the microstructure of which changes under the thermal influence of the subsequent beads. In areas 1, 5–8 and 20, the metal is coarse-grained and brittle, and the dominant factor is peak temperature around 1600 K. In areas 2, 9–11 and 19, the metal is fine-grained and highly tough. In areas 3, 13–16, there is a risk of intercritical embrittlement. Area 16 is affected particularly adversely, initially being coarse-grained and then partially turning into

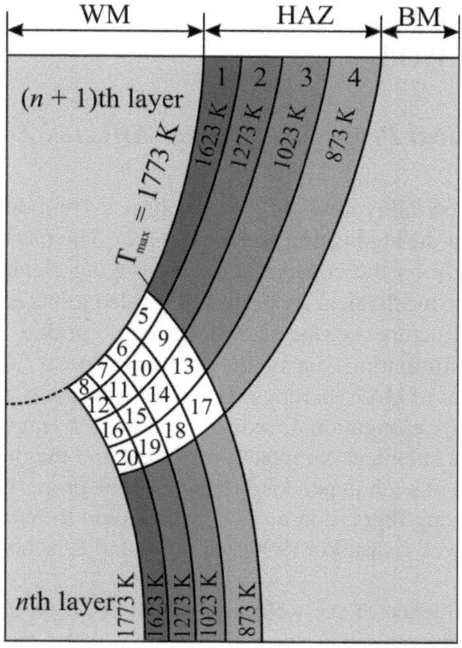

Fig. 13.14 Heat affected zones in multi-pass weld: the region independent of subsequent weld passes: 1—coarse grain zone; 2—fine grain zone; 3—zone of intercritical heating; 4—zone of subcritical heating; the region dependent on subsequent weld passes: 5–8—coarse grain zone; 9–11—fine grain zone; 12—recrystallisation zone; 13, 14—zone of intercritical heating; 15—zone of intercritical heating with fine grains; 16—zone of intercritical heating with coarse grains; 17—zone of subcritical heating; 18—zone of intercritical and subcritical heating; 19—zone of subcritical heating with fine grains; 20—zone of subcritical heating with coarse grains (Buchmayr 1991; Degenkoble et al. 1984)

austenite. This may further lead to dissolution of sensitive particles (e.g. Nb (CN)). Fast cooling of carbon-enriched austenite leads to the formation of martensite islands (Buchmayr 1991; Devillers et al. 1995).

Thus, the HAZ and weld metal of the previous passes are subjected to heat treatment during multi-pass welding. Considering the chemical composition of the base and welding materials, it is theoretically possible to find such thermal cycles (optimise welding conditions) that would lead to the required microstructures and mechanical properties of the welded joint. Helical surface welding of cylindrical solids (with concurrent cylinder rotation and source movement along the generatrix) allows obtaining optimum thermal cycles similar to multi-pass welding (Yazovskikh and Belenky 2011a, b).

13.4 Mechanical Properties of Steels in the Heat Affected Zone and Weld

13.4.1 Mechanical Properties of Heat Affected Zone Metal

The HAZ metal of low-alloy steels is a multi-phase system due to the phase transformations that occur during heating (see Sect. 13.2). The phase composition of the metal depends mainly on the concentration of alloying elements and the thermal cycle of welding. The mechanical properties of welded joints are largely determined by the HAZ microstructure, i.e. the size of austenitic grains, state of their boundaries and phase transformations during the cooling of metal (Kasatkin 1990). Local mechanical properties of HAZ (hardness HV, ultimate tensile strength σ_u, 0.2% offset yield strength $\sigma_{0.2}$, elongation δ, reduction in area ψ, impact energy KCV and KCU) depend on the chemical composition of steel and the thermal cycle of welding. The mechanism of such dependencies is yet to be properly studied. Therefore, they are described using regression models, which tend to reflect basic quantitative relationships rather well (Kasatkin 1990; Seyffarth and Kuscher 1982; Seyffarth and Kasatkin 2002).

Normally, the influence of the welding thermal cycle on the properties of steel is described using diagrams that show the change of the phase composition and mechanical properties depending on cooling time $\Delta t_{8.5/5}$ in the temperature interval 1123–773 K (850–500 °C). The diagrams examined below are obtained using specimens made of low-alloy steels of the following composition, %: C $\leq$ 0.4, Mn $\leq$ 2, Si $\leq$ 0.8, Cr $\leq$ 2, Mo $\leq$ 1, Ni $\leq$ 1.5, V $\leq$ 0.3, Ti $\leq$ 0.06, Al $\leq$ 0.06, Nb $\leq$ 0.1, W $\leq$ 0.5, Cu $\leq$ 0.5. The used thermal cycles provide peak heating temperature T_{max} = 1623 K (1350 °C) and cooling time $\Delta t_{8.5/5}$ = 5–200 s. As a result of experimental data processing, the following interpolation models are obtained (Kasatkin and Seyffarth 1984):

$$\begin{aligned}\mathrm{HV} = {} & (390 + 680\mathrm{C} + 17.7\mathrm{Mn})M + (234 + 122\mathrm{C})B + (98 + 275\mathrm{C} \\ & + 15.4\mathrm{Mn})(F + P);\end{aligned} \tag{13.4.1}$$

$$\begin{aligned}\sigma_u = {} & (798 + 1100\sqrt{C})M + (590 + 960\mathrm{C} + 39.7\mathrm{Mn} + 200\mathrm{V})B \\ & + (297 + 1360\mathrm{C} + 60\mathrm{Mn} + 140\mathrm{V})(F + P), \mathrm{MPa};\end{aligned} \tag{13.4.2}$$

$$\begin{aligned}\sigma_{0.2} = {} & (662 + 1610\mathrm{C})M + (500 + 460\mathrm{C} - 120\mathrm{C}^2 + 150\mathrm{V})B \\ & + (187 + 926\mathrm{C} + 47\mathrm{Mn} + 90\mathrm{V})(F + P), \mathrm{MPa};\end{aligned} \tag{13.4.3}$$

$$\begin{aligned}\delta = {} & (12.2 - 67\mathrm{C}^2 - 1.5\mathrm{Mn} + 0.76\ln\Delta t_{8.5/5})M + (21.3 - 35.6\mathrm{C} \\ & - 4.0\mathrm{Mn} - 5.0\mathrm{V} + 1.84\ln\Delta t_{8.5/5})B + (36.5 - 127\mathrm{C} + 153\mathrm{C}^2 \\ & - 1.16\mathrm{Mn} + 8.0\mathrm{V} + 0.66\ln\Delta t_{8.5/5})(F + P), \%;\end{aligned} \tag{13.4.4}$$

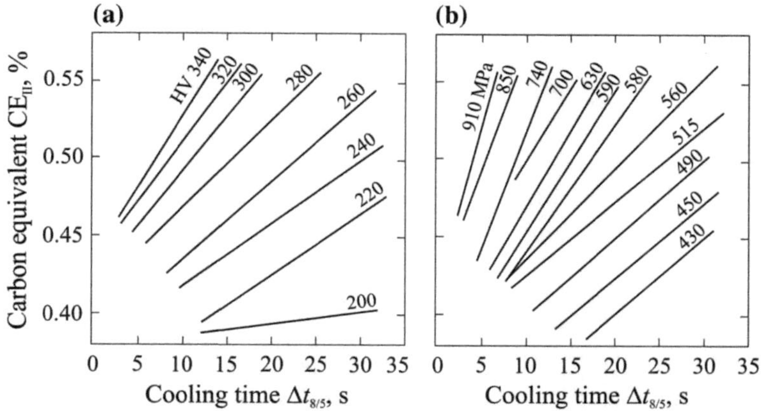

Fig. 13.15 HAZ iso-hardness (**a**) and iso-yield strength (**b**) for high strength low-alloy steels (Grong 1994; Kluken et al. 1992)

$$\psi = (48.5 - 158C + 116C^2 + 0.98 \ln \Delta t_{8.5/5})M + (53.3 - 132C + 103C^2 - 5.1Mn - 10V + 3.4 \ln \Delta t_{8.5/5})B + (65.4 - 88C - 82C^2 - 6.7Mn + 18V + 0.6 \ln \Delta t_{8.5/5})(F + P), \%; \quad (13.4.5)$$

$$\mathrm{KCU} = (1.06 - 2.8C + 1.3C^2 - 0.081Mn + 0.054 \ln \Delta t_{8.5/5})M + (1.3 - 1.6C - 0.08Mn)B + (1.47 - 1.8C + 0.80C^2 - 0.076Mn - 0.045 \ln \Delta t_{8.5/5})(F + P), \mathrm{MJ\,m^{-2}}. \quad (13.4.6)$$

Formulas for the calculation of martensite M, ferrite and pearlite $F + P$ and bainite B in the HAZ structure are provided in Sect. 13.2.

A simplified approach has become widespread in the welding practice, when the effect of the chemical composition of steel is described by a single parameter, namely, carbon equivalent. Exemplifying this approach, Fig. 13.15 shows lines of equal values of hardness HV and yield strength depending on cooling time $\Delta t_{8/5}$ and carbon equivalent CE_{II}, where (Yurioka et al. 1981)

$$CE_{II} = C + \frac{Si}{24} + \frac{Mn}{5} + \frac{Cu}{10} + \frac{Ni}{18} + \frac{Cr}{5} + \frac{Mo}{2.5} + \frac{V}{5} + \frac{Nb}{3}. \quad (13.4.7)$$

There are many publications that provide chemical compositions, welding TTT diagrams and the dependence of mechanical properties on cooling time or rate (Seyffarth 1978; Seyffarth and Kuscher 1982; Seyffarth et al. 1992; Shorshorov and Belov 1972). According to the calculated thermal cycle of welding, it is easy to determine the final phase composition and mechanical properties of the HAZ metal. As an example, Fig. 13.16 shows the cross-section of a natural and calculated weld and hardness distribution in a welded joint of 12Kh1MF steel. The chemical composition

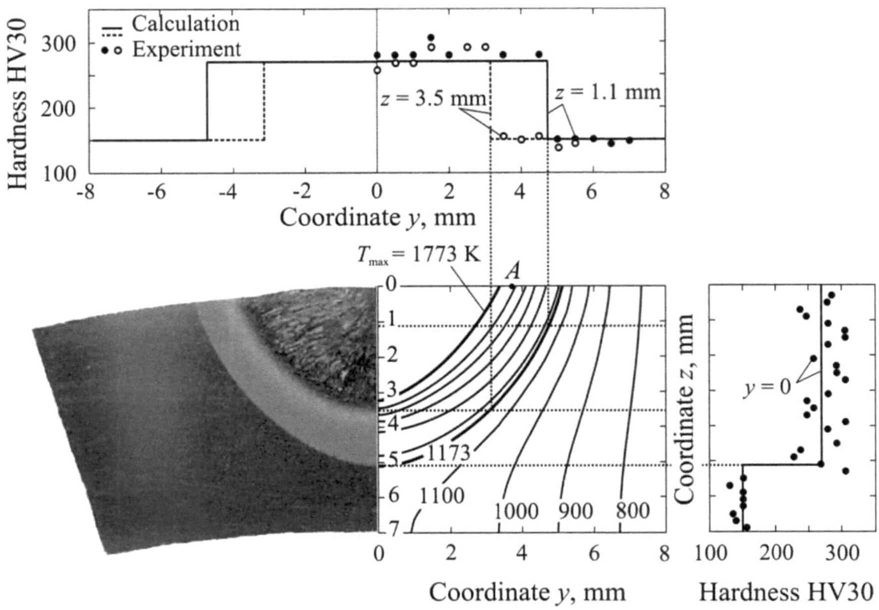

Fig. 13.16 Field of peak temperature T_{max} and distribution of hardness HV30 in the cross-section of a joint after gas tungsten arc welding of 12Kh1MF steel (I = 250 A, U = 12 V, v = 5 mm s^{-1} (Karkhin et al. 2006)

of steels 12Kh1MF and 12CrMo4.3 (Fig. 13.5) is similar. The temperature field is calculated using the inverse modelling method, while hardness distribution is calculated according to cooling time $\Delta t_{8.5/5}$ ($\Delta t_{8.5/5} = 5.6$ s) using Fig. 13.6. Figure 13.16 illustrates that the hardness of the HAZ metal almost doubles compared with the initial state. The metal consists of 76% bainite and 24% martensite (Karkhin et al. 2006). For hybrid laser-arc welding of the carbon steel, the mechanical properties of the HAZ metal was predicted with consideration for real weld shape (Karkhin et al. 2013a, b, 2015a, b).

Computer programs for calculating thermal processes of welding and cladding and predicting the microstructure and mechanical properties of the HAZ metal are developed (Muzhichenko et al. 1991; Muzhichenko 2000).

At complex thermal cycles (e.g., in low-frequency pulsed welding and multi-pass welding), the HAZ metal is heated and cooled multiple times, making it extremely difficult to predict the mechanical properties of metal in particular areas due to the lack of composition-properties' calculation models. In these cases, it is proposed to use an experimental-calculation method: a specimen of the base metal is subjected to a complex specified (calculated) thermal cycle on specialised installations in order to obtain the required microstructure, with further examination of the microstructure and properties of the obtained metal. This approach is particularly efficient if the width of the HAZ areas (e.g. the coarse grain area) is small, which makes it difficult to determine local mechanical properties (e.g. impact energy).

When using the experimental-calculation method to determine local mechanical properties, it is important to keep in mind that diffusion processes near the weld interface are not allowed for. For example, in the welding of different steels, an area with reduced carbon concentration ("soft interlayer") and, consequently, different mechanical properties, may be formed in the HAZ (Petrov 1963).

13.4.2 Mechanical Properties of Weld Metal

The mechanical properties of weld metal in the welding of low-alloy steels depend on the chemical composition of the weld metal and cooling time $\Delta t_{8.5/5}$ in the range 1123 K (850 °C)–773 K (500 °C). High-strength weld metal with the following composition is examined: 0.05–0.15% C, 0.5–2.0% Mn, 0.2–0.6% Si, $\leq$ 2.0 Cr, $\leq$ 2.0% Mo, $\leq$ 3.0% Ni, $\leq$ 0.2% V, $\leq$ 0.06% Ti, $\leq$ 0.06% Al, $\leq$ 0.2% Cu, 0.005–0.04% S, 0.005–0.04% P, 0.015–0.025% N, 0.01–0.06% O. At time $\Delta t_{8.5/5} = 7$–30 s, weld metal has the following properties (offset yield strength $\sigma_{0.2}$, ultimate tensile strength σ_u, elongation δ, reduction in area ψ) (Kasatkin 1984b, 1990; Kasatkin and Mikhoduy 1992):

$$
\begin{aligned}
\sigma_{0.2} = {} & 290 + 1660\mathrm{C} + 164\mathrm{Mn} + 17.4\mathrm{Si} + 103\mathrm{Cr} + 190\mathrm{Mo} + 92\mathrm{Ni} \\
& + 324\mathrm{V} - 120\mathrm{Ti} - 90\mathrm{Al} - 32\mathrm{Nb} + 13\mathrm{W} + 40\mathrm{Cu} + 960\mathrm{Zr} - 1200\mathrm{P} + 1740\mathrm{N} \\
& - 110\mathrm{O} - 32.3\mathrm{C}\cdot\mathrm{Mn} + 712\mathrm{C}\cdot\mathrm{Si} + 291\mathrm{C}\cdot\mathrm{Ni} + 470\mathrm{C}\cdot\mathrm{V} - 24.1\mathrm{Mn}\cdot\mathrm{Ni} \\
& + 67.1\mathrm{Si}\cdot\mathrm{Cr} - 64.2\mathrm{Si}\cdot\mathrm{Ni} + 201\mathrm{Si}\cdot\mathrm{V} - 24\mathrm{Mo}\cdot\mathrm{Ni} - 59.2\mathrm{Mo}\cdot\mathrm{V} + 15.3\mathrm{Ni}\cdot\mathrm{V} \\
& - 161\mathrm{C}^2 - 31.3\mathrm{Mn}^2 - 15\mathrm{Mo} - 4.5\mathrm{Ni}^2 - 275\mathrm{V}^2 - (27.5 + 350\mathrm{C} + 4.1\mathrm{Mn} \\
& + 4.3\mathrm{Cr} + 3.4\mathrm{Mo} + 2.2\mathrm{Ni} + 6.6\mathrm{V}) \ln \Delta t_{8.5/5}, \mathrm{MPa}
\end{aligned}
\tag{13.4.8}
$$

$$
\begin{aligned}
\sigma_u = {} & 350 + 1720\mathrm{C} + 171\mathrm{Mn} - 23.6\mathrm{Si} + 202\mathrm{Cr} + 147\mathrm{Mo} + 97\mathrm{Ni} + 426\mathrm{V} \\
& - 310\mathrm{Ti} - 70\mathrm{Al} - 65\mathrm{Nb} + 57\mathrm{W} + 43\mathrm{Cu} - 70\mathrm{Zr} - 200\mathrm{P} + 1900\mathrm{N} - 90\mathrm{O} \\
& + 170\mathrm{C}\cdot\mathrm{Mn} + 1330\mathrm{C}\cdot\mathrm{Si} + 870\mathrm{C}\cdot\mathrm{Cr} - 160\mathrm{C}\cdot\mathrm{Mo} + 310\mathrm{C}\cdot\mathrm{Ni} - 694\mathrm{C}\cdot\mathrm{V} \\
& - 23.5\mathrm{Mn}\cdot\mathrm{Ni} + 64.5\mathrm{Si}\cdot\mathrm{Cr} - 43\mathrm{Cr}\cdot\mathrm{Mo} - 25.5\mathrm{Cr}\cdot\mathrm{V} - 20.7\mathrm{Mo}\cdot\mathrm{Ni} - 50.7\mathrm{Mo}\cdot\mathrm{V} \\
& + 16.1\mathrm{Ni}\cdot\mathrm{V} + 86\mathrm{C}^2 - 17\mathrm{Mn}^2 - 50\mathrm{Cr}^2 - 4.5\mathrm{Mo}^2 + 5.4\mathrm{Ni}^2 - 255\mathrm{V}^2 - (23.0 \\
& + 290\mathrm{C} + 3\mathrm{Mn} + 2.5\mathrm{Si} + 14\mathrm{Cr} - 13\mathrm{Mo} + 10.8\mathrm{Ni} + 13\mathrm{V}) \ln \Delta t_{8.5/5}, \mathrm{MPa}
\end{aligned}
\tag{13.4.9}
$$

$$
\begin{aligned}
\delta = {} & 38.2 - 103\mathrm{C} - 3.94\mathrm{Mn} + 1.61\mathrm{Si} - 8.59\mathrm{Cr} - 9.0\mathrm{Mo} - 20.6\mathrm{V} \\
& + 12.4\mathrm{Ti} + 0.478\mathrm{Nb} - 0.85\mathrm{Cu} + 2.43\mathrm{Co} - 18\mathrm{S} - 18\mathrm{P} - 103\mathrm{N} - 18.5\mathrm{O} \\
& + 9.2\mathrm{C}\cdot\mathrm{Mn} + 11\mathrm{C}\cdot\mathrm{Cr} + 20\mathrm{C}\cdot\mathrm{Mo} - 1.1\mathrm{Mn}\cdot\mathrm{Si} + 1.64\mathrm{Mn}\cdot\mathrm{Mo} + 0.577\mathrm{Cr}\cdot\mathrm{Ni} \\
& + 0.826\mathrm{Mo}\cdot\mathrm{Ni} + 3.86\mathrm{Mo}\cdot\mathrm{V} + 78\mathrm{C}^2 + 0.27\mathrm{Cr}^2 + 0.686\mathrm{Mo}^2 + 2.12\mathrm{V}^2 + (8.4\mathrm{C}
\end{aligned}
$$

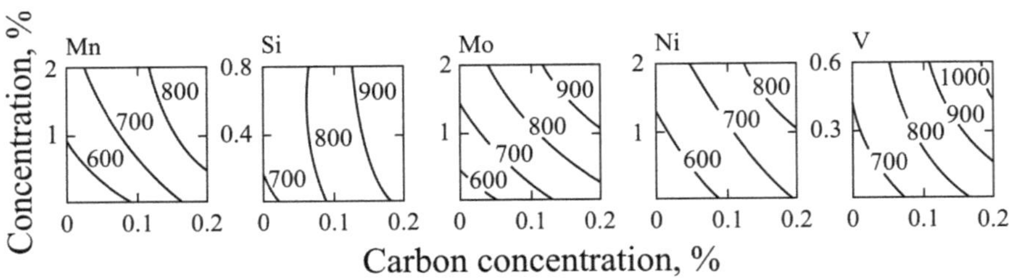

Fig. 13.17 Effect of alloying elements on the 0.2 percent offset yield strength of 10KhGN2 M steel (1% Mn, 0.3% Si, 1% Cr, 0.5% Mo, 1.5% Ni, 0.02% V, 0.02% S, 0.02% P, 0.01% N, 0.025% O) at cooling time $\Delta t_{8.5/5} = 20$ s (Kasatkin and Mikhoduy 1992)

$$+ 0.905\text{Cr} - 0.596\text{Ni} + 4.9\text{V} - 2\text{P}) \ln \Delta t_{8.5/5}, \% \quad (13.4.10)$$

$$\begin{aligned} \psi = {} & 93.4 - 213\text{C} + 2.86\text{Mn} + 10.5\text{Si} - 11.9\text{Cr} - 11.4\text{Mo} - 32.2\text{V} \\ & + 47.5\text{Ti} - 15.8\text{Al} - 3.5\text{Cu} + 4.7\text{Co} - 25\text{S} - 20\text{P} - 263\text{N} - 158\text{O} + 4.45\text{C} \cdot \text{Cr} \\ & - 1.38\text{Mn} \cdot \text{Mo} - 0.592\text{Mn} \cdot \text{Ni} - 2.76\text{Mn} \cdot \text{V} - 1.52\text{Ni} \cdot \text{V} + 110\text{C}^2 + 0.31\text{Cr}^2 \\ & + 4.1\text{V}^2 + (-2.83 + 18.8\text{C} - 1.76\text{Mn} - 4.36\text{Si} + 1.76\text{Cr} + 3.9\text{Mo} + 0.54\text{Ni} \\ & + 8\text{V} - 5\text{P}) \ln \Delta t_{8.5/5}, \% \end{aligned} \quad (13.4.11)$$

Regression models (13.4.8)–(13.4.11) show the non-additive and non-linear effects of alloying elements on the indicators of strength and ductility of weld metal. The degree of the effect of an individual element depends on the concentration of other elements.

Figure 13.17 illustrates the influence of major alloying elements on offset yield strength $\sigma_{0.2}$. As can be seen in the figure and the provided formulas, alloying with carbon, manganese, chromium, molybdenum and vanadium significantly increases the yield strength of weld metal, also reducing its ductility.

As the rate of weld metal cooling decreases (cooling time $\Delta t_{8.5/5}$ increases), so do its offset yield strength $\sigma_{0.2}$ and ultimate tensile strength σ_u. However, there is an increase in elongation δ and reduction in area ψ. Sensitivity of the mechanical properties of the weld metal to the thermal cycle grows with an increase in the concentration of major alloying elements.

There are also known regression models for assessing true tensile strength (Kasatkin 1984a), limit of proportionality (Kasatkin 1984b), impact energy KCV of low-alloy weld metal and its temperature dependence (Kasatkin 1990, 2005), resistance of weld metal to fatigue crack propagation (Kasatkin 1985), fracture toughness of low- and medium-alloy metal (Kasatkin 1994).

Formulas (13.4.8)–(13.4.11) make it possible to select a weld alloying system in the welding of low-alloy high-strength steels with allowance for welding conditions (time $\Delta t_{8.5/5}$, which can be calculated).

13.5 Microstructure and Mechanical Properties of Aluminium Alloys

Aluminium alloys are extensively used in welded structures due to their high strength, low density and high corrosion resistance. Cu, Mn, Si, Mg, Zn and other elements are used in alloying. Al–Mg (5XXX series) and Al–Mg–Si (6XXX series) alloys are used in the manufacturing of ships, pressure vessels, railroad coaches, automobiles, pipes, etc.

Al–Mn, Al–Si and Al–Mg alloys are non-heat treatable. If the welded metal is cold-worked, after heating the cold work of the metal in the HAZ is partially or completely removed, and the metal's properties become similar to the annealed metal (reduced hardness and strength, increased ductility) (Petrov and Tumarev 1977). Composition and microstructure of the weld metal is determined by composition of the filler metal, which is similar to the base metal in most cases.

Al–Cu, Al–Mg–Si and Al–Zn alloys are heat-treatable. They are subjected to homogenising (solution heat treatment) and quenching with further natural or artificial ageing. The composition of the filler metal is determined by the composition of the welded alloy. For example, in the welding of a 6005 alloy (0.8% Si, 0.5% Mg) it is recommendable to use 4043 alloy (4.5–6.0% Si, 0.8% Fe, 0.30% Cu) with reduced solidus temperature in order to avoid solidification cracking.

In the welding of heat-treatable alloys, the microstructure and local mechanical properties of the HAZ metal depend on the thermal cycle (peak temperature and heating and cooling time), which in turn depends on welding conditions, joint geometry and the thermal and physical properties of the metal (see Chap. 8).

Let us analyse the degree of softening the HAZ metal of the welded 6005–T6 Al–Mg–Si alloy, which has been subjected to homogenising and stabilisation prior to welding (Martikainen et al. 2011, 2013a). Deposition on a massive body is performed with preheating, $T_0 = 393$ K. Figure 13.18 shows the natural and calculated cross-section and thermal cycles of three points in the HAZ. Under heating, yield strength slumps as temperature increases (Fig. 13.19). Under cooling, yield strength grows as temperature drops. However, it still remains rather low if peak temperature is above 600 K. Such temperature dependencies of local mechanical properties serve as crucial input data for the solution to thermomechanical problems aimed at determining transient and residual welding stresses and distortion of welded structures.

13.6 Effect of Thermal Cycles on Mass Diffusion Processes in Welding

Microstructure and mechanical properties also depend on the chemical phenomena occurring in the weld and HAZ metal. It is necessary to take into account temperature history (thermal cycle) in the calculation of chemical kinetics (element diffusion, grain growth, formation and decomposition of chemical compounds, etc.). The

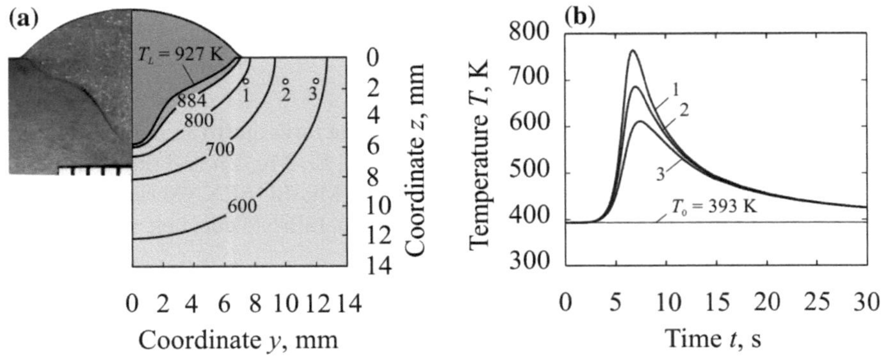

Fig. 13.18 Weld cross-section (**a**) and thermal cycles of points 1–3 in HAZ (**b**) (Martikainen et al. 2011)

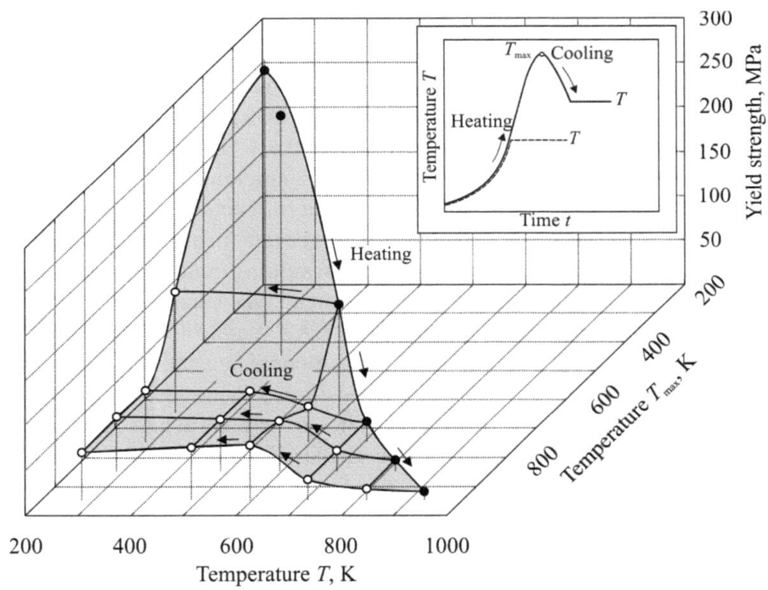

Fig. 13.19 The temperature dependence of the yield strength of 6005-T6 aluminium alloy under heating and cooling and schematic diagram of testing (Martikainen et al. 2011, 2013a, b)

solution to this problem is complicated by the fact that using the known analytical solutions obtained for isothermal conditions is problematic.

Based on the concept of equivalent time, it is possible to reduce non-isothermal diffusion to isothermal diffusion (Karkhin et al. 1978; Karkhin and Okhapkin 2011; Petrov 1963; Shewmon 1963).

Let us make the following assumptions: the diffusion flows are one-dimensional; the zone of active diffusion is relatively small and under isothermal conditions at any given time; the thermal cycle is known. Then element diffusion can be described by the following equation:

$$\frac{\partial C}{\partial t} = D(T(t))\left(\frac{\partial^2 C}{\partial x^2}\right), \tag{13.6.1}$$

where C is the concentration and D is the diffusion coefficient dependent on temperature T, which, in its turn, depends on time t:

$$D(T(t)) = D_0 \exp\left(-\frac{Q}{RT(t)}\right). \tag{13.6.2}$$

Here, D_0 is the multiplier, Q is the activation energy of diffusion and R is the gas constant. Values D_0 and Q depend on the microstructure and are assumed to be known.

Non-linear Eq. (13.6.1) with variable diffusion coefficient D can be reduced to a linear equation with constant coefficient D_{ref} taken relative to a reference constant temperature T_{ref} (Karkhin et al. 1978; Karkhin and Okhapkin 2011):

$$\frac{\partial C}{\partial t_{eq}} = D_{ref}\frac{\partial^2 C}{\partial x^2}, \tag{13.6.3}$$

where t_{eq} is the equivalent time determined according to the following formula:

$$t_{eq} = \frac{1}{D_{ref}}\int_0^t D(T(t'))dt'. \tag{13.6.4}$$

Equations (3.6.1) and (13.6.3) are identical. Formula (13.6.4) shows that thermal cycle T (t) is allowed for in the equivalent time.

Therefore, the solution to problem (13.6.1)–(13.6.2) with a variable diffusion coefficient can be reduced to the solution to problem (13.6.3)–(13.6.4) with constant coefficient D_{ref} in the form of function C (x, t_{eq}), and then the required function C (x, t) is built using Eq. (13.6.4).

The integral in (13.6.4) with allowance for (13.6.2) can be calculated using known numerical methods. Let us find it for a piecewiselinear function, which can be used to rather accurately approximate any thermal cycle of welding and heat treatment T (t) (Fig. 13.20). Temperature T_n in each n section is described by a linear function of local time Δt_n:

$$T_n(\Delta t_n) = a_n + b_n \Delta t_n, \tag{13.6.5}$$

where a_n и b_n are defined by two points: T_{ns}, t_{ns} and T_{nf}, t_{nf} (Fig. 13.20). Then equivalent time t_{eq} at any time t_n is determined by summation by sections:

Fig. 13.20 Thermal cycle $T(t)$ and dependence of equivalent time t_{eq} on real time t_n

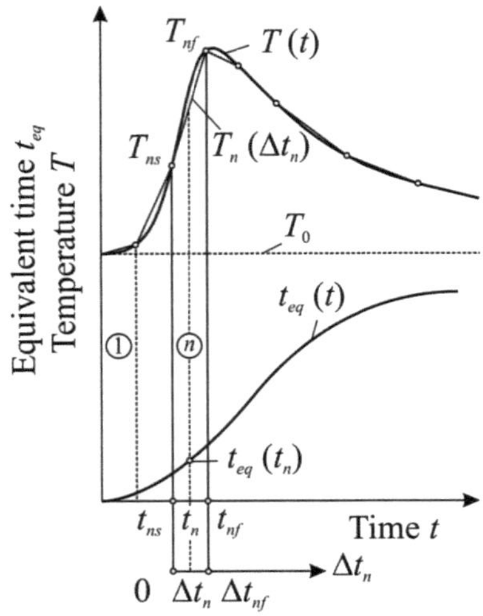

$$t_{eq}(t_n) = \begin{cases} \Delta t_{eq\,n}(\Delta t_n) & \text{for } n = 1; \\ \sum\limits_{i=1}^{n-1} \Delta t_{eq\,i}(\Delta t_{if}) + \Delta t_{eq\,n}(\Delta t_n) & \text{for } n > 1. \end{cases} \tag{13.6.6}$$

Let us assume that the pattern of function $D\,(T)$ is constant at each nth section (D_{0n} and Q_n are constant). Then equivalent time for the nth section follows from formulas (13.6.2) and (13.6.4):

$$\Delta t_{eq\,n}(\Delta t_n) = \frac{1}{D(T_{ref})}\int\limits_0^{\Delta t_n} D_{0n}\exp\left(-\frac{Q_n}{RT_n(t')}\right)dt' =$$
$$= \begin{cases} \frac{D_{0n}}{D_{0ref}}\exp\left(\frac{Q_{ref}}{RT_{ref}} - \frac{Q_n}{Ra_n}\right)\Delta t_n \quad \text{for } b_n = 0; \\ \frac{D_{0n}}{D_{0ref}}\frac{Q_n}{Rb_n}\exp\left(\frac{Q_{ref}}{RT_{ref}}\right)\left[\frac{R(a_n+b_n\Delta t_n)}{Q_n}\exp\left(-\frac{Q_n}{R(a_n+b_n\Delta t_n)}\right) - \right. \\ \left. -\frac{Ra_n}{Q_n}\exp\left(-\frac{Q_n}{Ra_n}\right) + \mathrm{Ei}\left(-\frac{Q_n}{R(a_n+b_n\Delta t_n)}\right) - \mathrm{Ei}\left(-\frac{Q_n}{Ra_n}\right) \quad \text{for } b_n \neq 0. \right. \end{cases} \tag{13.6.7}$$

Here, values D_{0ref} and Q_{ref} refer to the material at temperature T_{ref}, while values D_{0n} and Q_n refer to the material at the nth section.

Equation (13.6.7) can be represented in the following form:

$$\Delta t_{eq\,n} = \frac{D_{0n}}{D_{0ref}}\exp\left(\frac{Q_{ref}}{RT_{ref}}\right)\Theta(u, v)\Delta t_n, \tag{13.6.8}$$

where

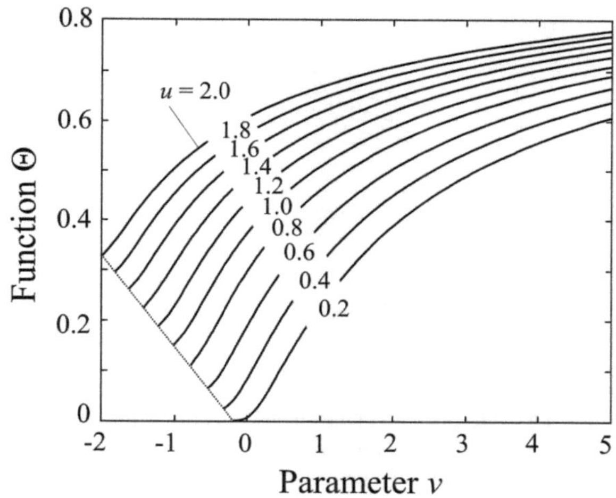

Fig. 13.21 Dependence of the function Θ on dimensionless parameters $u = Ra_n/Q_n$ and $v = Rb_n\Delta t_n/Q_n$

$$\Theta = \begin{cases} \exp\left(-\frac{1}{u}\right) & \text{for } v = 0; \\ \frac{1}{v}\left[(u+v)\exp\left(-\frac{1}{u+v}\right) - u\exp\left(-\frac{1}{u}\right) + \right. & \\ \left. +\mathrm{Ei}\left(-\frac{1}{u+v}\right) - \mathrm{Ei}\left(-\frac{1}{u}\right)\right] & \text{for } v \neq 0; \end{cases}$$

$$u = Ra_n/Q_n; \quad v = Rb_n\Delta t_n/Q_n; \quad u + v > 0. \tag{13.6.9}$$

Function Θ (u, v) is shown in Fig. 13.21. It can be seen that as temperature increases (along with coefficients a_n and b_n, and, consequently, parameters u and v), so does equivalent time $t_{eq\,n}$ (diffusion is activated).

Thus, the obtained Eqs. (13.6.6)–(13.6.8) make it possible to use the solution to diffusion problem (13.6.3) when the diffusion coefficient is constant, $D_{ref} = D\,(T_{ref})$, taking equivalent time t_{eq} for time. An analytical solution to diffusion Eq. (13.6.3) can be obtained using known methods.

Let us now find a correlation between equivalent time t_{eq} and actual time t for two typical schemes of calculating the thermal cycles in fusion welding: a rapidly moving point source on the surface of a massive body and a rapidly moving line source in a thin plate (Sect. 5.1.4). Let us define the diffusion zone by the specified peak temperature $T_{\max}$.

By substituting Eqs. (13.6.2), (5.1.121) and (5.1.129) into expression (13.6.4), we get the dependence of equivalent time t_{eq} on actual time t for a point source (Karkhin and Okhapkin 2011):

$$t_{eq}(t) = \frac{q/v}{\lambda(T_{\max} - T_0)} \frac{D_0}{D_{ref}} F(E, \Theta_0, \tau), \tag{13.6.10}$$

where

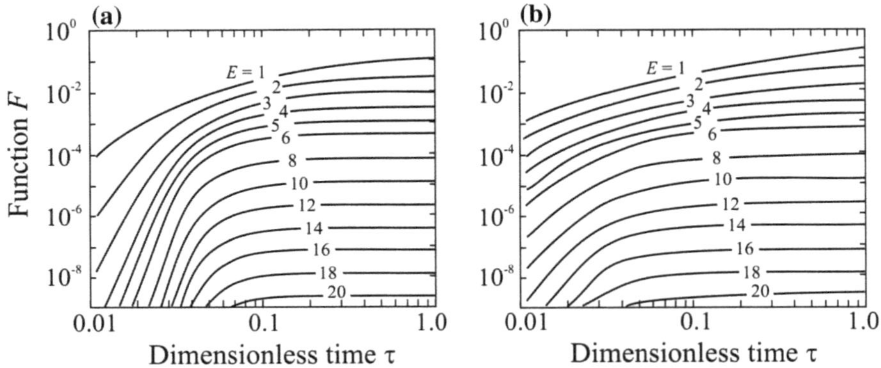

Fig. 13.22 Dependence of the integral function F on dimensionless real time τ at the dimensionless activation energy E and dimensionless initial temperature $\theta_0 = 0.2$: a rapidly moving point source on a thick solid (**a**) and a rapidly moving line source in a plate (**b**) (Karkhin and Okhapkin 2011)

$$F(E, \Theta_0, \tau) = \int_0^{\tau} \exp\left(-\frac{E}{\Theta_0 + (2\pi u)^{-1} \exp\left(-(2\pi e u)^{-1}\right)}\right) du;$$

$$\tau = \frac{\lambda(T_{\max} - T_0)}{q/v} t; \; E = \frac{Q}{R(T_{\max} - T_0)}; \; \Theta_0 = \frac{T_0}{T_{\max} - T_0}. \tag{13.6.11}$$

Here, τ is the dimensionless actual time; E is the dimensionless activation energy; Θ_0 is the dimensionless initial temperature.

Dependence $F(\tau)$ is shown in Fig. 13.22a. The greater energy E is, the lesser integral F and time t_{eq} as well as their growth will be. Function F remains unchanged after complete cooling; it reaches its maximum value $F_{\max}$. It can be assumed that condition $\tau = 1$ corresponds to the end of thermally active diffusion. The only exception is hydrogen ($E < 5$), which diffuses at low temperature as well. With the increase in energy E, function $F_{\max}$ (time t_{eq}) drops. A rise of initial temperature Θ_0 enhances diffusion and, consequently, function $F_{\max}$ and time t_{eq} proportionally to the increase in energy E. This also follows from Eq. (13.6.11): the values of the integrand and the integral increase with the growth of Θ_0. Thus, preheating and additional heating increase the equivalent diffusion time (Karkhin and Okhapkin 2011).

Similarly, by substituting Eqs. (13.6.2), (5.1.146) and (5.1.153) into expression (13.6.4), we get the dependence of equivalent time t_{eq} on actual time t for a line source (Karkhin and Okhapkin 2011):

$$t_{eq}(t) = \frac{1}{c\rho\lambda}\left[\frac{q/(vh)}{T_{\max} - T_0}\right]^2 \frac{D_0}{D_{ref}} F(\mathrm{E}, \Theta_0, \tau), \tag{13.6.12}$$

where

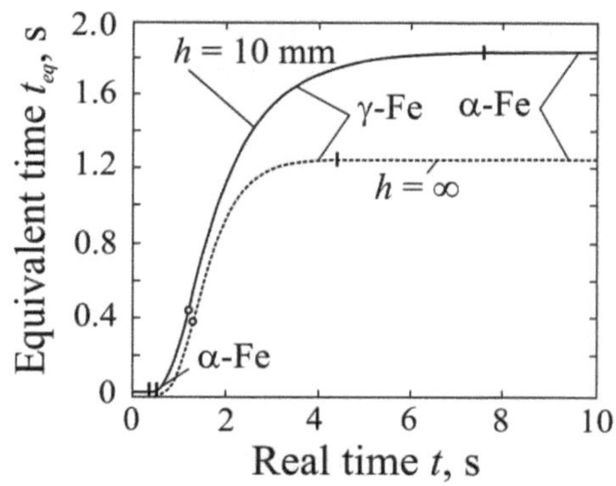

Fig. 13.23 Dependence of the equivalent time of diffusion of carbon on the real time in the HAZ in welding a 10 mm thick steel plate ($h = 10$ mm) and a thick solid ($h = \infty$) with the heat input of 1148 J mm^{-1} and the initial temperature of 293 K (Karkhin and Okhapkin 2011)

$$F(E, \Theta_0, \tau) = \int_0^{\tau} \exp\left(-\frac{E}{\Theta_0 + (4\pi u)^{-1/2}\exp\left(-(8\pi e u)^{-1}\right)}\right) du;$$

$$\tau = c\rho\lambda\left[\frac{T_{\max} - T_0}{q/(vh)}\right]^2 t. \tag{13.6.13}$$

Dependence $F(\tau)$ is also defined by two parameters: E and Θ_0. Comparison of Fig. 13.22a, b shows that the corresponding curves are qualitatively similar.

It should be noted that the diffusion zone can be characterised not by peak temperature, but by the distance from the diffusion zone to the weld centreline (Karkhin et al. 1978). Then the formulas change their form.

If during heating and cooling there is a number of microstructural transformations accompanied by changes in D_0 and Q, the entire section $0 - t$ in Eqs. (13.6.11) and (13.6.13) is divided into separate stages with uniform microstructure (constant D_0 and Q).

Example 13.6.1 Required: a correlation between equivalent time t_{eq} and actual time t for carbon in the metal heated to temperature $T_{\max} = 1758$ K in the welding of a thick steel slab. Given: welding conditions $q/v = 1148$ J mm^{-1} and $T_0 = 293$ K ($\Theta_0 = 0.2$) provide a weld of width 10 mm and 5 mm in thickness (Petrov 1963).

Let us refer diffusion to temperature $T_{ref} = T_{\max}$. Figure 13.23 (curve $h = \infty$) shows that diffusion at the stages of heating (curve beginning) and cooling (curve end) is weakly expressed (equivalent time changes slightly). The most intense diffusion occurs in γ-Fe at $T_{\max} = 1758$ K (depicted as a circle on the curve). At the cooling stage, diffusion is more intense and longer than at the heating stage. The complete thermal cycle corresponds to equivalent time $t_{eq} = 1.27$ s. This time is longer for a plate, $t_{eq} = 1.86$ s (curve $h = 10$ mm), which means that diffusion processes in a plate are more intense, since the plate metal heats faster and cools slower that the metal of a massive body (Karkhin and Okhapkin 2011).

13.7 Optimisation of Inert Gas Tungsten Arc Welding Conditions for Stainless Steel

Optimisation of welding conditions requires a mathematical thermal process model that would correlate welding parameters with the characteristics of the temperature field (Fig. 11.1). So far, we have considered this model to represent a conventional heat conduction problem (partial differential equation of heat conduction equation, initial and boundary conditions). Technically, this model can be described in a simplified form, e.g. by a polynomial regression equation. Naturally, the regression equation should be obtained using statistical processing of the results of numerous experiments (numerical and natural). Let us illustrate the latter approach through the example of a gas tungsten arc welding technique.

Preliminary application of a thin coating of activating fluxes on the surface of the material makes it possible to increase significantly the depth of fusion (Parshin 2011, 2013). For example, single-pass gas tungsten arc (A-TIG) welding of 304LN stainless steel ($\leq$0.03% C, 18–20% Cr, 8–12% Ni, $\leq$2% Mn, $\leq$1% Si) provides the depth of fusion up to 10–12 mm (Vasudevan et al. 2007).

There is a known regression equation that directly correlates the depth of fusion $h_{p,}$ weld width W and weld overfill height h_{of} with the following parameters of the tungsten arc welding of 304LN steel along the activating flux layer: current I = 80–280 A, speed v = 1.00–3.33 mm s^{-1}, arc length l = 0.75–1.50 mm, argon comsumption 0.17 l s^{-1}, thoriated electrode diameter 3 mm, electrode tip angle 45°, multi-component flux. In tungsten arc welding without filler metal, weld overfill (reinforcement) forms resulting from residual transverse plastic shortening strains. Regression equations for 304LN steel have the following form (Vasudevan et al. 2007):

$$\begin{aligned} h_p = {} & 0.27946 + 0.04961 \cdot I + 0.11002 \cdot U - 0.21162 \cdot v - 0.53303 \cdot l \\ & - 0.00141 \cdot I \cdot U - 0.014 \cdot I \cdot v + 0.00316 \cdot I \cdot l \\ & + 0.0000437177 \cdot I^2 + 0.01065 \cdot l^2; \end{aligned} \tag{13.7.1}$$

$$\begin{aligned} W = {} & 0.38168 + 0.0854 \cdot I + 0.28924 \cdot U - 0.1794 \cdot v + 0.2496 \cdot l \\ & - 0.0002042 \cdot I \cdot U - 0.0005326 \cdot I \cdot v + 0.0009386 \cdot I \cdot l \\ & + 0.00003310 \cdot I^2 + 0.00664 \cdot l^2; \end{aligned} \tag{13.7.2}$$

$$\begin{aligned} h_{of} = {} & -0.30796 + 0.0004187 \cdot I + 0.10023 \cdot U - 0.01046 \cdot v \\ & - 0.28719 \cdot l - 0.0007915 \cdot I \cdot U + 0.00001245 \cdot I \cdot v + 0.00202 \cdot I \cdot l \\ & + 0.00003797 \cdot I^2 + 0.00004965 \cdot v^2. \end{aligned} \tag{13.7.3}$$

Here, U is the arc voltage. It is possible to add an arc equation that would correlate voltage U with current I and arc length l, if it is known. The optimum values of the variable parameters I, U, v and l can be found by solving Eqs. (13.7.1)–(13.7.3)

Fig. 13.24 Weld shape produced by A-TIG welding of 304LN austenitic stainless steel

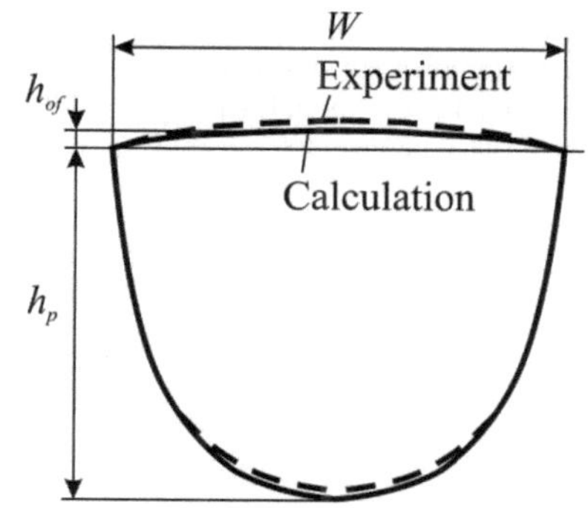

for these four parameters with allowance for additional physical constraints. This problem can be formulated as an optimisation problem by taking objective function F as a sum of squares of relative deviations of the calculated parameters from the required parameters (marked by superscripts m):

$$F(I, U, v, l) = \left(\frac{h_p^m - h_p}{h_p^m}\right)^2 + \left(\frac{W^m - W}{W^m}\right)^2 + \left(\frac{h_{of}^m - h_{of}}{h_{of}}\right)^2 \rightarrow \min. \quad (13.7.4)$$

Equation (13.7.4) with allowance for the corresponding constraints on welding parameters in the form of equations and inequations can be solved using the methods described in Chap. 11 or using a genetic algorithm (Vasudevan et al. 2007). Genetic algorithm is a heuristic search algorithm used for solving optimisation and modelling problems by randomly selecting, combining and varying the required parameters using mechanisms reminiscent of biological evolution. The fundamentals of this theory can be found in specialised literature. The genetic algorithm has found use in solutions to welding problems (Kim et al. 2002; Kim and Rhee 2003; Kumar et al. 2005; Vasudevan et al. 2007).

Figure 13.24 and Table 13.1 show weld geometry and the corresponding welding parameters. The comparison of the calculated and experimental weld size and the corresponding welding parameters shows satisfactory calculation accuracy. The weld is represented by two semi-ellipses despite the actual shape of the weld being more complex (Vasudevan et al. 2007). The accuracy of the weld geometry calculation directly depends on accuracy (adequacy) of the regression equation.

Table 13.1 Comparison between predicted and actual welding process variables and weld shape parameters

Parameter	Calculation	Experiment
W (mm)	6.14	6.16
h_p (mm)	3.87	3.83
h_{of} (mm)	0.26	0.33
I (A)	136	140
U (V)	15.0	13.6
v (mm s^{-1})	1.33	1.33
l (mm)	1.0	1.5

References

Adams, C. (1958). Cooling rates and peak temperatures in fusion welding. *Welding Journal, 37*(5), 210-s–215-s.

Berkhout, C. F., & van Lent, P. H. (1968). Anwendung von Spitzentemperatur-Abkuhlzeit (STAZ)-Schaubildern beim Schweissen hochfester Stahle. *Schweissen und Schneiden, 6*, 256–260 (in German).

Buchmayr, B. (1991). *Computer in der Werkstoff – und Schweisstechnik. Anwendung von mathematischen Modellen* (436 pp.). Duesseldorf: DVS – Verlag (in German).

Buchmayr, B., & Cerjak, H. (1988). Mathematical description of HAZ behaviour of low-alloyed structural steels. In *Proceedings of the International Conference on Improved Weldment Control with Special Reference to Computer Technology* (pp. 43–51). Vienna.

Degenkoble, J., Uwer, D., & Wegmann, H. (1984). Characterisation of weld thermal cycles with regard to their effect on the mechanical properties of weld joints by the cooling time $t_{8/5}$ and its determination (17 pp.). IIW Doc. IX-1336-84.

Devillers, L., Kaplan, D., & Testard, P. (1995). Predicting the microstructures and toughness of weld HAZs. *Welding International, 9*(2), 128–138.

Frolov, V. V. (Ed.). (1988). *Theory of welding processes* (559 pp.). Moscow: Vysshaya Shkola (in Russian).

Gliha, V. (2005). The microstructure and properties of materials at the fusion line. *Metalurgija, 44*(1), 13–18.

Grong, O. (1994). *Metallurgical modelling of welding* (581 pp.). London: The Institute of Materials.

Karkhin, V. A., Homich, P. N., Ivanov, S. Yu., & Karimi J. (2013a). Prediction of microstructure and mechanical properties of weld metal in hybrid laser-arc welding. In *Proceedings of the 7th International Scientific and Technical Conference on Beam Technologies and Laser Application,* 18–21 September 2013, (pp. 38–45). St. Petersburg, Russia: St. Petersburg Polytechnic University Publishing.

Karkhin, V. A., Khomich, P. N., Ivanov, S. Yu., Michailov, V. G., & Kah, P. (2013b). Prediction of the microstructure and mechanical properties of metal in welded joints with consideration for real weld geometry. International Institute of Welding. IIW RD305. Rev 3-3/7 (7 pp.).

Karkhin, V. A., Khomich, P. N., Ivanov, S. Yu., & Martikainen, J. (2015a). Prediction of microstructure and mechanical properties of heat affected zone in hybrid laser-arc welding. *Welding and Diagnostics, 3*, 9–12 (in Russian).

Karkhin, V. A., Homich, P. N., Ivanov, S. Yu., Michailov, V. G., & Panchenko, O. V. (2015b). Prediction of microstructure and mechanical properties of weld metal in hybrid laser-arc welding. *Advanced Materials Research, 1120–1121*, 1292–1296.

Karkhin, V. A., Khomich, P. N., & Michailov, V. G. (2006). Prediction of microstructure and mechanical properties of weld metal with consideration for real weld geometry. In W. Lucas & V. I. Makhnenko (Eds.), *Proceedings of Joint International Conference "Computer Technology in Welding and Manufacturing (16th International Conference) and Information Technologies in Welding and Related Processes (3rd International Conference)"* (pp. 162–166). Kiev.

Karkhin, V. A., Mnushkin, O. S., & Petrov, G. L. (1978). An approximate calculation of hydrogen redistribution in welded joints. In *Proceedings of Leningrad Polytechnic Institute, No. 364 "Welding Production"* (pp. 3–8) (in Russian).

Karkhin, V. A., & Okhapkin, K. A. (2011). Evaluation of the equivalent time of non-isothermal diffusion processes in the heat-affected zone in fusion welding. *Welding International, 25*(8), 629–632.

Kasatkin, O. G. (1984a). Dependence of ultimate strength and true rupture strength of weld metal on alloying and welding thermal cycle. *Automatic Welding, 9*, 1–5 (in Russian).

Kasatkin, O. G. (1984b). Dependence of yield strength of weld metal on concentration of alloying elements. *Automatic Welding, 8*, 11–12 (in Russian).

Kasatkin, O. G. (1984c). Interpolation models for evaluation of phase composition in arc welding of low-alloyed steels. *Automatic Welding, 1*, 7–11 (in Russian).

Kasatkin, O. G. (1985). Calculation of weld metal resistance to fatigue crack propagation. *Automatic Welding, 12*, 1–4 (in Russian).

Kasatkin, O. G. (1990). Mathematical modelling of composition-property relationships for welded joints and development of calculation-experimental system for optimisation of main technological factors of welding of low-alloyed structural steels. Doctoral thesis, Kiev: The Paton Welding Institute (in Russian).

Kasatkin, O. G. (2005). Estimation of impact energy of low-alloyed weld metal. *Automatic Welding, 1*, 57–58 (in Russian).

Kasatkin, O. G., & Mikhoduy, L. I. (1992). Selection of alloying system for welding of low-alloyed high strength steels. *Automatic Welding, 5*, 19–25 (in Russian).

Kasatkin, O. G., & Seyffarth, P. (1984). Influence of chemical and phase composition of the heat affected zone on its mechanical properties during arc welding of low-alloy steels. *Automatic Welding, 2*, 5–10 (in Russian).

Kasatkin, O. G., & Seyffarth, P. (1994). Dependence of impact energy of low-alloy and alloy weld metal on its composition and microstructure. *Automatic Welding, 3*, 52–66 (in Russian).

Kim, D., & Rhee, S. (2003). Optimisation of GMA welding process using the dual response approach. *International Journal of Production Research, 41*(18), 4505–4515.

Kim, D., Rhee S., & Park H. (2002). Modelling and optimisation of a GMA welding process by genetic algorithm and response surface methodology. *International Journal of Production Research, 40*(7), 1699–1711.

Kluken, A. O., Ibarra, S., Liu, S., & Olson, D. L. (1992). Use of predictive equations for arctic steel heat affected zone properties. In *Proceedings of 11th International Conference on Offshore Mechanics and Arctic Engineering* (Vol. A, pp. 1–7). Calgary, Canada, June 1992. Publ. ASME.

Kumar, A., Zhang, W., Kim, C.-H., & DebRoy, T. (2005). A smart bi-directional model of heat transfer and free surface flow in gas metal arc fillet welding for practicing engineers. In H. Cerjak, H. K. D. H. Bhadeshia, & E. Kozeschnik (Eds.), *Mathematical modelling of weld phenomena* (Vol. 7, pp. 3–37). Graz: Verlag der Technischen Universitaet Graz.

Martikainen, J., Hiltunen, E., Brhane, F., Karkhin, V., & Ivanov, S. (2011). Prediction of liquation crack initiation in Al-Mg-Si alloys welded joints. In J. Lippold, T. Boellinghaus, & C. E. Cross (Eds.), *Hot cracking phenomena in welds III* (pp. 71–86). Springer.

Martikainen, J., Hiltunen, E., Karkhin, V., & Ivanov, S. (2013a). Numerical analysis of liquation cracking in aluminium alloy welded joints. In C. Sommitsch & N. Enzinger (Eds.), *Mathematical modelling of weld phenomena* (Vol. 10, pp. 401–411). Graz: Verlag der Technischen Universitaet Graz.

Martikainen, J., Hiltunen, E., Karkhin, V. A., & Ivanov, S. Yu. (2013b). A method for evaluating the liquation cracking susceptibility of welded joints in Al-Mg-Si alloys. *Welding International, 2*, 139–143.

Michailov, V., Karkhin, V., & Petrov, P. (2016). *Principles of welding*. St. Petersburg: Polytechnic University Publishing.

Muzhichenko, A. F. (2000). Software for prediction of microstructure and mechanical properties of HAZ metal in welding of structural steels. *Automatic Welding, 6*, 40–43 (in Russian).
Muzhichenko, A. F., Demchenko, V. F., & Romanenko A. V. (1991). Software for personal computers to calculate thermal processes in welding and surfacing. *Automatic Welding, 6*, 73–74 (in Russian).
Odanovic, Z., & Nedeljkovic, L. (2001). Numerical modelling of microstructure in heat affected zone of GMA welded HY-100 steel. In H. Cerjak (Ed.), *Mathematical modelling of weld phenomena* (Vol. 5, pp. 381–392). London: IOM Communications.
Ossenbrink, R., & Michailov, V. (2007). Thermomechanical numerical simulation with the maximum temperature austenisation cooling time model (STAAZ). In H. Cerjak, H. K. D. H. Bhadeshia, & E. Kozeschnik (Eds.), *Mathematical modelling of weld phenomena* (Vol. 8, pp. 357–372). Graz: Verlag der Technischen Universitaet Graz.
Parshin, S. G. (2011). Use of ultradispersed particles in activating flux to increase productivity of MIG/MAG welding of steels. *Welding Production, 6*, 28–32 (in Russian).
Parshin, S. G. (2013). *Nanostructured and activating materials for arc welding* (624 pp.). St. Petersburg: St. Petersburg Polytechnic University Publishing (in Russian).
Petrov, G. L. (1963). *Inhomogeneity of weld metal* (206 pp.). Leningrad: Sudpromgiz (in Russian).
Petrov, G. L., & Tumarev, A. S. (1977). *Theory of welding processes* (2nd ed., 392 pp.). Moscow: Vysshaya Shkola Publishing (in Russian).
Radaj, D. (1992). *Heat effects of welding. Temperature field, residual stress, distortion* (348 pp.). Berlin: Springer.
Seyffarth, P. (1978). *Schweiss – ZTU – Schaubilder. Atlas* (151 pp.). Rostock: Wilhelm – Pieck – Universitaet, Band 1.; Band 2 (167 pp.) (in German).
Seyffarth, P., & Kassatkin, O. G. (1979). Mathematisch – statistische Beschreibung der Austenitumwandlung in der Waermeeinfusszone. *Schweisstechnik. Berlin, 3*, 117–119 (in German).
Seyffarth, P., & Kasatkin, O. G. (2002). Calculation models for evaluating mechanical properties of HAZ metal in welding low-alloyed steels. In V. I. Makhnenko (Ed.), *Proceedings of International Conference on Mathematical Modelling and Information Technologies in Welding and Related Processes* (pp. 103–106). Kiev: Paton Welding Institute Publishing (in Russian).
Seyffarth, P., & Kuscher, G. (1982). *Schweiss – ZTU – Schaubilder* (233 pp.). Berlin: VEB Verlag Technik (in German).
Seyffarth, P., Meyer, B., & Scharff, A. (1992). *Grosser Atlas Schweiss – ZTU – Schaubilder* (176 pp.). Duesseldorf: DVS – Verlag (in German).
Shewmon, P. G. (1963). *Diffusion in solids* (200 pp.). New York: McGraw-Hill Book Co.
Shorshorov, M. H., & Belov, V. V. (1972). *Phase transformations and changes of steel properties in welding* (220 pp.). Moscow: Nauka (in Russian).
Vasudevan, M., Bhaduri, A. K., & Raj, B. (2007). Genetic algorithm for optimizing the A-TIG welding of austenitic stainless steels. In H. Cerjak, H. K. D. H. Bhadeshia, & E. Kozeschnik (Eds.), *Mathematical modelling of weld phenomena* (Vol. 8, pp. 23–35). Graz: Verlag der Technischen Universitaet Graz.
Yazovskikh, V. M., & Belenky, V. Ya. (2011a). Thermal processes in surfacing of solid cylinders. *Welding and Diagnostics, 3*, 27–31 (in Russian).
Yazovskikh, V. M., & Belenky, V. Ya. (2011b). Thermal processes in surfacing of cylindrical solids with strip electrode. *Welding Production, 12*, 20–24 (in Russian).
Yurioka, N., Ohshita, S., & Tamehiro, H. (1981). Study on carbon equivalents to assess cold cracking tendency and hardness. In *Proceedings of International Symposium on Pipeline Welding in the '80s*, 18 March 1981, (pp. 1–15). Publication of Australian Welding Research Association.

Index

V. A. Karkhin, *Thermal Processes in Welding*, Engineering Materials,
https://doi.org/10.1007/978-981-13-5965-1

FSC
www.fsc.org
MIX
Papier aus verantwortungsvollen Quellen
Paper from responsible sources
FSC® C105338

If you have any concerns about our products,
you can contact us on
ProductSafety@springernature.com

In case Publisher is established outside the EU,
the EU authorized representative is:
Springer Nature Customer Service Center GmbH
Europaplatz 3, 69115 Heidelberg, Germany

Printed by Libri Plureos GmbH
in Hamburg, Germany